Biomolecular Action of Ionizing Radiation

Series in Medical Physics and Biomedical Engineering

Series Editors: John G Webster, E Russell Ritenourt, Slavik Tabakov,
and Kwan-Hoong Ng

Other books in the series:

An Introduction to Rehabilitation Engineering
R A Cooper, H Ohnabe, and D A Hobson

The Physics of Modern Brachytherapy for Oncology
D Baltas, N Zamboglou, and L Sakelliou

Electrical Impedance Tomography
D Holder (Ed)

Contemporary IMRT
S Webb

The Physical Measurement of Bone
C M Langton and C F Njeh (Eds)

**Therapeutic Applications of Monte Carlo Calculations
in Nuclear Medicine**
H Zaidi and G Sgouros (Eds)

Minimally Invasive Medical Technology
J G Webster (Ed)

Intensity-Modulated Radiation Therapy
S Webb

Physics for Diagnostic Radiology
P Dendy and B Heaton

Achieving Quality in Brachytherapy
B R Thomadsen

Medical Physics and Biomedical Engineering
B H Brown, R H Smallwood, D C Barber, P V Lawford, and D R Hose

Monte Carlo Calculations in Nuclear Medicine
M Ljungberg, S-E Strand, and M A King (Eds)

Introductory Medical Statistics 3rd Edition
R F Mould

Ultrasound in Medicine
F A Duck, A C Barber, and H C Starritt (Eds)

Design of Pulse Oximeters
J G Webster (Ed)

October 9, 2007

Thank you for your purchase of *Biomolecular Action of Ionizing Radiation*
(Shirley Lehnert).

Figure 4.5 on page 59 was corrupted at time of printing. The corrected Figure is
shown below:

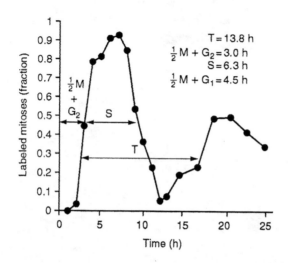

FIGURE 4.5 Determination of cell cycle phase durations by the fraction of
labeled mitosis method. Mouse myoblasts were pulse-labeled for 15 min
with 3H-thymidine. At indicated times cells were incubated for 15 min with
Colcemid, fixed to microscope slides and autoradiographed. The fraction of
labeled mitotic cells was plotted as a function of time after the pulse label.
(From Merrill, G., Methods Cell Biol., 57, 229, 1998. With permission.)

We sincerely regret any inconvenience this may have caused you. Please let us
know if we can be of assistance regarding this or any other title that Taylor &
Francis publishes.

Taylor & Francis Group, LLC

#IP164/9780750308243

Series in
Medical Physics and Biomedical Engineering

Biomolecular Action of Ionizing Radiation

Edited by

Shirley Lehnert
Montreal General Hospital/McGill University
Canada

Taylor & Francis
Taylor & Francis Group
New York London

Taylor & Francis is an imprint of the
Taylor & Francis Group, an **informa** business

CRC Press
Taylor & Francis Group
6000 Broken Sound Parkway NW, Suite 300
Boca Raton, FL 33487-2742

© 2008 by Taylor & Francis Group, LLC
CRC Press is an imprint of Taylor & Francis Group, an Informa business

Library of Congress Cataloging-in-Publication Data

Lehnert, Shirley.
    Biomolecular action of ionizing radiation / Shirley Lehnert.
        p. ; cm. -- (Series in medical physics and biomedical engineering)
    "A CRC title."
    Includes bibliographical references and index.
    ISBN-13: 978-0-7503-0824-3 (hardcover : alk. paper)
    ISBN-10: 0-7503-0824-9 (hardcover : alk. paper)
    1. Ionizing radiation--Physiolgical effect. 2. Molecular radiobiology I. Title. II. Series.
    [DNLM: 1. Radiation, Ionizing. 2. Cells--radiation effects. 3. DNA--radiation effects. WN 105 L523b 2007]

QP82.2.I53L44 2007
612'.01448--dc22   2007015244

Visit the Taylor & Francis Web site at
http://www.taylorandfrancis.com

and the CRC Press Web site at
http://www.crcpress.com

Series Blurb

The *Series in Medical Physics and Biomedical Engineering* describes the applications of physical sciences, engineering, and mathematics in medicine and clinical research.

The series seeks (but is not restricted to) publications in the following topics:

- Artificial organs
- Assistive technology
- Bioinformatics
- Bioinstrumentation
- Biomaterials
- Biomechanics
- Biomedical engineering
- Clinical engineering
- Imaging
- Implants
- Medical computing and mathematics
- Medical/surgical devices
- Patient monitoring
- Physiological measurement
- Prosthetics
- Radiation protection, health physics, and dosimetry
- Regulatory issues
- Rehabilitation engineering
- Sports medicine
- Systems physiology
- Telemedicine
- Tissue engineering
- Treatment

The *Series in Medical Physics and Biomedical Engineering* is an international series that meets the need for up-to-date texts in this rapidly developing field. Books in the series range in level from introductory graduate textbooks and practical handbooks to more advanced expositions of current research.

The *Series in Medical Physics* is the official book series of the International Organization for Medical Physics.

Series Preface

The International Organization for Medical Physics

The International Organization for Medical Physics (IOMP), founded in 1963, is a scientific, educational, and professional organization of 76 national adhering organizations, more than 16,500 individual members, several corporate members, and four international regional organizations.

IOMP is administered by a council, which includes delegates from each of the adhering national organizations. Regular meetings of the council are held electronically as well as every three years at the World Congress on Medical Physics and Biomedical Engineering. The president and other officers form the executive committee, and there are also committees covering the main areas of activity, including education and training, scientific, professional relations, and publications.

Objectives

- To contribute to the advancement of medical physics in all its aspects
- To organize international cooperation in medical physics, especially in developing countries
- To encourage and advise on the formation of national organizations of medical physics in those countries which lack such organizations

Activities

Official journals of the IOMP are *Physics in Medicine and Biology* and *Medical Physics and Physiological Measurement*. The IOMP publishes a bulletin *Medical Physics World* twice a year, which is distributed to all members.

A World Congress on Medical Physics and Biomedical Engineering is held every three years in cooperation with IFMBE through the International Union for Physics and Engineering Sciences in Medicine (IUPESM). A regionally based international conference on medical physics is held between world congresses. IOMP also sponsors international conferences, workshops, and courses. IOMP representatives contribute to various international committees and working groups.

The IOMP has several programs to assist medical physicists in developing countries. The joint IOMP Library Programme supports 69 active libraries in 42 developing countries, and the Used Equipment Programme coordinates equipment donations. The Travel Assistance Programme provides a limited number of grants to enable physicists to attend the world congresses.

The IOMP Web site is being developed to include a scientific database of international standards in medical physics and a virtual education and resource center.

Information on the activities of the IOMP can be found on its Web site at www.iomp.org.

This book is dedicated to my family, who have had to put up with me while it was being written. To John, Tim, Heather, Nadya, Miko, Dominic, Joanne, Tabitha, and recent addition Spencer.

Contents

Acknowledgments

I am extremely grateful to the following friends and colleagues who read chapters of this book relating to their own expertise:

Alex Almasan, Ed Azzam, Eric Bernhard, Doug Boreham, Rob Bristow, Tony Brookes, Don Chapman, Slobodan Devic, JoAnna Dolling, Christina Haston, Mike Joiner, Susan Lees-Miller, Jian-Jian Li, Ron Mitchel, Ann Moons, Dave Murray, Peggy Olive, Tej Pandita, Guy Poirier, Stephane Richard, Jan Seuntjens, Frank Verhaegen, Mike Weinfeld, and Don Yapp.

They invariably provided constructive and helpful criticism and occasionally corrected a blatant error. Most importantly, they gave encouragement and support.

In addition, I thank the innumerable members of the scientific community with whom I have had the privilege to interact over many years for their generosity in sharing information and insights.

Finally, I am happy to acknowledge Ildiko Horvath for her work on the illustrations, and Anne Forrest, proofreader and English consultant who, among other things, was in charge of the distribution of semicolons.

Abbreviations

^{123}I-IAZA	^{123}I-labeled iodoazomycin arabinoside
^{18}F-EF5	2-(2-nitro-1[H]-imidazol-1-yl)-N-(2,2,3,3, 3-pentafluoropropyl)-acetamide
4E-BP1	eukaryotic translation initiation factor binding protein
5′dRP	deoxyribose-phosphate residue
^{60}CuATSM	^{60}Cu-labeled diacetyl-bis(N4-methylthiosemicarbazone)
6PGD	6-phosphogluconate dehydrogenase
A	(1) adenine; (2) adenosine
AP	(1) apurinic; (2) apyrimidinic
Apaf-l	apoptosis activating factor-l
AR	adaptive response
ARD1	arrest defective 1
ARF	alternate reading frame
Asmase	acid sphingomyelinase
ASM	airway smooth muscle
AT	ataxia telangiectasia (syndrome)
ATM	mutated in ataxia telangiectasia
ATF-2	activating transcription factor 2
ATF-3	activating transcription factor 3
ATP	adenosine triphosphate
BAEC	bovine aortic endothelial cells
Bcl-2	B-cell lymphoma gene 2
BER	base excision repair
bFGF	basic fibroblast growth factor
bp	base pair
BrdU	bromodeoxyuridine
C′	carboxy (terminus of protein)
C	(1) cytidine; (2) cytosine
CAD	caspase-activated deoxyribonuclease
CAK	CDK activating kinase
CAM	cell adhesive matrix
cAMP	cyclic adenosine monophosphate
CBP	cyclic AMP response binding protein
cDNA	complementary DNA
CDKN1A	cyclin-dependent kinase inhibitor-1A
CDK	cyclin-dependent kinase
CdkI	CDK inhibitor

COX-2	cyclooxygenase 2
CRE	cAMP response element
CREB	cAMP response element-binding
CT	computed tomography
dA	deoxyadenosine
DAG	diacyl glycerol
DAPI	4′6′-diamidino-2-phenylindole
DD	death domain
DDB2	DNA-damage binding protein 2
dG	deoxyguanosine
Diablo	direct IAP-binding protein with low pI
DIE	death-inducing factor
DISC	death-inducing signaling complex
DNA	deoxyribose nucleic acid
DPC	DNA-protein cross-links
DPI	diphenyliodonium
DR	cell surface death receptors
ds	double stranded (DNA or RNA)
DSB	DNA double-strand break
DTT	dithiothreitol
ECM	extracellular matrix
EGF	epidermal growth factor
EGF-R	EGF receptor
Egr-1	early growth response 1 transcription factor
eIF-4E	eukaryotic translation initiation factor 4E
EMSA	electrophoretic shift mobility assay
ERK1/2	extracellular signal-regulated kinase 1 and 2
eV	electron volt
FACS	fluorescence activated cell sorting
FAD	flavin adenine dinucleotide
FADD	Fas-associated death domain
FaPy	2,6-diamino-4-hydroxy-5-formamidopyrimidine
FasL	ligand of the Fas death receptor
FdG	18-fluorodeoxyglucose
FGF	fibroblast growth factor
FISH	fluorescence in situ hybridization
FMN	flavin mononucleotide
Ftase	farnesyl transferase
FTIs	farnesyl transferase inhibitors
GADD45A	growth arrest and DNA-damage-inducible, alpha
GAP	GTPase-activating proteins
GBM	glioblastoma multiforme
G-CS	granulocyte colony stimulating factor
GCV	gancyclovir
GDP	guanosine diphosphate
GEF	guanine nucleotide exchange factors

GF	growth factor
GFP	green fluorescent protein
GG-NER	global genome repair
GGT	gamma glutamyl transferase
GGTI	gamma glutamyl transferase inhibitors
GGTase	geranylgeranyltransferase
GI	gastrointestinal
GM-CSF	granulocyte-macrophage colony stimulating factor
GPx	glutathione peroxidases
GSNO	S-nitrosoglutathione adduct
GTP	guanine nucleotide triphosphate
GTPase	enzyme that cleaves GTP, generating GDP
Gy	gray
^{3}H	tritium
HATS	histone acetyltransferases
HbO$_2$	oxyhemoglobin
HDAC	histone deacetylase
HGF	hepatocyte growth factor; also called scatter factor (SF)
HIF-1	hypoxia-inducible factor-1
HLH	helix-loop-helix
HNC	head and neck cancer
HNSCC	head and neck squamous cell carcinoma
HRS	low-dose hypersensitivity
HSP	heat shock protein
HSV-tk	thymidine kinase gene
HTH	helix-turn-helix
IAP	inhibitor of apoptosis
ICAD	inhibitor of CAD
ICAM-1	intracellular adhesion molecule-1
IEG	immediate early gene
IFP	interstitial fluid pressure
IFN	interferon
IGF-1R	insulin-like growth factor-1 receptor
IGR-1	insulin-like growth factor 1
IKK	IκB kinases
IL	interleukin
iNOS	nitric oxide synthase
IP$_3$	inositol triphosphate
IR	ionizing radiation
IRR	increased radioresistance
Jak	Janus kinase
JNK/SAPK	c-Jun NH$_2$ terminal kinases or stress-activated kinase
kb	kilobase
kbp	kilobase pair
kDa	kilodalton
keV	kiloelectron volts

LAP	latency associated peptide
LCM	laser capture microdissection
LET	linear energy transfer
LOH	loss of heterozygosity
LPS	lipopolysaccharides
LQ	linear quadratic
M	mitosis
MoAb	monoclonal antibody
MAPK	mitogen activated protein kinase
MAPKK	kinase that phosphorylates MAPKK
MAPKKK	kinase that phosphorylates MAPKKK
MAP-KAP2	mitogen-activated protein kinase-activated protein-2 kinase
mBAND	multicolor chromosome banding
Mdm2	mouse double minute chromosome
MDS	multiply damaged sites
MEF2	myocyte enhancer factor-2
MEK 1/2	MAPK/Erk kinase 1/2
MEKK	MAPK/Erk kinase kinase
mFISH	multifluor FISH
μm	micrometer
MKK4/7	mitogen activated protein kinase kinase 4/7
MNK1	MAP kinase-interacting kinase 1
MnSOD	manganese superoxide dismutase
MORT-l	mediator of receptor-induced toxicity
MRI	magnetic-resonance imaging
MRI-BOLD	blood oxygen level dependent
mRNA	messenger RNA
mTOR	mammalian target of rapamycin
N′	N (amino) terminus
NAC	*N*-acetylcysteine
NADH	reduced form of nicotinamide adenine dinucleotide
NADPH	nicotinamide adenine dinucleotide phosphate
NBS	Nijmegen break syndrome
NER	nucleotide excision repair
NF-κß	nuclear factor kappa B
nM	nanomolar (10^{-9} molar)
NMR	nuclear magnetic resonance
NOS	nitric oxide synthase
NSCLC	nonsmall cell lung cancer
Oct-1	octamer-binding protein
OER	oxygen enhancement ratio
OFH	oxygen-fixation hypothesis
p	short arm of chromosome
p	probability of an event
PAI-1	plasminogen activator inhibitor-1

PAK	p21-activated kinase 2
PARP	poly(ADP-ribose) polymerase
PCC	premature chromosome condensation
PCNA	proliferating cell nuclear antigen
PDGF	platelet-derived growth factor
PDK1	phosphoinositide-dependent kinase 1
PET	positron emission tomography
PET-CT	combination PET CT imaging
PH	Pleckstrin homology
pHe	extracellular pH
pHi	intracellular pH
Pi	phosphate
PI3K	phosphatidylinositol 3-kinase
pK_a	acid-base ionization/dissociation constant
PKA	protein kinase A
PKB	protein kinase B (= Akt)
PKC	protein kinase C
PLDR	repair of potentially lethal damage
pO_2	oxygen tension
Prx	peroxiredoxin
PTB	phosphotyrosine binding
PtdIns(3,4,5)	$P_3$3-phosphorylated lipid phospatidylinositol-3,4,6 triphosphate
PTEN	phosphatase and tensin homolog deleted on chromosome 10
q	long arm of a chromosome
RAG	recombination activating gene
Rb	retinoblastoma
RBE	relative biological effectiveness
RCC	renal cell carcinoma
RCE	retinoblastoma control elements
RNS	reactive nitrogen species
ROS	reactive oxygen species
RTK	tyrosine kinase receptor
SAPK/JNK	stress-activated protein kinase
SCC	squamous cell carcinoma
SCE	sister chromatid exchange
SCF	stem cell factor
SCID	severe immunodeficiency syndrome
SER	sensitizer enhancement ratio
SF	scatter factor (see HGF)
SH1	Src homology 1 domain
SH2	Src homology 2 domain
SH3	Src homology 3 domain
SI	sphingosine
SIE	sis-inducible element

SKY	spectral karyotyping
Smac	second mitochondria-derived activator of caspase
SOS	son of sevenless
SPECT	single photon emission computed tomography
SRE	serum response element
SNP	single nucleotide polymorphism
SOD	superoxide dismutase
SRF	serum response factor
SSA	single-strand annealing
SSB	single-strand break
STAT	signal transducer and activator of transcription
T	(1) thymine; (2) thymidine
TCR	(1) T cell receptor; (2) transcription coupled DNA repair
TERT	telomerase reverse transcriptase
TGFα	transforming growth factor α
TGF-ß1	transforming growth factor β1
TK	tyrosine kinase
TKI	tyrosine kinase inhibitor
TNF	tumor necrosis factor
TNFerade	adenovector carrying the transgene for human tumor necrosis factor α
TNFR-1	TNF-alpha receptor 1
tRNA	transfer RNA
TRADD	TNF receptor-associated death domain
TPA	12-O-tetradecanoylphorbol-13 acetate
TRAIL	tumor necrosis-factor related apoptosis-inducing ligand
Trx	thioredoxin
TUNEL	TdT-mediated dUTP nick end labeling
U	(1) uracil; (2) uridine
uPA	urokinase-type plasminogen activator
uPAR	urokinase plasminogen activator receptor
UV	ultraviolet
VEGF	vascular endothelial growth factor
VHL	von Hippel-Lindau protein
XIAP	X-linked inhibitor of apoptosis tumor necrosis factor
Z	atomic number

1

Introduction

This book originated when it was suggested to me by a publisher that since I had been involved in teaching radiation biology for a number of years, I might be interested in writing a text on the subject. To this I naturally replied that there was nothing to be gained from writing a radiation biology textbook because there already was one, which was frequently updated [1]. (In fact, there is more than one [2,3].) The suggestion, however, caused me to think around the subject and particularly about the fact that—as have all branches of biomedicine—the radiation sciences have been overtaken and irreversibly transformed by the concepts and technology of molecular biology.

An axiom about the effect of ionizing radiation is that the primary target for cell killing is DNA and the distribution of radiation dose is random, not constrained by pharmacology. The finding that cell killing correlates well with unrepaired double-strand DNA breaks indicated that DNA is the prime target. Results of many recent studies have shown, however, that there are radiation-induced changes in an array of cellular targets. These include signal transduction pathway changes, mitochondrial changes, and apoptotic and cell cycle changes. Some of the changes are in response to DNA damage; however, non-DNA-mediated targets including the cell membrane and mitochondria can be critically involved in cell killing or can be the primary targets. New biological concepts for the molecular radiation oncology era are the possibility to modulate normal tissue response based on the understanding that late radiation injury is the result of chronic persistent inflammatory and cytokine-mediated processes.

Almost coincidental with the development of molecular radiobiology, mechanistic and molecular studies of the effects of low doses of radiation have resulted in radical changes in the perception of radiation effect. Bystander effects occur in non-hit cells as well as in cells where energy has been deposited. Genomic instability, which might be the result of radiation exposure, has been shown to play a role in the development of cancer. Finally, recent studies have shown that radiation-induced changes in gene expression can occur at very low doses and that these may be involved in protective or adaptive responses. These nontargeted effects, which have major implications for dosimetry, treatment planning, and risk assessment, are also discussed in this book.

Molecular medicine is providing new insights, which have major implications for diagnosis and treatment. As information on the molecular processes induced by radiation emerges, it creates unique opportunities to improve the use of radiation as a cytotoxic and cytostatic agent and to develop treatments which will target molecular intermediates or signaling molecules involved in radiation-induced processes in such a way as to be protective or sensitizing. The high-throughput techniques such as microarray and proteomics are a means to identify a molecular signature of the cell, which will enable treatment to be tailored to individual patients in terms of their tumor and normal tissue radiosensitivities. Although not specifically covered in this book, it should be remembered that the new techniques of molecular or functional imaging arise from an understanding of tumor molecular biology combined with sophisticated imaging techniques to allow noninvasive localization of key molecules.

Radiation science is by its nature interdisciplinary and multidisciplinary. Many of the earlier radiation biologists were trained as medical physicists and brought a rigorous quantitative approach to the wealth of data derived from basic radiation biology models. The radiation scientists of the twenty-first century should have a thorough grounding in all the basic sciences involved. The combined concepts and tools of radiation/molecular biology are essential for determining the mechanism of cell and tissue responses to radiation and in applying relevant findings to radiation oncology, carcinogenesis, mutagenesis, and other radiation effects. Quantification has always been an important part of the radiation sciences and should ideally carry over into analyzing the findings of molecular, cellular, and tissue research and to the design of treatments based on the results of this research. In effect, the goal should be to integrate the strengths of the molecular, mechanistic, and the phenomenological approaches to create a powerful consortium for research and development.

One of the features accompanying the advent of molecular biology in radiation medicine has been the coining of terms to describe the perceived subcategories of radiation biology. Thus, classical radiobiology has been compared with modern radiobiology or molecular radiobiology. At one meeting in Canada, I heard the term "nouveau radiobiology" used presumably in deference to both an emerging scientific field and our bilingual environment. Recently, these descriptions have been heard less frequently and, indeed, unnecessarily divide the field rather than underscoring its depth and breadth.

Recent years have seen extraordinary developments in the hardware and software involved in the delivery of ionizing radiation for the treatment of cancer. Techniques for anatomically directed radiation including intensity-modulated radiation therapy, stereotactic techniques, tomotherapy, proton therapy, and brachytherapy, available in the major medical centres in the developed world, have revolutionized the field. While this has been occurring, the other vital part of the treatment process, understanding the biology of radioresponse, has appeared to lag behind. In fact, this perception is mistaken; the last few years have witnessed the accumulation of a huge body of knowledge as to the effects of ionizing radiation on molecular and cellular biology, which continues at an ever-increasing pace. It is true that in most cases this understanding is not yet applied to the treatment of disease, but inevitably these discoveries move toward clinical application for the treatment of cancer and radiation injuries implying that radiation oncologists must understand the effects of ionizing radiation at all levels including the molecular processes underpinning the observed phenomena.

The target audiences for this book are residents in radiation oncology, medical physicists, and other students and graduate students who have an interest in the radiation sciences and in cancer treatment. Many of the chapters are concerned with one aspect of cellular or molecular biology, for instance apoptosis, and how this is affected when the cell is exposed to ionizing radiation. In fact, the separation of the material in this way is a convenient device but not representative of the fact that nothing happens in the cell without ultimately impacting every other cellular process and the organism of which it is a part. In effect, the initial energy deposition, which results from ionizing exposure, is like throwing a stone into a pond. The ripples engendered when the stone breaks the surface of the water extend to every other part of the pond and beyond.

The accumulation of knowledge about the molecular processes underlying the radiation response has generated both confusion and clarity. Confusion arises in the minds of those who, for instance, attend seminars where the effect of radiation on signal transduction pathways is described. Those whose expertise is not in molecular biology often become hopelessly lost in a maze of cross talking signal pathways and molecules identified only by inscrutable acronyms. These individuals form part of the target audience for this book; I hope it will help. The clarity comes when, for instance, the mechanisms underlying some hitherto inexplicable effect of radiation are revealed.

This book is by no means comprehensive. The important topics of radiation carcinogenesis, radioprotection, and radiation risk to human populations in occupational and other contexts have received only peripheral attention. My personal bias in radiation research has always been toward understanding that part of radiation biology which would contribute to improvements in the treatment of cancer with ionizing radiation. I consider this to be the most important application of this discipline, and in most of the chapters consideration is given at some point to the implications of the subject matter for the practice of radiation oncology.

References

1. Hall EJ and Giaccia AJ. *Radiobiology for the Radiologist*, 6th Edition, Lippincott Williams & Wilkins, Philadelphia, 2006.
2. Steele GG. *Basic Clinical Radiobiology*, 3rd Edition, Arnold, London, 2002.
3. Tannock IF, Hill RP, Bristow RG, and Harrington L, Eds. *The Basic Science of Oncology*, 4th Edition, McGraw Hill, New York, 2005, (chapters 14 and 15).

2

Basic Radiation Physics and Chemistry

The biological effects of irradiation are the end product of a series of events which are set in motion by the passage of radiation through the medium. The initial events are ionizations and excitations of atoms and molecules of the medium along the tracks of the ionizing particles. These physical perturbations lead to physicochemical reactions, then chemical reactions, and finally the biological effect. The physical and chemical reactions that initiate the process are briefly described in this chapter.

2.1 Ionization and Excitation

Ionizing radiation is radiation, either particulate or electromagnetic (EM), which is of sufficient energy to eject one or more orbital electrons from the atoms or molecules with which it interacts, thus causing the atom to be ionized. The characteristic which defines ionizing radiation is the localized release of large amounts of energy. Whenever the energy of a particle or photon exceeds the ionization potential of a molecule, a collision with the molecule can lead to ionization. The dose of radiation to biological material is defined in terms of the amount of energy absorbed per unit mass. 1 Gy is equivalent to 1 J/kg (Table 2.1).

An interaction that transfers energy, but does not completely displace an electron, produces an "excited" atom or molecule and is called an excitation. Excitation raises an electron in an atom or molecule to a higher energy level without actual ejection of an electron.

2.2 Types of Ionizing Radiation

2.2.1 Electromagnetic Radiation

Experiments with biological systems and most clinical applications involve either x-rays or gamma (γ)-rays, two forms of EM radiation with similar properties. The designations, x-ray or γ-ray, depend on the way by which they are produced. x-Rays are produced in an x-ray tube when accelerated

TABLE 2.1

Units Used in Radiation Research

Radiation (Physical Quantities)	Unit	Conversion to Other Systems
Absorbed dose	Gray (Gy) Joules/kg	1 rad = 0.01 Gy
Dose equivalent	Sievert (Sv) Joules/kg	1 rem = 0.01 Sv
Energy	Joule (J) N m	$1 \text{ eV} = 1.6 \times 10^{-19} \text{ J}$
Dose rate	Gy/s	
Exposure	Röentgen (R)	$1 \text{ Röentgen} = 2.58 \times 10^{-4} \text{ C/kg}$
Electric charge	Coulomb (C)	
Activity	Becquerel (Bq)/s	$1 \text{ Curie (Ci)} = 3.7 \times 10^{10} \text{ Bq}$

electrons hit a tungsten target and then decelerate, emitting a spectrum of bremsstrahlung radiation as part of the kinetic energy (KE) of the electrons which is converted into x-rays. (Bremsstrahlung are x-rays resulting from the interaction of an electron with a target nucleus.) The resulting spectrum is filtered and otherwise modulated to produce a clinically useful x-ray beam. γ-Rays, in contrast, are produced spontaneously. They are emitted by radioactive isotopes and represent excess energy that is given off as the unstable nucleus breaks up and decays as it reverts to a stable form.

x-Rays and γ-rays are part of the continuous spectrum of EM radiation that includes radio waves, heat, and visible and ultraviolet (UV) light (Figure 2.1).

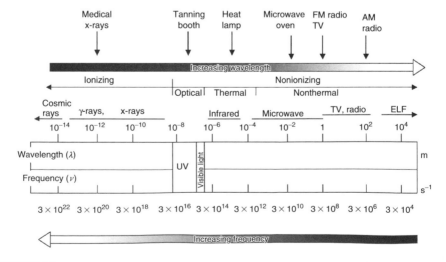

FIGURE 2.1
Electromagnetic spectrum showing relationship between wavelength (m) and frequency (s⁻¹). Bands in the spectrum and their characteristics are indicated.

Electromagnetic radiation can be considered as a series of waves described by the relation:

$$\nu\lambda = c = 3 \times 10^8 \text{ m/s}$$

where
  c = speed of light
  ν = frequency (s^{-1})
  λ = wavelength (m)

Ionizing radiation comes from the extreme end of the EM spectrum characterized by short wavelength (usually expressed in Ångström units) and high frequency.

EM radiation can also be viewed as moving packets or quanta of energy called photons. Photons travel at velocity c and the energy (in joules) of each quantum is equal to the frequency (ν) multiplied by Planck's constant (η).

$$E = \eta\nu$$

The amount of energy per photon is defined by the position in the EM spectrum, x- or γ-ray photons are at the high-energy end of the EM spectrum, and individual photons have enough energy to completely displace an electron from its orbit around the nucleus of an atom, leaving the atom or molecule ionized with a net positive charge. Typical binding energies for electrons in biological material are around 10 eV and photons with energies greater than 10 eV are considered to be ionizing radiation, while photons with energies of 2–10 eV are in the UV range and are nonionizing, but can cause excitation.

The concept of x-rays being composed of photons is very important to the understanding of the biological effects of radiation because it implies that when x-rays are absorbed, the energy is deposited in the tissues and cells unevenly, in discrete packets. Thus, for a beam of x-rays the energy is quantized into large individual packets, each of which is big enough to break a chemical bond and initiate the chain of events that culminates in biological damage. The critical difference between nonionizing and ionizing radiations is the size of the individual packets of energy, not the total energy involved. Energy delivered in the form of heat or mechanical energy is absorbed homogenously and much greater quantities of energy in this form are required to produce damage in biological material.

2.2.2 Particulate Radiations

Other types of radiation that occur in nature and also are used experimentally and in some cases clinically are electrons, protons, α-particles, neutrons, negative π-mesons, and heavy charged ions.

- Electrons are small, negatively charged particles that can be accelerated to high energy to a speed close to that of light by means of an electrical device such as a betatron or linear accelerator. They are used in the treatment of cancer.

- Protons are positively charged particles having a mass 1860 times greater than that of an electron. Because of their mass they require complex and expensive equipment, such as a cyclotron, to accelerate them to useful energies. Naturally occurring protons originating from the Sun represent part of the natural background radiation. The surface of the Earth is protected by the atmosphere and the Earth's magnetic field which deflects charged particles, but protons still remain a major hazard to long-range space missions.

- α-Particles are nuclei of helium atoms consisting of two protons and two neutrons. They have a net positive charge and can therefore be accelerated with very large machines similar to those used to accelerate protons. α-Particles are also emitted during the decay of naturally occurring heavy radionuclides, such as uranium and radium.

- Neutrons are uncharged particles with a mass similar to that of a proton. Since they are electrically neutral, they cannot be accelerated in an electrical device but they can be produced artificially when a charged particle, such as a deuteron, is accelerated to high energy and strikes a suitable target material (a deuteron is a nucleus of deuterium and consists of a closely associated proton and a neutron). Neutrons are emitted when radioactive isotopes of heavy elements undergo fission, and consequently are present in large quantities in nuclear reactors, and they are also emitted by some artificially synthesized heavy radionuclides. Neutrons are important components of space radiation and contribute significantly to the exposure of passengers and crew of high-flying aircraft.

2.3   Processes of Energy Absorption

The process by which x-ray photons are absorbed depends on the energy of the photons concerned and the chemical composition of the absorbing material. When x-ray photons interact with tissue, they give up energy by one of three processes: the photoelectric effect, the Compton effect, or by pair production. All three of these interactions result in the production of energetic electrons that, in turn, lose energy by exciting and ionizing target atoms and molecules and setting more electrons in motion. If an x-ray beam is absorbed by tissue, the net result is the production of a large number of fast electrons, many of which can ionize other atoms of the absorber and are sufficiently energetic to break chemical bonds.

In the photoelectric process, the x-ray photon interacts with a bound electron in the K, L, or M shell of an atom of the absorbing material. The photon gives up all of its energy to the electron. Part of the transferred energy acts to overcome the binding energy of the electron and releases it from its orbit, while the remainder contributes to the KE of the electron. The KE of the ejected electron is given by the expression

$$KE = h\nu - EB$$

in which $h\nu$ is the energy of the incident photon and EB is the binding energy of the electron in its orbit (Figure 2.2A).

Ejection of an electron leaves a vacancy that must be filled by another electron falling in from an outer shell of the same atom or by a conduction electron from outside the atom. Movement of the negatively charged electron from a loosely bound to a tightly bound shell is a decrease of potential energy, which is balanced by the emission of an EM photon characteristic with low energy, typically 0.5 kV which is of little biological consequence.

In the energy range most widely used in radiotherapy (100 keV–25 MeV), the Compton effect is the most important mechanism leading to deposition of energy in tissue. This energy-transfer process involves a collision between the

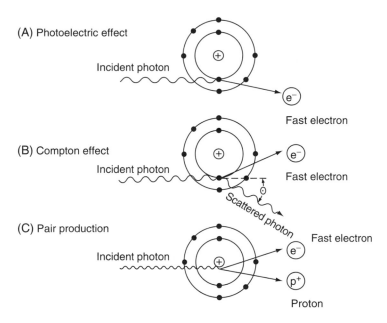

FIGURE 2.2

Absorption of a photon of x- or γ-rays by (A) photoelectric process—the photon gives up its energy entirely and the electron is ejected with energy equal to that of the incident photon minus the binding energy of the electron; (B) the Compton process—part of the energy of the photon is transferred to a loosely bound planetary electron as kinetic energy and the photon is deflected and proceeds with reduced energy; (C) pair production—γ-ray photons with energy greater than 1.02 MeV interact with a nucleus to form an electron–positron pair. Excess energy is carried away equally by these two particles which produce ionization as they travel through the medium. The positron is eventually captured by an electron and annihilation of the two particles occurs resulting in the release of two photons each of 0.51 MeV annihilation radiation.

photon and an outer orbital electron of an atom, with partial transfer of energy to the electron and scattering of the photon into a new direction. The photon interacts with what is usually referred to as a "free" electron, an electron whose binding energy is negligibly small compared with the photon energy. The electron (and the photon) can then undergo further interactions, causing more ionizations and excitations, until its energy is dissipated (Figure 2.2B).

For lower photon energies such as those used in diagnostic radiology, both Compton and photoelectric absorption processes occur, with Compton absorption dominating at the higher end of the energy range and the photoelectric effect being more important at lower energies.

The Compton and photoelectric absorption processes differ in several respects that are important for the application of x-rays to diagnosis and therapy. For the Compton process, the mass absorption coefficient is independent of the atomic number (Z) of the absorbing material, whereas the mass absorption coefficient for photoelectric absorption varies with Z. As a consequence, x-rays are absorbed to a greater extent by bone because bone contains elements, such as calcium, with high atomic numbers. This differential absorption in materials of high Z is the basis of the diagnostic utility of radiography. On the other hand, for radiotherapy, high-energy photons in the megavoltage range are preferable, because in that case the Compton process is predominant and, since the absorbed dose is about the same in soft tissue, muscle, and bone, the differential absorption in bone is avoided.

Pair production requires higher-energy photons (>1.02 MeV, usually >5 MeV). γ-Ray photons with energy greater than 1.02 MeV may interact with a nucleus to form an electron–positron pair. This amount of energy is just sufficient to provide the rest masses of the electron and positron (0.51 MeV). Excess energy will be carried away equally by these two particles which produce ionization as they travel through the medium. The positron is eventually captured by an electron, and annihilation of the two particles occurs. This results in the release of two photons, each of 0.51 MeV, known as annihilation radiation. These two photons then lose energy by Compton scattering or the photoelectric effect (Figure 2.2C).

Differences among the various absorption processes are of practical importance in radiology, but the consequences for radiobiology are minimal since in all cases most of the energy of the absorbed photon is converted into the KE of a fast electron.

2.4   Direct and Indirect Action of Radiation

Any form of radiation, photons, neutrons, or charged particles, may interact directly with target structures causing ionization or excitation, thus initiating the chain of events that leads to a biological change. The direct action of radiation is the dominant process for radiations with high linear energy transfer (LET), such as neutrons or α-particles.

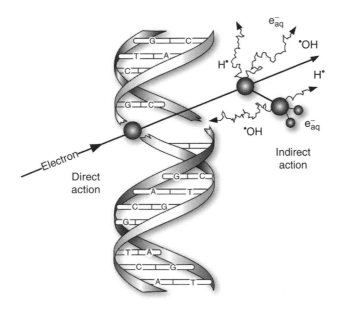

FIGURE 2.3
Direct and indirect actions of radiation. In direct action, a secondary electron reacts directly with the target to produce an effect. In indirect action, free radicals produced as a result of the radiolysis of water interact with the target to produce target radicals.

Radiation can also interact with other atoms or molecules in the cell (particularly water) to produce radicals, more frequently called free radicals that are able to diffuse over distances sufficient to interact with, and cause damage to, critical targets; this is called indirect action of radiation (Figure 2.3). The term radical is used to describe an atom or group of atoms containing an unpaired electron, which results in great chemical reactivity (Figure 2.4). Free radicals are usually denoted by a dot (•) placed to the left or right of the chemical symbol depending on the position of the unpaired electron. Most of the energy deposited in cells is absorbed initially in water (the cell contains 70%–80% water), leading to the rapid (within 10^{-5}–10^{-4} s) production of oxidizing and reducing reactive radical intermediates (the OH radical [•OH], an oxidizing agent, is probably the most damaging), which in turn can react with other molecules in the cell.

In model conditions, such as the irradiation of DNA in solution, the overwhelming contribution of free-radical damage to DNA will be caused by the free radicals generated by the radiolysis of water and only a negligible part will arise from the absorption of the energy of the ionizing radiation by DNA itself. The situation in the cellular environment is more complex than in aqueous solution, involving, in the case of DNA, bound water and endogenous free-radical scavengers. The cell contains naturally occurring thiol compounds such as glutathione and cysteine, which contain sulfhydryl (SH) groups that can react chemically with the free radicals to

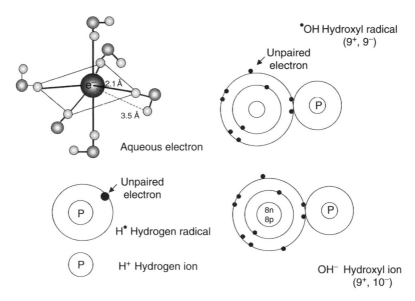

FIGURE 2.4
Reactive species produced by the ionization of water.

decrease their damaging effects. Other antioxidants include vitamins C and E and intracellular manganese superoxide dismutase (MnSOD) (see Chapter 5). Nevertheless, given that the cellular environment is 70%–80% water, the contribution of free-radical processes (generated by the radiolysis of water) to biological damage for sparsely ionizing radiation far exceeds that of direct action.

2.5 Radiolysis of Water

When the energy loss of high-energy electrons is absorbed by water, approximately 100 eV of energy is deposited per ionization event (on average) to generate a water radical and an electron (Equation 2.1). The latter may still contain enough energy to cause further ionizations in the very near neighborhood. These areas containing a number of ionization and, occasionally also, excitation events (Equation 2.2) are called spurs (Figure 2.5). In the case of sparsely ionizing radiation, these spurs do not overlap, but for densely ionizing radiations they may form cylinders of spurs (called tracks) from which some δ-rays (medium-energy electrons) may branch off.

$$H_2O \xrightarrow[\text{radiation}]{\text{ionizing}} H_2O^{\bullet+} + e^- \tag{2.1}$$

$$H_2O \xrightarrow[\text{radiation}]{\text{ionizing}} H_2O^* \tag{2.2}$$

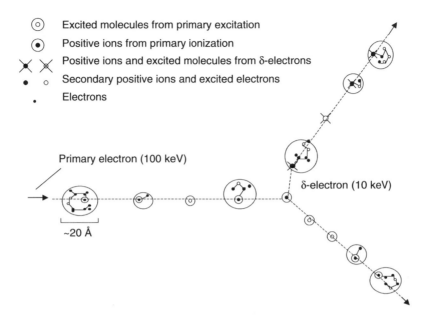

Excited molecules from primary excitation

Positive ions from primary ionization

Positive ions and excited molecules from δ-electrons

Secondary positive ions and excited electrons

Electrons

Primary electron (100 keV)

~20 Å

δ-electron (10 keV)

FIGURE 2.5
The spur model of energy deposition. An ionizing particle passing through absorbing matter energy is not deposited uniformly but in small packages called spurs.

The water radical cation produced in Equation 2.1 is a very strong acid and immediately loses a proton to neighboring water molecules, thereby forming •OH (Equations 2.3 and 2.4). The electron becomes hydrated by water (Equation 2.5). Electronically excited water can decompose into •OH and H• (Equation 2.6). As a consequence, three kinds of free radicals are formed side by side in the spurs •OH, e_{aq}^-, and H•. To match the charge of the electrons an equivalent amount of H^+ is also present.

$$H_2O^{\bullet+} + H_2O \rightarrow H_3O^+ + {}^\bullet OH \tag{2.3}$$

$$H_2O^{\bullet+} \rightarrow {}^\bullet OH + H^+ \tag{2.4}$$

$$e^- + nH_2O \rightarrow e_{aq}^- \tag{2.5}$$

$$H_2O^* \rightarrow {}^\bullet OH + H^\bullet \tag{2.6}$$

Since a spur can contain more than one free-radical pair, there is the possibility that they will interact with one another. While processes such as the reactions of •OH with the e_{aq}^- (Equation 2.7), or with H• (Equation 2.8), will not lead to measurable final products, the self-termination of two •OH radicals produces H_2O_2 (Equation 2.9), while H_2 is formed in Equations 2.10 and 2.11.

$$^\bullet OH + e_{aq}^- \rightarrow OH^\bullet \tag{2.7}$$

$$\text{\textbullet OH} + \text{H} \rightarrow \text{H}_2\text{O} \tag{2.8}$$

$$2\text{\textbullet OH} \rightarrow \text{H}_2\text{O}_2 \tag{2.9}$$

$$2\text{H}^\bullet \rightarrow \text{H}_2 \tag{2.10}$$

$$2\text{e}^-_{aq} + 2\text{H}^+ \rightarrow \text{H}_2 \tag{2.11}$$

To study the reaction of $\text{\textbullet OH}$ without the contribution of e^-_{aq}, e^-_{aq} can be converted into further $\text{\textbullet OH}$ by saturating the solution with N_2O.

$$2\text{e}^-_{aq} + \text{N}_2\text{O} \rightarrow \text{\textbullet OH} + \text{N}_2 + \text{OH}^- \tag{2.12}$$

Saturation of the solution with O_2 converts $\text{H}^\bullet$ and e^-_{aq} into $\text{O}_2^{\bullet-}$ (superoxide radical) and $\text{HO}_2^\bullet$ (hydroperoxyl radical).

$$\text{e}^-_{aq} + \text{O}_2 \rightarrow \text{O}_2^{\bullet-} \text{ (superoxide radical)} \tag{2.13}$$

$$\text{H}^\bullet + \text{O}_2 \rightarrow \text{HO}_2^\bullet \text{ (hydroperoxyl radical)} \tag{2.14}$$

$$\text{HO}_2^\bullet \rightleftarrows \text{H}^+ + \text{O}_2^{\bullet-} \tag{2.15}$$

Thus, the overall reaction for the radiolysis of water in the presence of oxygen can be written as

$$\text{H}_2\text{O} \rightarrow \text{\textbullet OH}, \text{e}^-_{aq}, \text{H}^\bullet, \text{O}_2^{\bullet-}, \text{H}_2\text{O}_2, \text{H}_2 \tag{2.16}$$

2.5.1 Haber–Weiss Reaction

The OH radical can be formed from hydrogen peroxide in a reaction catalyzed by metal ions (Fe^{2+} or Cu^+) (the Fenton reaction).

$$\text{H}_2\text{O}_2 + \text{Fe}^{2+}/\text{Cu}^+ \rightarrow \text{Fe}^{3+}/\text{Cu}^{2+} + \text{\textbullet OH} + \text{OH}^- \tag{2.17}$$

Superoxide also has a role in recycling metal ions.

$$\text{Fe}^{3+}/\text{Cu}^{2+} + \text{O}_2^{\bullet-} \rightarrow \text{Fe}^{2+}/\text{Cu}^+ + \text{O}_2 \tag{2.18}$$

The sum of these two reactions is known as the Haber–Weiss reaction.

$$\text{O}_2^{\bullet-} + \text{H}_2\text{O}_2 \rightarrow \text{H}_2\text{O} + \text{O}_2 + \text{\textbullet OH} \tag{2.19}$$

This reaction is also discussed in Chapter 5.

2.5.2 Reactions of the Primary Radiolytic Products of Water with Target Molecules

Reactions of the primary radiolytic products of water with target molecules are as follows:

2.5.2.1 Hydrogen Abstraction

Examples:

$$R - H + H^\bullet \rightarrow R^\bullet + H_2$$
$$R - H + {}^\bullet OH \rightarrow R^\bullet + H_2O$$

(H represents a reactive hydrogen atom and $R^\bullet$ the remainder of the molecule.)

The result of the reaction of the solute molecule with either $H^\bullet$ or ${}^\bullet OH$ is the formation of a solute radical $R^\bullet$ (e_{aq}^- does not extract hydrogen atoms from C–H bonds). The creation of solute radicals by a process of hydrogen abstraction is believed to be a major contributor to the damage of biological molecules.

2.5.2.2 Dissociation

Examples:

$$R - NH_3^+ + e_{aq}^- \rightarrow R^\bullet + NH_3$$
$$R - NH_2 + H^\bullet \rightarrow R^\bullet + NH_3$$

2.5.2.3 Addition Reactions

Example:

${}^\bullet OH$ reacts with C–C double bonds at close to diffusion-controlled rates but is highly selective largely due to its electrophilic nature. For example, the C(5)–C(6) double bond in the pyrimidines is preferentially attacked at the electron-richer C(5). (Radiation damage to purines and pyrimidines is described in more detail in Chapter 6.)

2.5.3 Solute Radicals Form Stable Products

Solute radicals can react with each other or with other reactants in the milieu to create stable end products. The most frequent types of reactions are

$$\text{Dimerization and addition:} \quad R_1^\bullet + R_2^\bullet \rightarrow R_1 - R_2$$
$$R^\bullet + R^\bullet \rightarrow R - R$$

Addition of oxygen: $R^\bullet + O_2 \rightarrow RO_2^\bullet \rightarrow RO_2H$

$R^\bullet + HO_2^\bullet \rightarrow RO_2H$

Hydrogen transfer: $R^\bullet + P - H \rightarrow R - H + P^\bullet$

The addition of O_2 to target radicals to form stable hydroperoxides is very important from the standpoint of the biological effects of radiation since it irreversibly fixes the radiation-induced lesion. Molecular oxygen is a highly effective and almost universally available radiosensitizer that acts at the free-radical level. (This is discussed at greater length in Chapter 16.)

The converse of the addition of oxygen is restoration of the target molecule by H donation. This is also an important reaction at the cellular level and is one mode of action of radioprotective compounds, especially thiols.

2.6   Linear Energy Transfer

As a charged particle moves through matter, it transfers energy by a series of interactions occurring randomly. Thus, the rate at which particles lose energy along their track (dE) is proportional to their energy. The energy loss dE along a portion of track dx is also dependent on the particle's velocity (v), charge Z, and the electron density of the target ρ, according to the relationship

$$\frac{dE}{dx} \propto \frac{Z^2\rho}{v^2}$$

The important difference between different types of radiation is the average density of energy loss along the path of the particle. The average energy lost by a particle over a given track length is known as the LET and the units of LET are given in terms of energy lost per unit path length (i.e., keV/μm). Some representative values of LET for different particles are given in Table 2.2.

LET is a simplistic way to describe the quality of different types of radiation since it fails to address the size of the individual energy-loss events that occur along the track of a particle. It is, however, useful as a crude index of radiation quality. For a given type of particle or ray, the higher the energy, the lower the LET (Table 2.2). At first sight, this may seem counterintuitive, but in fact as the radiation becomes more energetic, the beam penetrates deeper into tissue because it is sparsely ionizing, deposits less energy along its track, and hence has lower LET and relative biological effectiveness (RBE).

As a charged particle slows down, it loses energy more and more rapidly and reaches a maximum rate of energy loss (the Bragg peak) just before it comes to rest (Figure 2.5). This pattern of energy-loss distribution applies to

TABLE 2.2

Linear Energy Transfer of Various
Radiations

Radiation	LET (keV/μm)
Photons	
^{60}Co (~1.2 MeV)	0.3
200 keV x-ray	2.5
Electrons	
1 MeV	0.2
100 keV	0.5
10 keV	2
1 keV	10
Charged particles	
Proton 2 MeV	17
Alpha 5 MeV	90
Carbon 100 MeV	160
Neutrons	
2.5 MeV	15–80
14.1 MeV	3–30

all charged particles including the electrons displaced by EM photons. Photon irradiation sets in motion electrons which are easily deflected due to their small mass and whose track through the tissue is convoluted. Each electron track has a Bragg peak at its termination and a range of LET values along its track, but both the initiation and termination points of the electron tracks occur at random in the tissue, so that the LET spectrum is similar at all depths. In contrast, for a beam of heavy charged particles (e.g., high LET protons or heavy nuclei) the tracks are much straighter because their larger mass reduces the probability that they will be deflected. The Bragg peak then occurs at a similar depth in tissue for all particles and there is a region where a relatively large amount of energy is deposited, a feature which makes these particles potentially attractive as a treatment modality.

The concept of an average LET does not address the size of the individual energy-loss events that occur along the track of a particle. These fluctuations are important because they are the reason why equal doses of different radiations produce effects of different magnitude, i.e., difference in RBE. While the absorbed dose determines the average number of energy deposition events, each cell as an individual entity will react to the actual energy deposited in it. The average response of a system of cells should therefore depend on the energy distribution on a scale that is at least as small as the dimensions of the cell. In addition, the pattern of energy deposition at the subcellular level is also of importance because radiosensitive components occupy only a portion of the cell.

A picture of the submicroscopic pattern of ionizations within a cell nucleus can be realized using special techniques for measuring ionization in very small volumes combined with computer simulations of dose distribution, a field of study known as microdosimetry. Microdosimetric calculations for the tracks of γ-rays compared with high LET radiation indicate why one type of track leads to more severe damage. γ-Rays deposit much more of their energy as single, isolated ionizations creating damage with higher probability of repair by intracellular mechanisms. About 1000 of these sparse tracks are produced per Gy of absorbed radiation. The α-particles produce fewer tracks, but each track leads to more severe (non-repairable) damage where the track intersects vital structures. In terms of LET, the γ-rays have an LET of about 0.3 keV/μm (low LET radiation) whereas α-particles have an LET of about 100 keV/μm and are high LET radiation [1].

2.7 Relative Biological Efficiency

As LET increases radiation, it causes more cell kill per Gy; in other words, it increases the biological effectiveness of the radiation. The relative biological efficiency (RBE) of a radiation is defined as

$$\text{RBE} = \frac{\text{Dose of reference radiation}}{\text{Dose of test radiation}}$$

to give the same biological effect. The reference low LET radiation has historically been 250 kV x-rays. The RBE is not constant but depends on the end point and the level of biological damage, and hence on the dose level. RBE rises to a maximum at an LET of about 100 keV/μm and then falls at higher levels of LET due to "overkill" (Figure 2.6). Low LET radiation of very dense ionization (LET > 100 keV/μm) is inefficient because more energy than necessary is deposited at critical sites and energy is wasted because the ionizing events are too close together. Since RBE is a ratio of doses to produce an equal biological effect, this densely ionizing radiation has a lower RBE than does radiation of lower energy but optimal LET.

FIGURE 2.6
Schematic illustration of the energy deposition by a charged particle along its track in tissue. The particle has a high velocity at the left hand side of the figure, but as it loses energy, it slows down until it comes to rest in the region of the Bragg peak.

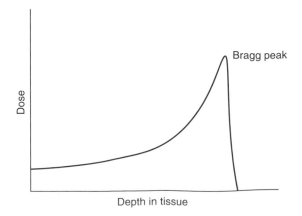

2.8 Summary

x-Rays and γ-rays are indirectly ionizing; the first step in their absorption is the production of fast recoil electrons. Neutrons are also indirectly ionizing; their absorption results in the production of fast recoil protons and other particles. Biological effects of x-rays are due to direct action (recoil electrons interact with target molecule) or indirect action (recoil electrons interact with the solvent to produce free radicals which interact with the target). The direct action of radiation is the dominant process for radiations with high LET, such as neutrons or α-particles.

The most important products of the radiolysis of water, which mediate indirect action, are $^\bullet OH$, $H^\bullet$, and e_{aq}. About two-thirds of biological damage by sparsely ionizing radiation is due to indirect action, and $OH^\bullet$ is considered to be the most damaging radical species. Damage produced by indirect action can be modified by chemical sensitizers and protectors.

The radicals produced by the radiolysis of water react with target molecules to produce solute radicals by processes of hydrogen abstraction, dissociation, and addition. Solute radicals revert to stable products by processes of dimerization or addition, disproportionate reactions, or by reaction with hydrogen or oxygen.

LET is the energy transferred by photon or particle per unit length of track. It is expressed as an average quantity, keV/μm. In fact, energy is not deposited homogenously and the study of microdosimetry has developed to describe the distribution of dose on a microscopic scale. It is based on the principle that biological effect is related to the quantity of energy deposited in elemental structures of very small size. Relative biological efficiency (RBE) is a ratio of doses D_{x250}/D_r to produce the same biological effect. RBE increases with LET up to 100 keV/μm and then decreases due to overkill. RBE varies according to the tissue or end point studied.

Reference

1. Goodhead DT. Spatial and temporal distribution of energy. *Health Phys.* 55: 231–240, 1988.

Bibliography

1. Steele GG. *Basic Clinical Radiobiology*, 3rd Edition, Arnold, London, 2002.
2. Altman K, Gerber GB, and Okada S. *Radiation Biochemistry*, 1st Edition, vol. 1, Academic Press Inc., New York, 1970.

3. Hall EJ and Giaccia AJ. *Radiobiology for the Radiologist*, 6th Edition, Lippincott Williams & Wilkins, Philadelphia, 2006.
4. Tubiana M, Dutreix J, and Wambersie A. *Introduction to Radiation Biology*, Taylor and Francis, London, New York, Philadelphia, 1990.
5. von Sonntag C. *Free-Radical-Induced DNA Damage and Its Repair. A Chemical Perspective*, Springer, Berlin, New York, 2006.

3

Basic Cell Biology and Molecular Genetics

This chapter is concerned with the basic cell biology and molecular genetics, which will be helpful to know when reading this book. It is obviously impossible to provide a thorough background in the scope of one chapter, but there are a number of excellent books and Web sites dealing with this subject, some of which are referenced at the end of the chapter. In addition, chapters include background on specific topics.

3.1 Basic Cell Biology

Eukaryotic cells are characterized by organization and compartmentalization of the cell, which is necessary for the orderly progression of the multitude of reactions that are taking place simultaneously in the cell. Eukaryotic cells are large compared with prokaryotes and have a cytoskeleton that gives the cell mechanical strength and controls its shape. The cell is enclosed in a plasma membrane. By definition, eukaryotic cells keep their DNA in a separate internal compartment; the nucleus and the nuclear membrane are only one of an elaborate set of internal membranes, structurally similar to the plasma membrane, enclosing different types of spaces within the cell and dedicated to different cellular processes. In this section the structures and organelles to be found in the mammalian cell are briefly described (Figure 3.1).

3.1.1 Cell Membrane

The cell membrane or plasma membrane is a selectively permeable bilayer, which comprises the outer layer of a cell. The dimensions of the plasma membrane (65–100 angstroms [Å] in thickness) are below the limits of the light microscope, but can be visualized by electron microscopy. The membrane largely consists of phospholipid and protein molecules. The basic functions of the plasma membrane are to enclose the components of the cell and to serve as a boundary through which substances must pass to enter or exit the cell. Plasma membranes are described as semipermeable, differentially permeable, or selectively permeable since they are usually freely permeable to water but act as partial barriers to the movement of almost all other substances.

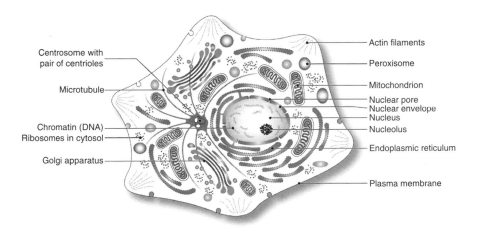

FIGURE 3.1
Major features of a typical mammalian cell.

The basic composition and structure of the plasma membrane is the same as that of the membranes that surround organelles and other subcellular compartments. The foundation is a phospholipid bilayer, and the membrane as a whole is often described as a fluid mosaic—a two-dimensional fluid of freely diffusing lipids, dotted or embedded with proteins, which may function as channels or transporters across the membrane, or as receptors for specific ligands. Some of these proteins simply adhere to the membrane (extrinsic or peripheral proteins), whereas others may be said to reside within it or to span it (intrinsic proteins). Glycoproteins in the membrane have carbohydrates attached to their extracellular domains. Cells may vary the variety and the relative amounts of different lipids to maintain the fluidity of their membranes despite changes in temperature and, in eukaryotes, regulation is assisted by cholesterol molecules.

As already noted, the cell membrane is semipermeable. This means that only some molecules can pass unhindered in or out of the cell—mostly those molecules which are either small or lipophilic. Factors that influence the passage of molecules through the membrane are as follows:

(i) Size of molecules: Large molecules cannot pass through the plasma membrane.

(ii) Solubility in lipids: Substances that dissolve easily in lipids pass through the membrane more readily than other substances.

(iii) Charge on ions: The charge on an ion attempting to cross the plasma membrane can determine how easily the ion enters or leaves the cell.

(iv) The availability of carriers: Plasma membranes contain special molecules called carriers that are capable of attracting and transporting substances across the membrane regardless of size, ability to dissolve in lipids, or membrane charge.

The main processes by which substances can be moved across the plasma membrane are as follows:

- Passive transport moves different chemical substances across membranes through diffusion of hydrophobic (nonpolar) and small polar molecules, or facilitated diffusion of polar and ionic molecules, which relies on a transport protein to provide a channel or bind to specific molecules. This spontaneous process decreases free energy and increases entropy in a system, but, unlike active transport, it does not involve any chemical energy (from ATP).

- Diffusion is the net movement of material from an area of high concentration of that material to an area with lower concentration. The difference of concentration between the two areas is often termed a concentration gradient, and diffusion will continue until this gradient has been eliminated.

- Facilitated diffusion is movement of molecules across the cell membrane via special transport proteins that are embedded within the cellular membrane. Many large molecules, such as glucose, are insoluble in lipids and too large to fit through the membrane pores. In this case the molecule will bind to its specific carrier proteins, and the complex will then be bonded to a receptor site and moved through the cellular membrane. Facilitated diffusion is a passive process and the solutes are still moving down a concentration gradient.

- Filtration is movement of water and solute molecules across the cell membrane due to hydrostatic pressure generated by the cardiovascular system. Depending on the size of the membrane pores, only solutes of a certain size may pass through.

- Active transport can be of two main types: primary and secondary. In primary transport, energy is directly coupled to the movement of a desired substance across a membrane independent of any other species. Secondary transport concerns the diffusion of one species across a membrane to drive the transport of another.

Primary active transport directly uses energy to transport molecules across a membrane. Most of the enzymes that perform this type of transport are transmembrane ATPases. A primary ATPase universal to all cellular life is the sodium–potassium pump, which helps maintain the cell potential.

In secondary active transport, there is no direct coupling of ATP; instead, the electrochemical potential difference created by pumping ions out of cells is used. The three main forms of this are uniport, counter-transport (antiport), and cotransport (symport). In counter-transport two species of an ion or other solutes are pumped in opposite directions across a membrane. One of these species is allowed to flow from high to low concentration, which yields the entropic energy to drive the transport of the other solute from a low concentration region to a high one. An example is the sodium–calcium exchanger or antiporter, which allows three sodium ions into the cell to transport one calcium ion out. Cotransport also uses the flow of one solute species from high to low concentration to move another molecule against its preferred direction of flow. An example is the glucose symporter, which cotransports two sodium ions for every molecule of glucose it imports into the cell.

3.1.2 Cytoplasm

The cytoplasm is the matrix or ground substance surrounding the various subcellular components. Cytoplasm is a thick, semitransparent, elastic fluid containing suspended particles. Chemically, cytoplasm is approximately 90% water plus solid components, mostly proteins, carbohydrates, lipids, and inorganic substances. The inorganic substances and most carbohydrates are soluble in water and are present as a true solution. The majority of organic compounds, however, are present as colloids—particles that remain suspended in the surrounding ground substance. The particles of a colloid bear electrical charges that repel each other and, as a consequence, they remain suspended and separate from each other. Functionally, cytoplasm is the milieu in which chemical reactions occur. The cytoplasm receives raw materials from the external environment and converts them into usable energy by decomposition reactions, and cytoplasm is also the site where new substances are synthesized for cellular use. In the cytoplasm, chemicals are packaged for transport to other parts of the cell or to other cells of the body and the excretion of waste materials is facilitated. Despite the myriad chemical activities occurring simultaneously in the cell, there is little interference of one reaction with another. This is possible because, within the cytoplasm there is a system of compartmentalization provided by structures, collectively called organelles that are specialized portions of the cell with specific roles in growth, maintenance, repair, and control.

3.1.3 Nucleus

The nucleus is generally a spherical or oval organelle. In addition to being the largest structure in the cell, the nucleus contains hereditary factors of the cell, called genes, which control cellular structure and direct many cellular activities. The nucleus is separated from the cytoplasm by a double membrane, the nuclear membrane or envelope. Between the two layers of the nuclear membrane is a space called the perinuclear cisterna. This arrangement of each of the nuclear membranes resembles the structure of the plasma membrane. Minute pores in the nuclear membrane allow the nucleus to communicate with a membranous network in the cytoplasm called the endoplasmic reticulum (ER), and substances entering and exiting the nucleus are believed to pass through the pores. The available space in the nucleus is filled with a gel-like fluid called nucleoplasm containing one or more spherical bodies, the nucleoli, which are composed primarily of RNA and have a function in protein synthesis. The genetic material consists principally of DNA, packaged as chromatin, which when the cell is not dividing looks like a ball of thread. Before cellular reproduction the chromatin shortens and coils into rod-shaped bodies called chromosomes. The structures of chromatin and chromosomes are described in detail in Chapters 9 and 10.

3.1.4 Mitochondria

3.1.4.1 Mitochondrial Structure

Mitochondria are small, spherical, rod-shaped, or filamentous structures, ranging from 1 to 10 μm in size, which occur throughout the cytoplasm. A mitochondrion has an elaborate internal organization consisting of a double-unit membrane, each unit of which is similar in structure to the plasma membrane. The outer mitochondrial membrane is smooth, but the inner membrane is arranged in a series of folds called cristae. The inner and outer membranes have different properties and, in fact, five distinct compartments are delineated within the mitochondria: the outer membrane, the inter-membrane space (between the outer and inner membranes), the inner membrane, the cristae space (formed by infoldings of the inner membrane), and the matrix (space within the inner membrane). The center of the mitochondrion is called the matrix. Because of the nature and arrangement of the cristae, the inner membrane provides an enormous surface area for chemical reactions. Enzymes involved in energy-releasing reactions that form ATP are located on the cristae. The outer mitochondrial membrane, which encloses the entire organelle, contains numerous integral proteins called porins, which surround a relatively large internal channel (about 2–3 nm) that is permeable to all molecules of 5000 Da or less. Larger molecules can only traverse the outer membrane by active transport.

The inner mitochondrial membrane contains proteins that carry out a number of functions, including oxidation reactions of the respiratory chain. ATP synthase, which makes ATP, is found in the matrix; mitochondria are frequently called the "powerhouses of the cell" because they are the sites for the production of ATP, and active cells such as muscle and liver cells have a large number of mitochondria because of their high energy expenditure. Other important proteins of the inner mitochondrial membrane are specific transport proteins that regulate the passage of metabolites into and out of the matrix and the protein import machinery.

The mitochondrial matrix, which is the space enclosed by the inner membrane, contains a highly concentrated mixture of hundreds of enzymes, in addition to mitochondrial ribosomes, transfer RNA (tRNA), and several copies of the mitochondrial DNA genome. Major functions of the matrix enzymes include oxidation of pyruvate and fatty acids, and the citric acid cycle. Mitochondria possess their own genetic material, and the machinery to manufacture their own RNAs and proteins. This nonchromosomal DNA encodes a small number of mitochondrial peptides (13 in humans) that are integrated into the inner mitochondrial membrane, along with polypeptides encoded by genes that reside in the host cell nucleus.

3.1.4.2 Mitochondrial Function

The main functions of mitochondria are to convert organic materials into cellular energy in the form of ATP. However, mitochondria also play an important role in many other metabolic tasks, including apoptosis (programmed cell death), cellular proliferation, regulation of the cellular redox state, heme synthesis, and steroid synthesis.

A dominant role for the mitochondria is the production of ATP by oxidizing the major products of glycolysis: pyruvate and NADH that are produced in the cytosol. This process of cellular respiration, also known as aerobic respiration, is dependent on the presence of oxygen. When oxygen is limiting, the glycolytic products will be metabolized by anaerobic respiration, a process that is independent of the mitochondria.

Each pyruvate molecule produced by glycolysis is actively transported across the inner mitochondrial membrane, and into the matrix where it is oxidized and combined with coenzyme A to form CO_2, acetyl-CoA, and NADH. The acetyl-CoA is the primary substrate to enter the citric acid cycle, also known as the tricarboxylic acid (TCA) cycle or Krebs cycle. The enzymes of the citric acid cycle are located in the mitochondrial matrix with the exception of succinate dehydrogenase, which is bound to the inner mitochondrial membrane. The citric acid cycle oxidizes the acetyl-CoA to carbon dioxide and in the process produces reduced cofactors (three molecules of NADH and one molecule of $FADH_2$) that are a source of electrons for the electron transport chain, and a molecule of GTP (that is readily converted to an ATP).

The redox energy from NADH and $FADH_2$ is transferred to oxygen in several steps via the electron transport chain. Protein complexes in the inner membrane (NADH dehydrogenase, cytochrome *c* reductase, and cytochrome *c* oxidase) perform the transfer and the incremental release of energy is used to pump protons (H^+) into the intermembrane space. This process is efficient, but a small percentage of electrons may prematurely reduce oxygen, forming the toxic free-radical superoxide. This can cause oxidative damage in the mitochondria and may contribute to the decline in mitochondrial function associated with the aging process.

As the proton concentration increases in the intermembrane space, a strong electrochemical gradient is established across the inner membrane. The protons can return to the matrix through the ATP synthase complex and their potential energy is used to synthesize ATP from ADP and inorganic phosphate (P_i).

Mitochondria replicate their DNA and divide mainly in response to the energy needs of the cell; their growth and division is not linked to the cell cycle. At cell division, mitochondria are distributed to the daughter cells more or less randomly during the division of the cytoplasm. Mitochondria divide by binary fission similar to bacterial cell division but, unlike bacteria, mitochondria can also fuse with other mitochondria. Mitochondrial genes are not inherited by the same mechanism as nuclear genes, in that the mitochondria and the mitochondrial DNA come from the egg only and mitochondria are inherited through the female line.

3.1.5   Endoplasmic Reticulum and Ribosomes

In the cytoplasm, there is a system of paired parallel membranes enclosing narrow cavities of varying shapes. This is the endoplasmic reticulum or ER. The ER is, in effect, a network of channels (cisternae) running through the entire cytoplasm. These channels are continuous with both the plasma membrane and the nuclear membrane. It is believed that the ER provides a surface area for chemical reactions, a pathway for transporting molecules within the cell, and a storage area for synthesized molecules.

Parts of the ER are covered with ribosomes (where amino acids are assembled into proteins). Their rough appearance under the electron microscope led to their being called rough ER (rER); other parts, called smooth ER (sER), are free of ribosomes. The rough and smooth ER differs in both appearance and function but are contiguous with each other. The ribosomes on the surface of the rough ER insert the freshly produced proteins directly into the ER, which processes them and then passes them on to the Golgi apparatus, whereas the smooth ER has functions in several metabolic processes, including synthesis of lipids, metabolism of carbohydrates, and detoxification of drugs and toxic substances.

The ER is both a protein-sorting and transport pathway. Proteins that are transported by the ER and from the ER to other parts of the cell have a labeling sequence on the N-terminus of the polypeptide chain that targets them to the correct location. Proteins that are destined for places outside the ER are packed into transport vesicles and moved along the cytoskeleton toward their destination.

3.1.6   Golgi Complex

Another structure found in the cytoplasm is the Golgi complex that consists of four to eight flattened bag-like channels, stacked upon each other with expanded areas at their ends. Like those of the ER, the stacked elements are called cisternae and the expanded, terminal areas are vesicles. Generally, the Golgi complex is located near the nucleus and is directly connected, in places, to the ER. An important function of the Golgi complex is the secretion and packaging of proteins. Proteins synthesized by the ribosomes associated with rough ER are transported into the ER cisternae and then migrate along the ER cisternae until they reach the Golgi complex. As proteins accumulate, the cisternae of the Golgi complex expand to form vesicles that pinch off from the cisternae after a certain size is reached. The protein and its associated vesicle, referred to as a secretory granule, then moves toward the surface of the cell where the protein is secreted.

The primary function of the Golgi apparatus is to process proteins targeted to the plasma membrane, lysosomes, or endosomes, and sort them within vesicles. Vesicles that leave the smooth ER are transported to the Golgi apparatus, where they are modified, sorted, and shipped toward their final destination. The Golgi apparatus is present in most eukaryotic cells, but tends to be more prominent where there are many substances, such as proteins, being secreted.

3.1.7   Cytoskeleton

The cytoskeleton is composed of a number of different components:

- Actin filaments are approximately 7 nm in diameter and are composed of two actin chains oriented in a helicoidal shape. They are mostly concentrated just beneath the plasma membrane, as they keep cellular shape, form cytoplasmatic protuberances (like pseudopodia and microvilli), and participate in some cell-to-cell or cell-to-matrix junctions and in the transduction of signals. They are also important for cytokinesis and, along with myosin, muscular contraction.

- Intermediate filaments are 8–11 nm in diameter and form the more stable (strongly bound) and heterogeneous constituents of the cytoskeleton. They organize the internal tridimensional structure of the cell (as structural components of the nuclear envelope and the sarcomeres) and they also participate in some cell–cell and cell–matrix junctions.

- Microtubules are hollow cylinders of about 25 nm, formed by 13 protofilaments which, in turn, are polymers of α- and β-tubulin. They have a very dynamic behavior, binding GTP for polymerization. They are organized by the centrosome.

Cytoskeletal components play key roles in intracellular transport, in transport organelles like mitochondria or vesicles in the axoneme of cilia and flagella and in the mitotic spindle.

3.1.7.1   Centrosome and Centrioles

A dense area of cytoplasm, generally spherical and located near the nucleus, is called the centrosome or centrosphere. Within the centrosome is a pair of cylindrical structures called centrioles. Each centriole is composed of a ring of nine evenly spaced bundles, and each bundle in turn consists of three microtubules. The two centrioles are situated so that the long axis of one is at right angles to the long axis of the other. Centrioles have a role in cell division, and certain cells, such as most mature nerve cells which do not have a centrosome, do not divide.

3.1.8 Lysosomes

When viewed under the electron microscope, lysosomes appear as membrane-enclosed spheres. Lysosomes have only a single membrane and lack detailed structure; they contain powerful digestive enzymes capable of breaking down many kinds of molecules and are also capable of digesting bacteria that enter the cell. White blood cells, which ingest bacteria by phagocytosis, contain a large number of lysosomes. It has been questioned why these powerful enzymes do not also destroy their own cells. The answer may be that the lysosome membrane in a healthy cell is impermeable to enzymes, so they cannot move out into the cytoplasm, but when a cell is injured, the lysosomes release their enzymes that digest the cell down to its chemical constituents, which are then either reused by the body or excreted.

3.1.9 Extracellular Materials

Extracellular materials are not components of the cell, but they are important for the functioning of the cell in multicellular organs and tissues. They include body fluids (interstitial fluid and plasma) that provide a medium for dissolving, mixing, and transporting substances. Extracellular materials also include secreted inclusions like mucus and the special substances that form the matrix in which some cells are embedded. The matrix materials are produced by cells and deposited outside their plasma membranes. They act to support the cells, bind them together, and give strength and elasticity to the tissue. Some matrix materials are amorphous (hyaluronic acid and chondroitin sulfate) and have no specific shape, whereas others (collagen and collagenous fibres) are fibrous or threadlike and provide strength and support for tissues. Collagen is found in all types of connective tissue, especially in bones, cartilage tendons, and ligaments. Elastin, found in elastic fibers, gives elasticity to skin and to tissues forming the walls of blood vessels.

3.1.10 Summary: Cell Biology

This section of the chapter is summarized in Table 3.1.

3.2 Molecular Genetics

All living things on this planet store their hereditary information in the form of double-stranded molecules of DNA. Furthermore, all cells transcribe the information from DNA into molecules of the related polymer RNA by the same mechanism of templated polymerization that is used in the replication of DNA. RNA is the template for the translation of genetic information into proteins, the principal catalysts of almost all chemical reactions in the cell. The relationship between the synthesis of DNA, RNA, and protein can be described as follows:

DNA directs the synthesis of RNA, which then directs the synthesis of protein; specialized proteins catalyze the synthesis of both DNA and RNA. The flow of information from DNA to RNA to protein occurs in all cells and has been called the "central dogma of molecular biology." The understanding of biology and genetics at the molecular level, which went hand-in-hand with the development of recombinant DNA technology, described in the next chapter, has revolutionized radiation biology to the same extent that it has impacted all other branches of biomedical science. This part of the chapter is designed to describe some of the basic principles of molecular genetics.

3.2.1 DNA Structure

The DNA molecule consists of two long polynucleotide chains composed of four types of nucleotide subunits. Hydrogen bonds between the base portions of the nucleotides hold the two chains together. Nucleotides are composed of a five-carbon sugar to which are attached one or more phosphate

TABLE 3.1

Structure and Function of Components of the Eukaryotic Cell

Organelle or Structure	Function	Structure	Notes
Plasma membrane	Separates the cell interior from its surroundings. Controls the passage of materials in and out of the cell by diffusion, facilitated diffusion, and active transport	Selectively permeable phospholipid bilayer with intrinsic and peripheral proteins, which may function as channels or transporters or as receptors	Fluid mosaic—a two-dimensional fluid of freely diffusing lipids, dotted or embedded with proteins
Nucleus	DNA maintenance and transcription to RNA	Double-membrane compartment	Contains most of the genome
Mitochondria	Energy production	Double-membrane compartment	Contains some genes
Golgi apparatus	Sorting and modification of proteins	Single-membrane compartment	
Endoplasmic reticulum	Modification and folding of new proteins and lipids	Single-membrane compartment	
Cytoskeleton	Gives the cell shape and capacity for directed movement. Microtubules form mitotic spindle and are involved in cytokinesis	Spatially organized network of protein filaments (actin, intermediate filaments), or hollow tubes of α- and β-tubulin (microtubules)	

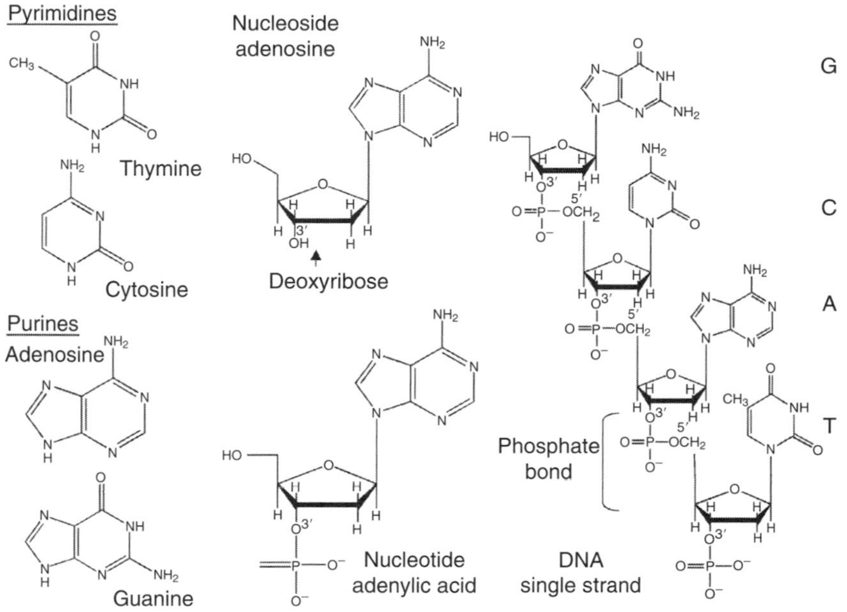

FIGURE 3.2
Building blocks of DNA. The bases are nitrogen containing ring compounds, either pyrimidines or purines. A base linked to a sugar by an N-glycosidic bond is a nucleoside. Nucleotides consist of a base, a five-carbon sugar, and one or more phosphate groups. DNA is made up of four types of nucleotides that are linked together to form a polynucleotide chain.

groups and a nitrogen-containing base. In the case of DNA, the sugar is deoxyribose (hence the name deoxyribonucleic acid), which is attached to a single phosphate group and the base can be adenine (A), cytosine (C), guanine (G), or thymine (T). The nucleotides are covalently linked together in a chain through the sugars and phosphates, forming a backbone of alternating sugar–phosphate–sugar–phosphate (Figure 3.2). (The symbols A, C, G, and T are also used to denote the four corresponding nucleotides, i.e., the bases with their attached sugar and phosphate groups.)

The way in which the nucleotide subunits are linked together gives the DNA strand a chemical polarity. Each sugar has a phosphate at the 5′ position on one side and a hydroxyl at the 3′ on the other, and each completed chain is formed by linking 5′ phosphate to 3′ hydroxyl so that all of the subunits are lined up in the same orientation. This polarity of the DNA chain is defined by referring to one end as the 3′ end and the other as the 5′ end.

The three-dimensional structure of DNA, the double helix, arises from the chemical and structural features of the two polynucleotide chains. Because the two chains are held together by hydrogen bonding between the bases on the different strands, all the bases are on the inside of the double helix, and the sugar–phosphate backbones are on the outside (Figure 3.3). In each case, a bulkier two-ring base (a purine) is paired with a single-ring base

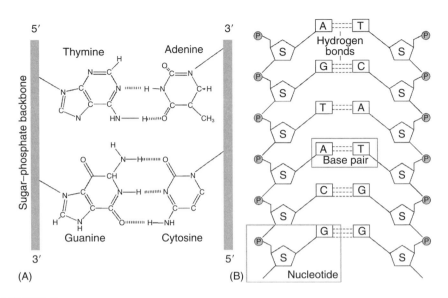

FIGURE 3.3

(A) and (B) Hydrogen bonding between A and T and between G and C, molecules that can be brought close enough together without distorting the double helix. Two hydrogen bonds form between A and T and three form between G and C. The bases can pair in this way only if the polynucleotide chains are antiparallel.

(a pyrimidine), A being paired with T, and G with C. This complementary base-pairing enables the base pairs to be packed in the energetically most favorable arrangement in the interior of the double helix. Each base pair is of similar width, thus the sugar–phosphate backbones are an equal distance apart along the DNA molecule. The efficiency of base-pair packing is maximized by having the two sugar–phosphate backbones wind around each other to form the double helix, with one complete turn every 10 base pairs (Figure 3.4A). The members of each base pair can fit together within the double helix only if the two strands of the helix are antiparallel, i.e., if the polarity of one strand is oriented opposite to that of the other strand. Thus, each strand of a DNA molecule contains a sequence of nucleotides that is exactly complementary to the nucleotide sequence of the opposing strand.

3.2.2 DNA Structure Is the Basis for Heredity

The units of hereditary, the genes, carry biological information that must be copied accurately for transmission to the next generation each time a cell divides to form two daughter cells. The information is encoded through the order or sequence of the nucleotides along each strand. Each base, A, C, T, or G, can be considered as a letter in a four-letter alphabet that spells out biological messages through the chemical structure of the DNA, and the

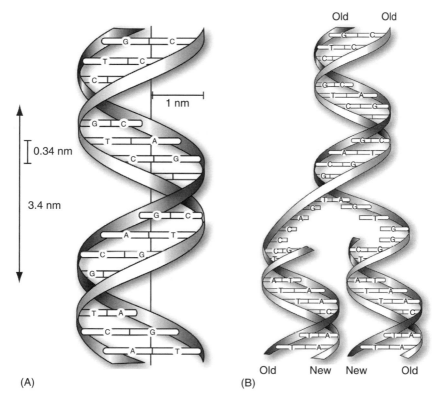

(A)   (B)

FIGURE 3.4

(A) The DNA molecule is made of the four types of nucleotides linked covalently in a polynucleotide chain (a DNA strand) with sugar–phosphate backbone from which the bases (A, C, G, and T) extend. The two strands are held together in a double-helical configuration DNA replication by hydrogen bonding. (B) DNA replication is semiconservative. Each of the two strands of DNA is used as a template for the formation of a complementary DNA strand. The original strands remain intact through many cell generations.

DNA message in turn encodes proteins (Table 3.2). The properties of a protein, which are responsible for its biological function, are determined by its three-dimensional structure, and its structure is in turn determined by the linear sequence of the amino acids with which it is composed. The complete set of information in an organism's DNA is called its genome, and it carries the information for all the proteins the organism will ever synthesize.

The genetic information in DNA can be accurately copied by a simple and economical process in which one strand separates from the other and each separated strand then serves as a template for the production of a new complementary partner strand that is identical to its former partner. The ability of each strand of a DNA molecule to act as a template for producing a complementary strand enables a cell to copy, or replicate, its genes before passing them on to its descendants.

TABLE 3.2

The Genetic Code

| 1st Position | 2nd Position | | | | 3rd Position |
(5' End)	U	C	A	G	(3' End)
U	Phe	Ser	Tyr	Cys	U
	Phe	Ser	Tyr	Cys	C
	Leu	Ser	**STOP**	**STOP**	A
	Leu	Ser	**STOP**	Trp	G
C	Leu	Pro	His	Arg	U
	Leu	Pro	His	Arg	C
	Leu	Pro	Gln	Arg	A
	Leu	Pro	Gln	Arg	G
A	Ile	Thr	Asn	Ser	U
	Ile	Thr	Asn	Ser	C
	Ile	Thr	Lys	Arg	A
	Met	Thr	Lys	Arg	G
G	Val	Ala	Asp	Gly	U
	Val	Ala	Asp	Gly	C
	Val	Ala	Glu	Gly	A
	Val	Ala	Glu	Gly	G

3.2.3 Mechanism of DNA Replication

All organisms must duplicate their DNA with extraordinary accuracy before each cell division, while duplicating DNA at rates as high as 1000 nucleotides per second. This accuracy is achieved by DNA templating, the process by which the nucleotide sequence of a DNA strand (or selected portions of a DNA strand) is copied by complementary base-pairing (A with T, and G with C) into a complementary DNA sequence (Figure 3.4B). Separation of the two strands of the DNA helix allows the hydrogen-bond donor and acceptor groups on each DNA base to become exposed for base-pairing with the appropriate incoming free nucleotide and to be aligned for enzyme-catalyzed polymerization into a new DNA chain by the nucleotide-polymerizing enzyme, DNA polymerase.

3.2.3.1 DNA Replication Fork

During DNA replication each of the two old DNA strands serves as a template for the formation of a complete new strand. Thus each of the two daughters of a dividing cell inherits a new DNA double helix containing one old and one new strand, and the DNA double helix is said to be replicated semiconservatively by DNA polymerase. A localized region of replication called the replication fork because of its Y-shaped structure moves progressively along the parental DNA double helix. At the replication fork, new daughter strands are synthesized by a multienzyme complex that contains the DNA polymerase. The simplest mechanism of DNA replication would be the continuous growth of both new strands, nucleotide by nucleotide, at

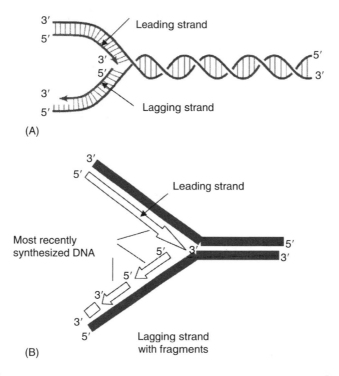

FIGURE 3.5
(A) and (B) Two views of the DNA replication fork. Because both daughter strands are synthesized in the 5′-to-3′ direction, the DNA synthesized on the lagging strand must be made initially as a series of short DNA molecules.

the replication fork as it moves from one end of a DNA molecule to the other. However, because of the antiparallel orientation of the two DNA strands in the double helix, such a mechanism would require one daughter strand to polymerize in the 5′-to-3′ direction and the other in the 3′-to-5′ direction. This does not occur; instead one DNA daughter strand, the leading strand, is synthesized continuously slightly ahead of the other daughter strand (the lagging strand), which is synthesized discontinuously. In the lagging strand, replication intermediates 100–200 nucleotides long are polymerized only in the 5′-to-3′ chain direction and joined together after their synthesis to create long DNA chains. For the lagging strand, the direction of nucleotide polymerization is opposite to the overall direction of DNA chain growth. The replication fork therefore has an asymmetric structure (Figure 3.5).

3.2.3.2 Proofreading Mechanisms
The fidelity of copying DNA during replication is such that only about one mistake is made for every 10^9 nucleotides copied. This is much higher than would be expected on the basis of the accuracy of complementary base

pairing, since the standard complementary base pairs are not the only ones possible. With small changes in helix geometry, two hydrogen bonds can form between G and T in DNA while rare, tautomeric forms of the four DNA bases occur transiently in ratios of 1 part to 10^4 or 10^5, and these forms can mispair without a change in helix geometry.

In fact, the high fidelity of DNA replication depends on several proof-reading mechanisms that act sequentially to correct any initial mispairing that might have occurred. The first proofreading step is carried out by the DNA polymerase and occurs just before a new nucleotide is added to the growing chain. The correct nucleotide has a higher affinity for the moving polymerase than does the incorrect nucleotide, because only the correct nucleotide can correctly base-pair with the template and this predisposes the polymer to reject the incorrect nucleotide.

Another error-correction reaction known as exonucleolytic proofreading takes place immediately after an incorrect nucleotide is covalently added to the growing chain. DNA polymerase enzymes cannot begin a new poly-nucleotide chain by linking two nucleoside triphosphates together. Instead, they require a base-paired 3′–OH end of a primer strand on which to add further nucleotides. Those DNA molecules with a mismatched (improperly base-paired) nucleotide at the OH end of the primer strand are not effective as templates because the polymerase cannot extend such a strand. In this case a 3′-to-5′ proofreading exonuclease clips off any unpaired residues at the primer terminus, continuing until enough nucleotides have been removed to regenerate a base-paired 3′–OH terminus that can prime DNA synthesis.

3.2.3.3 *Proteins Involved in DNA Replication*

DNA replication requires the cooperation of many proteins. These include in addition to DNA polymerase

- DNA primase, to catalyze nucleoside triphosphate polymerization
- DNA helicases and single strand DNA binding proteins, to help in opening up the DNA helix so that it can be copied
- DNA ligase and an enzyme that degrades RNA primers, to seal together the discontinuously synthesized lagging-strand DNA primers
- DNA topoisomerases, to help relieve helical winding and DNA tangling problems

Many of these proteins associate together at the replication fork to form a highly efficient replication machine through which the activities and spatial movements of the individual components are coordinated.

The proteins that initiate DNA replication bind to DNA sequences at the replication origin to catalyze the formation of a replication bubble with two outward moving replication forks. The process begins when an initiator protein–DNA complex is formed that subsequently loads a DNA helicase onto the DNA template. Other proteins are then added to form the multienzyme replication machine that catalyzes DNA replication at each replication fork. The DNA replication in eukaryotes takes place in only one part of the cell cycle and eukaryote chromosomes require many replication origins to complete their replication in a typical 8 h S phase. The problem of replicating the ends of chromosomes is solved by a specialized end structure, the telomere which requires a special enzyme telomerase. Telomerase extends the chromosome by using an RNA template that is an integral part of the enzyme itself, producing a highly repetitive DNA sequence that extends for 10,000 nucleotide pairs or more at each chromosomal end.

3.2.4 Transcribing DNA to RNA

The first stage in the read out of genetic instructions is to copy a specific portion of its DNA nucleotide sequence, i.e., the gene, into an RNA nucleotide sequence, a process called transcription. The information in RNA is written in essentially the same language as is that of DNA since, like DNA, RNA is a linear polymer made of four different types of nucleotide subunits linked together by phosphodiester bonds (Figure 3.6). It differs from DNA

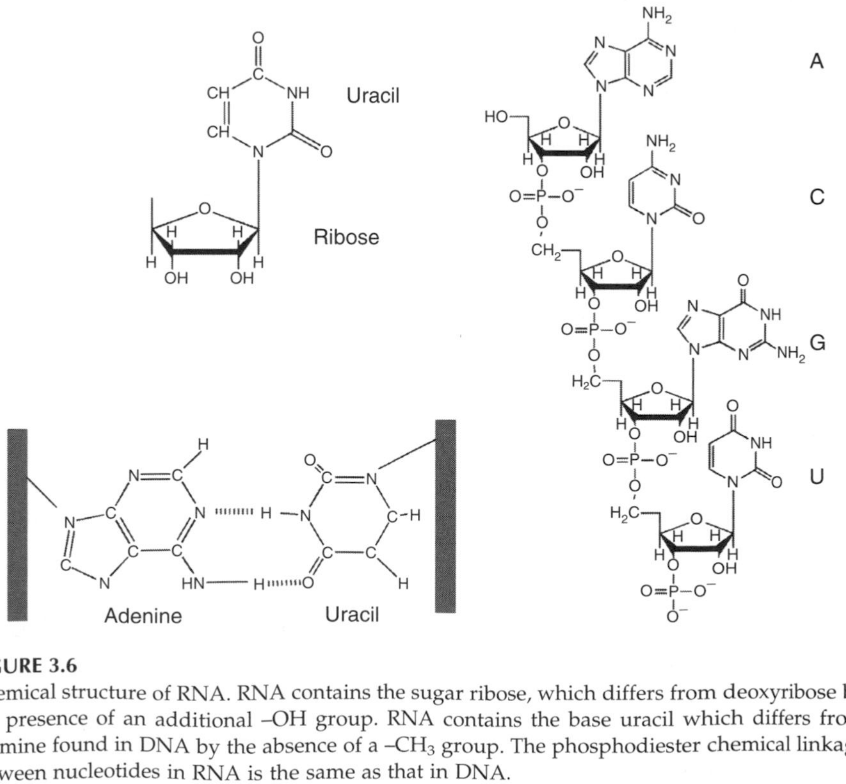

FIGURE 3.6
Chemical structure of RNA. RNA contains the sugar ribose, which differs from deoxyribose by the presence of an additional –OH group. RNA contains the base uracil which differs from thymine found in DNA by the absence of a –CH$_3$ group. The phosphodiester chemical linkage between nucleotides in RNA is the same as that in DNA.

chemically in two respects, the nucleotides in RNA are ribonucleotides, that is, they contain the sugar ribose rather than deoxyribose, and RNA contains the base uracil (U) instead of the thymine (T) found in DNA. Since uracil like thymine can base-pair by hydrogen-bonding with adenine, the complementary base-pairing properties of DNA apply also to RNA. Where DNA and RNA differ radically is in overall structure. While DNA always occurs in cells as a double helix, RNA is single-stranded and RNA chains can fold up into a variety of shapes. The ability to fold into complex three-dimensional shapes allows some RNA molecules to have structural and catalytic functions.

3.2.4.1 One DNA Strand Is Transcribed into RNA

All of the RNA in a cell is made by DNA transcription, a process that has certain similarities to the process of DNA replication already described. Transcription begins with the opening and unwinding of a small portion of the DNA double helix to expose the bases on each DNA strand (Figure 3.7). One of the two strands of the DNA double helix then acts as a template for the synthesis of an RNA molecule. As in DNA replication, the nucleotide sequence of the RNA chain is determined by the complementary base-pairing between incoming nucleotides and the DNA template. When a correct match is made, the incoming ribonucleotide is covalently linked to the growing RNA chain in an enzymatically catalyzed reaction. The RNA chain produced by transcription, the transcript, is therefore elongated one nucleotide at a time, and it has a nucleotide sequence that is exactly complementary to the strand of DNA used as the template. Transcription, however, differs from DNA replication in several crucial ways. Unlike a newly formed DNA strand, the RNA strand does not remain hydrogen-bonded to the DNA template strand. Instead, just behind the region where the ribonucleotides are being added, the RNA chain is displaced and the DNA helix re-forms. Thus, the RNA molecules produced by transcription are released from the DNA template as single strands. In addition, because they are copied from only a limited region of the DNA, RNA molecules are much shorter than DNA molecules. Most RNAs are no more than a few

FIGURE 3.7
DNA transcription by messenger RNA. A characteristic sequence of nucleotides marks the beginning of a gene on the DNA strand, and binds to a promoter protein that initiates mRNA synthesis. RNA polymerase binds to the promoter and starts to unwind DNA strands. One of the strands serves as a template for RNA formation. The polymerase reads the DNA template from 3′ to 5′ and produces an RNA transcript from 5′ to 3′.

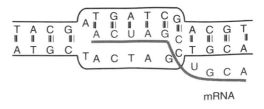

thousand nucleotides long whereas a DNA molecule can be up to 250 million nucleotide pairs long.

The enzymes that perform transcription are called RNA polymerases and, like the DNA polymerase, RNA polymerases catalyze the formation of the phosphodiester bonds that link the nucleotides together to form a linear chain. The RNA polymerase moves stepwise along the DNA, unwinding the DNA helix just ahead of the active site for polymerization to expose a new region of the template strand for complementary base-pairing. In this way, the growing RNA chain is extended by one nucleotide at a time in the 5'-to-3' direction. The substrates are nucleoside triphosphates (ATP, CTP, UTP, and GTP) and, as is the case for DNA replication, the hydrolysis of high-energy bonds provides the energy needed to drive the reaction forward. The almost immediate release of the RNA strand from the DNA as it is synthesized means that many RNA copies can be made from the same gene in a relatively short time and the synthesis of the next RNA molecules can be started before the first RNA is completed. In eukaryotes, over a thousand transcripts can be synthesized in an hour from a single gene.

3.2.4.2   RNA and RNA Polymerases

The majority of genes specify the amino acid sequence of proteins; the RNA molecules that are copied from these genes are called messenger RNA (mRNA) molecules. For a small number of genes the final product is the RNA itself. These RNAs, like proteins, serve as enzymatic and structural components for a wide variety of processes in the cell. Some small nuclear RNA (snRNA) molecules direct the splicing of pre-mRNA to form mRNA, ribosomal RNA (rRNA) molecules form the core of ribosomes, and tRNA molecules form the adaptors that select amino acids and hold them in place on a ribosome for incorporation into protein. Overall, RNA makes up a few percent of the dry weight of the cell and most of the RNA in cells is rRNA. mRNA comprises only 3%–5% of the total RNA in a typical mammalian cell and the mRNA population is made up of tens of thousands of different species with only 10–15 molecules of each species of mRNA being present in each cell.

Eukaryotic nuclei have three types of RNA polymerase: RNA polymerase I, RNA polymerase II, and RNA polymerase III. RNA polymerases I and III transcribe the genes encoding tRNA, rRNA, and various small RNAs, whereas RNA polymerase II transcribes the vast majority of genes, including all those that encode proteins.

3.2.4.3   Transcription Stop Signals

To transcribe a gene accurately, RNA polymerase must recognize where on the genome to start and where to finish. The initiation of transcription is an important step in gene expression because it is the main point at which the cell regulates which proteins are to be produced and at what rate. The DNA sequence involved in the initiation of transcription is called the promoter. The promoter comprises a region of DNA extending 150–300 base pairs upstream of the transcription site that contains binding sites for RNA polymerase and a number of other proteins that regulate the rate of transcription of the adjacent gene.

3.2.4.4 General Transcription Factors

Eukaryotic RNA polymerase II requires general transcription factors that help to position the RNA polymerase correctly at the promoter, aid in pulling apart the two strands of DNA to allow transcription to begin, and release RNA polymerase from the promoter into the elongation mode once transcription has begun. These proteins are called general because they assemble on all promoters used by RNA polymerase II. They consist of a set of interacting proteins, designated as TFII (for transcription factor for polymerase II) and are listed as TFIIA, TFIIB, etc.

The assembly process starts with the binding of the general transcription factor TFIID to a short double-helical DNA sequence primarily composed of T and A nucleotides known as the TATA sequence or TATA box. The TATA box is typically 25 nucleotides upstream from the transcription start site. It is not the only DNA sequence that signals the start of transcription, but for most the polymerase II promoters it is the most important. Other factors are then assembled along with DNA polymerase II to form a complete transcription initiation complex.

Once RNA polymerase II is guided to the promoter DNA it must gain access to the template strand at the transcription start point. This step is facilitated by another transcription factor, TFIIH, which contains a DNA helicase. Polymerase II remains at the promoter synthesizing short lengths of RNA until it is released by a conformational change to start transcribing a gene. A key step in the release is the addition of phosphate groups to the tail of the RNA polymerase (the C-terminal domain). This phosphorylation is also catalyzed by TFIIH, which in addition to the helicase contains a protein kinase as one of its subunits. The polymerase then disengages from the general transcription factors, undergoes a series of conformational changes that tighten its interaction with DNA, and acquires a new set of proteins that allow it to transcribe for long distances without dissociating. Most of the general transcription factors are released so that they are available to initiate another round of transcription with another polymerase molecule.

The model of transcription just described was established by studying the action of RNA polymerase II and the general transcription factors on purified DNA templates in vitro. In fact, DNA of eukaryotic cells is packaged in nucleosomes and further packaged in higher order structures (see Chapter 9). Transcription under these circumstances requires many proteins that must assemble to initiate transcription. The order of assembly of these proteins may differ for different genes and, in fact, some protein assemblies may be preformed away from DNA.

3.2.4.5 *Processing the mRNA Molecule*

In eukaryotic cells the RNA molecule produced by transcription alone (sometimes referred to as the primary transcript) contains both coding (exon) and noncoding (intron) sequences. The typical eukaryotic gene is present in the genome as short blocks of protein coding sequence (exons) separated by long sequences of introns and RNA splicing is the important step that brings together different portions of a coding sequence. Before it can be translated into protein, the two ends of the RNA are modified, the introns are removed by an enzymatically catalyzed RNA splicing reaction, and the resulting mRNA is transported from the nucleus to the cytoplasm. RNA splicing provides higher eukaryotes with the ability to synthesize several proteins from the same gene. The modifications of the ends of eukaryotic mRNA are capping on the 5' end and polyadenylation of the 3' end (Figure 3.8).

Interphase chromosomes occupy discrete territories in the nucleus, and transcription and pre-mRNA splicing must take place within these territories. It has been proposed that, although a typical mammalian cell may be expressing in the order of 15,000 genes, transcription and RNA splicing may be localized to only a few thousand sites in the nucleus. These sites themselves are highly dynamic and may comprise an association of transcription

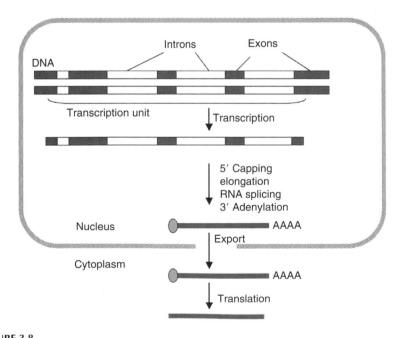

FIGURE 3.8
Posttranscriptional processing results in a fully functional mRNA ready for export to the cytoplasm and ribosomes. Steps involved are splicing RNA to remove the introns, elongation, capping at the 5' end, and adenylation at the 3' end.

and splicing components to create small assembly lines where the local concentration of these components is very high.

3.2.5 From RNA to Protein

This section describes how the cell converts the information carried in an mRNA molecule into the amino acid sequence of a protein molecule.

3.2.5.1 Decoding mRNA

Once an mRNA has been produced by transcription and processing, the information present in its nucleotide sequence is used to synthesize a protein. Unlike transcription, where DNA can act as a direct template for the synthesis of RNA by complementary base-pairing, the conversion of the information in the base sequence of RNA into protein represents a translation of the information into another language that uses different symbols. Moreover, since there are only 4 different nucleotides in mRNA and 20 different types of amino acids in a protein, this translation cannot be accounted for by a direct one-to-one correspondence between a nucleotide in RNA and an amino acid in protein. The nucleotide sequence of a gene, through the medium of mRNA, is translated into the amino acid sequence of a protein by a set of rules, known as the genetic code that was deciphered in the early 1960s. The sequence of nucleotides in the mRNA molecule is read consecutively in groups of three. RNA is a linear polymer of four different nucleotides, so there are $4 \times 4 \times 4 = 64$ possible combinations of three nucleotides: the triplets AAA, AUA, AUG, and so on. However, only 20 different amino acids are commonly found in proteins. It has been shown that the code is redundant, some amino acids being specified by more than one triplet, as is shown by the completely deciphered genetic code in Table 3.3. Each group of three consecutive nucleotides in RNA is called a codon and each codon specifies either one amino acid or a stop to the translation process. This genetic code is found in all organisms alive today. (Mitochondria have their own transcription and protein synthesis systems that operate independently from those of the rest of the cell and some minor differences from the universal genetic code are found in the small genomes of mitochondria.) Since each amino acid is specified by a sequence of three nucleotides, an RNA sequence can, in principle, be translated in anyone of three different reading frames, depending on where the decoding process begins. However, only one of the three possible reading frames in an mRNA encodes the required protein. A special punctuation signal at the beginning of each RNA message sets the correct reading frame at the start of protein synthesis.

The translation of mRNA into protein depends on adaptor molecules that recognize and bind both to the mRNA codon and, at another site on their surface, to the corresponding amino acid. These adaptors consist of a set of small molecules known as tRNAs each about 80 nucleotides in

TABLE 3.3

Amino Acids and Codons

Amino Acid	Symbol	1 Letter Code	Codon
Alanine	Ala	A	GCA GCC GCG GCU
Cysteine	Cys.	C	UGC UGU
Aspartic acid	Asp	D	GAC GAU
Glutamic acid	Glu	E	GAA GAG
Phenylalanine	Phe	F	UUC UUU
Glycine	Gly	G	GGA GGC GGG GGU
Histidine	His	H	CAC CAU
Isoleucine	Ile	I	AUA AUC AUU
Lysine	Lys	K	AAA AAG
Leucine	Leu	L	UUA UUG CUA CUC CUG CUU
Methionine	Met	M	AUG
Asparagine	Asn	N	AAC AAU
Proline	Pro	P	CCA CCC CCG CCU
Glutamine	Gln	Q	CAA CAG
Arginine	Arg	R	AGA AGG CGA CGC CGG CGU
Serine	Ser	S	AGC AGU UCA UCC UCG UCU
Threonine	Thr	T	ACA ACC ACG ACU
Valine	Val	V	GUA GUC GUG GUU
Tryptophan	Trp	W	UGG
Tyrosine	Tyr	Y	UAC UAU

length. RNA molecules can fold up into precisely defined three-dimensional structures and, in the case of tRNA molecules, four short segments of the folded tRNA are double-helical. More folding results in a compact L-shaped structure that is held together by additional hydrogen bonds between different regions of the molecule (Figure 3.9). Two regions of unpaired nucleotides situated at either end of the L-shaped molecule are crucial to the function of tRNA in protein synthesis. One of these regions forms the anticodon, a set of three consecutive nucleotides that pairs with the complementary codon on an mRNA molecule. The other is a short single-stranded region at the 3' end of the molecule; this is the site where the amino acid that matches the codon is attached to the tRNA. Since the genetic code is redundant, several different codons can specify a single amino acid. This might mean that there is more than one tRNA for many of the amino acids or that some tRNA molecules can base-pair with more than one codon and, in fact, both these situations occur. Some amino acids are encoded by more than one tRNA while some tRNAs are constructed so that they require accurate base-pairing only at the first two positions of the codon and can tolerate a mismatch (or wobble) at the third position. The wobble in base-pairing explains why many of the alternative codons for an amino acid differ only in their third nucleotide (Table 3.3).

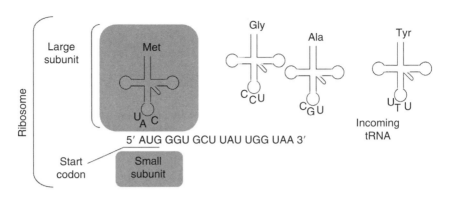

FIGURE 3.9

Transfer RNAs (tRNAs) have distinctive three-dimensional structures consisting of loops of single-stranded RNA connected by double-stranded segments. This structure is further wrapped into an assembly that binds the amino acid at the end of one arm and a characteristic anticodon region at the other end. The anticodon consists of a nucleotide triplet that is the complement of the amino acid's codons. The ribosome accepts an mRNA molecule, binding initially to a characteristic nucleotide sequence at the 5'-end ensuring that polypeptide synthesis starts at the right codon. A tRNA molecule with the appropriate anticodon then attaches at the starting point and this is followed by a series of adjacent tRNA attachments, peptide bond formation, and shifts of the ribosome along the mRNA chain to expose new codons.

3.2.5.2 RNA Message Is Decoded on Ribosomes

The translation of the nucleotide sequence of an mRNA molecule into protein takes place in the cytoplasm on a large ribonucleoprotein assembly called a ribosome, a complex catalytic structure made up of more than 50 different proteins and several RNA molecules. The amino acids used for protein synthesis are first attached to a family of tRNA molecules, each of which recognizes, by complementary base-pairing, particular sets of three nucleotides in the mRNA (codons). The sequence of nucleotides in the mRNA is then read from one end to the other in sets of three according to the genetic code. To initiate translation, a small ribosomal subunit binds to the mRNA molecule at a start codon (AUG) that is recognized by a unique initiator tRNA molecule. A large ribosomal subunit then binds to complete the ribosome to begin the elongation phase of protein synthesis. During this phase, aminoacyl tRNAs, each bearing a specific amino acid, bind sequentially to the appropriate codon in mRNA by forming complementary base pairs with the tRNA anticodon. Each amino acid is added to the C-terminal end of the growing polypeptide by means of a cycle of three sequential steps: aminoacyl-tRNA binding, peptide bond formation, and ribosome translocation. The mRNA molecule progresses codon by codon through the ribosome in the 5' to 3' direction until one of three stop codons is reached. A release factor then binds to the ribosome, terminating translation and releasing the completed polypeptide (Figure 3.10).

The rRNA has the dominant role in translation, determining the overall structure of the ribosome, forming the binding sites for the tRNAs, matching

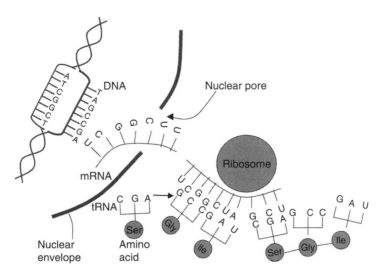

FIGURE 3.10

Translating an mRNA molecule. Following transcription, mRNA is exported from the nucleus into the cytoplasm where protein synthesis occurs. The mRNA couples with the protein synthesis apparatus (the ribosome). Another type of RNA, tRNA, brings free amino acids to the ribosome. The anticodon present on the tRNA recognizes the codon present on the mRNA, and the ribosome adds the amino acid to the growing chain of linked amino acids (polypeptides), cleaving it away from the tRNA. As the polypeptide chain grows, it folds to form a protein.

the RNAs to codons in the rRNA, and providing the peptidyl transferase enzyme activity that links amino acids together during translation. In the final steps of protein synthesis, two distinct types of molecular chaperones guide the folding of polypeptide chains. These chaperones, known as hsp60 and tP701, recognize exposed hydrophobic patches on proteins and act to prevent the protein aggregation that would otherwise compete with the folding of newly synthesized proteins into their correct three-dimensional conformations. This protein folding process must also compete with a highly elaborate quality control mechanism that destroys proteins with abnormally exposed hydrophobic patches. In this case, ubiquitin is covalently added to a misfolded protein by a ubiquitin ligase, and the resulting multi-ubiquitin chain is recognized by the proteasome and moved to the interior of the proteasome for proteolytic degradation. (The proteosomal mechanism of protein degradation is described in Chapter 8.)

3.2.5.3 Regulation of Gene Expression

The genome of a cell contains in its DNA sequence the information to make many thousands of different protein and RNA molecules. A cell typically expresses only a fraction of its genes, and the different types of cells in multicellular organisms arise because different sets of genes are expressed. Moreover, cells can change the pattern of genes they express in response to changes in their environment, such as signals from other cells.

There are many steps in the pathway leading from DNA to protein, and all of them can in principle be regulated. Thus a cell can control the proteins it makes by

(i) Controlling when and how often a given gene is transcribed (transcriptional control)

(ii) Controlling how the RNA transcript is spliced or otherwise processed (RNA processing control)

(iii) Selecting which completed mRNAs in the cell nucleus are exported to the cytosol and determining where in the cytosol they are localized (RNA transport and localization control)

(iv) Selecting which mRNAs in the cytoplasm are translated by ribosomes (translational control)

(v) Selectively destabilizing certain mRNA molecules in the cytoplasm (mRNA degradation control)

(vi) Selectively activating, inactivating, degrading, or compartmentalizing specific protein molecules after they have been made (posttranslational control) (Figure 3.11)

The general transcription factors have already been described. In addition, other transcription factors or gene regulatory proteins recognize short stretches of double-helical DNA of defined sequence and thereby determine which of the thousands of genes in a cell will be transcribed. Although each of these regulatory proteins has unique features, most bind as homodimers

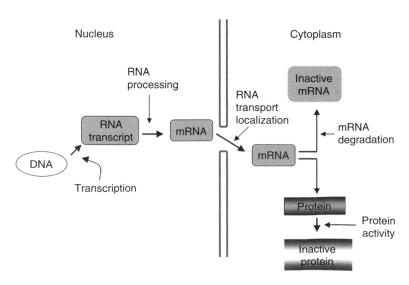

FIGURE 3.11
Steps at which eukaryotic gene expression can be controlled.

or heterodimers and recognize DNA through one of a small number of structural motifs (described in greater detail in Chapter 11). The precise amino acid sequence that is folded into a motif determines the particular DNA sequence that is recognized and hetero-dimerization increases the range of sequences that can be recognized by gene regulatory proteins.

3.2.6 Proteins

Proteins are made of amino acids arranged in a linear chain and joined by peptide bonds. By means of the information transfer pathway just described, a DNA sequence, the gene, specifies the sequence of amino acids in a protein. Most proteins fold into unique three-dimensional structures unassisted simply through the structural propensities of their component amino acids; others require the aid of molecular chaperones to efficiently fold to their native states.

There are four distinct aspects of a protein's structure. The primary structure is the amino acid sequence. The secondary structure consists of regularly repeating local structures stabilized by hydrogen bonds, the most common examples being the α helix and β sheet. Secondary structures are local and many regions of different secondary structures can be present in the same protein molecule. The tertiary structure results from the spatial relationship of the secondary structures to one another giving the overall shape to the single protein molecule. Tertiary structure (or protein folding) is generally stabilized by nonlocal interactions, most commonly the formation of a hydrophobic core, but also through salt bridges, hydrogen bonds, disulfide bonds, and posttranslational modifications. The quaternary structure is the shape or structure that results from the interaction of more than one protein molecule or protein subunits that function as part of the larger assembly or protein complex.

Proteins can be informally divided into three main classes, which correlate with typical tertiary structures: globular proteins, fibrous proteins, and membrane proteins. Almost all globular proteins are soluble and many are enzymes. Fibrous proteins are often structural, while membrane proteins often serve as receptors or provide channels for polar or charged molecules to pass through the cell membrane.

The best-known role of proteins is as enzymes, which catalyze chemical reactions including most of the reactions involved in metabolism and catabolism as well as DNA replication, DNA repair, and RNA synthesis. About 4000 reactions are known to be catalyzed by enzymes. Although enzymes can consist of hundreds of amino acids, it is usually only a small fraction of the residues that come in contact with the substrate and an even smaller fraction: 3–4 residues on average form the active site that binds the substrate and contains the catalytic residues.

Many proteins are involved in the process of cell signaling and signal transduction. Some proteins, such as insulin, are extracellular proteins, and others are membrane proteins that act as receptors whose main function is to bind a signaling molecule and induce a biochemical response in the cell. Many receptors are membrane proteins that have a binding site exposed on the cell surface and an effector domain within the cell, which may have enzymatic activity or may undergo a conformational change detected by other proteins within the cell.

Antibodies are protein components of the adaptive immune system, whose main function is to bind antigens, or foreign substances in the body, and target them for destruction. While enzymes are limited in their binding affinity for their substrates by the necessity of conducting their reaction, antibodies have no such constraints. An antibody's binding affinity to its target is extraordinarily high.

Structural proteins confer stiffness and rigidity to otherwise fluid bio-logical components. Most structural proteins are fibrous proteins; for example, actin and tubulin are globular and soluble as monomers but polymerize to form long, stiff fibers that comprise the cytoskeleton, which allows the cell to maintain its shape and size. Collagen and elastin are critical components of connective tissue such as cartilage, and keratin is found in hard or filamentous structures. Other proteins that serve structural functions are motor proteins such as myosin, kinesin, and dynein, which are capable of generating mechanical forces. These proteins are crucial for cellular motility and they also generate the forces exerted by contracting muscles.

3.2.7 Summary: Molecular Genetics

Genetic information is carried in the linear sequence of nucleotides in DNA. Each molecule of DNA is a double helix, formed from two complementary strands of nucleotides held together by hydrogen bonds between G–C and A–T base pairs. Duplication of the genetic information occurs by the use of one DNA strand as a template for formation of a complementary strand. The genetic information stored in the DNA contains the instructions for all the proteins the organism will ever synthesize.

DNA replication takes place at a Y-shaped structure called a replication fork. A self-correcting DNA polymerase enzyme catalyzes nucleotide poly-merization in a 5'-to-3' direction, copying a DNA template strand with a high degree of fidelity. Since the two strands of a DNA double helix are antiparallel, this 5'-to-3' DNA synthesis can take place continuously on only one of the strands (the leading strand). On the lagging strand, short DNA fragments synthesized in the 5'-to-3' direction are ligated together. In addition to DNA polymerase, DNA replication requires the cooperation of many proteins associated with each other at a replication fork to form "replication machines."

Before the synthesis of a particular protein can begin, the corresponding mRNA molecule is produced by transcription. mRNA is synthesized by RNA polymerase II, one of three RNA polymerases found in eukaryotic cells. This enzyme requires a series of additional proteins, termed the general transcription factors, to initiate transcription on a purified DNA template and still more proteins (including chromatin-remodeling complexes and histone acetyltransferases) to initiate transcription on its chromatin template inside the cell. During the elongation phase of transcription, the nascent RNA undergoes three types of processing events: a special nucleotide is added to its 5' end (capping), intron sequences are removed from the middle of the RNA molecule (splicing), and the 3' end of the RNA is generated (cleavage and polyadenylation). Genes for which RNA is the end product are usually transcribed by either RNA polymerase I or III. RNA polymerase I synthesizes ribosomal RNAs as large precursors that are chemically modified, cleaved, and assembled into ribosomes in the nucleolus.

The translation of the nucleotide sequence of an mRNA molecule into protein takes place in the cytoplasm on a large ribonucleoprotein assembly called a ribosome. The amino acids to be used for protein synthesis are attached to a family of tRNA molecules, each of which recognizes, by complementary base-pair interactions, particular sets of three nucleotides in the mRNA (codons). The sequence of nucleotides in the mRNA is then read from one end to the other in sets of three according to the genetic code. Translation is initiated when the small ribosomal subunit binds to the mRNA molecule at a start codon (AUG) recognized by a unique initiator tRNA molecule. Aminoacyl tRNAs, each bearing a specific amino acid, bind sequentially to the appropriate codon in mRNA by forming complementary base pairs with the tRNA anticodon. Each amino acid is then added to the C-terminal end of the growing polypeptide. The mRNA molecule progresses codon by codon through the ribosome in the 5'-to-3' direction until a release codon binds to the ribosome, terminating translation. The genome of a cell contains in its DNA sequence the information to make many thousands of different protein and RNA molecules, but the cell typically expresses only a fraction of its genes. The different types of cells in multicellular organisms arise because different sets of genes are expressed. Cells can also change the pattern of genes they express in response to changes in their environment. The initiation of RNA transcription is the most important point of control, but genes can be regulated at a number of points between transcription and expression of protein activity.

The base sequence of the gene specifies the sequence of amino acids in a protein, and the three-dimensional conformation of the protein molecule, which defines its function, is determined by its amino acid sequence. The structure of the folded protein is stabilized by non-covalent interactions between different parts of the polypeptide chain. Many proteins are enzymes that catalyze biochemical reactions and are vital to metabolism. Enzyme function and that of some other proteins is determined by the chemical properties of their surfaces, which form binding sites for ligands or the active sites of enzymes. Other proteins have structural or mechanical functions, such as the proteins in the cytoskeleton, which forms a system of scaffolding that maintains cell shape. Proteins are also important in cell signaling, immune responses, cell adhesion, and the cell cycle.

Bibliography

1. Alberts B, Johnson A, Lewis J, Raff M, Roberts K, and Walter, P. *Molecular Biology of the Cell*, 4th Edition, Garland Science, New York and London, 2002.
2. Tortora GJ. *Principles of Anatomy and Physiology*, 11th Edition, John Wiley, Hoboken, New Jersey, 2003.
3. Cooper GM and Haussman RE. *The Cell: A Molecular Approach*, 4th Edition, ASM Press, Washington DC.
4. Turner P, McLennan A, Bates A, and White M. *Molecular Biology*, 3rd Edition, Taylor and Francis, New York.
5. There are a multitude of Web sites which deal with topics discussed in this and other chapters. The best strategy is probably to search a specific topic (e.g., mitochondria). The following are a small sample of Web sites which cover many aspects of cell and molecular biology.
   - www.cytochemistry.net/cytochemistry.
     A group of Web sites providing basic information about cell biology topics.
   - micro.magnet.fsu.edu/cells/animalcells.html.
   - www.web-books.com/MolBio/.
     Molecular biology Web book.
   - www.rothamsted.bbsrc.ac.uk/notebook/course/guide.
     Beginners guide to molecular biology.

4

Methods of Cell and Molecular Radiobiology

4.1 Methods of Classical Radiobiology

4.1.1 Cell Survival In Vitro: The Clonogenic Assay

The clonogenic assay that celebrates its golden jubilee in 2006 is the cornerstone of classical radiobiology. This technique evolved from one that had long been used by bacteriologists for determining bacterial survival; the "plating" of diluted suspensions of bacteria on nutrient agar in a petri dish and counting the number of bacterial colonies that became visible after a short period of incubation. Puck and Marcus [1] developed an equivalent technique using a nutrient medium that would support the growth of mammalian cells. Single cells were plated in petri dishes; a proportion of the cells attached to the glass and those which were viable proliferated to form macroscopic colonies that could be counted after 7–10 days incubation. The first paper to describe this technique showed that radiation reduced the viability of mammalian cells in terms of clonogenic survival (Figure 4.1).

4.1.1.1 Colony Formation as an Index of Survival

A cell that retains unlimited ability to proliferate after radiation has survived treatment, whereas one which has lost the ability to generate a clone or colony is reproductively dead even though it may complete a few cell divisions and remain intact and metabolically active in the cell population for some time. Assessing survival of cells after radiation thus depends on the demonstration that they retain the ability to produce a large number of progeny (i.e., to produce a colony). Colony formation following irradiation is an important end point for radiobiologists and radiation oncologists, since it reflects the ability of the cell to repopulate normal or tumor tissues following exposure to ionizing radiation.

4.1.1.2 Clonogenic Assay: Procedure

For the clonogenic assay, cells grown in culture are irradiated either before or after preparation of a suspension of single cells and plated at low density

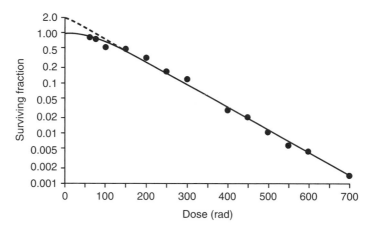

FIGURE 4.1
The first published radiation survival curve for mammalian cells. Survival of the reproductive capacity of HeLa cells. (From Puck, T.T. and Markus, P.I., *J. Exp. Med.*, 103, 653, 1956. With permission.)

in tissue-culture dishes. The cells are incubated for a number of days and those that retain proliferative capacity divide and grow to form discrete colonies of cells. At the end of the incubation period, the colonies are fixed and stained so that they can be counted easily. Cells that do not retain proliferative capacity following irradiation (i.e. are killed) may divide a few times but form only very small abortive colonies. A colony is considered significant if it contains more than 50 cells, that is, it must have resulted from at least six cell-division cycles indicating that the cells of the colony are capable of continued growth.

The plating efficiency (PE) of the cell population is calculated by dividing the number of colonies formed by the number of cells plated. The ratio of the PE for the irradiated cells to the PE for control cells gives the fraction of cells that have survived. If a range of radiation doses is used, then these cell-survival values can be plotted to give a survival curve (Figure 4.2).

4.1.1.3   Clonogenic Assay: Results
Results obtained using the clonogenic assay laid the foundation of much of the cellular radiation biology and by extension of the basic science under-pinning the treatment of cancer by radiation. Generally these studies have used established cancer cell lines of rodent or human origin, which are grown rapidly and are immortal in tissue culture.

Specific results will be described in the relevant chapters, but some examples are given here of the application of the clonogenic assay to a number of experimental situations (Figure 4.3A through 4.3D). The cell survival curve (Figure 4.3A) is a graph of surviving fraction against dose, which is almost invariably plotted on semi-log coordinates. The reasons for

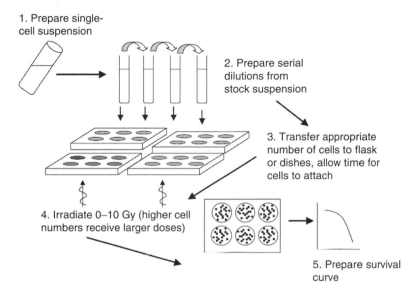

1. Prepare single-cell suspension

2. Prepare serial dilutions from stock suspension

3. Transfer appropriate number of cells to flask or dishes, allow time for cells to attach

4. Irradiate 0–10 Gy (higher cell numbers receive larger doses)

5. Prepare survival curve

FIGURE 4.2
Procedure to generate a single-cell survival curve. A single-cell suspension is prepared by trypsinization and known numbers of cells are transferred to T25 flasks or 6-well plates. The number of cells/well is dependent on the dose of radiation to be given, larger numbers of cells are plated for higher doses. Plates are incubated for 1–2 weeks after which the medium is removed and the colonies attached to the plate are fixed, stained, and counted. The number of colonies in the plate which received 0 Gy represents the plating efficiency (PE) (i.e., the proportion of cells which attach and form colonies in the absence of radiation). PE irradiated/PE nonirradiated = surviving fraction. Surviving fractions are plotted against dose on semi-log coordinates to give a survival curve.

this are twofold. Firstly, cell killing by radiation is random; consequently, survival is an exponential function of dose, which will be a straight line on a semi-log plot; and secondly, a logarithmic scale allows an easier comparison of effects at very low survival levels. In Figure 4.3A, the survival curve for cells grown under hypoxic conditions is also shown indicating the enhanced radioresistance of hypoxic cells.

For most mammalian cells the survival curve is characterised by a shoulder at low dose levels. Recovery from sublethal damage (SLD) was demonstrated by showing that the shoulder of the radiation survival curve returned when a second dose was delivered 1–2 h after an initial dose (Figure 4.3B). Elkind and Sutton [2] demonstrated that the shoulder on the survival curve represented accumulation of SLD that could be repaired. The term sublethal damage was used to describe the kind of damage that was not lethal by itself, but could interact with additional radiation-induced lesions to cause a lethal event. The repair of cellular damage between radiation doses is the mechanism that underlies the observation that a larger total dose can be tolerated when the radiation dose is fractionated.

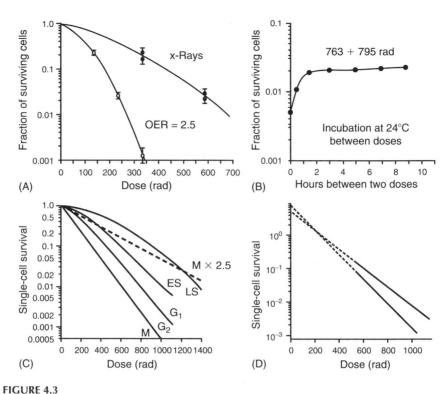

FIGURE 4.3
Important results in radiation biology which have been obtained using the clonogenic assay.
(A) Survival curves in air and nitrogen demonstrating the radio-sensitizing effect of oxygen.
(From Barendsen, G.W., Beusker, T.L.J., Vergroesen, A.J., and Budke, L., *Radiat. Res.*, 13, 841, 1960.
With permission.) (B) Survival of Chinese hamster cells exposed to two fraction of radiation and
incubated at room temperature between fractions. Demonstrates the repair of sublethal damage
between fractions. (From Elkind, M.M., Sutton-Gilbert, H., Moses, W.B., Alescio, T., and Swain,
R.B., *Radiat. Res.*, 25, 359, 1965. With permission.) (C) Cell survival curves for Chinese hamster
cells at different phases of the cell cycle. M—mitosis, ES—early S, LS—late S. Demonstrates
the variation in radiosensitivity throughout the cell cycle. (From Sinclair, W.K., *Radiat. Res.*,
33, 620, 1968. With permission.) (D) Survival curves for density-inhibited stationary phase
cells, subcultured either immediately or at 6–12 h after irradiation. Demonstrates the repair of
potentially lethal damage when cells are held under suboptimal conditions for growth following
radiation. (From Little, J.B., Hahn, G.M., Frindel, E., and Tubiana, M., *Radiology* 106, 689, 1973.
With permission.)

In delayed plating experiments, cells are irradiated in a nongrowing state
and left for increasing periods of time before plating in fresh medium.
Under these circumstances, an increase in survival is frequently observed.
This is termed recovery from potentially lethal damage (PLD) (Figure 4.3C).

Finally, using clonogenic assay it was shown that the radiosensitivity of
cells varies considerably as they pass through the cell cycle. This has not
been studied in a large number of cell lines but there is a general tendency
for cells in the latter part of S phase to be most resistant and for cells in G_2

and mitosis to be most sensitive. The classic results of Sinclair and Morton [3] are shown in Figure 4.3D.

4.1.1.4 Clonogenic Assay Adapted for Study of Response to Low Doses

The clonogenic assay involves the plating of an aliquot of cell suspension that on average will contain a known number of cells, but the actual number of cells in a given aliquot will vary according to Poisson statistics. More precision is required for comparison of the effects of low radiation doses where the differences are small, and this can be achieved only by knowing the actual number of cells plated, or even better, the actual number of cells that have attached to the plate and are potentially clonogenic. Methods have been developed that have improved the accuracy of cell survival measurement by determining exactly the number of cells "at risk" in a colony-forming assay. This is achieved by using either a fluorescence-activated cell sorter (FACS), which can plate a precisely known number of cells [3], or by microscopic scanning after plating to locate and count all the cells that have attached [4]. The latter technique uses equipment called a dynamic microscopic image processing scanner (DMIPS). Cell survival curves obtained using these methods and the processes underlying low dose survival will be discussed in Chapter 14.

4.1.2 Non-Clonogenic Assays

Non-clonogenic assays can be completed in a short period and are useful for those cells that do not produce well-defined colonies. Many of these assays are growth assays, which determine the number of viable cells at various times following irradiation in untreated or treated cell cultures. Cellular growth can be measured by a variety of methods including the MTT assay, in which cellular viability is measured by colorimetric assessment of the reduction of a tetrazolium compound.

One growth assay that has been optimized for determining the radio-sensitivity of human tumor cells is the cell-adhesive matrix (CAM) assay in which primary human tumor cells are plated directly onto culture dishes coated with a combination of cell-adhesive proteins [5]. Although clonogenic survival remains preferable for determining the reproductive potential of cells following irradiation, data from growth assays can be informative provided that the acquisition and interpretation of the data are rigorous [6].

Non-clonogenic assays have been used to estimate the relative radio-sensitivity of cells, but assays that estimate cell growth for a short period (24–48 h) after irradiation, including the MTT assay, can only predict clono-genic survival for those cancer cell lines which are killed uniformly by a process of apoptosis. Overall, short-term assays are of very limited value for radiosensitivity studies because it is rarely possible to assess more than one decade of cell kill, and in fact, the results usually do not correlate with results of the clonogenic assay, which remains the gold standard for determining the radiosensitivity of cells in vitro.

4.1.3   Methods of Cell Synchronization

As noted above, the radiation response varies throughout the cell cycle. Cell kinetic factors are important in the response of cell populations to ionizing radiation both in vitro and in vivo. For these reasons, the use of synchronized cell populations has been an important tool in radiobiological research. A number of methods have been developed, each of which has advantages and disadvantages (reviewed by Merrill [7]).

4.1.3.1   Synchrony by Release from G_0 Arrest

Cultured cells deprived of growth factors withdraw from the cell cycle with a $2N$ complement of DNA and the technique is made more effective by using both serum withdrawal and contact inhibition to induce cell cycle arrest. Synchrony is then achieved by passaging cells to serum-rich medium at a lower cell density. The advantage of this technique is that it requires no drugs or mechanical aids to achieve a large population of synchronized cells. The disadvantage is that reentry after a G_0 arrest is often not very synchronous. For many established cell lines, cells begin to enter S phase within 3 h of growth factor restoration, but can take as long as 15 h for the labeling index to reach maximal levels.

4.1.3.2   Synchrony by Release from M-Phase Block

Several agents or conditions that reversibly disrupt microtubule function have been used to synchronize cells by inducing a temporary mitotic arrest. The microtubule destabilizers colchicine, demecolcine (Colcemide), and vinblastine arrest cells in M phase by preventing formation of a mitotic spindle. Methyl-(5-[2-thienylcarbonyl]-1H-benzimidazol-2-yl) carbamate (Nocodazole) arrests cells in M phase by inhibiting assembly of a microtubule subpopulation required for chromosome separation. Nocodazole has become the drug of choice for inducing mitotic synchrony because its effects are generally more readily reversed.

4.1.3.3   Synchrony by Release from S-Phase Block

Several agents or conditions that prevent DNA replication have been used to prepare synchronized cell populations. Hydroxyurea and high concentrations of thymidine prevent replication by inhibiting ribonucleotide reductase, whereas aphidicolin blocks replication by directly binding and inhibiting DNA polymerase. Regardless of whether the blocking agent works by preventing DNA precursor synthesis or by blocking DNA polymerization, two incubations in the presence of the blocking agent are necessary to induce synchrony. During the first incubation, the agent is added to exponentially growing cells. Cells in S phase stop replicating and remain distributed throughout S phase whereas cells in G_2, M, and G_1 continue traversing the cell cycle until they enter S phase, where they arrest with a small percentage of their DNA replicated. When the blocking agent is removed, the cells arrested at the start of S, as well as those cells distributed throughout S, resume replication. After allowing all cells sufficient time to complete replication, the blocking agent is restored. All cells should be in G_2, M, or G_1 at this point and will continue traversing the cell cycle until they enter S phase and arrest. The whole cell population is now arrested in early S phase and when the blocking agent is removed again the cells synchronously complete DNA replication and proceed through the cell cycle.

4.1.3.4 Synchrony by Mitotic Detachment

A minimally disruptive way of synchronizing monolayer cells is by the mitotic shake-off method. As most monolayer cells enter mitosis, they round up and become only tenuously attached to the substratum. Mitotic cells can be isolated by shaking the dish vigorously and transferring the medium containing the detached cells to a fresh culture vessel. Shortly after transfer, the cells complete mitosis, attach to the new dish, and enter G_1. A drawback of the method is the low yield of mitotic cells. Some biochemical experiments, such as those involving Northern blot analysis of specific RNAs or immunoblot analysis of specific proteins, require more than 10^5 cells per time point. To generate larger number of cells, a number of plates are used and reused at regular intervals. Cells can be collected by centrifugation and resuspended in smaller volume.

4.1.3.5 Synchrony by Centrifugal Elutriation

Centrifugal elutriation is a useful method of fractionating cells according to their size. The system consists of a specialized centrifuge rotor in which the centrifugal force on a cell population can be countered by bulk medium flow in the opposite direction. When these antagonizing forces are exactly balanced, cells form a gradient in the elutriation chamber, with smaller cells at the top and larger cells at the bottom. As either the rotor speed is decreased or the rate of medium flow is increased, the gradient of size-separated cells is pushed toward the top of the elutriation chamber. Eventually, the small cells at the top of the gradient are pushed out of the elutriation chamber and into a collection vessel. With further incremental decreases in rotor speed or increases in medium flow, progressively larger cells are pushed out of the elutriation chamber. As early G_1 cells are roughly half the size of mitotic or late G_2 cells, centrifugal elutriation can be used to fractionate cells according to their position in the cell cycle. Centrifugal elutriation requires specialized equipment but offers a number of advantages over other methods of synchronizing cells in that it does not involve the use of drugs or serum deprivation to collect cells in a uniform cell cycle stage, it gives a relatively high yield of synchronized cells and it can be used to synchronize both monolayer and suspension culture cells.

4.1.4 Determination of Duration of Phases of the Cell Cycle

4.1.4.1 Microscopy

4.1.4.1.1 Cell Cycle Time and Duration of Mitosis

If the mean generation time of a population is known, the duration of M phase can be calculated by determining the frequency of mitotic cells or the mitotic index.

4.1.4.1.2 Fraction of Labeled Mitoses Method

This technique depends on labeling cells with tritiated thymidine (^{3}H-thymidine), which is incorporated into DNA and identifying the labeled cell by autoradiography (Figure 4.4). Exponentially growing cells are pulse-labeled with ^{3}H-thymidine, and at intervals thereafter cells are incubated with Colcemid (15 min) fixed to microscope slides and autoradiographs are prepared. The fraction of labeled mitotic cells is plotted as a function of time after the pulse label. Cell cycle phase durations are calculated from the following measurements:

- $^{1}/_{2}$ M + G$_{2}$ is the time required for labeling index to rise to half maximal levels.

- The duration of the S phase is the time between 50% labeling on the ascending curve and 50% labeling on the descending curve.

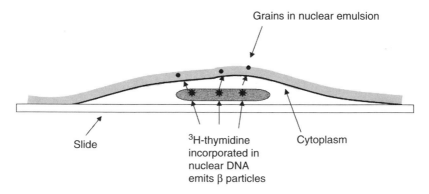

FIGURE 4.4
Autoradiography. Cells labeled with a radioactive DNA precursor (^{14}C- or ^{3}H-thymidine) attached to a glass microscope slide are coated with a layer of photographic emulsion and kept in cold and dark conditions for several weeks. β-Particles emitted during the decay of ^{14}C- or ^{3}H interact with the silver halide in the emulsion and appear as black grains when the emulsion is developed and fixed. (The same principal applies when a radioactive oligonucleotide is used to probe a gel in Southern or Northern blotting. A piece of x-ray film is applied to the gel and black band is seen at the point where the probe is bound when the film is developed.)

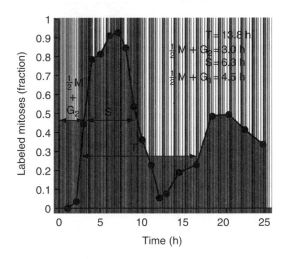

FIGURE 4.5

Determination of cell cycle phase durations by the fraction of labeled mitosis method. Mouse myoblasts were pulse-labeled for 15 min with ^{3}H-thymidine. At indicated times cells were incubated for 15 min with Colcemid, fixed to microscope slides and autoradiographed. The fraction of labeled mitotic cells was plotted as a function of time after the pulse label. (From Merrill, G., *Methods Cell Biol.*, 57, 229, 1998. With permission.)

- The time between half maximal labeling of first and second peak corresponds to the length of the entire division cycle:

$$G_1 + 1/2\,M = T - (S + 1/2\,M + G_2)$$

- The length of the M phase estimated from the mitotic index is a minor component of the G_1 and G_2 duration times (Figure 4.5).

4.1.4.2   Flow Cytometry

Flow cytometry is a general method for rapidly analyzing large numbers of cells individually using light scattering, fluorescence, and absorbance measurements. The power of this method lies both in the wide range of cellular parameters that can be determined and in the ability to obtain information on how these parameters are distributed in the cell population. Flow cytometric assays have been developed to determine both cellular characteristics such as size, membrane potential, and intracellular pH, and the levels of cellular components such as DNA, proteins, surface receptors, and calcium.

4.1.4.2.1   Principles of Flow Cytometry

A flow cytometry system consists of five main operating units: a light source (mercury lamp or laser), flow cell, optical filter units for specific wavelength detection over a broad spectral range, photodiodes or photomultiplier tubes for sensitive detection of the signals of interest, and a data processing and

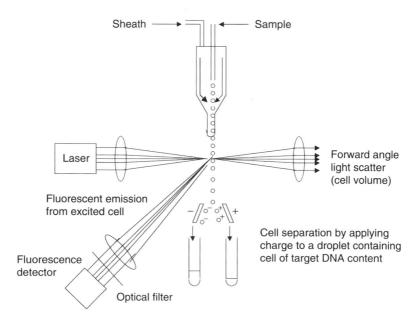

FIGURE 4.6

A fluorescence-activated cell sorter. A cell passing through a laser beam is monitored for fluorescence. Droplets containing cells are given a positive or negative charge depending on the fluorescence of the cell. The cells are then deflected by an electric field into a collection tube depending on their charge.

operating unit. A cell suspension is injected into the flow cell where the cells pass one after another across a laser beam (or mercury lamp light) that is orthogonal to the flow. Using this technology, it is possible to detect up to 10,000 cells/s. Some intracellular compounds have an intrinsic fluorescence (e.g. NAD(P)H), in other cases staining with specific fluorescent dyes allow certain cell components to be selectively assayed. The resulting fluorescent emission is processed through the photomultiplier to the data processing system; and the data are analyzed by the cytometer software. The scattering and fluorescence signals can be combined in various ways that allow all subpopulations to be observed. In most cases, dual parameter plots are used to visualize cytometric data, but recently, multivariate data analysis methods have been developed to improve the extraction of information from data obtained by multiparameter analysis (Figure 4.6).

4.1.4.2.2   Estimation of Duration of Cell Cycle Phases

If the mean generation time for a given population is known, the fraction of proliferative cells is high, the frequency of cell loss is low, and cells are reasonably homogenous with respect to their cell cycle kinetics, the lengths of G_1, S, and G_2 can be estimated by flow cytometry. Asynchronously growing cells are stained with a DNA-intercalating dye such as propidium iodide, and fluorescence intensity is used to determine the frequency of cells

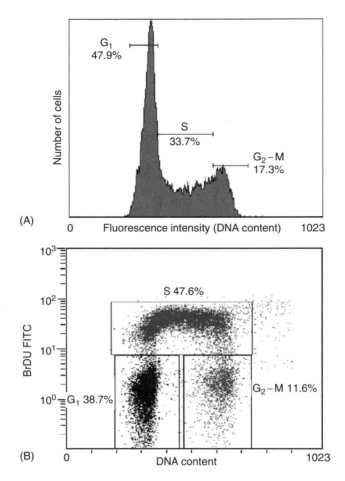

FIGURE 4.7
Results obtained by flow cytometry. (A) Cell cycle distribution analyzed on the basis of DNA content by flow cytometry DNA is stained with propidium iodide (PI) and the profile is analyzed by appropriate software to give the proportional cell numbers in G_1-S-G_2/M phases. (B) Double labeling: Bromodeoxyuridine (BrDU) incorporated into newly synthesized DNA is labeled by a fluorescein conjugated monoclonal antibody. Bulk DNA is labeled with PI. This protocol gives a more accurate quantification of S-phase cells.

with a 2N, intermediate, or 4N complement of DNA (Figure 4.7A). Staining with bromodeoxyuridine (BrDU) in addition to PI identifies newly synthesized DNA and is a more accurate method to identify cells entering the S phase (Figure 4.7B).

4.1.4.2.3 Cell Sorting

Flow cytometers can be combined with sorting units that offer the possibility of separating selected subpopulations. Most sorters are based on a sorting unit that breaks the cell stream into droplets and ejecting the focused stream of cells through a nozzle into air. By vibrating the nozzle with a transducer a stable stream of droplets can be produced and adjusting the relative flow of sample and sheath fluids makes it possible for the majority of cells to be placed individually in droplets separated by a number of droplets not containing cells. The sheath fluid is usually phosphate buffered saline (PBS) or a similar electrolyte so that the whole stream can be electrically charged (+, −, or 0) just before a droplet separates, enabling the three populations to be selected after passing through an electric field. As the droplet leaves the stream, it passes between deflection plates carrying a high voltage and the droplet is deflected depending on the charge that it was given. Uncharged droplets which pass through undeflected and deflected droplets are collected in tubes.

4.1.5   Measuring Cell Survival In Vivo

Cell survival studies in vitro are rapid, convenient, and easily quantified; however, the natural milieu of a mammalian cell is not a plastic dish containing culture medium but a three-dimensional structure composed of many different types of cells organized into interactive multicellular components. A number of techniques have been devised, which model the response of cells irradiated in situ as part of a solid tumor or normal tissue.

4.1.5.1   Transplantable Solid Tumors in Experimental Animals

Techniques that have been described for measuring survival of individual tumor cells irradiated in situ can use two different procedural approaches: firstly, irradiation of the tumor and measurement of tumor response in situ on the basis of noninvasive measurements of tumor growth; or secondly, irradiation of the tumor in situ and measurement of survival of tumor cells transplanted to a nonirradiated recipient. The details of some of these techniques are summarized in Table 4.1.

In the first category is tumor growth measurement that is possibly the simplest end point to use, and tumor control, the TCD$_{50}$, which is the model with the most obvious relevance to clinical radiotherapy. The tumor cell transplantation techniques involve irradiation of tumor in situ and transplantation to recipient animals (limiting-dilution technique and the lung colony assay) or into tissue culture (the in vivo/in vitro technique or excision assay).

The transplantation techniques are attractive in that the results are in the form of a surviving fraction of tumor cells irradiated and radiation response curves can be plotted similar to those prepared by the in vitro clonogenic assay. There are, however, some drawbacks to methods that require transplantation of cells since these depend on disaggregation and production of single-cell suspension from the donor tumor. Tumor tissues differ in the ease with which they can be disaggregated and the methods used are potentially damaging to cells since they involve the use of proteolytic enzymes (trypsin, collagenase, and pronase) or mechanical disaggregation.

TABLE 4.1

Methods for Assay of Tumor Cell Survival

Type of Assay	Method	Procedure	Comments
In situ tumor irradiation, in situ measurement of response	Tumor growth measurement	Uniform size transplanted tumors show a temporary shrinkage after radiation followed by regrowth. Response is scored on time to regrow to pretreatment volume or a specified multiple of pretreatment volume.	Growth delay increases as a function of dose. Simplest end point to use. Relevance to clinical radiotherapy.
	Tumor control (TCD$_{50}$) assay	Group of tumors of uniform size are irradiated locally with graded doses and subsequently observed for tumor recurrence. The TCD$_{50}$, the dose at which 50% of the tumors are locally controlled, is determined by statistical analysis.	Tumor control is the model with the most obvious relevance to clinical radiotherapy.
In situ tumor irradiation. Response determined by transfer to recipient animal or to tissue culture	Limiting dilution assay	Different concentrations of cells from an irradiated tumor implanted into recipients. Surviving clonogenic cells form tumors in vivo. Number of cells required to transmit the tumor to 50% of recipient is the TD$_{50}$.	Requires preparation of single-cell suspension from solid tumor. First in vivo survival curve [35]. Uses large number of animals.
	Lung colony assay	Survival of cells solid tumor irradiated in situ assayed by injecting single-cell suspension into recipient animals and counting the number of lung colonies which develop.	Requires preparation of single-cell suspension from solid tumor.
	Excision assay	Solid tumor irradiated in situ and a single-cell suspension put into tissue culture for clonogenic assay.	Combines validity of in situ irradiation with speed and economy of in vitro end point assay.
Xenografts		Human tumor cells implanted in immunodeficient recipient animals are irradiated in situ and response assayed by growth delay or cell survival techniques.	Cells retain some characteristics of the individual human tumors. Stroma of mouse origin so xenografts are no better than murine tumors in cases where vascular supply is a factor in treatment response.

For the excision assay, some tumor cells will grow attached to plastic tissue-culture dishes while others require a feeder layer of lethally irradiated connective tissue or tumor cells to be laid down first. For tumor samples taken directly from patients or animals, the sample will inevitably be contaminated with normal tissue fibroblasts that may grow better than the tumor cells under conditions which permit anchorage dependent growth. This can be avoided by using semisolid medium containing agar or methylcellulose, which inhibits the growth of anchorage-requiring cells but allows growth of anchorage-independent tumor cells.

Xenograft tumors (i.e., human tumor cells planted in immunodeficient rodent hosts) have become possible because animal strains have been developed that are congenitally immune-deficient. Best known are nude mice, which are hairless and lacking a thymus. More recently nude rats have been developed, and SCID mice that suffer from the severe combined immunodeficiency syndrome and are deficient in both B-cell and T-cell immunity have been used. The survival of xenografted tumor cells irradiated in situ is usually determined by growth delay method or by cell survival determination after transplantation to tissue culture.

Methods for measuring survival of tumor cells irradiated in situ are summarized in Table 4.1.

4.1.5.2 Assays of Radiation–Dose Relationships for Cells from Normal Tissues

A number of experimental techniques are available to obtain dose–response relationships for the cells of normal tissues. As was the case for the tumor cell assays, techniques to measure normal tissue response can be divided into those which depend on scoring survival in situ and those which require cells from an irradiated donor to be transplanted to a recipient in order to score survival. The techniques developed by Withers and his collaborators [8–11], which include assay systems for survival of skin stem cells, testes stem cells, kidney tubule epithelium, and regenerating crypts in the jejunum, are based on the observation of a clone of cells regenerating in situ in irradiated tissue. The assay systems for the stem cells in the bone marrow or cells of the thyroid and mammary glands, on the other hand, depend on the observation of the growth of clones of cells taken from a donor animal and transplanted into a different tissue in a recipient animal.

A disadvantage of the intestinal crypt, testes, and kidney tubule systems is that experiments can be done only at doses that deplete cell survival sufficiently for individual surviving stem cells to be identified in situ, whether as regenerating crypts or as the epithelial lining of testes or kidney tubules. This problem can be overcome if the dose is delivered as a number of small fractions. As long as the total dose results in enough cell kill to be able to identify and score the survivors, the shape of the entire survival curve can be reconstructed from the multi-fraction data, if it is assumed that each dose fraction produces the same amount of cell killing [12].

The best-known survival assay based on the ability of cells to form colonies in situ following transplantation from an irradiated donor is the spleen-colony method, which has been used to assess the radiation and drug sensitivity of bone marrow stem cells [13]. When mouse bone marrow cells are injected intravenously into syngeneic recipients that have received sufficient whole-body irradiation to suppress endogenous hemopoiesis, colonies are produced in the spleen that are derived from the stem cells in the graft. The colonies vary in morphology (erythroid, granulocyte, or mixed) indicating that these stem cells are pluripotent. Initially their precise identity was not known and they were called colony-forming units (CFU). The first survival curve for bone marrow cells demonstrating their extreme radiosensitivity was obtained using this assay. Subsequently, clonogen transplant assays for epithelial cells of the mammary and thyroid glands were devised. Assays for normal cell survival following irradiation in situ are summarized in Table 4.2.

4.2 Methods for Detecting Damage to DNA

This section is based in part on a review by Olive (1997) and the papers referenced therein [14].

The random nature of energy deposition events means that radiation-induced changes can occur in any molecule in the cell. However, because of its biological role, even relatively small amounts of DNA damage can lead to lethality and a number of lines of evidence have identified DNA as the most important target of the lethal action of radiation. Further, results of a number of assays of DNA double-strand breaks (DSBs) suggest that cell survival following radiation is correlated with the residual level of DNA double-strand breaks [15]. For these reasons, determination of the extent of DNA strand breakage and the rate at which breaks are repaired were of interest even before any of the other manifestations of radiation insult at the molecular level were investigated.

Early experiments on the radiation chemistry of DNA demonstrated that ionizing radiation could cause loss of viscosity in DNA solutions. This was shown to result from breakage of DNA, which could involve one strand only (single-strand break [SSB]) or both strands (double-strand break [DSB]). Extraction of DNA from cells involves chemical and mechanical processes that themselves introduce a lot of damage so that the purified DNA used in these experiments was already reduced in size by the large numbers of strand breaks introduced during extraction. To increase this number by an amount sufficient to produce measurable differences in viscosity required very high doses of radiation. A variety of methods were subsequently developed to detect DSBs without excessive background interference from exogenously introduced damage and with increased sensitivity to lower doses.

TABLE 4.2

Methods for Assay of Survival of Cells of Normal Tissues Following Radiation

Type	Method	Procedure	Comments
In situ irradiation, in situ measurement of response	Skin stem-cell colonies	A test dose is given to island of intact skin which has been isolated by a massive dose to surrounding area. Nodules of regrowing skin are scored some days later.	Range of doses is restricted by the area of skin. Split doses can be used [8].
	Small intestine crypt stem cells	Mice given total-body radiation are sacrificed and the jejunum sectioned after 3 days. Number of regenerating crypts/circumference of the jejunum is scored. Crypt survival does not reflect cell survival unless dose sufficient to reduce the stem-cell survivors to <1/crypt.	Single doses must be >10 Gy but whole survival curve can be reconstructed from multi-fraction data [9].
	Testes stem cells	Mouse testes are sectioned and histology done 6 weeks postirradiation. The proportion of tubules containing spermatogenic epithelium is scored.	High single doses are necessary, but complete survival curve can be reconstructed from multi-fraction data [10].
	Kidney stem cells	One kidney irradiated and removed for histology at 60 weeks. Number of regenerated tubules scored.	First clonal assay for a late-responding tissue [11].
In situ irradiation, response determined by transfer to recipient	Lung colony assay	Suspension of bone marrow cells injected into previously supra-lethally irradiated recipients. Spleen colonies counted at 9 days.	Using this assay Till and McCulloch obtained the first survival curve for bone marrow cells [13]. [36,37].
	Mammary cells, thyroid cells	Mammary or thyroid-gland cells in the rat, are irradiated before the gland is removed from the donor and processed to a single-cell suspension. Known numbers of cells are then injected into the inguinal or interscapular white fat pad of recipient animals where they give rise to mammary/thyroid structures that can be counted.	

4.2.1   Strand Break Assays

A number of methods measure strand breaks in bulk DNA on the basis of direct or indirect measurement of the extent to which the size of the DNA molecule has been changed by radiation.

4.2.1.1   Sucrose Gradient Centrifugation

An important advance occurred in 1966 when McGrath and Williams [16] introduced alkaline sucrose gradient sedimentation as a method to measure DNA SSBs in irradiated cells. This method reduced the amount of damage introduced into DNA by lysing the cells directly on top of the sucrose gradient through which they were then centrifuged. Alkaline sucrose gradient sedimentation is notable among methods for the measurement of DNA damage in that the physical principle on which it is based is well understood. Unfortunately, the original method proved insensitive to detection of strand breaks produced by biologically relevant doses in mammalian cells, largely because conditions required for ideal sedimentation required that DNA already contain a significant number of strand breaks.

4.2.1.2   Filter Elution

For this technique, cells containing radio-labeled DNA are lysed on the top of a filter, the pore size of which is many times larger than the DNA filament. The flow of elution buffer carries the DNA fragments through the membrane and the rate of elution of DNA is related to the size of the DNA molecules on the membrane. To measure DNA DSBs, the elution buffer is at pH 7.4 or 9.6. The use of pH 9.6 not only increases the sensitivity of the technique but also allows the detection of alkali-labile sites that are not actual DSBs at physiological pH. Increasing pH to 12.3 leads to separation of the two strands of the DNA and allows the measurement of DNA SSBs.

Alkali filter elution detects DNA pieces of different sizes as they elute from polycarbonate filters: small pieces elute early, whereas large pieces elute after a longer period. In the case of neutral filter elution, interpretation of the results is complicated by the fact that analysis of the sizes of DNA fragments that elute from the filter has shown that the pieces are of similar size, regardless of the time at which they elute [17,18]. It is hypothesized that similar-sized fragments are produced by the shear forces of the eluting solution and it is the number of pieces available for elution at any time, and not their size, that determines the amount of DNA eluted.

4.2.1.3   Electrophoretic Methods

Fragments of DNA carry a net negative charge, and when an electric field is applied to DNA incorporated into an agarose gel, the fragments migrate at a speed that is inversely related to their size. For analysis of DNA size by electrophoresis, about 10^6 cells are embedded in low-gelling-temperature agarose plugs before being lysed and the plugs are then sealed into the wells of a preformed gel and subjected to electrophoresis for various times. DNA is detected in the gel by fluorescence or sectioning of gels containing radio-labeled DNA. The method of embedding intact cells in agarose before lysis reduces the background level of strand breaks, increasing the sensitivity of the method.

With careful choice of parameters, constant-field gel electrophoresis can demonstrate sensitivity to modest doses of radiation, but it fails to separate DNA pieces on the basis of size once they exceed about 50 kb. The separation of fragments is improved by pulsing and alternating the direction of the electric field, a process called pulsed field gel electrophoresis (PFGE). The theory behind PFGE is that by alternating the direction of the electric field at regular intervals, DNA is forced to reorient in the new field direction, and smaller DNA pieces will reorient faster than larger ones because of the physical resistance of the agarose matrix. This technique overcomes problems of anomalous movement of large DNA molecules in an electric field, so that their separation can be translated into a measure of strand breakage produced by small radiation doses. DNA of known molecular weight (for instance intact yeast chromosomes) is used to calibrate the movement of irradiated DNA in the gels.

A number of different electrode configurations, represented by a lexicon of four and five letter acronyms have been used: orthogonal field agarose gel electrophoresis (OFAGE), transverse alternating field electrophoresis (CHEF), field inversion gel electrophoresis (FIGE), to name but a few. These have similar dose–response relationships when the amount of DNA to migrate out of the well (FAR) is used as an end point. However, using this simple end point (amount of DNA smaller than a defined exclusion size) PFGE does not present much advantage over constant-field electrophoresis in terms of sensitivity. A more sophisticated approach requires analysis of fragment size but the method cannot be applied in a uniform manner to measurements over the complete range of potential DNA fragment sizes since a given combination of voltage, pulse time, and duration of electrophoresis are generally applicable only to a specific size range. Another factor influencing on PFGE results is DNA replicative status since DNA from cells in S phase migrates 3–4 times slower than DNA from cells in other phases [19,20]. In fact, this is a limitation on all DSB assays although when DNA is detected by radiolabeling methods, a chase time for several hours in isotope-free medium following radiolabeling minimizes the problem.

4.2.1.4 Nucleoid Assays

The lysis of cells at neutral pH in the presence of high salt concentration and a nonionic detergent allows the interphase nucleus to open up and reveal the tangled mass of chromatin. The resulting structures, often called nucleoids, consist of supercoiled DNA still retaining attachment to residual protein structures. The sedimentation of these structures in a sucrose gradient is influenced by the induction of SSB, which allows the domains to relax, and therefore to enlarge. One adaptation of this technique, the halo method, assesses the expansion of nucleoids by incorporating a fluorescent dye (usually ethidium bromide) into the DNA and measuring the size of halos by microscopy. The concentration of the intercalating dye greatly influences the degree of unwinding of the domains, and the relationship between halo size and dye concentration gives information about the chromatin structure in the nucleoid. Nucleoids and the relationship between radiation sensitivity and the kinetics of nucleoid unwinding are discussed at greater length in Chapter 9.

4.2.2 Measurement of DNA Damage and Repair in Individual Mammalian Cells

4.2.2.1 *Single-Cell Gel Electrophoresis or Comet Assay*

The comet assay (so-called because of the appearance of individual nuclei in this method) that evolved from the method of Ostling and Johanson [21] is a powerful technique, because it can be used to identify subpopulations differing in damage by relatively small amounts. Single cells embedded in low-density agarose are lysed to remove proteins and then exposed to an electric field (Figure 4.8). The domain structures unwind and when the preparations are subjected to electrophoresis, the broken DNA migrates

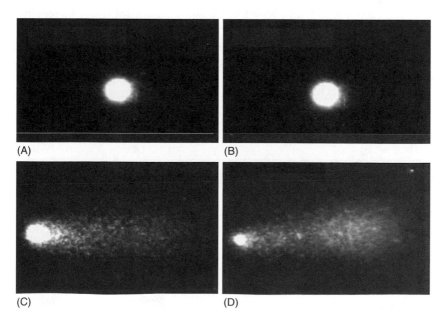

(A) (B) (C) (D)

FIGURE 4.8
Single-cell electrophoresis (comet assay). (A) Normal cell, (B–D) increasing amounts of DNA damage.

away from the general mass of DNA in the nucleus. The amount of DNA that migrates is proportional to the number of strand breaks. Once DNA is stained using a fluorescent DNA-binding drug, such as propidium iodide, damage can be quantified using an image analysis system [22]. Variations in the lysis conditions allow the measurement of different lesions (e.g., alkaline conditions assess SSB while neutral conditions allow the assessment of DSB) and the effect of chromatin structure can also be manipulated to some degree in a manner similar to nucleoid sedimentation. This assay has the advantages of high sensitivity to SSB (though not to DSB) and a requirement for only small numbers of cells. Irradiation of solid tumors combined with analysis using the alkaline comet assay can be used as an indicator of the presence of hypoxic cells [23], since the latter show threefold fewer DNA strand breaks after irradiation than aerobic cells.

4.2.2.2 Micronucleus Test

Micronuclei are cytoplasmic chromatin-containing bodies formed when acentric chromosome fragments or chromosomes lag during anaphase and fail to become incorporated into daughter-cell nuclei during cell division. Because the genetic damage that results in chromosome breaks, structurally abnormal chromosomes, or spindle abnormalities leads to micronucleus formation, the incidence of micronuclei serves as an index of these types of damage. It has been established that essentially all agents that cause double-strand chromosome breaks induce micronuclei. Enumeration of micronuclei is faster and less technically demanding than is scoring of chromosomal aberrations.

If cells are cultured in the presence of cytochalasin-B, which blocks cytokinesis, binucleate cells are seen after cell division, thus allowing nuclei that have undergone one posttreatment division to be identified (Figure 4.9). Micronuclei are scored as small extranuclear bodies in binucleate cells, their frequency is related to radiation dose and gives a measure of radiation

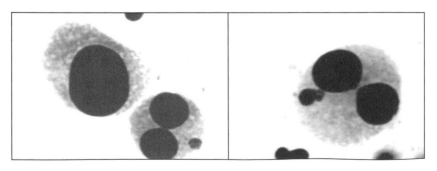

FIGURE 4.9
Micronuclei in cytokinesis-blocked lymphocytes. Stained with Geimsa. (Courtesy of Dr Frank Verhaegen. With permission.)

sensitivity [24]. The reliability is limited by the fact that diploid, polyploid, and aneuploid cells may differ in their tolerance of genetic loss, and therefore of micronucleus formation.

4.2.2.3 Chromosome Aberrations, Premature Chromosome Condensation, and Fluorescent In Situ Hybridization

One of the most obvious cytological effects of irradiation is the production of damage to chromosomes. The morphological signs of chromosome damage are apparent in metaphase chromosomes when cells undergo mitosis. Cells irradiated early in the cell cycle take some hours to enter mitosis, and therefore the chromosome breaks that are observed are those that have failed to rejoin during the time between radiation and mitosis. However, if an interphase cell is fused with a mitotic cell, it is found that the interphase chromatin undergoes a process of premature chromosome condensation (PCC) by which its chromosomes become visible. The mitotic cell can be of a different cell type and its chromatin can be labeled with BrUdR so that, within the binucleate fusion product, it is possible to identify the chromosomes of the target cell. This technique enables breaks in chromatin to be scored within 10–15 min of irradiation and the speed of their rejoining to be determined [25,26].

The analysis of both metaphase chromosome aberrations and PCC has been greatly facilitated by the development of chromosome-specific lengths of DNA (i.e. probes) that can be used to identify specific gene sequences. In this technique, the chromosomes are spread and fixed on microscope slides, then heated so that much of their DNA becomes single stranded. The specimens are incubated with labeled probe DNA and the probe binds to those regions of the chromosomal DNA with which it is homologous. If the probe is tagged with a fluorescent ligand, it can be detected by fluorescence microscopy. The use of this technique (called fluorescent in situ hybridization [FISH]) has made the individual chromosomes and translocations between chromosomes much easier to identify and has led to the identification of highly complex aberrations that were not previously detectable. The formation and dose relationships of chromosome aberrations are discussed at greater length in Chapter 10.

4.2.2.4 Measurement of the Fidelity of Rejoining

Although there are several techniques that provide information about DNA damage and repair in terms of changes in the sizes of DNA fragments, a more refined approach is required to determine if repair processes have faithfully recreated the DNA sequence present before radiation damage. The methods used to answer this question depend largely on molecular biology techniques described in the next section of this chapter and will only be briefly described here.

Thacker et al. [27] developed an integrating DNA vector, encoding a selectable marker gene that could be broken in defined sites in the coding sequence by restriction endonucleases. Plasmids were created with two selectable markers, a *gpl* gene, which is cut with restriction enzymes (*Kpn*l or *Eco*RV representing staggered and blunt cutters), and a *neo* gene as a marker of transfection. The fidelity of the rejoining process could then be followed by measuring survival of vector-transformed cells in a medium that would only support growth of cells containing the functional *gpl* gene. Powell and McMillan [28] used this method to probe differential cell repair capacity in a series of human tumor cell lines varying in intrinsic radiosensitivity.

4.2.3 Summary

This section is summarized in Table 4.3.

4.3   Tools and Techniques of Molecular Biology

This section is intended to provide an overview of some of the techniques of molecular biology most relevant to this book and to be an indication of what is possible. Biotechnology moves at a furious pace. Every day new techniques become available and better, quicker, and more expensive ways of performing familiar assays become standard. There are a number of comprehensive books on the subject which are frequently updated and a multitude of ever-changing Web sites. Many procedures can be accomplished using kits with which the manufacturer includes detailed instructions and recipes.

4.3.1   Hybridization of Nucleic Acids

Many of the techniques of molecular biology depend on the ability of single-stranded nucleic acids to hybridize or renature, with complementary sequences. Under conditions of high temperature or pH, double-stranded DNA is denatured by breakage of the hydrogen bonds between the base pairs, separating the complementary strands into single-stranded DNA. Upon return to appropriate conditions, the original duplex DNA molecule can be reassembled by hybridization of the complementary strands. Complementary regions of DNA can join together or hybridize to form a double-stranded molecule and hybridization can take place between two complementary single-stranded DNA molecules or between one DNA molecule and one RNA molecule.

Many important techniques in molecular biology depend on DNA hybridization including Southern blotting and the screening of bacteria to isolate cloned DNA, both of which require the construction of a specific DNA probe. A single-stranded oligonucleotide hybridized to a single-stranded template will act as a primer for DNA synthesis if it is incubated in a reaction

TABLE 4.3

Detection of DNA Damage Induced by Ionizing Radiation

Assay	Dose Range	Method	Comments
Sucrose velocity sedimentation	SSB > 5 Gy DSB > 15 Gy	Larger DNA fragments sediment more rapidly.	Insensitive to low doses.
Filter elution	SSB > 1 Gy (alkaline elution) DSB > 5 Gy (neutral elution)	Smaller DNA fragments elute more rapidly through filter.	Influenced by cell conformation, cell cycle, cell number, and lysis conditions.
Pulse-field gel electrophoresis (PFGE)	DSB > 5–10 Gy	DSB quantified by relative migration in the gel.	Influenced by cell conformation and number of cells in S phase.
Nucleoid sedimentation	SSB 1–20 Gy	Alterations of DNA supercoiling in the nucleus reflect strand breakage.	Uncertain which DNA lesions are detected.
Comet assay	SSB > 1 Gy (alkaline lysis) SSB > 1 Gy (neutral lysis)	Cells lysed; nuclei electrophoresed. Fragmented DNA moves away from the nucleus to form the comet tail. Quantitated as measure of DNA damage.	Image analysis system required to quantify DNA damage. Influenced by number of cell in S phase.
Chromosome aberrations detected with FISH	Dose > 1 Gy	Chromosome-specific probes visualized by fluorescent ligand. Good for detection of translocations.	Complicated by translocations present before radiation.
Premature chromosome condensation (PCC)	Dose > 1 Gy	Irradiated interphase cell fused to a mitotic cell causing chromosome of the mitotic cell to condense rendering chromosome damage visible.	Chromosome aberrations may be present before irradiation.
γ-H2AX intranuclear foci	Dose > 0.05 Gy	Immunofluorescence microscopy or flow cytometry using antibody to γ-H2AX phosphoprotein.	Requires image analysis system.

mixture containing DNA polymerase, the four nucleotides, and appropriate cofactors. The enzyme (DNA polymerase I) adds nucleotides to the 3′-hydroxyl (3′-OH) end of the oligonucleotide leading to synthesis of a complementary new strand of DNA. By including radioactive nucleotides in the reaction mixture, the complementary copy of the template can be labeled and used as a radioactive probe (Figure 4.9).

4.3.2 Restriction Enzymes

A key early event in the evolution of molecular biology was the discovery of restriction enzymes. Restriction enzymes are endonucleases found in bacteria that cut DNA at exactly the same site within a specific nucleotide sequence and always cut the DNA at that sequence. These enzymes were first discovered in bacterial cells, where they are believed to protect the cell against infection by viruses and to participate in DNA repair and recombination. Figure 4.10 shows some commonly used restriction enzymes together with the sequence of nucleotides that they recognize and the position at which they cut the sequence. Some restriction enzymes create ends that can be joined together because the DNA is cut in different places on the two strands, creating an overhang of single-stranded DNA. This small single-stranded section can hybridize to other fragments with compatible sequences, which is to fragments that have been produced by digestion with the same restriction enzyme. This allows the DNA to be cut and the fragments rejoined in specific sequences.

Type II restriction enzymes are those which have endonuclease activity only. More than a thousand type II enzymes have been isolated of which about 70 are commercially available. The names of restriction enzymes are based on the genus, species, and strain of the organism from which they are derived and the order in which they were discovered.

Most restriction recognition sites have symmetry. For example, *Eco*RI recognizes the sequence 5' GAATTC 3'; the complementary strand is also 5' GAATTC 3'. *Eco*RI cuts the DNA between the G and A on each strand, leaving 5' single-strand sequences of AATT on each strand (line 1 of Figure 4.10), which are complementary. Thus, all DNA fragments generated with *Eco*RI are complementary and will hybridize with each other (Figure 4.11).

4.3.3 Gel Electrophoresis and Blotting Techniques

An important analytical tool in molecular biology is gel electrophoresis, which separates pieces of DNA on the basis of size. As described in Section 4.2,

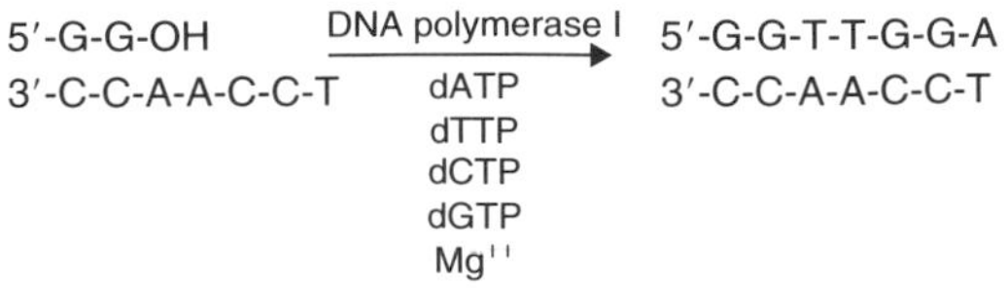

```
5'-G-G-OH        DNA polymerase I   5'-G-G-T-T-G-G-A
3'-C-C-A-A-C-C-T      dATP          3'-C-C-A-A-C-C-T
                      dTTP
                      dCTP
                      dGTP
                      Mg⁺⁺
```

FIGURE 4.10
Synthesis of a complementary strand of DNA. The substrate is a 3'-hydroxyl end of a primer hybridized to a single-stranded template. The primer is a short synthetic single-stranded oligonucleotide. All four nucleotides and magnesium are required for DNA polymerase to function.

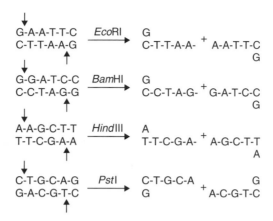

FIGURE 4.11
Restriction enzymes. The nucleotide sequences recognized by four endonucleases are shown. On the left side are the recognized sequence and the site where the enzymes cut (arrows). On the right side are the two fragments produced by digestion with the specific restriction enzyme. Each recognition sequence is a palindrome. In all the examples shown the fragment produced by digestion has a single-stranded tail. This tail allows fragments cut with the same restriction enzyme to anneal with each other.

DNA samples are pipetted into wells at one end of a slab of solid agarose. The gel is immersed in electrolyte, and under the influence of an electric field, DNA molecules migrate toward the positive pole and are sorted by size since, in a given time, small fragments move farther through the gel than do large fragments. Dye is pipetted into the wells at the same time as DNA to allow progress through the gel to be visualized (Figure 4.12). The concentration of the agarose is varied according to the size of the DNA fragment to be separated and visualized, higher concentrations for smaller fragments and lower concentrations for larger fragments. After separation is complete, the gel is soaked in ethidium bromide, which intercalates into DNA and fluoresces under ultraviolet light to visualize the position of the DNA.

4.3.3.1 Southern and Northern Blotting

Southern blotting, named after its discoverer [29], is widely used for analyzing DNA structure (Figure 4.12). DNA to be analyzed is cut into defined lengths using a restriction enzyme and the fragments are separated by electrophoresis in agarose gel. Pieces of DNA of known size are electrophoresed at the same time to provide a molecular weight scale. A piece of nylon membrane is laid on top of the gel and fluid is drawn through the gel into the membrane by means of a vacuum pump causing the DNA to pass from the gel into the membrane where it is immobilized.

One application of Southern blotting is to determine the size of the fragments of DNA which carry a particular gene. For such an analysis, a cloned gene that has been made radioactive by hybridization, as described above, is used as a probe. The nylon membrane to which the electrophoresed DNA

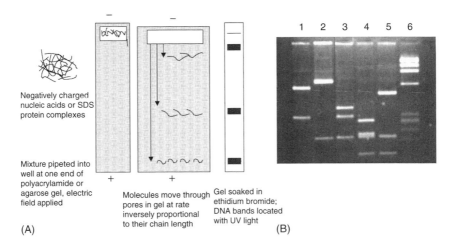

FIGURE 4.12
Gel electrophoresis. DNA, which is negatively charged, migrates toward the anode in an electrical field. During electrophoresis DNA fragments sort by size, smaller molecules moving farther than larger molecules.

fragments have been transferred is incubated with the radioactive probe, which anneals with homologous DNA sequences present on the membrane. Single-stranded unbound fragments are washed off and the position of the bound probe is located by exposing a piece of x-ray film to the membrane for several hours in the cold. When the film is developed, dark bands indicate the position of the radioactive probe on the membrane (Figure 4.13).

A very similar procedure known as Northern blotting can be used to characterize mRNA and to evaluate the expression patterns of genes. Northern blotting differs from Southern blotting in that the DNA probe is hybridized to RNA rather than to DNA. RNA is separated by electrophoresis, transferred to nylon membranes, and probed with labeled cloned fragments of DNA. The RNA is already single stranded and does not need previous digestion with restriction enzymes. Northern blotting is the technique of choice to monitor mRNA abundance and turnover. The intensity of the bands on the autoradiograph reflects the abundance of message and its expression in the cell or tissue.

4.3.4 Polymerase Chain Reaction

The application of blotting techniques is limited by the fact that the amount of DNA or RNA available for hybridization analysis may be so small that the signal generated is too weak for visualization. This problem of sensitivity can be overcome by use of the polymerase chain reaction (PCR), which uses enzymatic amplification to increase the number of copies of a DNA fragment (Figure 4.14). The principle is based on primer extension by DNA polymerases, which was described in Section 4.3.1 for the synthesis of radioactive probes. The reaction depends on a DNA polymerase enzyme

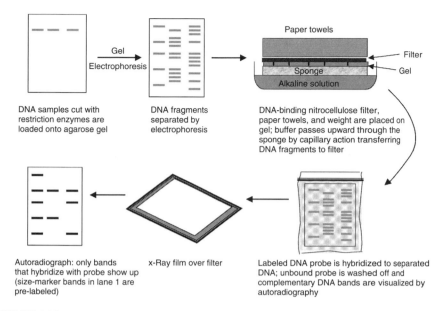

FIGURE 4.13

Detection of specific RNA or DNA molecules by gel-transfer hybridization. A mixture of either single-stranded RNA molecules (Northern blotting) or the double-stranded DNA molecules created by restriction nuclease treatment (Southern blotting) is separated according to length by electrophoresis. A nitrocellulose nylon membrane is laid over the gel and the separated RNA or DNA fragments are transferred to the sheet by blotting. The membrane is peeled off the gel and sealed in a plastic bag with a buffered salt solution containing a radioactive DNA probe and incubated under conditions favoring hybridization. Finally, the sheet is removed from the bag and washed to remove unhybridized probe molecules. Autoradiography is performed and the DNA that has hybridized to the labeled probe shows up as dark bands on the autoradiograph.

which is resistant to denaturation at high temperatures and the enzyme most frequently used is a heat-stable Taq DNA polymerase derived from *Thermus aquaticus,* a bacterium that lives in hot springs. The other requirement is for specific oligonucleotide primers based on a precise knowledge of the sequences flanking the region of interest. These two short oligonucleotides complementary to the flanking regions are synthesized or bought. All the components of the reaction, target DNA, the primers, deoxynucleotides, and Taq polymerase are combined in a small tube and the reaction sequence is accomplished by changing the temperature of the reaction mixture in a cyclical way. During each cycle the sample is heated to about 94°C to denature the DNA strands, then cooled to about 50°C to allow the primers to anneal to the template DNA, and then heated to 72°C, the optimal temperature for Taq polymerase activity. Repetition of this cycle permits more rounds of amplification. Each cycle takes only a few minutes and 20 cycles can theoretically produce a millionfold amplification. PCR products can then be sequenced or subjected to other methods of genetic analysis.

PCR can also be used to study gene expression or to screen for mutations in RNA. In this case, reverse transcriptase is used to make a complementary

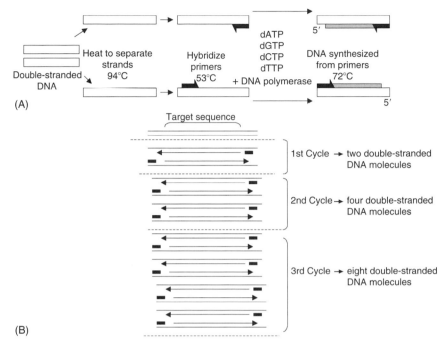

(A)

(B)

FIGURE 4.14

Amplification of DNA using the PCR technique. The known DNA sequence to be amplified is used to design two synthetic DNA oligonucleotides, each complementary to the sequence on one strand of the DNA double helix at opposite ends of the region to be amplified. These oligonucleotides serve as primers for in vitro DNA synthesis, which is performed by a DNA polymerase, and they determine the segment of the DNA that is amplified. PCR starts with a double-stranded DNA and each cycle of the reaction begins with a brief heat treatment to separate the two strands. After strand separation, cooling of the DNA in the presence of a large excess of the two primer oligonucleotides allows these primers to hybridize to complementary sequences in the two DNA strands. This mixture is incubated with DNA polymerase and the four deoxyribonucleoside triphosphates so that DNA is synthesized, starting from the two primers. The entire cycle is then begun again by a heat treatment to separate the newly synthesized DNA strands. As the procedure is performed over and over again the newly synthesized fragments serve as templates in their turn and, within a few cycles, the predominant DNA is identical to the original bracketed sequence.

single-strand DNA copy (cDNA) of an mRNA before performing the PCR. The cDNA is used as a template for a PCR reaction as described above. This technique, which is usually called reverse transcription PCR (RT-PCR), allows amplification of cDNA thereby providing a convenient source of DNA that can be screened for mutations. The RT-PCR technique can also provide approximate quantitation of expression of a particular gene.

4.3.4.1   Quantitative Real-Time Polymerase Chain Reaction

The ability to detect low copy numbers of DNA sequences is greatly increased by using PCR; however, the sequences cannot always be accurately quantified by this means. The next development, real-time quantitative PCR, has eliminated the variability associated with PCR, and has allowed the quantitation of low amounts of template. Similarly, quantitative RT-PCR provides a sensitive method to detect low levels of mRNA (often obtained from small samples or microdissected tissues) and to quantify gene expression.

Different chemistries are available for real-time detection of RNA and DNA. A very specific 5' nuclease assay uses a fluorogenic probe for the detection of reaction products after amplification, while a more economical assay uses a fluorescent dye (SYBR Green I) for the detection of double-stranded DNA products. In each case the fluorescence emission from each sample is collected by a charge-coupled device-camera and the data are automatically processed and analyzed by computer software. Quantitative real-time PCR using fluorogenic probes gives both amplification and analysis of samples, and multiple genes can be analyzed simultaneously within the same reaction without need for post-PCR processing. Simultaneous analysis of approximately 100 samples within 2 h is achievable.

4.3.5 Putting New Genes into Cells: DNA-Mediated Gene Transfer

The function of a gene can often be most readily studied by putting the gene into a cell where it is not normally expressed. For example, a gene tentatively identified as being involved in radioresistance might be inserted into a radiosensitive cell line. Expression of a radioresistant phenotype by the transfected cell line would be interpreted as contributing to proof that the gene did in fact encode a protein involved in radioresistance.

Mammalian cells do not take up foreign DNA naturally, in fact they try to protect themselves from invading DNA. Consequently, various stratagems must be used to bypass natural barriers. These include the following:

(i) Microinjection: This is the most direct but also the most difficult procedure. DNA is injected, one cell at a time, directly into the nucleus through a fine glass needle.

(ii) Calcium phosphate precipitation: This is the classic technique for introducing DNA into cells. Calcium phosphate causes DNA to precipitate in large aggregates, and for unknown reasons, some cells take up large amounts of DNA presented in this form.

(iii) Electroporation: This technique is useful for cells that are resistant to transfection by calcium phosphate precipitation. Cells in suspension are subjected to a brief electrical pulse that causes holes to open transiently in the membrane, allowing foreign DNA to enter.

Another approach is the use of viral vectors that can be targeted to a variety of cells and can infect nondividing cells. The genetic material of a retrovirus is RNA and when retroviruses infect mammalian cells, their RNA genomes are converted to DNA by the viral enzyme reverse transcriptase. The viral DNA is incorporated efficiently into the host genome, replicating along with the host DNA at each cell cycle. If a foreign gene has been incorporated into the retrovirus, it will be permanently maintained in the infected mammalian cell. A drawback of retrovirus is that only relatively small pieces of DNA (up to 10 kb) can be transferred by this means. Adenovirus, on the other hand, transfects very efficiently and can take larger inserts. Nonviral vectors such as liposomes are also used for transient expression of introduced DNA. Lipofectin is a commonly used reagent in which plasmid DNA is complexed with a liposome in suspension and added directly to cells in tissue culture.

Whatever the method of transfection, it is usually necessary to select for the cells which have retained the gene. If a selectable gene, such as one encoding resistance to an antibiotic is transfected along with the gene of interest, the cell will usually take up both genes during the same transfection. Cells which have been successfully transfected can then be selected on the basis of antibiotic resistance.

4.3.6 Generation of a Cloned Probe or DNA Library

When the target gene to be cloned has been identified, the DNA segment of interest is inserted into a vector (i.e., a self-replicating DNA molecule that has the ability to transport a foreign DNA molecule into a host cell). The most commonly used vectors are plasmids or bacterial viruses. Plasmids are circular DNA molecules that are found in bacteria and replicate independent of the host chromosome. A piece of foreign DNA can be inserted into a plasmid, which is then introduced into a bacterium where it replicates as the bacterial population increases. If the plasmid is also equipped with a gene conferring resistance to an antibiotic, transfected bacteria can be selected since only those that have taken up a plasmid will survive and replicate in culture medium containing the antibiotic. The limitation of plasmids is that they are useful only for relatively small DNA inserts and transfection into bacteria is relatively inefficient. Bacterial viruses or bacteriophage can also be engineered for drug resistance; they can infect the host at greater efficiency than a plasmid and can accommodate a wider range of different sized DNA fragments.

The vector DNA is prepared with the same restriction enzyme that was used to prepare the cloned gene, so that all the fragments have compatible single-strand sequences that can be spliced back together. The spliced fragments can be sealed with the enzyme DNA ligase, and the reconstituted molecule can be introduced into bacterial cells (Figure 4.15). Because, by taking up the vector, the bacteria have become drug resistant; they can be isolated and propagated in large numbers. In this way, large quantities of a gene can be obtained (i.e., cloned) and labeled with either radioactivity or biotin for use as DNA probes for analysis of Southern or Northern blots or used directly for nucleotide sequencing or transfer into other cells.

If the starting DNA is a mixture of all the different restriction fragments derived from the cell, representing the entire genome, a large number of different DNA fragments can be inserted into vector population and then introduced into bacteria. This constitutes a DNA library, which can be plated

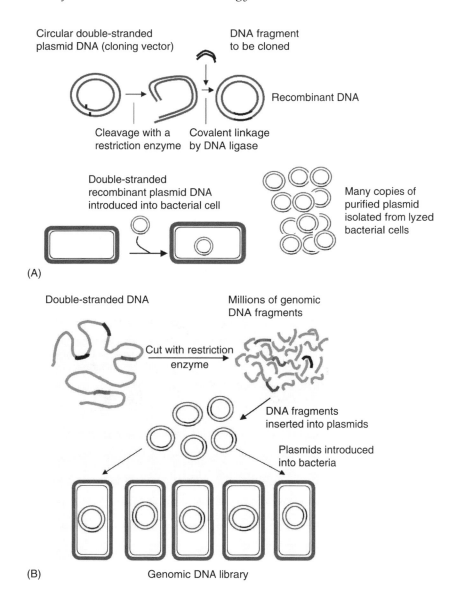

Circular double-stranded
plasmid DNA (cloning vector)

DNA fragment
to be cloned

Recombinant DNA

Cleavage with a   Covalent linkage
restriction enzyme by DNA ligase

Double-stranded
recombinant plasmid DNA
introduced into bacterial cell

Many copies of
purified plasmid
isolated from lyzed
bacterial cells

(A)

Double-stranded DNA

Millions of genomic
DNA fragments

Cut with restriction
enzyme

DNA fragments
inserted into plasmids

Plasmids introduced
into bacteria

(B)   Genomic DNA library

FIGURE 4.15

(A) Purification and amplification of a specific DNA sequence by DNA cloning in a bacterium.
The plasmid is cut open with a restriction nuclease (in this case one that produces cohesive ends)
and is mixed with the DNA fragment to be cloned (which has been prepared with the same
restriction nuclease), DNA ligase and ATP. The cohesive ends base pair and DNA ligase seals the
nicks in the DNA backbone producing a complete recombinant DNA molecule. (B) Construction
of a genomic DNA library. The genomic library is stored as a set of bacteria each carrying a
different fragment of human DNA. Only a few such fragments are shown in the diagram.

out and screened by hybridization with specific probes. An individual
recombinant DNA clone can be isolated from this library and used for other
applications.

The final step in the cloning of a gene is to screen the library to locate the gene of interest. This can be done in a number of ways, one of which is functional complementation. This technique depends on the DNA segment producing its corresponding protein within the cell, thereby giving the host a specific and detectable phenotype, for instance, radioresistance. Another approach is to, if possible, obtain a partial or complete amino acid sequence of the protein

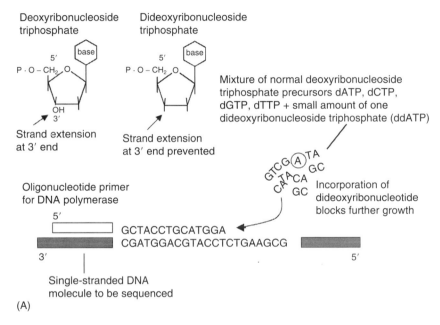

(A)

FIGURE 4.16

(A) and (B) The enzymatic or dideoxy method of sequencing DNA. The method relies on the use of dideoxyribonucleoside triphosphates, derivatives of the normal deoxyribonucleoside triphosphates that lack the 3′ hydroxyl group. Purified DNA is synthesized in a mixture that contains single-stranded molecules of the DNA to be sequenced, DNA polymerase, a short primer DNA, to enable the polymerase to start DNA synthesis and the four deoxyribonucleoside triphosphates (dATP, dCTP, dGTR, and dTTP). If a dideoxyribonucleotide analog of one of these nucleotides is also present in the nucleotide mixture it can become incorporated into a growing DNA chain, but because this chain now lacks a 3′ OH group, the addition of the next nucleotide is blocked and the DNA chain terminates at that point. Note: To determine the complete sequence of a DNA fragment, the double-stranded DNA is first separated into its single strands and one of the strands is used as the template for sequencing. Four different chain-terminating dideoxyribonucleosid triphosphates (ddATR, ddCTP, ddGTP, and ddTTR) are used in four separate DNA synthesis reactions on copies of the same single-stranded DNA template. Each reaction produces a set of DNA copies that terminate at different points in the sequence. The products of these four reactions are separated by electrophoresis in four parallel lanes of a polyacrylamide gel. The newly synthesized fragments are detected by a label (either radioactive or fluorescent). In each lane, the bands represent fragments that have terminated at a given nucleotide but at different positions in the DNA. By reading off the bands in order, starting at the bottom of the gel and working across all lanes, the DNA sequence of the newly synthesized strand can be determined.

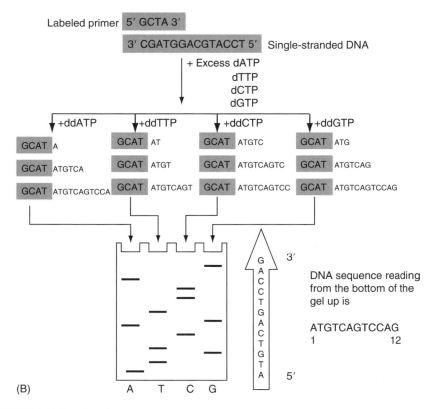

FIGURE 4.16 (continued)

encoded by the gene under study, then the coding sequence derived from these amino acids can be used to synthesize a short DNA sequence or oligonucleotide that can be used as a probe for the gene of interest. A third stratagem for screening a cDNA expression library is the use of antibody probes.

4.3.7   Sequencing of DNA

The usual method for characterizing genes and the proteins that they encode is to determine the sequence of the DNA by the dideoxy-chain termination method. The DNA to be sequenced may be amplified using PCR (Section 4.3.5). The DNA sequencing procedure is analogous to setting up an in vitro DNA replication reaction except that normal nucleotides are replaced with dideoxynucleotide triphosphates (ddNTPs). In fact, four separate reactions each containing one of the four ddNTPs (i.e., ddATP, ddCTP, ddGTP, or ddTTP) together with the other three normal nucleotides are set up. In each reaction the sequencing primers bind and start the extension of the chain at the same place. The extension of the primers, however, terminates at different points depending on where dideoxynucleotides are incorporated (Figure 4.16), producing fragments of different sizes with termination points corresponding to each nucleotide. The fragments are analyzed by polyacrylamide gel electrophoresis (PAGE), which allows fragments differing by a single base to be separated. Using this method a sequence of 200–500 bases can be read from a single gel.

4.3.8 Single Nucleotide Polymorphisms

DNA sequences can differ at single nucleotide positions within the genome. These single nucleotide polymorphisms (SNPs) can occur as frequently as 1 out of every 1000 base pairs and when SNPs occur in exons, they can affect protein structure and function. Polymorphisms are the result of point mutations, deletions, insertions, or variation in numbers of copies of a DNA fragment (tandem repeats).

Southern blotting can be used to detect DNA polymorphisms, using a probe that hybridizes to the polymorphic region of the DNA molecule. Another technique is the detection of restriction fragment length polymorphisms. These are seen when a particular restriction enzyme that is used to cut DNA does so at a polymorphic locus to yield restriction fragments of different sizes. Even a point mutation can be detected by this means if the resultant change in sequence removes or adds a new recognition site at which a restriction endonuclease cuts.

Larger scale polymorphisms such as deletions, insertions, or tandem repeats involving more than about 30 nucleotides can be detected as recognizable shifts in the Southern blot hybridization pattern.

A single base-pair difference between two short single-stranded DNA molecules may result in a difference in conformation between the two strands. This single-stranded conformation polymorphism can be detected by a difference in the molecule's electrophoretic mobilities on a neutral polyacrylamide gel. This change would not be undetected if electrophoresis were carried out under denaturing conditions where the strands are separated only according to size, and not base composition.

4.3.9 Functional Inactivation of Genes

Cloning of genes does not necessarily imply knowledge of their function. Clues as to the function of a gene can be provided by the occurrence of regions in the amino acid sequence which are similar to sequences in other proteins of known function. One way of testing the putative function of such a sequence is to see whether a mutation within the critical site causes loss of function.

4.3.9.1 *Site-Directed Mutagenesis*

Site-directed mutagenesis permits the introduction of mutations at a precise point in a cloned gene, resulting in a specific change in the amino acid sequence and hence in the secondary structure of the encoded protein.

By site-directed mutagenesis, amino acids can be deleted, altered, or inserted but usually not in such a way as to alter the reading frame or disrupt protein continuity.

There are two main ways of introducing a mutation into a cloned gene. If a restriction enzyme site occurs within the region to be altered, a few nucleotides may be inserted or deleted at this site when the gene is digested with the restriction endonuclease, by ligating a small oligonucleotide complementary to the sticky end of the restriction site. The second method requires more manipulation but does not depend on finding a restriction enzyme site in a critical location. The gene is first obtained in a single-stranded form by cloning into a vector such as a bacteriophage. A short oligonucleotide is synthesized containing the desired nucleotide change but otherwise complementary to the region to be mutated. The oligonucleotide will anneal to the single-stranded DNA but contains a mismatch at the site of mutation. The hybridized oligonucleotide–DNA duplex is then treated with DNA polymerase I in a reaction mixture containing four nucleotides and buffers, which will synthesize and extend a complementary strand with perfect homology at every nucleotide except at the site of mismatch in the primer used to initiate DNA synthesis. The double-stranded DNA is then introduced into bacteria where, because of the semiconservative nature of DNA replication, 50% of the vector (bacteriophage) produced will contain mutation (Figure 4.17). The bacteriophage containing artificially generated mutations are identified and used for gene transfer to cells in culture or to transgenic mice.

4.3.9.2 Antisense Oligonucleotides

Another approach to functional inactivation of a gene is to introduce a DNA or RNA sequence that will specifically inactivate the expression of a target gene. This can be achieved by introducing DNA or RNA molecules with a sequence that is homologous to that contained within a target gene but where the order of the bases is opposite to that of the complementary strand (i.e., 3′:5′ instead of 5′:3′). DNA molecules can combine in vitro specifically with their homologous sequences in mRNA and interfere with the expression of the gene.

4.3.9.3 siRNA

An evolution from antisense technology is the understanding that complementary RNA can directly interfere with gene expression (RNA interference: RNAi), leading to the specific disappearance of the selected gene products [30,31]. RNAi interferes with the stability of the mRNA transcript by initiating a degradation process of transcripts arising from the targeted gene. Specific gene inactivation in this way has the potential for therapy of tumors, for example, by inhibiting the expression of an oncogene. A limitation of all the above technologies is that once the nucleic acids enter a cell,

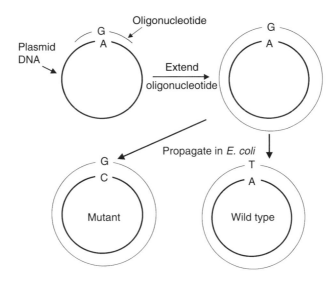

FIGURE 4.17
The use of a synthetic oligonucleotide to modify the protein-coding region of a gene by site-directed mutagenesis. A recombinant plasmid containing a gene insert is separated into its two DNA strands. A synthetic oligonucleotide primer corresponding to part of the gene sequence, but containing a single altered nucleotide at a predetermined point, is added to the single-stranded DNA under conditions that permit less than perfect DNA hybridization. The primer hybridizes to the DNA, forming a single mismatched nucleotide pair. The recombinant plasmid is made double-stranded by in vitro DNA synthesis starting from the primer and sealed by DNA ligase. The double-stranded DNA is introduced into a cell, where it is replicated. Replication using one strand of the template produces a normal DNA molecule, but replication using the other (the strand that contains the primer) produces a DNA molecule carrying the desired mutation. Only half of the progeny cells will end up with a plasmid that contains the desired mutant gene, but a progeny cell that contains the mutated gene can be identified, separated from other cells, and cultured to produce a pure population of cells, all of which carry the mutated gene.

they are vulnerable to a variety of cellular nucleases. To be biologically effective, a high concentration of molecules must be efficiently delivered to all cellular targets and must persist inside the cells for a prolonged period.

4.3.9.4   Transgenic and Knockout Mice

An approach to investigating the effect of gene expression on the function of the whole organism is to transfer genes directly into the germline to generate transgenic mice. The preferred method of gene transfer is microinjection, usually into the male pronucleus of a single-cell embryo where the gene integrates into a host chromosome to become part of the genome of the developing organism. Incorporation of the gene into the germline in this way creates a mouse that can become the progenitor of a line of transgenic mice all of which carry the newly introduced gene. Each transgene will have a unique location in the host chromosome and be transmitted in the same way as a naturally occurring gene. The activity of adjacent genes can influence the expression of a transgene so the site of integration is of importance. The reverse may also apply in that the transferred gene can alter the expression of endogenous genes. This observation has been exploited to develop a gene-targeting stratagem by which specific endogenous genes could be inactivated or knocked out, creating a so-called knockout mouse. Mice with inserted transgenes or with knocked-out endogenous genes are valuable tools for the study of effects of gene in the whole organism.

A mutation can be targeted to a specific endogenous gene by in vivo site-directed mutagenesis. In this case, a cloned gene fragment is inserted, not at random, but to a particular site in the genome by homologous recombination (described in Chapter 7). The cloned mammalian gene or DNA fragment is preferentially incorporated into the genome of the normal cell at the site on the chromosome where it occurs naturally by homologous recombination replacing the endogenous gene. The mutation that has been introduced can result in the knockout of gene expression and facilitate the study of gene function.

4.3.10 Genomic Methods of Tumor Analysis

Global expression analysis methods have been developed using microarrays or molecular cytogenetic methods that allow for the simultaneous analysis of thousands of genes at the DNA, RNA, or protein levels.

4.3.10.1 *Fluorescence In Situ Hybridization*

The use of FISH and multi-fluor fluorescence in situ hybridization (M-FISH) to characterize radiation-induced chromosome aberrations is described earlier in this chapter (Section 4.2.2.3) and in Chapter 10.

Chromosome aberrations can be detected by this method using specific centromere probes that give two signals from normal nuclei, one signal when there is only one copy of the chromosome (monosomy), or three signals when there is an extra copy (trisomy), whereas chromosome deletions can also be detected by using probes from the deleted region and counting the signals. If the probes used for FISH are close to specific translocation break points on different chromosomes, they will appear joined as a result of the translocation generating a color fusion signal.

4.3.10.2 *Spectral Karyotyping and Multi-Fluor Fluorescence In Situ Hybridization*

Universal chromosome painting techniques have been developed by which it is possible to analyze all chromosomes simultaneously. Two essentially similar approaches have been developed: spectral karyotyping (SKY) [32] and M-FISH [33]. Both techniques are based on the differential display of colored fluorescent chromosome-specific paints, which provide a complete analysis of the chromosomal complement in a given cell. If a total of five spectrally distinct hybridization fluors are used and two or three of these fluors are directed against a given chromosome, each of 24 different chromosome types in the human karotype is represented by a unique combinatorial signature. Computer software then recognizes these signatures and arbitrarily assigns convenient pseudo colors on a pixel by pixel basis. Combinatorial painting can produce a 24 color karyotype that is capable of revealing multiple radiation-induced interchromosome rearrangements simultaneously. Abnormal chromosomes can be identified by the pattern of color distribution along them and rearrangements between different chromosomes will lead to a distinct transition from one color to another at the position of the breakpoint. This technology is particularly suited to solid tumors where the complexity of the karyotypes may often mask the presence of subtle chromosomal aberrations.

4.3.10.3 *Microarray Analysis of Genes*

Microarray analysis involves the production of DNA arrays or chips on solid supports for large-scale hybridization experiments. There are two types of chip: in one, DNA probe targets are immobilized to a solid inert surface such as glass and exposed to a set of fluorescently labeled sample DNAs; in the second, an array of different oligonucleotide probes is synthesized in situ on the chip. The array is exposed to labeled DNA samples, and hybridized complementary sequences are determined by digital imaging. Microarray technology allows for the simultaneous analysis of the differential expression of thousands of genes at once for large-scale gene discovery expression, mapping, and sequencing studies and for detection of mutations or polymorphisms. This technology is having a major impact on understanding the dynamics of gene expression in cancer cells.

Microarrays of cDNAs can be used to study differential gene expression. In this case, mRNA from the sample of interest can serve as a template for producing complementary DNA (cDNA) in the presence of a reverse transcriptase enzyme. This cDNA is then fluorescently labeled and hybridized to the target gene sequences spotted on the cDNA microarray slide. The fluorescent intensity of each hybridized sequence in the array is read using a scanner linked to custom digital image analysis software, which produces a color-coded image of the array, and a quantitative value is recorded for each target gene (Figure 4.18). Intensity of fluorescence correlates with expression of the gene for which the spot codes. Gene expression results can be clustered using different methods of data analysis such as hierarchical clustering, a method which is widely applied to databases of gene expression profiles. This procedure merges individual datapoints into a tree structure called a dendrogram, based on their similarity.

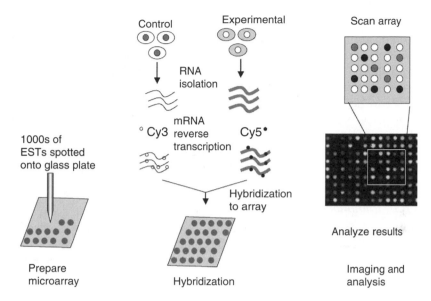

Control **Experimental** **Scan array**

RNA
isolation

mRNA
° Cy3 reverse Cy5•
transcription

1000s of
ESTs spotted
onto glass plate

Hybridization
to array

Analyze results

Prepare
microarray Hybridization Imaging and
analysis

FIGURE 4.18
Using DNA microarrays to monitor the expression of thousands of genes simultaneously. To prepare the microarray, DNA fragments, each corresponding to a gene, are spotted onto a slide. In the example mRNA is collected from two different cell samples for a direct comparison of their relative levels of gene expression. These samples are converted to cDNA and labeled one with a red fluorochrome, the other with a green fluorochrome. The labeled samples are mixed and then allowed to hybridize to the microarray. After incubation, the array is washed and the fluorescence scanned. Red spots indicate that the gene in sample 1 is expressed at a higher level than the corresponding gene in sample 2; green spots indicate that expression of the gene is higher in sample 2 than in sample l. Yellow spots show genes that are expressed at equal levels in both cell samples. Dark spots indicate little or no expression in either sample of the gene.

4.3.11 Analysis of Proteins

4.3.11.1 Western Blotting

Western blotting or immunoblotting is a gel electrophoretic procedure for separation of proteins. The name Western blot is derived by analogy with the Southern blot for DNA and the Northern blot for RNA. A complex mixture of proteins is separated on the basis of charge and size using PAGE. The separated proteins are transferred to a nitrocellulose or nylon membrane in an electric field maintaining their specific orientation on the filter. The membrane is incubated with a primary antibody against the protein of interest and then with a secondary antibody that is specific to the primary antibody and conjugated to horse radish peroxidase (HRP) or biotin. The protein–antibody conjugate can be detected by exposure to chemiluminescent detection agents, and detection of emitted fluorescent is by short exposure of a photographic film allowing the bands of interest to be identified. The secondary antibody can also be radio-labeled.

4.3.11.2   Two-Dimensional Electrophoresis

Two-dimensional polyacrylamide (2-D PAGE) is a technique for studying posttranslational modification of proteins. Every protein has a distinct electrochemical charge that is a result of its unique amino acid composition. Using an electrophoresis technique known as isoelectric focusing (IEF), proteins can be separated by charge in tube gels or paper strips. The IEF tubes or strips are then laid across the top of an SDS-PAGE gel and separated by size. The gel is stained to visualize proteins that appear as slightly elongated spots. The identity (and modification) of the proteins in the spots can be determined by Western blotting or mass spectrometry (Figure 4.19).

4.3.11.3   Immunoprecipitation

Immunoprecipitation is used to detect specific protein–protein interactions. An antibody against a specific protein is incubated with a sample containing the target protein and forms an immune complex. Small polystyrene or agarose beads containing staphylococcal protein A are added to the mixture. Protein A binds to antibodies and centrifugation of the mixture sediments protein A, the antibody and the specific protein to which it is bound. The unbound proteins in the supernatant are removed and the bound immune complex is dissociated and analyzed by Western blotting.

4.3.11.4   Enzyme-Linked Immune-Absorbant Assay

For the enzyme-linked immune-absorbant assay (ELISA) two independent proteins that bind to the protein of interest with high specificity and affinity

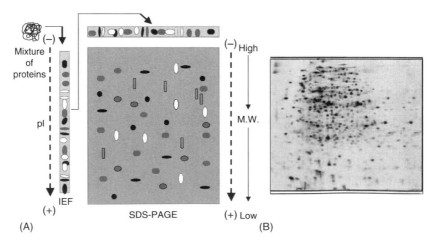

FIGURE 4.19
Two-dimensional electrophoresis (2DE). Proteins are first separated according to their isoelectric points (1st dimension) and subsequently according to their molecular weights (second dimension). Separation in two different dimensions (pI and Mw) makes 2DE capable of separating hundreds to thousands of proteins at high resolution.

are required. One of these antibodies is coupled to a plate that is incubated with the protein mix to be analyzed. The specifically bound protein is retained on the plate and is detected with a second antibody generated against the target protein that is coupled to an enzyme or protein allowing quantitation of the bound protein. ELISA assays are very sensitive and can detect picomole amounts of proteins.

4.3.11.5 Far Western Blotting

Far Western blotting uses a specific purified recombinant protein to probe a membrane of electrophoresed cellular protein. The recombinant protein binds to specific cellular proteins in the membrane and is detected by autoradiography or a specific antibody. Different protein chemistry techniques can be used to identify the interacting proteins on the membrane.

4.3.11.6 Fluorescent Proteins

A small number of naturally occurring fluorescent proteins have been discovered, which can act as reporters. The two most commonly used fluorescent proteins are green fluorescent protein (GFP) from a jellyfish and red fluorescent protein (RFP) from a reef coral. These proteins have widely different spectral properties and are ideal reagents for two-color labeling of cells. The usual approach is to fuse GFP to one protein of interest and RFP to another to monitor intracellular protein activity either in cells in culture or in mice.

GFP has also been engineered to fluoresce in the cyan (CFP) and yellow (YFP) spectra. The emission spectrum of CFP overlaps with the excitation spectrum of YFP, and by fusing CFP to a protein of interest and YFP to another protein it is possible to observe interactions by exciting CFP and seeing YFP fluoresce. This method, called fluorescence resonance transfer (FRET) is used to observe dynamic interactions of proteins in the cell.

4.3.12 Production of Monoclonal Antibodies

Techniques such as Western blotting and its histological equivalent, immunohistochemistry, depend on the specificity of the antibody probe used to localize and quantitate the protein. This specificity is assured by the use of monoclonal antibodies, immunoglobulin molecules of single epitope specificity in contrast to polyclonal antibodies that are a mixture of immunoglobulin antibodies, each recognizing a different epitope secreted against a specific antigen. Kohler and colleagues showed in 1976 [34] that it was possible to stimulate hybridization between malignant tumor cells maintained in continuous culture and immune lymphoid cells. Hybrid cells that grew in culture and produced antibodies with the single defined specificity of the immune lymphoid cell could then be selected by cloning (Figure 4.20).

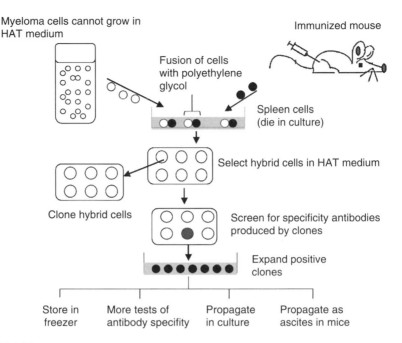

FIGURE 4.20

Major steps in the production of monoclonal antibodies by hybridoma technology. A normal B cell (a type of lymphocyte) obtained from the immunized mouse is hybridized with a myeloma cell. This union results in the formation of cloned hybridomas (hybrid cells) that have the cancer cell's trait of dividing endlessly and the B cell's ability to produce a specific type of antibody. Because neither cell type used for the initial fusion can grow on its own, only the hybrid cells survive.

Spleen cells from an animal that has been immunized with foreign antigens are placed in culture with a continuously growing myeloma cell line in the presence of polyethylene glycol to stimulate cell fusion. The myeloma cell line used is a mutant that does not secrete immunoglobulin and has been selected for an enzyme deficiency, which prevents its growth in medium containing hypoxanthine, aminopterin, and thymine (HAT medium). Normal spleen cells do not grow in culture, thus only hybrid cells formed by fusion will grow, the missing enzyme being produced by the fused spleen cells. After selection in HAT medium, hybrid cells are cloned by placing in individual cells into single wells of a tissue-culture plate. Antibodies secreted by each clone of hybrid cells (known as a hybridoma) can be tested for specificity.

4.3.13   Proteomics: Analysis of Protein Structure and Function

Proteomics includes a range of techniques for analysis of protein composition and function. Recent advances in computing power and analytical instrumentation such as mass spectrometry, nuclear magnetic resonance

(NMR) spectroscopy and x-ray diffraction, as well as increased knowledge of primary structure from the completion of the sequencing of the human genome, have stimulated this developing field. Mass spectrometry is used to analyze the mass and charge of proteins. NMR and x-ray diffraction, which are used to determine the three-dimensional structures (the relative positions of all atoms) of proteins and other biomolecules, are the primary methods of structural biology.

4.3.14 Analysis of Tissue Sections and Single Cells

Analysis of tissue samples has been improved by development of tissue arrays, which contain multiple small circular pieces derived from punches of representative areas of paraffin sections from different tumors or normal tissues. The arrays can contain pieces from hundreds of different tumors represented as 0.5–2 mm diameter discs of tissue containing several thousand cells, which retain the morphologic features of the original specimen from which the tissue punch was obtained. Tissue arrays are processed in the same way as regular paraffin sections containing only one tissue sample. Tissue in situ hybridization techniques rely on the hybridization of a specifically labeled nucleic acid probe to the cellular RNA in individual cells or tissue sections.

4.3.15 Laser Capture Microdissection

A problem associated with the molecular genetic analysis of small numbers of cancer cells is that substantial numbers of normal cells are often present in the tumor. This contamination may be scattered throughout a tumor section, and cannot be removed by dissection. This problem has recently been circumvented by the use of laser capture microdissection (LCM), in which tumor sections are coated with a clear ethylene vinyl acetate (EVA) polymer before microscopic examination. Tumor cells can be captured for subsequent analysis by briefly pulsing the area of interest with an infrared laser. The EVA film becomes sticky and will selectively attach to the tumor cells directly in the laser path. When sufficient cells have been fused to the EVA film, it is placed into nucleic acid extraction buffers and used for PCR or other molecular analyses.

References

1. Puck TT and Markus PI. Action of x-rays on mammalian cells. *J. Exp. Med.* 103: 653–666, 1956.
2. Elkind MM and Sutton H. Radiation response of mammalian cells grown in culture. I Repair of x-ray damage in surviving Chinese hamster cells. *Radiat. Res.* 13: 556–593, 1960.

3. Sinclair W and Morton R. X-ray sensitivity during the cell generation cycle of cultured Chinese hamster cells. *Radiat. Res.* 29: 450–474, 1966.

4. Spadinger I and Palcic B. Cell survival measurements at low doses using an automated image cytometry device. *Int. J. Radiat. Biol.* 63: 183–189, 1993.

5. Baker F, Spitzer G, Ajani JA, and Brock WA. Drug and radiation sensitivity testing of primary human tumor cells using the adhesive-tumor-cell culture system (ATCCS). *Prog. Clin. Biol. Res.* 276: 105–117, 1988.

6. Price P and McMillan TJ. The use of non-clonogenic assays in measuring the response of cells in vitro to ionising radiation. *Eur. J. Cancer* 30A: 838–841, 1994.

7. Merrill G. Cell synchronization. *Methods Cell Biol.* 57: 229–249, 1998.

8. Withers HR. The dose-survival relationship for irradiation of epithelial cells of mouse skin. *Br. J. Radiol.* 40: 187–194, 1967.

9. Withers HR. Regeneration of intestinal mucosa after irradiation. *Cancer* 28: 75–81, 1971.

10. Withers HR, Hunter N, Barkley HT, and Reid BO. Radiation survival and regeneration characteristics of spermatogenic stem cells of mouse testis. *Radiat. Res.* 57: 88–103, 1974.

11. Withers HR, Mason KA, and Thames HD. Late radiation response of kidney assayed by tubule cell survival. *Br. J. Radiol.* 59: 587–595, 1986.

12. Thames HD and Withers HR. Test of equal effect per fraction and estimation of initial clonogen number in microcolony assays of survival after fractionated irradiation. *Br. J. Radiol.* 53: 1071–1077, 1980.

13. McCulloch EA and Till JE. The sensitivity of cell from normal mouse bone marrow to gamma irradiation in vitro and in vivo. *Radiat. Res.* 16: 822–832, 1962.

14. Olive PL. Molecular approaches for detecting DNA damage. In: Hoekstra MF, (Ed.) *DNA Repair in Higher Eukaryotes*, Vol. 2. Humana Press Inc., Totowa, New Jersey, 1997, pp. 539–557.

15. Nunez MI, McMillan TJ, Valenzuela MT, Ruiz de Almodovar JM, and Pedraza V. Relationship between DNA damage, rejoining and cell killing by radiation in mammalian cells. *Radiother. Oncol.* 39: 155–165, 1996.

16. McGrath RA and Williams RW. Reconstruction in vivo of irradiated *Escherichia coli* deoxynucleic acid: The rejoining of broken pieces. *Nature* 212: 534–535, 1966.

17. Blocher D and Iliakis G. Size distribution of DNA molecules recovered from non-denaturing filter elution. *Int. J. Radiat. Biol.* 59: 919–926, 1991.

18. Yang CS, Wang C, Minden MD, and McCulloch EA. Fluorescence labeling of nicks in DNA from leukemic blast cells as a measure of damage following cytosine arabinoside. *Leukemia* 8: 2052–2059, 1994.

19. Iliakis G, Cicilioni O, and Metzger L. Measurement of DNA double strand breaks in CHO cells at various stages of the cell cycle using pulsed field gel electrophoresis: Calibration by means of 125I decay. *Int. J. Radiat. Biol.* 59: 343–359, 1991.

20. Olive PL and Banath JP. Detection of DNA double strand breaks through the cell cycle after exposure to X-rays, bleomycin, etoposide, and [125]IUrd. *Int. J. Radiat. Biol.* 64: 349–358, 1993.

21. Ostling O and Johanson KJ. Microelectrophoretic study of radiation-induced DNA damages in individual mammalian cells. *Biochem. Biophys. Res. Commun.* 123: 291–298, 1984.

22. Fairbairn DW, Olive PL, Johnson P, and O'Neill KL. The comet assay: A comprehensive review. *Mutat. Res.* 339: 37–59, 1993.

23. Olive PL, Durand RE, Le Riche J, Olivotto I, and Jackson S. Gel electrophoresis of individual cells to quantify hypoxic fraction in human breast cancers. *Cancer Res.* 53: 733–736, 1993.

24. Muller WU and Streffer C. Change in frequency of radiation induced micronuclei during interphase of four-cell mouse embryos in vitro. *Radiat. Environ. Biophys.* 25: 195–199, 1986.

25. Cornforth MN and Bedford JS. X-ray-induced breakage and rejoining of human interphase chromosomes. *Science* 222: 1141–1143, 1983.

26. Bedford JS. Sublethal damage, potentially lethal damage, and chromosomal aberrations in mammalian cells exposed to ionizing radiations. *Int. J. Radiat. Oncol. Biol. Phys.* 21: 1457–1469, 1991.

27. Thacker J, Fleck EW, Morris T, Rossiter BJF, and Morgan TL. Localization of deletion points in radiation-induced mutants of the *hprt* gene in hamster cells. *Mutat. Res.* 232: 163–170, 1990.

28. Powell SN and McMillan TJ. The repair fidelity of restriction enzyme-induced double strand breaks in plasmid DNA correlates with radioresistance in human tumor cell lines. *Int. J. Radiat. Oncol. Biol. Phys.* 29: 1035–1040, 1994.

29. Southern EM. Detection of specific sequences among DNA fragments separated by gel electrophoresis. *J. Mol. Biol.* 98: 503–517, 1975.

30. Kittler R and Buchholz F. RNA interference: Gene silencing in the fast lane. *Semin. Cancer Biol.* 13: 259–265, 2003.

31. Yin JQ, Gao J, Shao R, Tian WN, Wang J, and Wan Y. siRNA agents inhibit oncogene expression and attenuate human tumor cell growth. *J. Exp. Ther. Oncol.* 3: 194–204, 2003.

32. Veldman T, Vignon C, Schrock E, Rowley JD, and Ried T. Hidden chromosome abnormalities in haematological malignancies detected by multicolour spectral karyotyping. *Nat. Genet.* 15: 406–410, 1997.

33. Speicher MR, Gwyn BS, and Ward DC. Karyotyping human chromosomes by combinatorial multi-fluor FISH. *Nat. Genet.* 12: 368–375, 1996.

34. Kohler G, Howe S, and Milstein C. Fusion between immunoglobulin-secreting and non-secreting myeloma lines. *Eur. J. Immunol.* 6: 292–295, 1976.

35. Hewitt HB and Wilson CW. A survival curve for mammalian leukemia cells irradiated in vivo. *Br. J. Cancer* 13: 69–75, 1959.

36. Gould MN and Clifton KH. The survival of mammary cells following irradiation in vivo: A directly generated single dose survival curve. *Radiat Res.* 72: 343–352, 1979.

37. Mulcahy TT, Gould MN, and Clifton KH. Survival of thyroid cells: In vivo radiation and in situ repair. *Radiat. Res.* 84: 523–528, 1980.

5

Ionizing Radiation Effects to the Cytoplasm

5.1 Oxidative Stress

In considering the interactions of ionizing radiation with mammalian cells, the first thought is usually directed towards damage to nuclear DNA and the signaling which is initiated by this damage. This will be described in ensuing chapters. Less attention is paid to cytoplasmic radiation interactions; however, a considerable body of published information indicates that irradiation of the cytoplasm perturbs intracellular metabolic oxidation/reduction (redox) reactions and that systems affected by this initial insult may remain perturbed for minutes, hours, or days. It would seem logical that these cellular redox reactions might contribute to the activation of protective or damaging processes that could impact upon the damaging effects of radiation. These processes include redox-sensitive signaling pathways, transcription factors, gene expression, and the metabolic activities that govern the formation of intracellular oxidants and reductants. While a great deal is known about the molecular changes associated with the initial production of free radicals at the time of irradiation, the contribution of perturbations in redox-sensitive metabolic processes to biological outcomes following radiation exposure are only becoming appreciated recently.

Ionizing radiation results in the formation of free radicals in living systems that are believed to persist for milliseconds and to result in oxidative damage to biomolecules such as DNA, proteins, and lipids that contribute to the biological effects of radiation. In the presence of oxygen, radiation leads to the formation of reactive oxygen species (ROS) such as the superoxide anion ($O_2^{\bullet -}$), hydrogen peroxide (H_2O_2), hydroxyl radical ($^{\bullet}OH$), and singlet oxygen (described in detail in Chapter 2). The generation of intracellular ROS (and reactive nitrogen species [RNS]) by radiation takes place against the backdrop of normal cellular oxidative metabolism since ROS are constantly generated in aerobic cells as a result of electron-transfer processes. Although low amounts of ROS are easily tolerated by the cell, abnormally high levels of ROS, that might result from disruption in the balance between oxidant production and antioxidant defense, produce a state of oxidative stress, which is characteristic of some pathological conditions. ROS are produced concomitant to oxidative metabolism and after exposure to ionizing radiation, chemotherapeutic agents, hyperthermia, inhibition of antioxidant enzymes, and depletion of cellular reductants such as NADPH and glutathione (GSH).

5.1.1 Metabolic Oxidative Stress

Metabolic oxidative stress has been reviewed by Nordberg and Arnér [1]. It has been known for many years that metabolism in mammalian cells primarily derives energy from the tightly controlled biochemical oxidation of substrates (i.e., carbohydrates, fats, and amino acids) from which it obtains the reducing equivalents (electrons) necessary for mitochondrial electron transport chain-mediated oxidative phosphorylation which produces ATP and for which O_2 acts as the terminal electron acceptor. In addition, reducing equivalents in the form of NADH or NADPH are essential for many cellular biosynthetic processes involved in replication, cell division, and macromolecular synthesis. Therefore, the ability to extract, store, and move electrons through complex biological structures via a series of biochemical oxidation/reduction (redox) reactions involving protein catalysts has been hypothesized to provide the essential "life force" for maintaining metabolic homeostasis in mammalian cells. This metabolic strategy, while extremely efficient, leads to the formation of ROS as by-products.

The stepwise reduction of molecular oxygen via one-electron transfers, producing and also connecting the ROS molecules can be summarized as follows:

$$O_2 \xrightarrow{e^\bullet} O_2^{\bullet-} \xrightarrow[2H^+]{e^\bullet} H_2O_2 \xrightarrow{e^\bullet} {}^\bullet OH + OH^- \xrightarrow[2H^+]{e^\bullet} 2H_2O$$

ROS include a number of chemically reactive molecules derived from oxygen. Some of those molecules are extremely reactive, such as the ${}^\bullet OH$, while some are less reactive ($O_2^{\bullet-}$ and H_2O_2). Intracellular free radicals (i.e., free, low molecular weight molecules with an unpaired electron) are classified as ROS. Free radicals and ROS can readily react with most biomolecules, starting a chain reaction of free-radical formation. In order to stop this chain reaction, a newly formed radical must either react with another free radical, eliminating the unpaired electrons, or react with a free-radical scavenger—a chain-breaking or primary antioxidant.

In Table 5.1, the most common intracellular forms of ROS are listed along with their main cellular sources of production and the relevant enzymatic antioxidant systems scavenging these ROS molecules. ROS formation and metabolism can be summarized as shown in Figure 5.1.

TABLE 5.1

Major ROS Molecules and Their Metabolism

ROS Molecule	Source	Enzyme Defenses	Products
Oxygen (O_2)			
Singlet oxygen $^1O_2$	Excited form of oxygen in which one of the electrons jumps to a superior orbital after absorption of energy		
Superoxide ($O_2^{\bullet-}$)	Leakage of electrons from electron transport chain. Activated phagocytes Xanthine oxidase Flavoenzymes	Superoxide dismutase (SOD)	$H_2O_2 + O_2$
Hydrogen peroxide (H_2O_2)	From $O_2^{\bullet-}$ via SOD NADPH-oxidase Glucose oxidase Xanthine oxidase	Glutathione (GSH) Peroxidase Catalases Peroxiredoxins (Prx)	$H_2O + GSSG$ $H_2O + O_2$ H_2O
Peroxyl radical ($O_2^{\bullet 2-}$)			
Hydroxyl radical ($^{\bullet}OH$)	From $O_2^{\bullet-}$ and H_2O_2 via transition metals (Fe, Cu)		
Nitric oxide ($^{\bullet}NO$)	NO synthases	GSH/TrxR	GSNO

5.1.1.1 Important ROS/RNS

5.1.1.1.1 Superoxide

The superoxide anion ($O_2^{\bullet-}$) created from molecular oxygen by the addition of an electron is, in spite of being a free radical, not highly reactive. It lacks the ability to penetrate lipid membranes and is therefore enclosed in the compartment where it was produced. The formation of $O_2^{\bullet-}$ takes place spontaneously, especially in the electron-rich aerobic environment in the vicinity of the inner mitochondrial membrane with the respiratory chain (Figure 5.1). $O_2^{\bullet-}$ (as well as H_2O_2) is also produced endogenously by flavoenzymes, e.g., xanthine oxidase and by other superoxide-producing enzymes, including lipoxygenases and cyclooxygenases.

5.1.1.1.2 Hydrogen Peroxide

Hydrogen peroxide (H_2O_2) is not a free radical but is nonetheless highly important, mostly because of its ability to penetrate biological membranes. It plays a radical-forming role as an intermediate in the production of more ROS molecules including HOCl (hypochlorous acid) by the action of myeloperoxidase, an enzyme present in the phagosomes of neutrophils and, very importantly, in the formation of $^{\bullet}OH$ via oxidation of transition

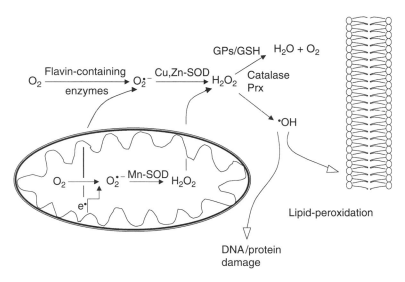

FIGURE 5.1

Schematic of oxidative and antioxidative systems in mammalian cells. Superoxide ($O_2^{\bullet-}$) is produced in significant amounts intracellularly, both in the cytosol via flavin-containing enzymes and in mitochondria due to escape of electrons from the respiratory chain. Two molecules of $O_2^{\bullet-}$ rapidly dismutate, either spontaneously or via superoxide dismutases (SOD) to dioxygen and hydrogen peroxide (H_2O_2), the latter permitting flux of ROS between cellular compartments. H_2O_2 can be enzymatically metabolized to oxygen and water by a number of different enzyme systems or converted to the hydroxyl radical ($^{\bullet}OH$), which is extremely reactive, via a chemical reaction catalyzed by transition metals.

metals. Another important function of H_2O_2 is its role as an intracellular signaling molecule. H_2O_2 is removed by at least three antioxidant enzyme systems: catalases, glutathione peroxidases (GPx), and peroxiredoxins (Prx) (Figure 5.1).

5.1.1.1.3 Hydroxyl Radical

Due to its strong reactivity with biomolecules, $^{\bullet}OH$ is probably capable of doing more damage to biological systems than any other ROS. The radical is formed from H_2O_2 in a reaction catalyzed by metal ions (Fe^{2+} or Cu^{+}) (the Fenton reaction).

$$H_2O_2 + Cu^+/Fe^{2+} \rightarrow {}^{\bullet}OH + OH^- + Cu^{2+}/Fe^{3+}$$

$O_2^{\bullet-}$ also plays an important role recycling the metal ions:

$$Cu^{2+}/Fe^{3+} + O_2^{\bullet-} \rightarrow Cu^+/Fe^{2++} + O_2$$

The sum of the last two reactions is known as the Haber–Weiss reaction. Transition metals, which play an important role in the formation of $^\bullet OH$ may be released from proteins such as ferritin by reactions with $O_2^{\bullet-}$.

5.1.1.1.4 Nitric Oxide

Nitric oxide ($^\bullet NO$) represents an odd member of the free-radical family and is similar to $O_2^{\bullet-}$ in that it does not readily react with most biomolecules despite its unpaired electron. On the other hand, it easily reacts with other free radicals (e.g., peroxyl and alkyl radicals), generating mainly less reactive molecules and in fact functioning as a free-radical scavenger. In this role $^\bullet NO$ has, for example, been shown to inhibit lipid peroxidation in cell membranes. If $O_2^{\bullet-}$ is produced in large amounts in parallel with $^\bullet NO$, both $O_2^{\bullet-}$ and $^\bullet NO$ react together to give $OONO^-$ (peroxynitrite), which is highly cytotoxic. Peroxynitrite can react directly with other biomolecules in one- or two-electron reactions; reaction with CO_2 for instance forms highly reactive nitroso-peroxocarboxylate ($ONOOCO_2^-$). Other products include peroxo-nitrous acid ($ONOOH$) and the results of peroxynitrite homolysis, $^\bullet OH$, and $^\bullet NO_2$ or rearrangement, NO_3. The rates of these different reactions of peroxynitrite will depend upon the pH, temperature and on the compounds present in the surrounding milieu. $^\bullet NO$ is synthesized enzymatically from L-arginine by $^\bullet NO$ synthase (NOS).

$$\text{L-arginine} + O_2 + NADPH \rightarrow \text{L-citrulline} + {}^\bullet NO + NADP^+$$

The complex enzymatic catalytic action of NOS involves the transfer of electrons from NADPH, via the flavins FAD and FMN in the carboxy-terminal reductase domain, to the heme in the amino-terminal oxygenase domain, where the substrate, arginine, is oxidized to citrulline and $^\bullet NO$. There are three main isoforms of the enzyme, neuronal NOS (nNOS), inducible NOS (iNOS), and endothelial NOS (eNOS), which differ in their respective expression and activities.

In physiologic concentrations, $^\bullet NO$ functions mainly as an intracellular messenger, stimulating guanylate cyclase and protein kinases and thereby relaxing smooth muscle in blood vessels. $^\bullet NO$ has the ability to cross cell membranes and can, as a result, transmit signals to other cells. When produced in larger amounts, as is the case when iNOS is induced by endotoxin and interferon γ, $^\bullet NO$ becomes an important factor in redox control of cellular function. Nitrosylation of proteins is known to regulate enzymatic activity including that of NOS itself.

Excessive production of $^\bullet NO$ is counteracted by its conjugation with GSH resulting in the S-nitrosoglutathione adduct (GSNO). GSNO can, in turn, be cleaved directly by mammalian thioredoxin reductase (TrxR) or by the complete thioredoxin system, which again liberates GSH and $^\bullet NO$. GSNO has also been shown to inhibit TrxR indicating a possible regulatory mechanism.

The total effect of NO on the redox status of cells is multifaceted since it functions as an antioxidant in some situations and as an oxidant in other situations.

5.1.1.2 Cellular Antioxidant Enzymes

The cellular antioxidant systems can be divided into the nonenzymatic, low molecular weight antioxidant compounds which act largely in conjunction with the thioredoxin system and the enzymatic antioxidant systems. The antioxidant enzyme systems include SOD, catalase, Prx, GPx, and other GSH-related systems and thioredoxin (Trx).

5.1.1.2.1 Superoxide Dismutases

Superoxide dismutases (SOD) were the first genuine ROS-metabolizing enzymes discovered. In eukaryotic cells, $O_2^{\bullet -}$ can be metabolized to H_2O_2 by two metal-containing SOD isoenzymes, an 80 kDa tetrameric Mn-SOD present in mitochondria and the cytosolic 32 kDa dimeric Cu/Zn-SOD (Figure 5.1). In the reaction catalyzed by SOD, two molecules of $O_2^{\bullet -}$ interact to form H_2O_2 and molecular oxygen. The reaction catalyzed by SOD is extremely efficient and is limited only by diffusion.

$$2O_2^{\bullet -} + 2H^+ \rightarrow H_2O_2 + O_2$$

In mitochondria, $O_2^{\bullet -}$ is formed in relatively high concentrations due to the leakage of electrons from the respiratory chain. The strictly mitochondrial Mn-SOD is obviously essential or near essential since no inherited diseases have been found in which Mn-SOD is deficient and knockout mice lacking Mn-SOD die soon after birth or suffer severe neurodegeneration. Using Mn-SOD-haploinsufficient mice (SOD2+/−), Mn-SOD is shown to be required for protecting ROS-induced injury in O_2 metabolism [2]. Reconstitution of the Mn-SOD efficiently suppresses ROS-induced damage and cell transformation [3]. Expression of Mn-SOD is, in contrast to Cu, Zn-SOD, induced by oxidative stress and also by Trx. Cytosolic Cu/Zn-SOD appears to be less important than Mn-SOD and transgenic animals lacking this enzyme are able to adapt so that the phenotype appears normal.

5.1.1.2.2 Catalases

Catalases are mainly heme-containing enzymes. The predominant subcellular localization in mammalian cells is in peroxisomes, where catalase catalyzes the dismutation of hydrogen peroxide to molecular oxygen and water:

$$2H_2O_2 \rightarrow O_2 + 2H_2O$$

Catalase also has functions in detoxifying different substrates, e.g., phenols and alcohols, via a coupled reduction of hydrogen peroxide:

$$H_2O_2 + R'H_2 \rightarrow R' + 2H_2O$$

One antioxidative role of catalase is to lower the risk of $^\bullet OH$ formation from H_2O_2 via the Fenton reaction catalyzed by Cu or Fe ions. Catalase also binds to NADPH protecting the enzyme from inactivation and increasing its efficiency.

5.1.1.2.3 Peroxiredoxins

Peroxiredoxins (Prx, thioredoxin peroxidases) are enzymes capable of directly reducing peroxides, such as H_2O_2 and different alkyl hydroperoxides. Trx (in mammalian cells) regenerates oxidized Prx formed in the catalytic cycle. In the mitochondria of mammalian cells the thioredoxin system is probably a specific reductant of Prx. Peroxiredoxins have been shown to inhibit apoptosis induced by p53 and by H_2O_2 on a level upstream of bcl-2.

5.1.1.2.4 Glutathione Peroxidases

There are at least four different glutathione peroxidases (GPx) in mammals (GPx1–4), all of them containing selenocysteine. GPx1 and GPx4 (or phospholipid hydroperoxide GPx) are cytosolic enzymes, abundant in most tissues. All GPx catalyze the reduction of H_2O_2 using GSH as substrate. They can also reduce other peroxides (including lipid peroxides in cell membranes) to alcohols

$$ROOH + 2GSH \rightarrow ROH + GSSG + H_2O$$

In this reaction, two molecules of GSH are oxidized to glutathione disulfide or oxidized glutathione (GSSG) that subsequently can be reduced by glutathione reductase (GR), the major mammalian GSSG-reducing enzyme.

5.1.1.2.5 Other Glutathione-Related Systems

Glutathione is the most abundant intracellular thiol-based antioxidant, prevalent in millimolar concentrations in all living aerobic cells. Its function is mainly as a sulfhydryl buffer, but GSH also serves to detoxify compounds either via conjugation reactions catalyzed by glutathione S-transferases (GST) or directly, as is the case with H_2O_2 in the GPx-catalyzed reaction. GSSG is reduced by the NADPH-dependent flavoenzyme GR.

Another class of proteins intimately related to GSH is the glutaredoxin (Grx), with functions overlapping those of Trx. A major qualitative difference between Grx and Trx is that Grx can be reduced by GSH and is capable of reducing GSH-mixed protein disulfides formed at oxidative stress, which play an important role in the total cellular antioxidant defense. Figure 5.2 summarizes the different GSH-related antioxidant systems and reactions involving GSH are shown in Figure 5.3.

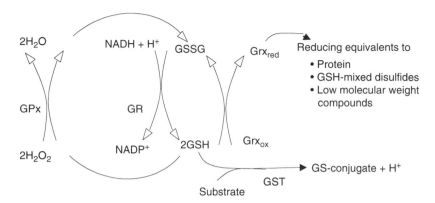

FIGURE 5.2
Summary of the major glutathione-associated antioxidant systems. Hydrogen peroxide (H_2O_2) is reduced by glutathione peroxidases (GPx) by oxidation of two molecules of glutathione (GSH) forming glutathione disulfide (GSSG) that subsequently can be reduced by glutathione reductase (GR) under consumption of NADPH. GSH also reduces glutaredoxins (Grx) that in their turn reduce various substrates. Specific for Grx is the reduction of GSH-mixed disulfides such as glutathionylated proteins. Glutathione *S*-transferases (GST) catalyze the conjugation of GSH with other molecules, acting as an intermediate step in the detoxification of miscellaneous toxins.

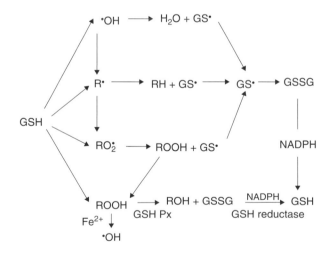

FIGURE 5.3
Reaction of glutathione with hydroxyl radicals (•OH), organic radicals (R•), peroxyl radicals (ROO•) and hydroperoxides (ROOH). GSH reacts with radicals to produce thiol radicals that self-associate to produce oxidized glutathione (GSSG). GSH is regenerated by NADPH generated from the pentose cycle. Reaction of Fe^{2+} with peroxide can lead to generation of •OH and increased production of malonaldehyde.

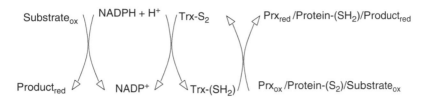

FIGURE 5.4
Enzymatic reactions of the thioredoxin system. Thioredoxin reductase (TrxR) reduces the active site disulfide in thioredoxin (Trx) and several other substrates directly under consumption of NADPH. Reduced Trx is highly efficient in reducing disulfides in proteins and peptides, including peroxiredoxins (Prx) and glutathione disulfide (GSSG).

5.1.1.3 Low Molecular Weight Antioxidant Compounds and Thioredoxin

A large number of low molecular weight compounds are considered to be antioxidants of biological importance, including vitamins C and E, different selenium compounds, lipoic acid, and ubiquinones and all of these interact with the mammalian thioredoxin system.

Thioredoxins are proteins with oxidoreductase activity and are ubiquitous in both mammalian and prokaryotic cells. The thioredoxin system consists of the two antioxidant oxidoreductase enzymes Trx and TrxR. The latter catalyzes the reduction of the active site disulfide in Trx using NADPH. Reduced Trx is a general protein disulfide reductant (Figure 5.4). Three distinct variants of human Trx encoded by separate genes have been cloned and characterized. The most intensely studied is the gene for the 12 kDa thioredoxin (Trx-1). The Trx-2 isoenzyme is located in mitochondria and includes a 60 amino acid N-terminal mitochondrial translocation signal. The third Trx, SpTrx, is a variant highly expressed in spermatozoa. The Trx of all organisms, including the human Trx-1, Trx-2, and SpTrx contain a conserved -Cys-Gly-Pro-Cys- active site, essential for the function as a general and potent protein disulfide oxidoreductase. Specific protein disulfide targets for reduction by Trx are ribonucleotide reductase and several transcription factors including p53, NF-κB, and AP-1. Trx is also a specific electron donor for many Prx and important for the reduction of peroxides.

5.2 Ionizing Radiation-Induced ROS/RNS

5.2.1 Demonstration of Radiation-Induced Intracellular ROS/RNS

Transient generation of ROS or RNS has been detected with dihydrochlorofluoroscein by fluorescence microscopy within minutes of exposing the cell to ionizing radiation [4]. In the 1–10 Gy dose range, the amount of ROS/RNS detected per cell was constant, but the percentage of producing cells increased with dose. The use of several inhibitors pointed to the mitochondria as the source for ROS/RNS, a conclusion which was supported by the observation of simultaneous transient depolarization of the mitochondrial membrane potential. In another study exposure to cobalt-60 γ radiation was found to enhance dose-dependent, cellular production of ROS/RNS in HepG2 cells with doses of 50–400 cGy, and was accompanied by a decrease in the level of reduced GSH [5]. Depletion of GSH before irradiation amplified the increase in ROS and radiosensitized the cells in terms of survival.

For high LET, radiation studies with fluorescent dyes demonstrated generation of ROS/RNS in cells within 15 min after passage of less than one α-particle per cell [6]. The intracellular production of ROS/RNS was 50-fold greater than the extracellular production. The effect was inhibited by diphenyliodonium (DPI), an inhibitor of flavoproteins, suggesting the involvement of a plasma membrane NADPH oxidase. This was a tentative conclusion since flavoproteins are also localized to the mitochondria and endoplasmic reticulum.

Important early investigations which implicated oxidative stress in radiation response were done by Oberley et al. [7] who found that modifying intracellular or extracellular SOD activity in bacteria reduced the oxygen enhancement ratio, implying that $O_2^{\bullet-}$ was at least partially responsible for the mechanisms leading to the effects of oxygen on radiosensitization. The investigation was extended to the whole body level by Petkau et al. [8] who showed that injection of Cu/Zn-SOD following 6 or 8 Gy whole body x-irradiation, significantly protected Swiss mice from lethality at 30 days. These studies clearly demonstrated that alterations in a metabolic enzyme with $O_2^{\bullet-}$ scavenging capability following radiation could result in radioprotection and implied that $O_2^{\bullet-}$ production (from some source) following radiation was participating in radiation-induced injury. These findings have been followed up more recently with observations that active SOD enzymes, SOD mimetic compounds and, in some instances, catalase can lead to inhibition of the deleterious effects induced by ionizing radiation in a wide variety of in vitro and in vivo experimental situations including transformation assays, bystander effects, normal tissue damage associated with inflammatory responses and fibrosis (reviewed by Spitz et al. [9]).

Recent studies of levels of ROS and RNS (derived from $^{\bullet}$NO) have shown that radiation exposure can induce increases in the metabolic production of these species for several minutes and hours post-irradiation [4,10,11] with indications that the source of this increase in radiation-induced pro-oxidant production involved mitochondria.

Indirect evidence for an extranuclear radical amplification mechanism has come from studies with high LET particles. In one case, bone marrow progenitor cells exposed to neutrons showed both an enhanced ability to oxidize a fluorescent probe for ROS/RNS and increased 8-hydroxy-2-deoxyguanosine levels indicative of oxidative DNA base damage [12].

5.2.2 Mechanisms of Generation and Amplification of ROS/RNS Following Irradiation of the Cytoplasm

The primary radical generated as a consequence of initial ionization events is $^{\bullet}OH$, which is short lived and only diffuses about 4 nm before reacting. The secondary ROS include $O_2^{\bullet-}$ and H_2O_2, which can react via Fenton chemistry with cellular metal ions to produce additional $^{\bullet}OH$. Taking into account secondary free-radical products resulting from the initial ionization, the calculated amount of ROS generated is relatively insignificant compared to the amount routinely produced by oxidative metabolism [13]. RNS might also be produced after irradiation, possibly as a consequence of radiation-induced stimulation of $^{\bullet}NO$ synthase activity in cells enriched in this enzyme. Reaction of $^{\bullet}NO$ with $O_2^{\bullet-}$ results in the formation of peroxynitrite, a membrane-permeant and relatively stable RNS. When peroxynitrite is protonated, it isomerizes to *trans*-peroxynitrous acid, which can cause protein and DNA damage similar to $^{\bullet}OH$.

The question that nevertheless arises is, how are the few primary ionizing events produced at clinically relevant doses ($\sim$2000/Gy/cell) sufficiently amplified to account for the activity that has been shown to be mediated by radiation-induced ROS/RNS, such as the rapid and robust activation of cellular signal transduction pathways? A partial answer to this question has come from recent studies that indicate that cells are endowed with cytoplasmic amplification mechanisms involving ROS/RNS and responsive to low doses of ionizing radiation. These mechanisms appear to be part of general cellular response pathways to oxidative stress (reviewed by Mikkelson and Wardman [11]).

The specific metabolic reactions responsible for alterations in ROS (or RNS) production following radiation exposure remain to be clarified. In general, the proposed metabolic sources of pro-oxidant production are ascribed to two possible systems which are not mutually exclusive: mitochondrial electron transport chains and oxidoreductase enzymes.

5.2.2.1 Increase of Pentose Cycle Activity

Increased pentose cycle activity is one of the best characterized metabolic alterations observed immediately following exposure to ionizing radiation and other oxidants [14,15]. The pentose cycle produces reducing equivalents in the form of NADPH and ribose-5-phosphate for RNA and DNA synthesis. In an emergency, under conditions of heightened oxidative stress, the pentose cycle is capable of producing large amounts of reducing equivalents, i.e., NADPH, dedicated to biosynthetic and repair processes, and bypassing the formation of ribulose-5-phosphate the precursor of ribose-5-phosphate.

The oxidative limb of the pentose cycle includes two $NADP^+$-dependent dehydrogenases glucose-6-phospate dehydrogenase (G6PD) and 6-phosphogluconate dehydrogenase (6PGD), which increases production of NADPH under conditions of oxidative stress. Increased production of NADPH is necessary for the reduction of GSSG and flavoproteins oxidized by xenobiotics. G6PD is the initial and rate-limiting enzyme of this metabolic pathway.

NADPH has a role as a cofactor in the reduction of hydroperoxides via the action of GPx and Prx. Radiation and oxidative stress-induced increases in glucose metabolism through the pentose cycle are believed to represent an immediate metabolic response to stress by attempting to offset increases in oxidative damage with increases in protective and reparative processes.

It has been shown that the oxidative pentose cycle protects cells from ionizing radiation-induced killing. CHO cells containing loss-of-function mutations in G6PD were significantly more sensitive to clonogenic cell death than were wild-type cells [16]. An elevated incidence of apoptosis was observed in G6PD$^-$ cells which accounted for reduced survival at doses lower than 2 Gy.

5.2.2.2 Role of NADPH Oxidase

As described in the preceding section, the pentose cycle supplies reducing equivalents for protection and repair under conditions of oxidative stress. Another possible way that radiation-induced changes in pentose cycle activity may be linked to radiation-induced processes including signal transduction and bystander effects is via the activity of NADPH-oxidase enzymes.

NADPH oxidases catalyze the production of $O_2^{\bullet-}$ by the one-electron reduction of oxygen, using NADPH as the electron donor:

$$2O_2 + NADPH \rightarrow O_2^{\bullet-} + NADP^+ + H^+$$

NADPH oxidases are flavin-containing enzymes with several different subunits which are broadly expressed in inflammatory cells but also in non-phagocytic cells, including fibroblasts and smooth muscle cells. Exposure of smooth muscle cells and fibroblasts to H_2O_2-mediated oxidative stress is known to lead to increases in non-phagocytic NADPH-oxidase activity contributing to H_2O_2-induced cell injury [17].

NADPH oxidases are also thought to be involved with signaling pathways leading to proliferation as well as fibrogenic responses. The NADPH-oxidase enzymes are dependent on NADPH to form ROS; consequently changes in pentose cycle activity could potentially impact upon the levels of ROS produced by these enzymes by affecting substrate availability. Results of several investigations indicate that inhibitors of flavin-containing enzymes such as NADPH oxidase are capable of inhibiting radiation-induced signal transduction and injury in cells adjacent to irradiated cells [18,19].

5.2.2.3 Mitochondrial Permeability and ROS/RNS Generation

Mitochondria are a major source of ROS in most cells and the total intracellular mitochondrial volume represents a substantial target volume. It is perhaps not surprising that studies with generation of cellular ROS following radiation showed that radiation-induced ROS was concomitant with Ca^{2+}-dependent propagation of a reversible permeability transition from one mitochondrion to another [4].

Reversible depolarization of the mitochondrial membrane potential, $(\Delta\psi)$, and decrease in fluorescence of a mitochondria-entrapped dye, calcein, were observed coincidentally with the stimulation of transient cellular generation of ROS/RNS following ionizing radiation in the therapeutic dose range. Radiation-induced ROS/RNS, $\Delta\psi$ depolarization, and calcein fluorescence are inhibited by the mitochondrial permeability transition-inhibitor, cyclosporin A but not the structural analogue cyclosporin H. Radiation stimulated generation of ROS/RNS was also inhibited by over-expressing the Ca^{2+}-binding protein, calbindin 28K or treating cells with an intracellular Ca^{2+} chelator. The ROS/RNS release observed with radiation was common to all cells investigated except those of an osteosarcoma cell line deficient in mitochondrial electron transport chain which showed neither radiation-induced ROS/RNS nor $\Delta\psi$ depolarization.

It has been proposed [10] that radiation damage to a few mitochondria could be transmitted via a reversible Ca^{2+} dependent mitochondrial permeability transition to adjacent mitochondria with resulting amplification of ROS/RNS generation. Measurements of radiation-induced mitogen-activated protein kinase activity indicate that this sensing/amplification mechanism is necessary for the activation of some cytoplasmic signaling pathways by low doses of radiation.

5.2.2.4 Nitric Oxide Synthase and Nitric Oxide

Nitric oxide is formed during the NOS catalyzed conversion of arginine to citrulline. Three isoforms of NOS with similar catalytic mechanisms have been described. The NOS-1 and NOS-3 isoforms are constitutively expressed in many cell types and their activities are Ca^{2+}/calmodulin dependent. NOS-1 and NOS-3 produce relatively low amounts of $^{\bullet}NO$ compared with the inducible NOS-2. NOS-2 tightly binds to calmodulin and its activity is largely Ca^{2+}-independent. Low doses of ionizing radiation transiently activate NOS-1.

RNS may thus have a significant role in the response of cells to radiation. $^{\bullet}NO$ reacts with $O^{\bullet-}$ at diffusion-controlled rates competing with endogenous SOD for substrate [20]. Inhibiting NOS activity or treating cells with $^{\bullet}NO$ scavengers stimulates oxidative distress [21]. The product of $O^{\bullet-} + {}^{\bullet}NO$, $ONOO^-$ mostly rearranges to form biologically inert nitrite or reacts with GSH to form the $^{\bullet}NO$ donor GSNO [20]. However, when the $^{\bullet}NO$ concentration approaches that of SOD the resulting high levels of $ONOO^-$ produce a number of cell damaging effects [22] including enhancement of tyrosine (Tyr) nitration of proteins, a cellular footprint of $ONOO^-$. Radiosensitization is observed with exogenous $^{\bullet}NO$ donors or after cytokine stimulation of inducible NOS-2 production [23].

The nature of the relationship between the radiation-induced reversible mitochondrial permeability transition described above and NOS-1 activation is unclear. One possibility is that the localized Ca^{2+} transients that propagate the permeability transition can simultaneously activate the Ca^{2+} dependent NOS-1. This mechanism is supported by the recent finding that NOS-1 can associate with mitochondria via its PDZ domain putting NOS-1 at the origin of mitochondria-released Ca^{2+}. Possibly also involved in this process is the ROS-induced ROS release that accompanies the mitochondrial permeability transition and that may be necessary for its propagation.

Another component of the mechanism stimulating ROS/RNS is activation by radiation of constitutive NOS, which has been shown to be comparable with that seen after treating the cells with a Ca^{2+} ionophore or purinergic receptor agonist. Downstream consequences of NOS activation included activation of •NO-dependent enzyme, soluble guanine cyclase and enhanced Tyr nitration, a footprint of $ONOO^-$. Radiation-induced Tyr nitration of Mn-SOD was inhibited by NOS inhibitors or by expression of dominant negative NOS-1 mutant. These results all demonstrate that ionizing radiation, at clinically relevant doses, effectively activates NOS-1.

The importance of •NO as a redox signaling molecule resides in part in its relative stability and the hydrophobic properties that permit its diffusion across cell membranes over several cell diameter distances, suggesting mechanisms by which an ionization event can be sensed in neighboring cells. A two-step mechanism has been suggested by Mikkelsen and Wardman [11] for cell signaling by radicals. The first step involves the nonspecific oxidative reaction of highly reactive radicals of limited diffusion distance, which are succeeded by a more stable radical with specific reactivity consistent with a signaling molecule. Highly reactive •OH, the initial product of ionizing radiation, has low target selectivity and short diffusion distance whereas, •NO and nitrite/nitrate are chemically inert with respect to most biological molecules but exhibit a high degree of reactant specificity.

5.2.3 Consequences of Radiation-Induced Generation of ROS/RNS

In many cases the action of radiation-induced ROS/RNS is detected on the basis of the occurrence of downstream reactions which occur at a distance, spatially or chemically from the initial ionizing events. These consequences are very important in themselves and they also provide markers for the detection of cytoplasmic consequences of radiation-generated ROS/RNS. The processes at this level, signal transduction, transcription, nuclear, and bystander effects are discussed in detail in succeeding chapters and will be described here in brief.

5.2.3.1 *Signal Transduction and Transcription Factor Activation*

Some ROS/RNS, including H_2O_2, •NO, and peroxynitrite, are membrane-permeant and are sufficiently stable to diffuse significant distances within cells, facilitating their interaction with biological molecules, including proteins, lipids, and DNA. Of particular interest for regulation of cytoplasmic signal transduction pathways is the targeting of protein SH residues. The interconversion of oxidized and reduced cysteine (Cys) represents a reversible controlling element in the regulation of protein functions.

RNS induce a number of reversible protein thiol modifications including S-nitrosylation and formation of sulfenic acids and both intramolecular (S–S) and mixed (S–SR) disulfides [24]. This is of special importance to radiation-induced activation of ERK1/2 (a signal transduction pathway described in Chapter 11) which has been shown to be RNS-dependent [10,25]. Both positive and negative regulatory components of ERK1/2 are potentially involved. The activity of the oncogene Ras is enhanced by nitrosylation of Cys118 resulting in enhanced downstream signaling and ERK1/2 activity [25]. On the other hand, the protein–Tyr phosphatase which inhibit ERK1/2 activation can be blocked by oxidation of Cys in the catalytic site inhibiting phosphatase activity [26]. Thus S-nitrosylation of either Ras or a protein–Tyr phosphatase could result in the enhanced ERK1/2 activity observed after radiation.

In addition to signal transduction pathways, transcription factors (AP-1, NF-κB, GADD153, p53) are also activated immediately following radiation exposure resulting in the transcription of downstream genes thought to be involved in the radiation response. It is now generally believed that these stress responsive transcription factors and their target genes form a pro-survival network to enhance cell resistance to radiation insults [9,27]. The exact mechanisms responsible for sensing radiation-induced free-radical production leading to the activation of these signaling and gene expression pathways are not, as yet, completely understood.

In one specific case, the link between radiation-induced changes in oxidative metabolism and activation of signaling leading to alterations in transcription factor activation is thought to be the pentose cycle. NADPH is the source of electrons for the reduction of Trx, which in the reduced form, is known to be involved with transcription factor activation [28,29]. It has been shown that TrxR passes electrons to Trx, resulting in Trx nuclear translocation and a subsequent interaction with redox factor-1 (Ref-1) leading to the activation of the AP1 transcription factor. This suggests that radiation-induced increases in pentose cycle activity can alter the availability of NADPH to donate electrons to TrxR initiating of the signaling cascade.

5.2.3.2 Mutagenic and Clastogenic Effects

Targeted irradiation of the cytoplasm, where many metabolic redox reactions occur, has been shown to induce mutations in nuclear DNA and manipulations of free-radical scavenging capability will modify radiation-induced nuclear DNA damage under these conditions. These results support the hypothesis that radiation-induced damage to cytoplasmic metabolic pathways that results in free-radical production contributes to heritable changes in the nuclear genome.

Direct evidence of cytoplasmic ionization events affecting nuclear processes has come from studies in which individual cells were irradiated with a microbeam of α-particles allowing selective irradiation of the cytoplasm or nucleus. Irradiation of the cytoplasm was shown to be mutagenic but not cytotoxic [30]. The major class of mutations was similar to those generated spontaneously which are believed to arise from DNA damage produced by endogenous ROS/RNS [31]. These mutations differ from those induced after direct irradiation of the nucleus.

5.2.3.3 Bystander Effects

A number of recent studies have challenged the traditional belief that the important biological effects of ionizing radiation are a result of DNA damage by its direct interaction with the nucleus. These findings indicate that irradiated and nonirradiated cells interact and that oxidative metabolism has an essential role in the signaling events leading to radiation-induced bystander effects. Preliminary evidence suggests that oxidants can modulate similar signal transduction pathways in irradiated and bystander cells, and thus contribute to the induction of bystander DNA damage. Genes that are directly responsive to oxidative stress have been identified in bystander cells (reviewed by Azzam et al. [19]). However, direct evidence explaining how these events occur is not yet available.

One scenario which has been proposed involves NAD(P)H-oxidase enzymes which are known to produce ROS in quantities capable of stimulating signaling pathways and are rapidly activated by a variety of soluble mediators and engagement of cell-surface receptors [32]. While $O_2^{\cdot-}$ anion is the major end product of NAD(P)H-oxidase activity, H_2O_2, the $^{\bullet}OH$, and hypochlorous acid are also formed as a result of its activation. These highly reactive species can be released into the cellular environment, where they can interact with closely neighboring cells or combine with serum proteins and lipids resulting in their oxidation and modulation of their function. Such activity would be consistent with a role for oxidizing diffusible factors in mediating the radiation-induced bystander response. Results from various laboratories support the hypothesis that a DPI-sensitive flavin-containing oxidase activity may represent a significant source of the ROS which are believed to mediate the bystander response produced in human fibroblast cultures exposed to low fluences of α-particles [19].

5.3   Effects of Ionizing Radiation on the Cell Membrane

The random nature of energy deposition characteristic of ionizing radiation ensures that the cell membrane is just as likely to be a site of radiation damage as is the cell nucleus. The physicochemical structure of biological membranes makes them peculiarly susceptible to oxidative damage, and consequently a target of radiation-generated ROS/RNS. Reaction of membrane lipids in the presence of oxygen results in lipid peroxidation, an effect that increases with decreasing dose rate and has profound effects on membrane structure and function.

5.3.1 Structure of the Cell Membrane

Composed chiefly of phospholipids, the cell surface is organized as a lipid bilayer with peripheral proteins which are restricted to the extracellular face of the lipid bilayer and transmembrane proteins which traverse the plasma membrane. Localization and orientation of membrane-resident proteins is not random, rather, proteins are synthesized and inserted into the plasma membrane in a defined and asymmetric fashion. Proteins are dynamically associated with lipid domains, including caveolae (having a unique coat containing caveolin-l), lipid rafts (enriched in cholesterol, sphingomyelin, and GM 1 ganglioside), and lipid shells (mobile cholesterol–phospholipid complexes) that surround proteins and that permit their translocation within the membrane. Incoming proliferation, differentiation, and apoptotic signals are transmitted from the extracellular environment into the cell by transmembrane cytokine receptors. The same growth regulatory signals are generated by the interaction of plasma membrane-anchored cytokines with their cognate receptors on neighboring cells. Contiguous cells may also communicate through gap junctions, water-filled channels that form between the plasma membrane of two adjacent cells the function of which is to permit direct passage of small molecules (e.g., cAMP) and ions (e.g., Ca^+) from the cytoplasm of one cell to that of its neighbor. Gap junctional associations permit the formation of a cell–cell communication network in which the cytoplasm of each cell in the network is linked to that of the others both electrically and metabolically.

The functional properties of the plasma membrane are principally dictated by proteins and their orientation within the membrane, but the biophysical properties of the cell-surface bilayer are imposed by lipids. Biophysical parameters, including membrane fluidity, resistance to changes in curvature, and membrane tension are controlled by the chemical nature and concentration of lipids which constitute the plasma membrane. These factors directly impact on the function of transmembrane proteins and, consequently on the capacity of the plasma membrane to regulate intercellular communication. Other transport mechanisms which do not involve protein participation, but rather large, local deformations of the plasma membrane (endocytosis, exocytosis, and membrane fusion events) are also determined by changes in the composition of lipids. Agents such as chemotherapeutic drugs and ionizing radiation that alter the concentration or chemical nature of plasma membrane lipids impact on plasma membrane function. In addition, ionizing radiation affects those functions which are mediated by transmembrane proteins by altering their expression or by changing the interaction(s) that normally takes place between membrane lipids and proteins.

5.3.2 Lipid Peroxidation in Plasma Membranes

Polyunsaturated lipids of the plasma membrane contain double bonds between some of their carbon atoms that are susceptible to attack by hydroxyl, hydrogen, and oxygen radicals. These radicals act on polyunsaturated fatty acids of cellular membranes and produce lipid peroxides. This section is based in part on reviews by Albanese and Dainiak [33], Stark [34], Rémita [35], and the articles referenced therein.

Initiation of lipid peroxidation is caused by attack upon a lipid of any species that has sufficient reactivity to abstract a hydrogen atom from a methylene ($-CH_2-$) group. Fatty acids with one or no double bonds are more resistant to such attack than are the polyunsaturated fatty acids (PUFAs). An adjacent double bond weakens the energy of attachment of the hydrogen atoms present on the next carbon atom (the allylic hydrogens), especially if there is a double bond on either side of the $-CH_2-$.

Hydroxyl radicals can readily initiate peroxidation:

$$-CH_2- + {}^{\bullet}OH \rightarrow -{}^{\bullet}CH- + H_2O$$

if they can reach the hydrocarbon side chain without interacting with something else first. ${}^{\bullet}OH$ generated outside the membrane can also attack extrinsic proteins such as cell-surface glycoproteins and "head groups" of phospholipids.

A mechanism for the formation of lipid peroxides is given in Figure 5.5. The initiating event leading to lipid peroxidation is abstraction of the allylic hydrogen from the polyunsaturated fatty acid by a hydroxyl or other radical creating a lipid radical, which subsequently undergoes molecular rearrangement to form a more stable conjugated lipid radical. In the presence of oxygen, a lipid peroxyl radical is formed when oxygen reacts with the conjugated lipid radical. Abstraction of a hydrogen atom from another lipid molecule then generates a lipid hydroperoxide and a new lipid radical. The lipid radical can then react with oxygen and a third lipid molecule and thus initiate an autocatalytic reaction that propagates the generation of lipid peroxides resulting ultimately in extensive damage to the membrane. Lipid peroxides are formed continually as the autocatalytic reaction proceeds until lipid radicals taking part in the chain reaction are eliminated by radical–radical interactions [34]. Structural alterations of the plasma membrane as a consequence of radiation-induced lipid hydroperoxides and lipid hydroperoxide breakdown products include unsaturated aldehydes such as malondialdehyde which alter plasma membrane fluidity, inducing a liquid crystal-to-gel phase transition [36]. The phase transition is a result of the formation of

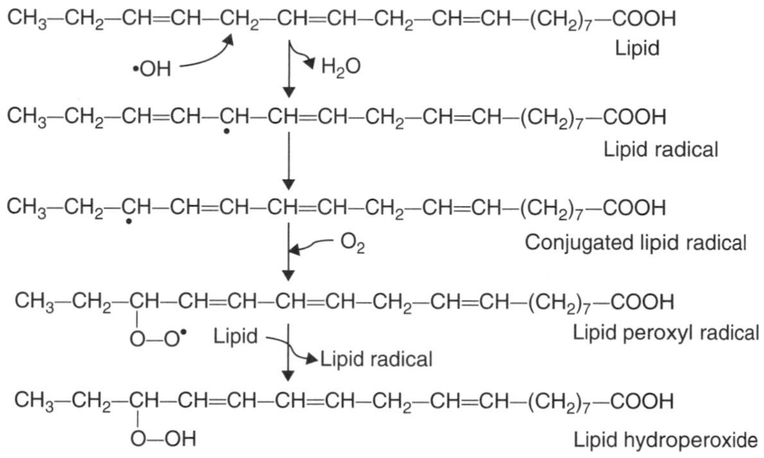

$$CH_3-CH_2-CH=CH-CH_2-CH=CH-CH_2-CH=CH-(CH_2)_7-COOH$$

•OH ⟶   ↘ H_2O Lipid

$$CH_3-CH_2-CH=CH-\underset{\bullet}{CH}-CH=CH-CH_2-CH=CH-(CH_2)_7-COOH$$

Lipid radical

$$CH_3-CH_2-\underset{\bullet}{CH}-CH=CH-CH=CH-CH_2-CH=CH-(CH_2)_7-COOH$$

↙ O_2 Conjugated lipid radical

$$CH_3-CH_2-\underset{\underset{O-O^\bullet}{|}}{CH}-CH=CH-CH=CH-CH_2-CH=CH-(CH_2)_7-COOH$$

Lipid ↘   Lipid peroxyl radical
↘ Lipid radical

$$CH_3-CH_2-\underset{\underset{O-OH}{|}}{CH}-CH=CH-CH=CH-CH_2-CH=CH-(CH_2)_7-COOH$$

Lipid hydroperoxide

FIGURE 5.5
Lipid peroxidation of plasma membrane lipids is initiated by a hydroxyl radical (•OH) abstracting a hydrogen atom from a polyunsaturated lipid molecule. This generates a lipid radical and shuffling of electrons leads to conjugated lipid radicals. In the presence of oxygen, a lipid peroxyl radical is formed which abstracts hydrogen from a nearby lipid to produce a lipid hydroperoxide molecule, regenerating the lipid radical. This can then react with another lipid molecule and begin the cycle again. Thus the •OH has initiated a self-perpetuating process of oxidative degeneration of polyunsaturated lipid molecules.

polar products that increase the dielectric constant within the hydrophobic core of the bilayer [34]. Increased polarity within the hydrophobic core facilitates crossing of the bilayer center and interdigitation of lipid fatty acyl chains with adjacent hydrocarbon chains in the opposing monolayer. An increase in membrane microviscosity occurs as interdigitation takes place, together with decrease in the width of the bilayer and increased ordering of the lipid chains. The mobility of hydrocarbon chains is also restricted by cross-links formed between lipid radicals, which in turn, further enhance rigidity of the plasma membrane [34,37]. These events are summarized in Figure 5.6.

5.3.3 Consequences of Damage to Plasma Membrane Lipids

Membrane lipid peroxidation results in increased membrane permeability to small molecules and ions [38]. In contrast, many transport systems are impeded as the cell-surface membrane becomes more rigid as a result of lipid peroxidation. Increased permeability to molecules and ions in spite of the decrease in membrane fluidity comes about because radiation not only causes a liquid crystalline-to-gel transition but also causes the formation of local protrusions and fenestrations at specific sites on the membrane [34]. An increase in plasma membrane microviscosity has been shown to result in

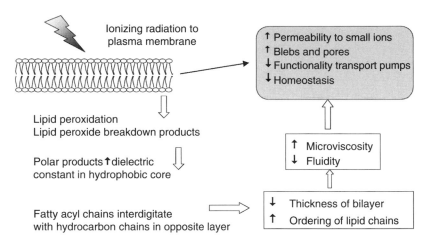

FIGURE 5.6

Biophysical changes induced in the plasma membrane by ionizing radiation.

the formation of "blebs" or protrusions on the cell surface and localized protrusions and pores are seen by transmission electron microscopy at the plasma membrane of lymphocytes as early as 15 min following exposure to x-rays (10 Gy) [39]. Presumably, the selective permeability of the plasma membrane is compromised by the creation of such openings, and molecules are allowed to pass through these structures in a nonspecific manner. Ultimately, the inability of the plasma membrane to maintain ionic homeostasis could result in cell death.

5.3.4   Plasma Membrane Is a Target for Ionizing Radiation-Induced Apoptosis

There is much evidence to support the concept that exposure to ionizing radiation initiates signals from the plasma membrane to activate pathways that culminate in apoptosis. Within seconds of irradiation, plasma membrane-associated sphingomyelinases are activated and catalyze the hydrolysis of sphingomyelin to generate ceramide [40,41]. The sphingomyelin pathway for induction of apoptosis is an ubiquitous, evolutionarily conserved signaling system which is initiated by the hydrolysis of sphingomyelin. Sphingomyelin-specific forms of the enzyme phospholipase C, called sphingomyelinases generate ceramide which acts as a second messenger to stimulate a cascade of kinases and transcription factors that are involved in a variety of cellular responses. Signals from the irradiated cell membrane mediated by ceramide lead to apoptosis via the stress-activated protein kinase (SAPK/JNK) cascade and terminate in the activation of transcription factors for apoptotic genes (described in detail in Chapter 12).

5.4 Summary

Ionizing radiation results in the formation of short-lived free radicals that cause oxidative damage to biomolecules. In the presence of oxygen, radiation leads to the formation of ROS and RNS. ROS are also constantly generated in aerobic cells by electron-transfer processes during normal cellular oxidative metabolism. Important ROS/RNS include $O_2^{\bullet-}$, H_2O_2, $^\bullet OH$, and $^\bullet NO$. $^\bullet NO$ is synthesized enzymatically from L-arginine by NOS and is unreactive with most biomolecules.

The antioxidant enzyme systems include SOD, catalase, Prx, GPx, other GSH-related systems and Trx. In the reaction catalyzed by SOD, two molecules of $O_2^{\bullet-}$ interact to form H_2O_2 and molecular oxygen while catalase catalyzes the dismutation of H_2O_2 to water and molecular oxygen. GSH is the most abundant intracellular thiol-based antioxidant; it functions as a sulfhydryl buffer and also acts to detoxify compounds either via conjugation reactions catalyzed by GST or directly, as is the case with H_2O_2 in the GPx-catalyzed reaction. The thioredoxin system consists of the two antioxidant–oxidoreductase enzymes: Trx and TrxR. Reduced Trx is a general protein disulfide reductant.

Early investigations which implicated oxidative stress in radiation response showed that modifying intracellular or extracellular SOD activity in bacteria reduced the oxygen enhancement ratio and that injection of SOD following whole body x-irradiation, significantly protected Swiss mice from lethality. More recently, intracellular ROS has been detected within minutes of exposing cells to low or high LET radiation.

In fact, the generation of ROS/RNS by radiation is relatively insignificant compared with the levels produced by oxidative metabolism, posing the question of how the few primary ionization events produced at clinically relevant doses are amplified to account for activity that has been shown to be mediated by radiation-induced ROS/RNS. There is evidence that metabolic sources of pro-oxidant production can involve two possible systems which are not mutually exclusive: mitochondrial electron transport chains and oxidoreductase enzymes. Low doses of ionizing radiation transiently activate NOS-1 and $^\bullet NO$ may thus have role in the response of cells to radiation. The importance of $^\bullet NO$ as a redox signaling molecule resides in its relative stability and the hydrophobic properties that permit its diffusion across cell membranes over several cell diameter distances, suggesting mechanisms by which an ionization event can be sensed in neighboring cells.

In fact, radiation-induced ROS/RNS are detected on the basis of downstream reactions which occur at a distance, spatially or chemically from the initial ionizing events and provide markers for the detection of cytoplasmic consequences of radiation-generated ROS/RNS. These reactions include activation of signal transduction intermediates and transcription factors and the signaling events leading to radiation-induced bystander effects including damage to nuclear DNA.

The physicochemical structure of biological membranes makes them peculiarly susceptible to oxidative damage and consequently a target of radiation-generated ROS/RNS. Composed chiefly of phospholipids, the cell surface is organized as a lipid bilayer with peripheral proteins restricted to the extracellular face of the bilayer and transmembrane proteins which traverse the plasma membrane. Polyunsaturated lipids of the plasma membrane contain double bonds between some of their carbon atoms that are susceptible to attack by hydroxyl, hydrogen, and oxygen radicals. These radicals act on polyunsaturated fatty acids, and irradiation of plasma membrane lipids in the presence of oxygen results in lipid peroxidation—an effect that increases with decreasing dose rate and has profound effects on membrane structure and function. Membrane lipid peroxidation results in increased membrane permeability to small molecules and ions and increased membrane rigidity. Exposure to ionizing radiation can also initiate signals from the plasma membrane to activate pathways that culminate in apoptosis. Plasma membrane-associated sphingomyelinases are activated by radiation to catalyze hydrolysis of sphingomyelin to ceramide which acts as a second messenger stimulating a signal transduction cascade culminating in apoptosis.

References

1. Nordberg J and Arnér ESJ. Reactive oxygen species, antioxidants, and the mammalian thioredoxin system. *Free Radic. Biol. Med.* 31: 1287–1312, 2001.
2. Williams MD, van Remmen H, Conrad CC, Huang TT, Epstein CJ, and Richardson A. Increased oxidative damage is correlated to altered mitochondrial function in heterozygous manganese superoxide dismutase knockout mice. *J. Biol. Chem.* 273: 28510–28515, 1998.
3. Zhong W, Oberley LW, Oberley TD, and St. Clair DK. Suppression of the malignant phenotype of human glioma cells by overexpression of manganese superoxide dismutase. *Oncogene* 14: 481–490, 1997.
4. Leach JK, Van Tuyle G, Lin PS, Schmidt-Ullrich RK, and Mikkelsen RB. Ionizing radiation-induced, mitochondria-dependent generation of reactive oxygen/nitrogen. *Cancer Res.* 61: 3894–3901, 2001.
5. Morales A, Miranda M, Sanchez-Reyes A, Biete A, and Fernandez-Checa J. Oxidative damage of mitochondrial and nuclear DNA induced by IR in human hepatoblastoma cells. *Int. J. Radiat. Oncol. Biol. Phys.* 42: 191–204, 1998.
6. Narayanan P, Goodwin E, and Lehnert B. Alpha particles initiate biological production of superoxide anions and hydrogen peroxide. *Cancer Res.* 57: 3963–3971, 1997.
7. Oberley LW, Lindgren AL, Baker SA, and Stevens RH. Superoxide ion as the cause of the oxygen effect. *Radiat. Res.* 68: 320–328, 1976.
8. Petkau A, Chelack WS, and Pleskach SD. Protection of postirradiated mice by superoxide dismutase. *Int. J. Radiat. Biol.* 29: 297–299, 1976.
9. Spitz DR, Azzam EI, Li JJ, and Gius D. Metabolic oxidation/reduction reactions and cellular responses to ionizing radiation: A unifying concept in stress response biology. *Cancer Metastasis Rev.* 23: 311–322, 2004.

10. Leach JK, Black SM, Schmidt-Ullrich RK, and Mikkelsen RB. Activation of constitutive nitric-oxide synthase activity is an early signaling event induced by ionizing radiation. *J. Biol. Chem.* 277: 15400–15406, 2002.

11. Mikkelsen RB and Wardman P. Biological chemistry of reactive oxygen and nitrogen and radiation-induced signal transduction mechanisms. *Oncogene* 22: 5734–5754, 2003.

12. Clutton S, Townsend K, Walker C, Ansell J, and Wright E. Radiation-induced genomic instability and persisting oxidative stress in bone marrow cultures. *Carcinogenesis* 17: 1633–1639, 1996.

13. Ward JF. The complexity of DNA damage: Relevance to biological consequences. *Int. J. Radiat. Biol.* 66: 427–432, 1994.

14. Tuttle SW, Varnes ME, Mitchell JB, and Biaglow JE. Sensitivity to chemical oxidants and radiation in CHO cell lines deficient in oxidative pentose cycle activity. *Int. J. Radiat. Oncol. Biol. Phys.* 22: 671–675, 1992.

15. Varnes ME. Inhibition of pentose cycle of A549 cells by 6-aminonicotinamide: Consequences for aerobic and hypoxic radiation response and for radiosensitizer action. *NCI Monogr.* 6: 199–203, 1988.

16. Tuttle SW, Stamato T, Perez ML, and Biaglow J. Glucose-6-phosphate dehydrogenase and the oxidative pentose phosphate cycle protect cells against apoptosis induced by low doses of ionizing radiation. *Radiat. Res.* 153: 781–787, 2000.

17. Li WG, Miller FJ, Zhang HJ, Spitz DR, Oberley LW, and Weintraub NL. H(2)O(2)-induced O(2) production by a non-phagocytic NAD(P)H oxidase causes oxidant injury. *J. Biol. Chem.* 276: 29251–29256, 2001.

18. Azzam EI, de Toledo SM, Spitz DR, and Little JB. Oxidative metabolism modulates signal transduction and micronucleus formation in bystander cells from alpha-particle irradiated normal human fibroblasts. *Cancer Res.* 62: 5436–5442, 2002.

19. Azzam EI, de Toledo SM, and Little JB. Oxidative metabolism, gap junctions and the ionizing radiation-induced bystander effect. *Oncogene* 22: 7050–7057, 2003.

20. Beckman JS and Koppenol WH. Nitric oxide, superoxide, and peroxynitrite: The good, the bad, and ugly. *Am. J. Physiol. Cell Physiol.* 271: C1424–C1437, 1996.

21. Janssen YMW, Soultanakis R, Steece K, Heerdt E, Singh RJ, Joseph J, and Kalyanaraman B. Depletion of nitric oxide causes cell cycle alterations, apoptosis, and oxidative stress in pulmonary cells. *Am. J. Physiol.* 275: L1100–L1109, 1998.

22. Durocq C, Blanchard B, Pignatelli B, and Ohshima H. Peroxynitrite: An endogenous oxidizing and nitrating agent. *Cell Mol. Life Sci.* 55: 1068–1077, 1999.

23. Ibuki Y and Goto R. Enhancement of NO production from resident peritoneal macrophages by in vitro gamma-irradiation and its relationship to reactive oxygen intermediates. *Free Radic. Biol. Med.* 22: 1029–1035, 1997.

24. Collins SP and Uhler MD. Cyclic AMP- and cyclic GMP-dependent protein kinases differ in their regulation of cyclic AMP response element-dependent gene transcription. *J. Biol. Chem.* 274: 8391–8404, 1999.

25. Lander HM, Ogiste JS, Teng KK, and Novogrodsky A. p21 as a common signaling target of reactive free radicals and cellular redox stress. *J. Biol. Chem.* 270: 21195–21198, 1995.

26. Barrett WC, DeGnore JP, Keng Y-F, Zhang Z-Y, Yim MB, and Chock PB. Roles of superoxide radical anion in signal transduction mediated by reversible regulation of protein–tyrosine phosphatase 1B. *J. Biol. Chem.* 274: 34543–34546, 1999.

27. Wang T, Zhang X, and Li JJ. The role of NF-kappaB in the regulation of cell stress responses. *Int. Immunopharmacol.* 2: 1509–1520, 2002.

28. Wei SJ, Botero A, Hirota K, Bradbury CM, Markovina S, Laszlo A, Spitz DR, Goswami PC, Yodoi J, and Gius D. Thioredoxin nuclear translocation and interaction with redox factor-1 activates the activator protein-1 transcription factor in response to ionizing radiation. *Cancer Res.* 60: 6688–6695, 2000.
29. Karimpour S, Lou J, Lin LL, Rene LM, Lagunas L, Ma X, Karra S, Bradbury CM, Markovina S, Goswami PC, Spitz DR, Hirota K, Kalvakolanu DV, Yodoi J, and Gius D. Thioredoxin reductase regulates AP-1 activity as well as thioredoxin nuclear localization via active cysteines in response to ionizing radiation. *Oncogene* 21: 6317–6327, 2002.
30. Wu L-J, Randers-Pehrson G, Xu A, Waldren CA, Geard CR, Yu Z, and Hei TK. Targeted cytoplasmic irradiation with alpha particles induces mutation in mammalian cells. *Proc. Natl. Acad. Sci. USA* 96: 4959–4964, 1999.
31. Rossman T and Goncharova T. Spontaneous mutagenesis in mammalian cells is caused mainly by oxidative events and can be blocked by antioxidants and metallothionein. *Mutat. Res.* 402: 103–110, 1998.
32. Babior BM. NADPH oxidase: An update. *Blood* 93: 1464–1476, 1999.
33. Albanese J and Dainiak N. Modulation of intercellular communication mediated at the cell surface and on extracellular, plasma membrane-derived vesicles by ionizing radiation. *Exp. Hematol.* 31: 455–464, 2003.
34. Stark G. The effect of ionizing radiation on lipid membranes. *Biochim. Biophys. Acta* 1071: 103–122, 1991.
35. Rémita S. De la peroxydation lipidique radioinduite: Les facteurs déterminante l'oxydabilité des lipides. *Can. J. Physiol. Pharmacol.* 79: 141–153, 2001.
36. Ianzini F, Guidoni L, Indovina PL, Viti V, Erriu G, Onnis S, and Randaccio P. Gamma-irradiation effects on phosphatidylcholine multilayer liposomes: Calorimetric, NMR, and spectrofluorimetric studies. *Radiat. Res.* 98: 154–166, 1984.
37. Chatterjee SN and Agarwal S. Liposomes as membrane model for study of lipid peroxidation. *Free Radic. Biol. Med.* 4: 51–72, 1988.
38. Stanimirovic DB, Wong J, Ball R, and Durkin JP. Free radical-induced endothelial membrane dysfunction at the site of blood-brain barrier: Relationship between lipid peroxidation, Na, K-ATPase activity, and 51Cr release. *Neurochem. Res.* 20: 1417–1427, 1995.
39. Chandra S and Stefani S. Plasma membrane as a sensitive target in radiation-induced cell injury and death: An ultrastructural study. *Int. J. Radiat. Biol. Relat. Stud. Phys. Chem. Med.* 40: 305–311, 1981.
40. Haimovitz-Friedman A. Radiation-induced signal transduction and stress response. *Radiat. Res.* 150: S102–S108, 1998.
41. Siskind LJ, Kolesnick RN, and Colombini M. Ceramide channels increase the permeability of the mitochondrial outer membrane to small proteins. *J. Biol. Chem.* 277: 26796–26803, 2002.

6

Damage to DNA by Ionizing Radiation

6.1 Mechanisms of DNA Damage: Physicochemical Relationships

Ionizing radiation deposits its energy, in the form of excitations and ionizations, in matter in amounts almost directly proportional to the mass of each molecular species present. Excitations do not appear to play a role in causing biologically significant damage but ionizations can occur in any molecule in the cell creating a cation radical.

$$RH \rightarrow RH^+ + electron$$

The electron produced by ionization can attach to another molecule or it can become solvated before reacting. The cation radical represents an alteration in the molecule ionized (a direct effect) and the radical site can be transferred to another molecule.

$$RH^+ + X \rightarrow RH + X^+$$

Cation radicals can also react by losing a proton to become a neutral radical, as is the case of the initial radical cation produced by the ionization of water. H_2O^+ reacts immediately (10^{-14} s) with a neighboring water molecule to produce the hydroxyl radical ($^\bullet OH$) as described in Chapter 2.

$$H_2O^+ + H_2O \rightarrow H_3O^+ + {}^\bullet OH$$

As a result of these reactions, cellular DNA can be damaged in several ways:

1. Direct ionization of the DNA
2. Reaction with electrons or solvated electrons
3. Reactions of $^\bullet OH$ or H_2O^+
4. Reactions with other radicals

6.1.1   Mechanisms of Damage Induction: Chemical End Points

In the early days of radiation chemistry, studies were carried out in two types of conditions, irradiating the DNA molecule itself (direct effect) or irradiating it in dilute aqueous solution (indirect effect). The direct effect was considered to result from the deposition of energy directly in the DNA, while the indirect effect resulted from reactions of radicals produced by ionizations of the solvent molecules. This basic approach using pure systems provided a great deal of information about mechanisms of formation of damage products, much of which could be transferred to the intracellular context. Examples of insights provided by these studies include the following:

1. The initial reactions of $^\bullet$OH radicals with DNA constituents were determined. The $^\bullet$OH radical adds to unsaturated bonds of the bases at almost diffusion-controlled rates and there seems to be little preference for one base over another [1,2]. Subsequent reactions of the DNA radicals, formed with oxygen or other radicals, lead to the final products.

2. At the same time, $^\bullet$OH radicals abstract hydrogen atoms from all sites on the deoxyribose moiety of DNA at a rate about 25% of the diffusion-controlled rate [1], again subsequent reactions with oxygen or other radicals lead to the final products.

3. Direct ionization of the DNA produces cation radicals, many of which were studied by electron spin resonance techniques. It is believed that the most probable cation radical in DNA is in guanine [3] either as a result of initial ionization of guanine or from radical transfer to guanine from other ionization sites.

A modification of the classification of the mechanisms by which damage is produced in DNA has been proposed by Becker and Sevilla [3] who suggested that three effects can be considered: direct, indirect, and quasi-direct. The first two terms have the same meaning as before while quasi-direct refers to an ionization that occurs initially in a molecule adjacent to DNA, but which then transfers its radical site to the DNA. For instance, ionizations transferred from adjacent water molecules close to DNA would be classified as quasi-direct although it is likely that many of the DNA radicals produced would be the same as those produced by direct ionization.

The radiochemical consequences of direct ionization of DNA and its associated water were investigated by Swarts and coworkers [4,5] who examined the effects of hydration on radiation-induced base release from DNA. (Base release is an indirect measurement of single-strand breakage [SSB].) On the basis of their results, a model was proposed of two mechanistically distinct compartments of water, a hydration layer of tightly bound waters with a water/deoxynucleotide ratio of 12–15, and more distant compartment of more loosely bound water (Figure 6.1). Based on this

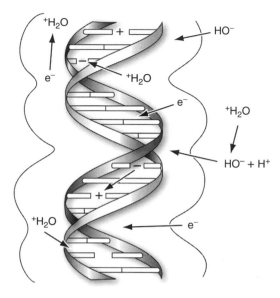

FIGURE 6.1
Schematic representation of free radicals produced by ionizing radiation and the two regions of hydration around the DNA molecules. The inner 15 water molecules are more tightly bound to DNA than the outer water molecules.

model, assuming the tightly bound water to be also present in cellular DNA, and the damage from this compartment to be nonscavengable, the yield of nonscavengable SSB was calculated for DNA within a mammalian cell to be 337/cell/Gy, which agrees well with cellular data, where the yield of SSB was found to be 1000/cell/Gy [6] of which two-thirds is caused by scavengable •OH radicals [7]. (Scavengers are compounds that react with radicals before they can reach their chemical or intracellular target.)

6.1.2 Mechanisms of Damage Induction: Cellular End Points

In paralleling radiation chemistry studies, work in cellular systems probed the source of the radiation damage causing cell death. The history of these explorations is described by Alper [8]. The first mechanistic investigation was that of Johansen [9] who showed that *Escherichia coli* was protected against killing by radiation by the addition to the medium of compounds which scavenge free radicals and that approximately 65% of the cell killing was caused by •OH radicals whereas no radioprotection was afforded when scavengers of electrons were used. Later, Roots and Okada [7], and Chapman and coworkers [10] carried out similar experiments in mammalian cells and came to the same conclusion that 65% of the cell killing was caused by •OH radicals. These studies have been extended to other systems and it has been shown that for low linear energy transfer (LET) radiation such as x-rays or γ-rays, high concentrations of •OH radical scavengers can, as well as protecting against cell killing, also reduce by a factor of 3, the yield of SSBs [7], double-strand breaks (DSBs) [11,12], mutations [11], and chromosome aberrations [13].

This data suggests another pragmatic classification of DNA damage origins in cells [14]. This assumes that the reduction of radiosensitivity achieved using radical scavengers reflects only the fraction of damage that originates from energy deposited in cellular water and that any damage that is a consequence of radical transfer from ionized water molecules next to DNA is not affected. On this basis, damage can simply be classified as scavengable and nonscavengable.

6.2 Types of DNA Damage

Ionizing radiation produces a wide range of damage in DNA and other macromolecules. Some of the possibilities are shown schematically in Figure 6.2 and listed in Table 6.1.

6.2.1 Simple Damages to DNA: Base Damage and Single-Strand Breaks

On the basis of assumed contributions from scavengable and nonscavengable reactions and extrapolated relative yields from studies in model systems, it was concluded that the total yield of base damage in mammalian cells should be on the order of 2.5–3 times the yield of SSBs. Thus with the

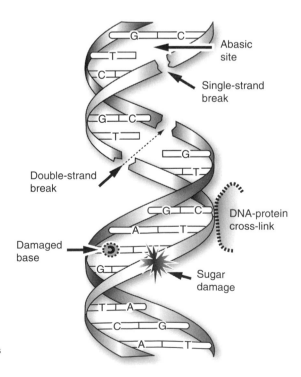

FIGURE 6.2
DNA lesions in eukaryotic cells
exposed to ionizing radiation.

TABLE 6.1

Estimation of the Number of Early Physical and
Biochemical Changes That Occur When Mammalian
Cells Are Irradiated with 1 Gy of Low LET Radiation

Initial Physical Damage	
Ionization in the cell nucleus	~1000,000
Ionization directly in DNA	~2000
Excitation directly in DNA	~2000
Selected Biochemical Damage	
Damaged bases	1000–2000
Damaged sugars	1200
DNA single-strand breaks (SSBs)	1000
Alkali-labile sites	250
Double-strand breaks (DSBs)	40
DNA-protein cross-links (DPC)	150
Selected Cellular Effects	
Lethal events	~0.2–0.8
Chromosome aberrations	~1
Hypoxanthine phosphoribosyltransferase (Hprt) aberrations	10^{-5}

yield of SSBs taken as 1000/cell/Gy, the yield of base damage would be 2500–3000/cell/Gy [15]. These values may not be representative of intracellular levels since results of several studies have indicated that the yields of base damage per strand break are higher for irradiated cells than for DNA irradiated in an aqueous solution [16].

As described in the previous section, 60%–70% of the cellular DNA damage produced by ionizing radiation is caused by hydroxyl radicals produced by the radiolysis of water. Chemical studies using nucleoside and nucleotide derivatives of DNA, have identified more than 100 free radical-induced products of DNA radiolysis including many examples of damage to the purine and pyrimidine rings, SSB, and sites of base loss (reviewed by Wallace [17]). A significant number of these damaged sites are stable, and thus when present in DNA they could contribute to the biological consequences of radiation damage. However, the damage produced by radiation-induced hydroxyl radicals is similar to that produced during oxidative metabolism [18], (although the relative yields of the individual products appear to be different). At the mean lethal dose of ionizing radiation for mammalian cells, the number of free radical-induced DNA-damaged sites is low compared to the spontaneous background of about 10,000–150,000 oxidative DNA damaged sites per human cell per day [19]. A DNA repair system, base excision repair (BER) (described in Chapter 7), which efficiently repairs lethal lesions, has evolved to counteract the effects of oxidative damage produced by free radicals so that in repair-proficient cells, all DNA damage produced by hydroxyl radicals is removed. Consequently, the radiation-induced DNA damage which results from free-radical attack on the DNA has not been thought to be important to the lethal and mutagenic consequences of ionizing radiation. Among prokaryotes, mutants devoid of BER are very radiosensitive [20] and in fact, are as radiosensitive as recombination-deficient mutants demonstrating that BER can be responsible for repairing potentially lethal irradiation-induced DNA lesions. As described below, a characteristic of radiation damage is the formation of more than one damaged moiety in proximity and this clustered damage has much greater potential for lethality than do single lesions.

Damage to bases by ionizing radiation has been extensively studied in vitro by irradiation of free bases, nucleosides, oligonucleotides, or DNA in aqueous solution under aerobic or anaerobic conditions, or in the solid state. The chemistry of lesion formation is fairly well understood and detailed accounts can be found in the literature (reviewed by von Sonntag [1], Téoule [21], and Wallace [17]). The structures of a number of stable free-radical damaged purine and pyrimidine bases are shown in Figure 6.3.

Many products and short-lived intermediates have been characterized for thymine. The •OH radical can attack the double bond of thymine at C-5 or C-6 and less frequently can abstract hydrogen from the methyl group. The 6-hydroxythymine intermediate can react with oxygen to yield thymine glycol (5,6-dihydroxy-5,6-dihydrothymine) a product which has been extensively studied (Figure 6.4). It has been suggested that the direct action of ionizing radiation may lead to the ejection of an electron from the unsaturated C-5 or C-6 position and the resulting cation radical may further

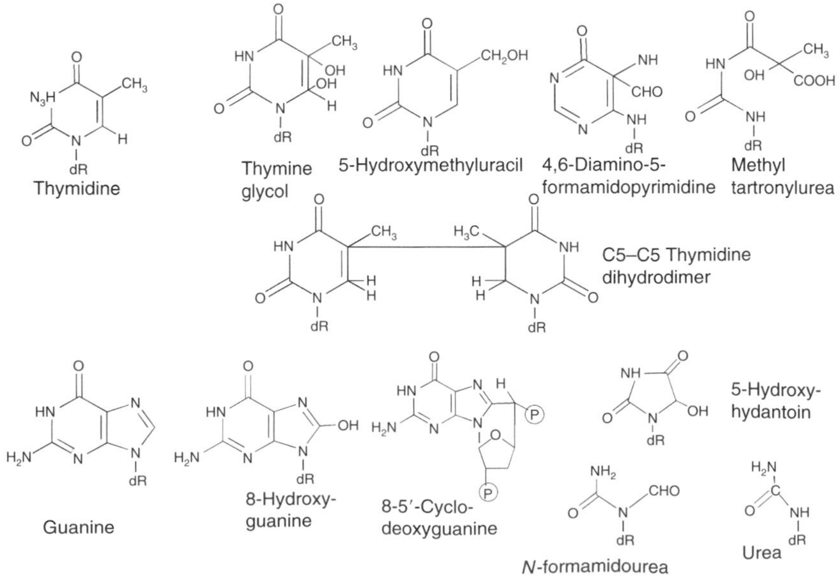

FIGURE 6.3
Examples of base damage induced by ionizing radiation.

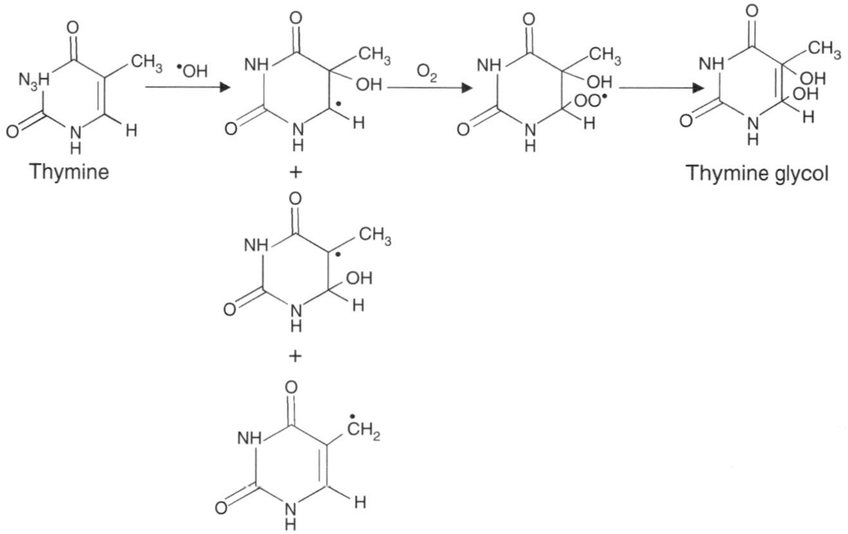

FIGURE 6.4
Formation of thymine glycol by •OH radical attack on thymine.

react with a hydroxyl ion [15]. Thus, direct and indirect radiation effects may result in identical reactive intermediates. The saturated ring can undergo further fragmentation and give rise to products such as methyltartronylurea, 5-hydroxyhydantoin (a five-ring derivative), N-formamidourea, and urea (Figure 6.3). A different spectrum of ring-saturated thymine derivatives is found in oxygen-free solutions; the products 5-hydroxythymine and 6-hydroxy-5,6-dihydrothymine detected under these conditions may be of particular biological importance [22]. 5-Hydroxymethyluracil is a less frequent lesion formed by attack with γ-rays or by radiolytic decay of [6-^{3}H]thymine in DNA [23].

There have been fewer studies on the effects of ionizing radiation-mediated damage to cytosine and other purines, although it has been established with respect to cytosine that the major site for attack by hydroxyl radicals is again the C-5, which is equal to C-6 double bond, as in the case of thymine [1,21]. Treatment of DNA with free-radical generating systems can result in the formation of an imidazole-opened ring derivative of guanine; 2,6-diamino-4-hydroxy-5-formamidopyrimidine (FaPy) as an abundant lesion. 4,6-Diamino-5-formamidopyrimidine is another biologically important purine derivative as is the mispairing lesion 8-hydroxyguanine (Figure 6.5). Purine residues can also undergo cyclization, giving rise to products such as 8,5′-cyclodeoxyguanosine or 8,5′-cyclodeoxyadenosine, which probably cause a distortion of the normal double helical structure of DNA [1,21]. Table 6.2 shows the yield of various single base products after γ irradiation of human cells.

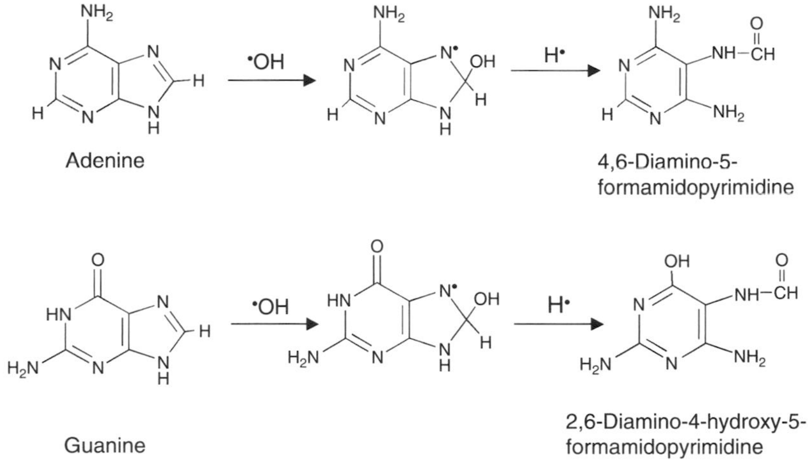

Adenine   → •OH →   → H• →   4,6-Diamino-5-
formamidopyrimidine

Guanine   → •OH →   → H• →   2,6-Diamino-4-hydroxy-5-
formamidopyrimidine

FIGURE 6.5

Imidazole ring opening in adenine and guanine following radical attack to yield the products shown.

6.2.2 Apurinic or Apyrimidinic Sites

Destabilization of the glycosyl bond and the formation of abasic deoxyribose residues can occur as a result of damage to DNA bases such as ring saturation. Base loss also results from radiation or spontaneous hydrolysis of *N*-glycosidic bonds to generate apurinic or apyrimidinic (AP) sites. AP sites are also generated by the first enzymatic step of BER and the repair of AP sites requires the completion of BER, which is a multistep process (Chapter 7). Sites of base loss are shown in Figure 6.6.

TABLE 6.2

Yields of DNA Base Products Following Irradiation
with 420 Gy γ-Rays

Product	No. of Molecules/10^5 DNA Bases	Fold Increase Over Background
FaPy guanine	34.4	13
8-Hydroxyguanine	23.3	3
5-Hydroxyhydantoin	23.2	2
Thymine glycol	10.2	6
FaPy adenine	10.0	3
8-Hydroxyadenine	5.5	2
2-Hydroxyadenine	4.9	2
5-Hydroxycytosine	4.7	2
5,6-dihydroxycytosine	4.1	13
5-Hydroxymethyluracil	2.8	4
5-Hydroxyuracil	1.8	5

Source: From Dizdaroglu, M., *Mutat. Res.*, 275, 331, 1992. With permission.

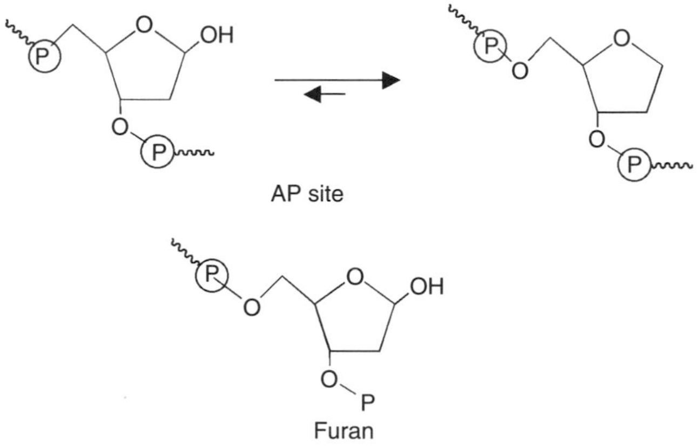

AP site

Furan

FIGURE 6.6
Sites of base loss resulting from cleavage of *N*-glycosyl bond by DNA-glycosylases, enzymes which recognize free-radical damaged purines or pyrimidines.

6.2.3 Modifiers of Radiation Effect

The biological effectiveness of ionizing radiation is markedly dependent on whether oxygen is present during radiolysis. Oxygen is a biradical species and, as such, reacts readily with many free radicals, including radiation-induced base-hydroxyl adduct radicals. In addition, oxygen is an avid scavenger of reducing radicals, such as the hydrated electron and the hydrogen atom. The manner in which oxygen sensitizes cells is related to its reactivity with DNA radicals leading to damage fixation. Intracellular radioprotectors, largely protein and nonprotein thiols, such as glutathione, can efficiently counteract damage induced by radicals at various levels. The amount of radical formed in deoxyribose or base residues that is amenable to the formation of stable, deleterious DNA damage is determined by the competition of thiol groups with oxygen [1,15]. In the absence of oxygen, it is believed that reducing species, such as the tripeptide glutathione, can react with radicals chemically repairing a molecule which has been damaged by hydrogen atom abstraction.

$$R^\bullet + XSH^- \rightarrow RH + XS^\bullet \text{ in the absence of oxygen}$$

as opposed to

$$R^\bullet + O_2 \rightarrow RO^\bullet \text{ in the presence of oxygen.}$$

The reaction of thiols with radical sites, however, does not necessarily reconstitute the parent molecule. The addition of a hydrogen atom to a base-hydroxyl adduct radical, for instance, forms a hydrate and not the original base and in the case of cytosine, the hydrate can break down to form uracil. In another case, hydrogen atom donation to a deoxyribose site from which a hydrogen has been abstracted can lead to the formation of an isomer of the sugar, posing another potential challenge to cellular repair processes [24].

6.2.4   DNA Strand Breaks

The earliest measurements of damage in the DNA of irradiated cells were of the induction of SSBs. Determinations were readily made from changes in the average length of single strands determined by velocity gradient sedimentation through alkaline sucrose gradients. Later, more sensitive techniques for the measurement of SSBs, such as alkaline unwinding and alkaline elution, were used (described in Chapter 5). These techniques measure DNA size, and are a means to accurately determine the number of breaks but provide little information about the nature of the damage induced or the mechanisms of its formation.

In order to cause a strand break directly, the site of damage must be the deoxyribose (Figure 6.7). Initial radical reactions with the deoxyribose moieties which lead to SSBs involve abstraction of a hydrogen atom from the deoxyribose moiety by an •OH radical [1]. Later, the radical can react with oxygen and form a peroxy radical. Possible mechanisms whereby strand breakage can arise from abstraction of any of the deoxyribose hydrogens include abstraction from the 2′-position of deoxyribose which is expected to be energetically less favorable than abstraction from the other sites and, as there are at least three products of deoxyribose damage, it is probable that

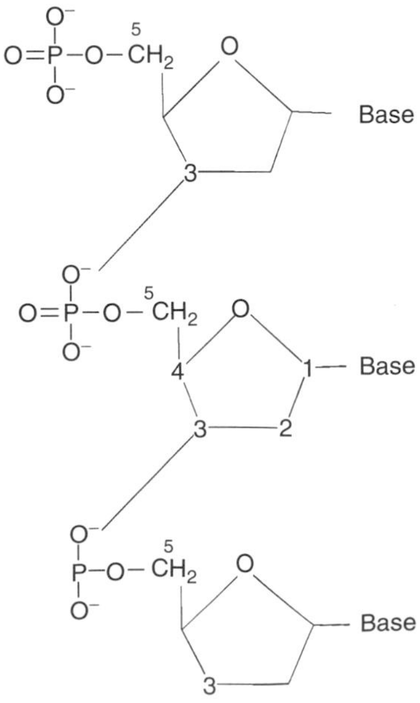

FIGURE 6.7
Breakage of the bond between the C-3–C-4 or between the C-4–C-5 can produce a break in the polynucleotide chain (single-strand break [SSB]).

attack on more than one site is involved, although the so-called 4'-mechanism is favored in the literature (reviewed by Ward [25]).

Damage to the DNA bases such as ring saturation resulting in destabilization of the *N*-glycosidic bond and the formation of abasic deoxyribose residues, (described above) and other lesions which are the result of direct attack by •OH radicals can be converted into strand breaks by hot-alkali treatment and are referred to as alkali-labile sites. Irradiation in the presence of oxygen results increase in the number of alkali-labile sites by a factor of 4 [26].

The majority of strand breaks induced by ionizing radiation are characterized by unusual or damaged termini which preclude repair by a simple DNA ligation step [1,27]. Although most of the 5' ends retain phosphate groups OH groups at 3' have been replaced by phosphate or phosphoglycolate groups [28]. The phosphate is also often entirely lost, and various fragmented sugar derivatives remain. In addition, the terminal base residue is frequently missing, and many strand breaks are in fact single nucleotide gaps [1,27] (Figure 6.8).

Many studies of radiation damage in DNA to individual sites have yielded much important data but have also brought the realization that these lesions when produced in mammalian cells at the levels induced by biologically relevant radiation doses are unimportant. This was demonstrated by experiments in which SSBs introduced into the DNA of mammalian cells by treatment with hydrogen peroxide at 0°C were found not to cause cell killing [29]. The production of these breaks is inhibited by the presence of an •OH radical scavenger, indicating that the intermediate reactive species causing the breaks is indeed an •OH radical.

In fact, lesions in cellular DNA cell present after a lethal dose of any genotoxic agent appear to be of two classes: those that cause singly

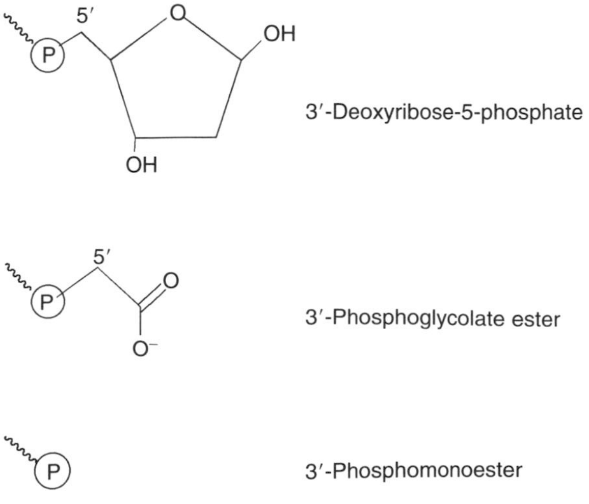

3'-Deoxyribose-5-phosphate

3'-Phosphoglycolate ester

3'-Phosphomonoester

FIGURE 6.8
Single-strand breaks produced in DNA. The major products are phosphate or phospoglycolate attached to the 3' side of the break.

damaged sites and those that produce multiply damaged sites (MDS) (such as DSBs). Whereas for singly damaged sites at a lethal dose, the numbers of lesions present per cell are of the order of 10^5–10^6; the corresponding number of MDS is less than 100, indicating the much greater biological effectiveness of MDS [25].

6.3   Double-Strand Breaks and Other Multiply Damaged Sites

As described in Chapter 2, ionizing radiation energy is neither deposited homogeneously nor is the amount deposited quantized. It is deposited in events that range in energies up to hundreds of eV (even for low LET radiation) with the average amount being 60 eV. Since the energy to form an ion pair (H_2O^+ and an electron) from water is ~20 eV, the average energy deposited per event is sufficient to produce approximately three ion pairs. The average half-life of the $^\bullet$OH radical in the mammalian cell is 8.5×10^{-9} s and during this time, the average distance that a scavengable $^\bullet$OH radical can move before it reacts with the DNA is 3.1 nm. If the $^\bullet$OH radical arose from an event in which several ion pairs were produced, it is likely that other radicals present in the vicinity will react within the same region of the DNA with the spacing between the positions at which two radicals can react varying up to about 10 nm [30]. The deposition of the energy in volumes of nanometer dimensions means that clusters of reactive species (radicals) are likely to be produced at high local concentrations. The realization that low LET radiation can produce hot spots in which clusters of ionizations may occur within a diameter of a few nanometers implies that a serious and non-repairable lesion could be produced if the localized energy deposition impinges on the DNA molecule (Figure 6.9). A simple example is the induction of a DSB by a single track of radiation as the result of a localized attack by two or more $^\bullet$OH radicals. In this case, an alternative mode of damage would be by a hybrid attack where one strand is damaged by the $^\bullet$OH radical and the other strand sustains direct damage within about 10 base pairs of the $^\bullet$OH attack.

Lesions which might involve one or more DSBs, several SSBs as well as base damage were originally called locally MDS and are now more usually referred to as MDS—a term coined by Ward in 1986 to describe this localized damage. The importance of MDS depends on the fact that unlike the majority of DNA lesions produced by endogenous or exogenous causes which are repaired, these lesions may fail to repair and hence be potentially lethal. MDS consist of mixtures of lesions, the individual types of damage within the MDS will be the same as those produced in singly damaged sites.

The distance in chromatin that $^\bullet$OH radicals can move before reacting with the DNA under in vivo conditions in the cell nucleus is believed to be controlled by the presence of low molecular compounds, histones, and polyamines. It was hypothesized that the major factor controlling the

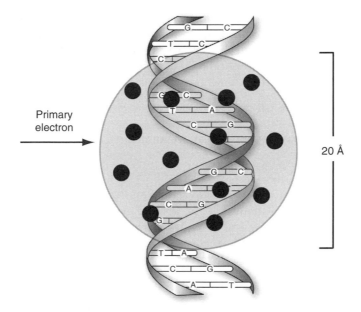

FIGURE 6.9
Schematic representation of potential multiply damaged site (MDS) produced by cluster of ionizations impinging on DNA.

distance in DNA that •OH can travel are the polyamines which act by neutralizing the negative charge on the macromolecule [31]. (For further discussion of the role of chromatin in the modulation of DNA damage, see Chapter 9.)

6.3.1 Distribution of MDS

A number of studies have considered whether differences in DSB distribution occur in different parts of the genome. The results have generally been in agreement that after low LET radiation, the overall distribution of DSBs within the genome is random (reviewed by Prise et al. [32]). In one case yields and distribution of breaks in the *Myc* gene relative to the whole genome were investigated [33]. No difference in the yield of DSBs was found but there was evidence of nonrandom distribution of breaks within the *Myc* gene. In particular it appeared that DNA close to the matrix attachment regions was more resistant to breakage than the rest of the gene. Significantly, differences were also observed in the repair kinetics with the fraction of residual breaks in the *Myc* gene being 30% higher than that in the genome overall. These variations, which are related to the fine structure of chromatin, are discussed further in Chapter 9.

In contrast, studies measuring distance between DSBs in total cellular DNA after exposure to high LET radiation by fragmentation analysis support the general conclusion that the higher order chromatin structure determines the distribution of DSBs for high LET particle tracks. A complex pattern of DSB distribution is closely related to the environment within the nucleus at the time of irradiation. Clustering of damage both at the level of the DNA helix (locally MDS) and over larger distances related to chromatin structure (regional MDS) appears to be involved.

6.3.2 Clustered Damage in DNA of Mammalian Cells

Until now, DSBs were the only clustered damage which could be measured in mammalian cells. Sutherland and coworkers [34] have developed and validated a method for measuring complex damages in large DNAs. By this method, clustered lesions were identified by treating irradiated DNA with lesion-specific enzymes, endonucleases, which make single strand cuts at oxidized bases *E. coli Nth* protein (endonuclease III) or abasic sites *E. coli Nfo* protein (endonuclease IV). At each cluster site, the enzyme produces two closely spaced single strand nicks on opposing strands. The frequency of DSBs (each resulting from a cluster site) was measured by quantitative gel electrophoresis to determine the number average molecular weight. By this method, it was shown that ionizing radiation produces clustered DNA damages containing abasic sites and oxidized purines and pyrimidines and that the frequencies of each of these cluster classes are equivalent to that of frank DSBs (Table 6.3). Clustered lesions were demonstrated by this method in DNA in solution and after low doses (0.1–1.0 Gy) of high LET irradiation in human cells.

The data showed that firstly, sparsely ionizing radiation like x-rays at realistic dose levels (1.0 Gy) induces clustered damages in addition to DSBs.

TABLE 6.3

Relative Clustered Damage Frequencies in DNA in Solution and in Human Cells

Complex Damage	Relative Cluster Frequencies			Percent Complex Damage
	DNA in PO₄ Buffer	DNA in Tris	Human Cells	Human Cells
Double-strand break	1	1	1	27.5
Fpg-oxypurine cluster	2	1.4	1	27.8
Nth-oxypyrimidine cluster	0.6	0.75	0.9	24.7
Nfo-abasic cluster	1.5	0.4	0.75	20

Source: From Sutherland, B.M., Bennett, P.V., Sutherland, J.C., and Laval, J., *Radiat. Res.*, 157, 611, 2002. With permission.

Note: DNAs from control and irradiated cells were isolated. Each sample was divided and half was treated with *E. coli* Nth protein while the other half was incubated without enzyme. In the irradiated cell, Nth protein released damaged bases (producing abasic sites) and its accompanying ligase produced a nick in the phosphodiester backbone at the abasic site. The samples were then analyzed on a neutral agarose gel, stained, and a quantitative electronic image is obtained from which the frequency of damages was calculated by number average length analysis of DNA distribution.

Secondly, the dose–response relationships for all the damages measured were linear indicating the clustered damage does not require multiple radiation hits but can originate from a single photon or particle and its resulting cloud of radicals. Thirdly, measured frequencies for all non-DSB clusters were in the same range as DSBs. The data also indicated that the relative frequencies of different cluster types are affected by the milieu surrounding the DNA.

6.4 DNA-Protein Cross-Links

The first reported studies of DNA-protein cross-links (DPC) used extremely high doses/supra–lethal doses of radiation (>30 Gy) and demonstrated the induction of DPCs in aerated cells [35]. It was later shown that ionizing radiation-induced DPCs in hypoxic cells more efficiently (from 1.5- to 5.5-fold) than in aerated cells [36]. Studies of DPCs induced by various agents showed a half-life from hours to days. Actively transcribing DNA regions were found to be more susceptible to DPC induction and cells in metaphase had a higher DPC background and a slower rate of induction by radiation. Recent studies have used sensitive methods to quantitate DPCs after lower doses of radiation (1–4 Gy). In a proteomic study, proteins which were covalently cross-linked were identified by mass spectrometry [37]. Twenty-nine proteins were identified as being cross-linked to DNA including structural proteins, regulators of transcription, stress response, and cell-cycle regulatory proteins. There was a linear dose–response relationship in the low dose range (0–1.5 Gy) for hamster and human cells and in these experiments no difference was seen in the induction of DPCs under aerated and hypoxic conditions.

Cross-linking of DNA to nuclear proteins will seriously affect DNA processing at the levels of replication, transcription, and repair but the role of radiation-induced DPCs in the overall radiation response has yet to be defined.

6.5 Summary

Ionizing radiation damages cellular DNA in a variety of ways, including direct ionization of the DNA, by reaction with electrons or solvated electrons, by reactions of $^{\bullet}OH$ or H_2O^+; and by reactions with other radicals. The $^{\bullet}OH$ radical adds to unsaturated bonds of the bases and also abstracts hydrogen atoms from all sites on the deoxyribose moiety of DNA; subsequent reactions with oxygen or other radicals lead to the final products of these reactions. Studies with bacterial and mammalian cell systems showed approximately 65% of cell kill to be caused by $^{\bullet}OH$ radicals. For low LET radiation the presence of high concentrations of •OH radical scavengers can, as well as protecting against cell killing, reduce by a factor of 3 the yield of SSBs DSBs mutations, and chromosome aberrations.

Chemical studies using nucleoside and nucleotide derivatives of DNA have identified more than 100 free radical-induced products of DNA radiolysis, including many examples of damage to the purine and pyrimidine rings, SSBs, and sites of base loss. However, the damage produced by radiation-induced hydroxyl radicals is similar to that produced during oxidative metabolism and at the mean lethal dose of ionizing radiation for mammalian cells the number of free radical-induced DNA-damaged sites is low compared to the spontaneous background. For both pyrimidines and purines, the C5–C6 double bond is the major site for •OH radical attack yielding a number of products and short-lived intermediates of which thymine glycol, (5,6-dihydroxy-5,6-dihydrothymine) has been the most extensively studied. Another example of free radical-generated damage is the formation of an imidazole-opened ring derivative of guanine; 2,6-diamino-4-hydroxy-5-formamidopyrimidine (FaPy). Destabilization of the glycosyl bond and base loss as a result of radiation or spontaneous hydrolysis of *N*-glycosidic bonds results in AP sites.

The biological effectiveness of ionizing radiation is dependent on the presence of oxygen during radiolysis. Oxygen sensitizes cells by reacting with radicals which formed in deoxyribose or base residues. The extent to which radicals that are converted to stable, deleterious DNA damage by oxygen fixation is determined by the competition for free-radical sites by endogenous radioprotective thiol groups.

Single-strand DNA breaks result from damage to the deoxyribose component of DNA. The majority of strand breaks induced by ionizing radiation are characterized by unusual or damaged termini which must be processed before repair can be completed by ligation. Singly damaged sites produced in cellular DNA by ionizing radiation are readily and accurately repaired after biologically significant radiation doses as would be expected since they are the same lesions as those produced by endogenous oxidation. The more significant radiation products are DSBs and other MDS that have several degrees of complexity and are problematic for the cell to repair. After low LET radiation, the overall distribution of DSBs within the genome is generally believed to be random. In contrast, studies measuring distance between DSBs in total cellular DNA after exposure to high LET radiation by fragmentation analysis support the general conclusion that the higher order chromatin structure determines the distribution of DSBs for high LET particle tracks. Clustering of damage both at the level of the DNA helix (locally MDS) and over larger distances related to chromatin structure (regionally MDS) appears to be involved.

The data from a recent study of clustered damage showed that sparsely ionizing radiation at realistic dose levels (1.0 Gy) induced clustered damages including DSBs. Dose–response relationships for all the damages measured were linear indicating the clustered damage does not require multiple radiation hits but can originate from a single photon or particle and its resulting cloud of radicals. The measured frequencies for all non-DSB clusters were in the same range as DSBs. This in vivo demonstration of the induction of clustered damage by x-rays confirms what was postulated on the basis of theoretical modeling, physicochemical measurement, and experiments with purified enzymes and synthetic oligonucleotides that damage by ionizing radiation has a unique capability for creating genotoxic damage.

DNA-protein cross-links were formerly studied using very high doses of radiation. Recently, using more sensitive methods, DPCs have been demonstrated to be formed by doses in the clinically significant dose range of 1–4 Gy. A proteomic study identified approximately 30 proteins as being cross-linked to DNA with a linear dose–response relationship in the low dose range (0–1.5 Gy) for hamster and human cells under aerated and hypoxic conditions. No difference was seen in the induction of DPCs under these conditions. Cross-linking of DNA to nuclear proteins will seriously affect DNA processing such as replication transcription and repair but the role of radiation-induced DPCs in the overall radiation response has yet to be defined.

References

1. von Sonntag C. *The Chemical Basis of Radiation Biology.* Taylor and Francis, London, 1987.
2. O'Niell P and Fielden EM. Primary free radical processes in DNA. *Adv. Radiat. Biol.* 17: 53–120, 1993.
3. Becker D and Sevilla MD. The chemical consequences of radiation damage to DNA. *Adv. Radiat. Biol.* 17: 121–180, 1993.
4. Swarts SG, Sevilla MD, Becker D, Tokar CK, and Wheeler KT. Radiation-induced DNA damage as a function of hydration. I. Release of unaltered bases. *Radiat. Res.* 129: 333–334, 1993.
5. Swarts SG, Becker D, Sevilla MD, and Wheeler KT. Radiation-induced DNA damage as a function of hydration. II. Oxidative base damage. *Radiat. Res.* 145: 304–314, 1996.
6. Elkind M and Redpath JL. Molecular and cellular biology of radiation lethality. In: Becker FF, (Ed.) *Cancer, a Comprehensive Treatise*, Vol. 6, Plenum Press, New York, 1977, pp. 51–99.
7. Roots R and Okada S. Protection of DNA molecules of cultured mammalian cells from radiation-induced single-strand scissions by various alcohols and SH compounds. *Int. J. Radiat. Biol.* 21: 329–342, 1972.
8. Alper T. *Cellular Radiobiology*, Cambridge University Press, Cambridge, 1979.
9. Johansen L. The contribution of water-free radicals to the X-ray inactivation of bacteria. In: *Cellular Radiation Biology*, Williams & Wlikins, Baltimore, Maryland, 1965, pp. 103–106.
10. Chapman JD, Reuvers A, Borsa J, and Greenstock CL. Chemical radioprotection and radiosensitization of mammalian cells growing *in vitro. Radiat. Res.* 56: 291–306, 1995.

11. Sapora OF, Barone M, Belli M, Maggi M, Quintilliani M, and Taocchini MA. Relationships between cell killing mutation and DNA damage in X-irradiated V79 cells: The influence of oxygen and DMSO. *Int. J. Radiat. Biol.* 60: 467–482, 1991.

12. Evans JW, Ward JF, and Limoli CL. Effects of dimethyl sulfoxide radioprotector on intracellular DNA. In: *33rd Radiation Research Society Meeting*. Los Angeles, California, 1985: Eh12.

13. Littlefield LG, Joiner EE, Colyer SP, Sayer AM, and Frome EL. Modulation of radiation-induced chromosomal aberration by DMSO, and OH radical scavenger. I. Dose response studies in human lymphocytes exposed to 220 kV X-rays. *Int. J. Radiat. Biol.* 53: 875–890, 1988.

14. Fielden EM and O'Niell P. *The Early Effects of Radiation on DNA*, Springer Verlag, Berlin, 1991.

15. Ward JF. DNA damage produced by ionizing radiation in mammalian cells: Identities, mechanisms of formation and repairability. *Prog. Nucleic Acids Res. Mol. Biol.* 35: 95–135, 1988.

16. Nackerdien ZOR and Dizdaroglu M. DNA base damage in chromatin of gamma-irradiated cultured human cells. *Free Radic. Res. Commun.* 16: 259–273, 1992.

17. Wallace S. In: Nicoloff JA and Hoekstra MF, (Eds.) *DNA Repair in Higher Eukaryotes*, Vol. 2. Humana Press Inc., Totowa, New Jersey, 1996.

18. Dizdaroglu M. Oxidative damage to DNA in mammalian chromatin. *Mutat. Res.* 275: 331–342, 1992.

19. Beckman KB and Ames BN. Oxidative decay of DNA. *J. Biol. Chem.* 272: 19633–19636, 1997.

20. Chen DS, Law C, and Keng PC. Reduction of radiation cytotoxicity by human apurinic endonuclease in a radiosensitive *Escherichia coli* mutant. *Radiat. Res.* 135: 405–410, 1993.

21. Téoule R. Radiation-induced DNA damage and its repair. *Int. J. Radiat. Biol.* 51: 573–589, 1987.

22. Breimer LH and Lindahl T. Thymine lesions produced by ionizing radiation in double-stranded DNA. *Biochemistry* 24: 4018–4022, 1985.

23. Demple B and Levin JD. Repair systems for radical-damaged DNA. In: Sies H, (Ed.) *Oxidative Stress: Oxidants and Antioxidants*, Academic Press Ltd., London, 1991, pp. 119–154.

24. Raleigh J. Secondary reaction in irradiated nucleotides: Possible significance for chemical "repair" mechanisms. *Free Radic. Res. Commun.* 6: 141–143, 1989.

25. Ward JF. Nature of lesions formed by ionizing radiation. In: Nickoloff JA, Hoekstra MF, (Eds.) *DNA Damage and Repair*, Vol. 2, Humana Press Inc., Totowa, New Jersey, 1998.

26. Hutchison F. Chemical changes produced in DNA by ionizing radiation. *Prog. Nucleic Acid Res.* 32: 115–154, 1985.

27. Obe G, Johannes C, and Schulte-Frohlinde D. DNA double strand breaks induced by sparsely ionizing radiation and endonucleases as critical lesions for cell death, chromosomal aberrations, mutations and oncogenic transformation. *Mutagenesis* 7: 3–12, 1992.

28. Henner WD, Rodriguez LO, Hecht SM, and Haseltine WO. Gamma ray-induced deoxyribonucleic acid strand breaks. *J. Biol. Chem.* 258: 711–713, 1983.

29. Ward JF. Biochemistry of DNA lesions. *Radiat. Res.* 104: S103–S111, 1985.

30. Roots R and Okada S. Estimation of life times and diffusion distances of radicals involved in X-ray-induced DNA strand breaks or killing of mammalian cells. *Radiat. Res.* 64: 306–320, 1975.

31. Newton GL, Aguilera JA, Ward JF, and Fahey RC. Polyamine-induced compaction and aggregation- a major factor in radioprotection of chromatin under physiologic conditions. *Radiat. Res.* 145: 776–780, 1996.
32. Prise KM, Pinto M, Newman HC, and Michael BD. A review of studies of ionizing radiation-induced double-strand break clustering. *Radiat. Res.* 156: 572–576, 2001.
33. Sak A, Stuschke M, Stapper N, and Streffer C. Induction of DNA double strand breaks by ionizing radiation at the c-myc locus compared to the whole genome: A study using pulse field gel electrophoresis and gene probing. *Int. J. Radiat. Biol.* 69: 679–685, 2001.
34. Sutherland BM, Bennett PV, Sidorkina O, and Laval J. Clustered DNA damages induced in isolated DNA and in human cells by low doses of ionizing radiation. *Proc. Natl. Acad. Sci. USA* 97: 103–108, 2000.
35. Ramakrishnan N, Chiu SM, and Oleinick NL. Yield of DNA-protein cross-links in gamma-irradiated Chinese hamster cells. *Cancer Res.* 47: 2032–2035, 1987.
36. Meyn RE, van Ankeren SC, and Jenkins WT. The induction of DNA-protein crosslinks in hypoxic cells and their possible contribution to cell lethality. *Radiat. Res.* 109: 419–429, 1987.
37. Barker S, Weinfeld M, Zheng J, Li L, and Murray D. Identification of mammalian proteins cross-linked to DNA by ionizing radiation. *J. Biol. Chem.* 280: 33826–33838, 2005.
38. Sutherland BM, Bennett PV, Sutherland JC, and Laval J. Clustered DNA damages induced by X-rays in human cells. *Radiat. Res.* 157: 611–616, 2002.

7

Repair of Radiation Damage to DNA

7.1 Overview of DNA Repair Mechanisms

The human genome comprises three billion base pairs coding for 30,000–40,000 genes. It is constantly under attack from endogenous mutagens, therapeutic drugs, and environmental stressors such as ionizing radiation, which threaten its integrity. Clearly the survival of the organism requires that the genome be under constant surveillance. This is accomplished by DNA repair mechanisms that have evolved to remove or tolerate pre-cytotoxic, pre-mutagenic, and pre-clastogenic lesions in an error-free or sometimes error-prone way. Defects in DNA repair cause hypersensitivity to DNA-damaging agents, accumulation of mutations in the genome, and ultimately to the development of cancer and metabolic disorders. The importance of DNA repair to normal function is illustrated by the sporadic occurrence of the DNA repair deficiency syndromes that are characterized by increased cancer incidence and multiple metabolic disorders.

There is an extensive array of enzyme systems in the cell, which can repair damage to DNA. Two recent papers compiled data of 130 human DNA repair genes, which had been cloned and sequenced, although not all of them have been characterized as to their function [1,2]. DNA-repair genes can be subgrouped into genes associated with signaling and regulation of DNA repair and into genes associated with distinct repair mechanisms such as mismatch repair (MMR), base excision repair (BER), nucleotide excision repair (NER), direct damage reversal, and DNA double-strand break (DSB) repair. Some of these are specific for a particular type of damage (e.g., a particular base modification), while others can handle a range of different damages. These systems also differ in the fidelity of repair, which is in the degree to which they are able to restore the wild-type sequence.

The eukaryotic cell has at least seven mechanisms to restore the structural integrity of DNA:

- Direct repair of defects such as O^6-alkylguanine formation
- NER, which fixes bulky lesions such as pyrimidine dimers produced by UV irradiation

- Base excision which repairs damaged bases and single-strand breaks (SSB)
- Mismatch repair which corrects mismatched nucleotides and small loops as they occur during replication
- Homologous recombination (HR) which repairs double-strand DNA breaks (DSBs)
- Nonhomologous end-joining (NHEJ) which repairs DSBs
- Poly(ADP-ribosyl) polymerase-1 (PARP) activity which, in conjunction with XRCC1 and ligase III, performs SSB repair and BCR.

Repair of O^6-alkylguanine, O^4-alkylthymine, and alkylphosphotriesters in DNA is an adaptive or inducible repair process involving several protein activities that recognize very specific modified bases, typically methylated, and transfers the modifying group from the DNA to themselves.

Excision repair involves the cutting out of damaged DNA sequences and replacement by repair synthesis. Aberrant bases, which are typically due to deamination of normal bases during oxidative metabolism and may also result from modification by chemicals or ionizing radiation, are removed by BER, an important part of the defense against irradiation damage to be considered in more detail below. NER works on bulky lesions such as base dimers and chemically induced intra-strand crosslinks. It involves an endonucleolytic cleavage near the dimer, followed by a polymerase with a $5'-3'$ exonucleolytic activity which cuts out the thymine dimer. This polymerase (the polA gene product) simultaneously synthesizes an appropriate matching strand. NER can proceed by two pathways termed global genome repair (GG-NER), which is transcription-independent and removes lesions from the entire genome, and transcription-coupled repair (TCR-NER). The GG-NER pathway surveys the entire genome for lesions that distort DNA. These lesions are removed rapidly. Pyrimidine dimers produced by UV irradiation are repaired slowly by GG-NER and are more efficiently removed from the transcribed strand of expressed genes by TCR-NER, which focuses on lesions that block the activity of RNA polymerases and overall transcriptional activity.

A third excision repair mode is mismatch excision repair or mismatch repair, a multienzyme system that recognizes inappropriately matched bases in DNA and replaces one of the two bases with one that "matches" the other. The major problem for this system is recognition of which of the mismatched bases is incorrect and must be excised and replaced. *Escherichia coli* discriminates between the old and the newly synthesized DNA strands on the basis of the fact that only the old strand of DNA is a methylated. An endonucleolytic attack can then take place on the new, unmethylated strand allowing removal of the incorrectly incorporated base (and everything else between the mutant site and the methylation site). The "half-sites" on the new strand are modified subsequently.

Recombination repair (or post-replication repair) repairs damage by a strand exchange from the other daughter chromosome. Since it involves homologous recombination (HR), it is largely error-free. This is the principal method for repair of radiation-induced DSBs in lower eukaryotes such as yeast and will be described in more detail below as will NHEJ, the chief mechanism for DSB repair in mammalian cells.

7.2 Repair of Radiation-Induced DNA Damage

Not all of the mechanisms are used in the repair of radiation-induced DNA damage. Those which are used are shown in Table 7.1.

7.2.1 Repair of Base Damage and Single-Strand DNA Breaks: Base Excision Repair

This section is based on reviews by Wallace [3], Friedman [4], and Christmann et al. [5] and the papers referenced therein. During BER, damaged DNA bases are excised from the genome and replaced by the normal nucleotide sequence. BER is initiated by DNA glycosylases and is so called because the chemically modified moieties are excised as free bases. BCR is responsible for removal of most of the base and for sugar phosphate lesions produced by oxidative damaging agents such as ionizing radiation. The action of a DNA-glycosylase is the first in a series of four enzymatic steps (summarized in Figure 7.1), which are required to repair a free radical-damaged purine or pyrimidine.

- The initial action on a damaged purine or pyrimidine is cleavage of the N-glycosyl bond by a DNA glycosylase, which catalyzes the hydrolysis of the N-glycosidic bonds that link chemically altered (or otherwise inappropriate) bases to the deoxyribose–phosphate

TABLE 7.1

Specific Pathways Involved in Repair of Radiation Damage

Repair Pathway	Type of Lesion	Repairs Radiation Damage
Base excision repair (BER)	Base damage, AP sites, single-strand breaks	Yes
Homologous recombination (HR)	Double-strand DNA breaks (DSBs)	Yes
Nonhomologous end joining (NHEJ)	DSBs	Yes
Nucleotide excision repair (NER)	Dimers, bulky adducts	No
Transcription coupled repair(TCR)	Dimers, bulky adducts	No
Mismatch repair (MMR)	Base mismatches	No

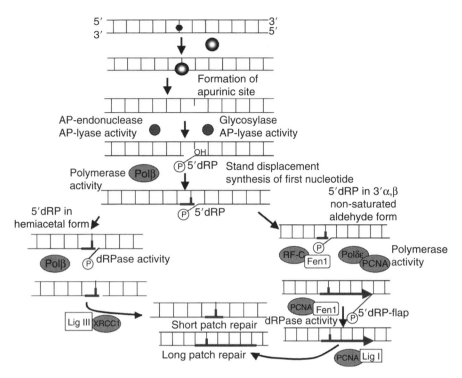

FIGURE 7.1
Mechanism of base excision repair (BER). Recognition of the DNA lesion occurs by a specific DNA glycosylase which removes the damaged base by hydrolyzing the N-glycosidic bond. The AP site is processed by AP endonuclease. Depending on the cleavability of the resulting 5'dRP by Polβ, repair is performed via the short- or long-patch BER pathway.

backbone. All the modified purine and pyrimidine bases depicted in Figure 6.3 and many others are substrates for DNA-glycosylases. DNA glycosylases that recognize free radical-damaged purines or pyrimidines have an associated DNA-lyase activity that nicks the DNA backbone 3' to the DNA lesion. The initial enzymatic event during BER generates another type of DNA damage, a site of base loss called an apurinic or apyrimidinic (AP) site. AP sites can also result from spontaneous hydrolysis of N-glycosidic bonds.

- Nicking the DNA backbone 3' to the DNA lesion leaves either a modified sugar or a phosphate residue attached to the 3' side of the resulting nick. This block must be removed before DNA polymerization, since DNA polymerases will only bind to a clean 3' hydroxyl end. The modified deoxyribose or phosphate residues attached to the 3' end are substrates for the diesterase or phosphatase activities of the 5' AP endonuclease. These same diesterase and phosphatase activities are also responsible for cleaning up SSBs produced in DNA by ionizing radiation that are frequently characterized by phosphate or phosphoglycolate groups attached to 3′ side of the SSB with a phosphate on the 5′ side of the break (see Figure 6.8). Thus, the second step in the repair of free radical-damaged purines or pyrimidines, the removal of the blocked 3′ terminus by a 5′ AP endonuclease, is also the first step in the repair of a radiation-induced SSB.

- Hydrolysis of the phosphodiester bond immediately 5′ to an AP site generates a 5′ terminal deoxyribose-phosphate residue (5′dRP). The removal of these residues during BER is by exonucleases that specifically initiate the degradation of DNA at free ends. Exonucleases are not repair-specific enzymes, they can degrade DNA with free ends in a variety of situations including repair, replication, and recombination. Those exonucleases whose action is restricted to the degradation of single-stranded DNA are sometimes referred to as DNA-deoxyribophosphodiesterases.

- As described in steps 1–3, the sequential action of a DNA glycosylase, a 5′ AP endonuclease, and a DNA deoxyribophosphodiesterase generates a single nucleotide gap in the DNA duplex. In short-patch BER, the 5′dRP is displaced by DNA polymerase β (Polβ), to create a single-base gap that is then filled in by Polβ using the opposing strand as the template for gap-filling synthesis. If the opposing strand is also damaged, repair may not be completed or attempted repair may result in more serious damage depending on the nature of the damages and their relative position.

The insertion of the first nucleotide is not dependent on the chemical structure of the AP site and up to this point Polβ is involved in both short-patch and long-patch BER. Besides polymerization activity, Polβ also has lyase activity and is able to catalyze the release of the hemiacetal form of 5′-dRP residues from incised AP sites. If however, the AP sites are oxidized or reduced (3′-unsaturated aldehydes, or 3′-phosphates) they are resistant to elimination by Polβ. In this case Polβ dissociates from damaged DNA and further processing occurs by PCNA (proliferating cell nuclear antigen)-dependent long-patch repair.

In contrast to short-patch repair, during which the DNA backbone is directly sealed following a single base insertion by Polβ, several additional steps occur during long-patch repair. After dissociation of Polβ and strand displacement, further DNA synthesis is accomplished by Polε or Polδ together with PCNA and RF-C resulting in longer repair patches of up to 10 nucleotides. The removal of the deoxyribosephosphate flap structure is executed by flap endonuclease FEN1 stimulated by PCNA.

The ligation step is performed by DNA ligases I and III. Ligase I interacts with PCNA and Polβ and participates mainly in long-patch BER, whereas DNA ligase III interacts with XRCC1, Polβ, and poly(ADP-ribose) polymerase-1 (PARP-1 and is involved only in short-patch BER).

If the initial damage is a site of base loss, it can be recognized by the lyase activity of the DNA glycosylases and processed by the same sequence of repair steps. Alternatively, a site of base loss can be recognized directly by a 5′ AP endonuclease that cleaves on the 5′ side of the deoxyribose, leaving the deoxyribose attached to the 5′ site of the nick. The deoxyribose needs to be removed by a phosphodiesterase activity specific for a 5′ attached sugar and the resulting single base gap is filled in by DNA polymerase and sealed by DNA ligase. Thus, the machinery of BER can encompass the repair of three different radiation-induced lesions: damaged bases, SSBs, and base loss.

7.2.2 Role of PARP

The material in this section is based on reviews by Lindahl et al. [6], D'Amours [7], Christmann et al. [5], and Vidakovic et al. [8]. An important role in the regulation of DNA repair is played by the members of the so-called poly(ADP-ribose) polymerase (PARP) family. These chromatin-associated enzymes modify several proteins by poly(ADP-ribosyl)ation. During this process, PARP consumes NAD^+ (nicotine adenine dinucleotide) to catalyze the formation of highly negatively charged poly(ADP-ribose) polymers of linear or branched structure with a length of 200–400 mono-mers, releasing nicotinamide as a by-product. The degradation of polymers is performed by poly(ADP-ribose) glycohydrolase (PARG). PARG exhibits endoglycosidic and exoglycosidic activity and produces a mono (ADP-ribosyl)ated protein plus mono(ADP-ribose).

To date, 18 different PARPs have been described, sharing a conserved catalytic domain responsible for poly(ADP-ribose) synthesis. Although PARP-1 plays an important role in DNA repair, the role of other family members is not yet completely understood. Besides its role in the regulation of DNA repair, PARP-1 has also been implicated in mammalian longevity [9] and is considered to be a master switch between apoptosis and necrosis. PARP-1 has a molecular weight of 113 kDa and comprises three different domains: the N-terminal DNA-binding domain consisting of two zinc fingers and the nuclear location signal, the C-terminal catalytic subunit which binds NAD+, and an internal domain that functions as acceptor site for poly(ADP-ribose). In response to DNA damage induced by ionizing radiation or alkylating agents, PARP-1 can specifically bind to SSBs. Upon binding to DNA, PARP-1 becomes auto-poly-(ADP-ribosyl)ated, which allows it to non-covalently interact with other proteins. It has been proposed that PARP-1 is involved in DNA repair by three different mechanisms:

- By direct interaction of automodified PARP-1 with XRCC1 and Polβ, which are both key proteins in BER. PARP-1 stimulates XRCC1 in vitro, together with FEN-1, strand displacement and DNA-repair synthesis by Polβ, thus stimulating long-patch BER. PARP-1-deficient mice are hypersensitive to methylnitrosurea (MNU)

and cell lines generated from these animals are hypersensitive to methyl methone sulfonate (MMS), showing reduced DNA strand-break resealing and increased apoptosis.

- By remodeling of chromatin structure upon the induction of DNA damage. It has been shown that automodified PARP-1 interacts with the 20s proteasome via ADP-ribose polymers (for a description of the proteasome see Chapter 8). The interaction and poly(ADP) ribosylation of the 20s proteasome results in increased proteolytic activity and the degradation of oxidatively damaged histones. Histone degradation leads to chromatin structure remodeling, giving DNA repair enzymes access to the site of DNA damage.

- A specific poly(ADP-ribose)-binding sequence motif was found in a number of DNA-repair and DNA-damage checkpoint proteins including p53, p21, XPA, DNA ligase III, XRCC1, DNA-PK$_{cs}$, Ku70, NF-κB, and telomerase. By poly(ADP) ribosylation of this motif, PARP-1 could potentially interfere with several functions of these proteins such as regulation of transcription, DNA repair, cell cycle regulation, and apoptosis (Figure 7.2).

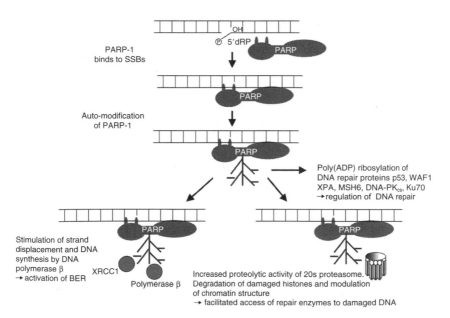

FIGURE 7.2

Role of PARP in the regulation of DNA repair. Binding of PARP-1 to SSBs results in auto-poly (ADP) ribosylation and increased activity. Activated PARP-1 is involved in DNA repair via (1) direct interaction and poly(ADP) ribosylation of XRCC1 and Polβ, leading to stimulation of BER; (2) poly(ADP) ribosylation and activation of the 20S proteasome, leading to relaxation of the chromatin structure; and (3) potential poly(ADP) ribosylation of DNA-repair proteins, thus modulating DNA repair.

There are a number of observations suggesting interaction between DNA-PK and PARP-1 (reviewed by Veuger et al. [10]). Hetero-modification and modulation of enzyme activity in vitro, i.e., poly (ADP-ribosylation) of DNA-PK by PARP-1 and phosphorylation of PARP-1 by DNA-PK have been demonstrated suggesting reciprocal regulation of enzyme activity. As noted above, a peptide binding motif for ADP-ribose polymers is found in DNA-PK and Ku70 and Ku forms a complex with PARP-1 in the absence of DNA, indicating an intimate association between the two enzymes.

- There is compelling evidence (reviewed later in this chapter) implicating BRCA1 in the signaling of DNA damage to facilitate repair by HR. It has been shown that some BRCA1- and BRCA2-deficient cells are extremely sensitive to PARP inhibitors and that the inhibition of poly(ADP-ribosyl)ation results in an increased level of DSBs. Auto-poly(ADP-ribosyl)ation of PARP-1 and PARP-2 is essential for the release of the enzymes from DNA strand-breaks and the recruitment of signal transduction and repair enzymes. Failure to auto-modify could cause PARP-1 and PARP-2 to stall on damaged DNA and hinder access of repair enzymes with the result that accumulating SSBs will generate DSBs at DNA replication forks. This will have drastic consequences in BRCA-deficient cells because DSBs associated with replication forks are predominantly repaired by HR. These findings have raised the idea of using PARP inhibitors as a clinical stratagem to target HR-deficient cancer cells (reviewed by Haince et al. [11]).

7.2.3 Processing of Multiply Damaged Sites by BER

The versatility of BER may create a situation in which attempted repair of a complex lesion can create as many problems as it solves. It has been shown that BER processing of multiply damaged sites (MDS), which are a unique feature of radiation damage, may convert otherwise nonlethal lesions into lethal DSBs or other types of MDS. MDS consist of closely spaced single lesions (oxidized purines and pyrimidines, abasic sites, and strand breaks) within one or two helical turns of the DNA molecule. DNA DSBs are the best known MDS, but they only represent a minor proportion of the radiation-induced MDS. It has been demonstrated that non-DSB clusters consisting of oxidized purines and pyrimidines and abasic sites comprise the majority of clustered lesions (described in Chapter 6).

A number of in vitro repair studies using model clustered damages and purified enzymes or cell extracts have demonstrated that processing of non-DSB bi-stranded damage clusters may actually convert nonlethal, easily reparable, or bypassable lesions into lethal DSBs [3,12,13]. The simplest MDS that can be envisaged consists of two closely opposed lesions that are subject to repair by BER and, intuitively, it would seem that if the two lesions on opposing strands are close together, say within 2–3 bases, there is increased potential for formation of a DSB as a result of initial damage or processing during repair of the lesion. In fact, the action of a repair enzyme within an MDS depends on the nature of the other damage and the juxta-position of the lesions, i.e., whether the second lesion alters DNA structure sufficiently to prevent the enzyme from binding to the DNA or from acting once bound [14]. The results of these constraints have been demonstrated using model systems. In one such system, a simple MDS consisting of one DNA strand containing a damaged pyrimidine, separated by one or more bases from a similar lesion on the opposite strand is cleaved on one strand by endo III (a bifunctional enzyme with glycosylase and AP lyase activity) but a break is not introduced into the opposing strand unless the lesion is separated by five to seven nucleotides from the damaged pyrimidine on the first strand [13]. This and other experiments demonstrated that a DNA glycosylase/AP lyase can only cleave the DNA backbone to form a DSB adjacent if the opposing lesions are separated by more than three nucleo-tides and in this case breaks are formed on both strands constituting a DSB. In the case of closely opposed lesions, the proximity of the second lesion interferes with enzymes binding to or processing the first lesion making one of the strands a poor substrate for cleavage. In this case a DSB is not formed and the lesions may be repaired sequentially with the possibility that, during repair of the first lesion, DNA polymerase may insert a nucleotide opposite the second lesion with potentially mutagenic consequences (shown schematically in Figure 7.3).

In a series of ingenious experiments, Wallace and coworkers regulated the expression of key base-excision enzymes, the major DNA glycosyla-ses/AP lyases, and measured the effect of radiation on clonogenic survival and DSB generation in mammalian cells. It was found that in H_2O_2-treated cells, where singly damaged sites predominate, over-expression of the DNA glycolsylases/AP lyase conferred resistance to the cytotoxic effects of H_2O_2 treatment, whereas cells with reduced enzyme expression were more sensitive. In contrast, for cells treated with radiation over-expression of the enzymes caused increased lethality, mutation frequency, and for-mation of DSB. Subsequently, it was shown using human B lymphoblast TK6 cells that targeting one of the DNA glycolsylases/AP lyases (hOGGI) by RNSi decreased radiation-induced lethality and DSB formation [15]. Targeting a different DNA glycolsylase/AP lyases (hNTH1) also resulted in decreased DSB formation at clustered damage sites. However, since in this case the cells have become deficient in the ability to process a major lethal single lesion, i.e., thymine glycol, they are more sensitive to the cytotoxic effects of radiation. In this situation the DNA glycosylase/AP lyase had a dual role in the processing of radiation damage, repairing potentially lethal single lesions while generating lethal DSBs at clustered damage sites.

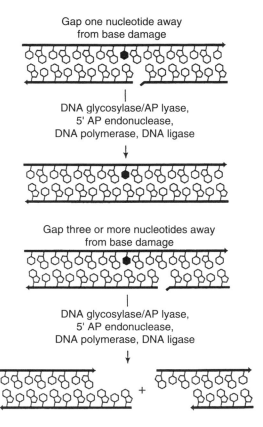

Gap one nucleotide away
from base damage

DNA glycosylase/AP lyase,
5' AP endonuclease,
DNA polymerase, DNA ligase

Gap three or more nucleotides away
from base damage

DNA glycosylase/AP lyase,
5' AP endonuclease,
DNA polymerase, DNA ligase

FIGURE 7.3
BCR processing of multiply damaged sites. (From Wallace, S.S., *Radiat. Res.*, 150, S60, 1998. With permission.)

7.3 Repair of DNA Double-Strand Breaks

As described in the preceding chapter, DNA DSBs are a common form of DNA damage resulting from ionizing radiation. DSBs can arise through the direct action of ionizing radiation and some chemicals and by a number of natural processes including replication, meiosis, and V(D)J rejoining. DSBs are highly potent inducers of genotoxic events (chromosome breaks and exchanges) and of cell death. The ability to repair DSBs and to ensure that repair is performed with sufficient fidelity is fundamental to genome protection. There are two main pathways of DSB repair, HR and NHEJ, which are error-free and error-prone, respectively. The pathways are conserved between *Saccharomyces cervisae* and mammalian cells although their relative importance differs. It is generally considered that HR is the predominant pathway of DSB repair in yeast, whereas in mammalian cells in a similar situation use a nonhomologous or illegitimate pathway. An overview of DNA DSB processes is shown in Figure 7.4.

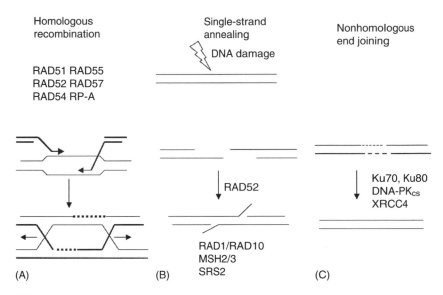

FIGURE 7.4

Schematic outline of homologous and nonhomologous double-stranded-break (DSB) repair pathways. (A) Homologous recombination: The extended 3′-single-stranded tails of molecules generated by exonucleolytic processing of DSBs invade homologous intact donor sequences. (B) Single-stranded annealing: Direct repeats on either side of the processed DSB are aligned and intervening sequences are removed. (C) Nonhomologous end joining.

DSB repair has been the subject of a number of excellent recent reviews; those which were used in the preparation of this chapter are listed at the end [5,16–25].

7.3.1 Homologous Recombination

Homologous recombination (HR) can repair a DSB, by using the undamaged sister chromatid as a template, and consequently HR generally results in the accurate repair of the DSB (shown schematically in Figure 7.5). During HR, the damaged chromosome enters into physical contact with an undamaged DNA molecule with which it shares sequence homology and which is used as template for repair. In organisms that range from yeast to mammals, HR is mediated through the so-called RAD52 epistasis group of proteins that includes RAD50, RAD51, RAD52, RAD54, and meiotic recombination 11 (MRE11). A partial list of the proteins involved in HR and NHEJ is shown in Figure 7.6.

HR is initiated by a nucleolytic resection of the DSB in the 5′–3′ direction by the MRE11–RAD50–NBS1 complex. The resulting 3′ single-stranded DNA is thereafter bound by a heptameric ring complex formed by RAD52 proteins, which protects it against exonucleolytic digestion. RAD52 competes with the Ku complex for the binding to DNA ends and this may

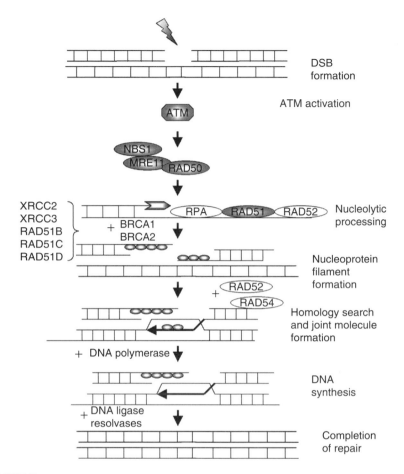

FIGURE 7.5
Double-stranded break repair through HR. Recognition of DNA ends occurs by the RAD52 protein. Nucleolytic processing of the DNA ends that requires the activity of the RAD50–MRE11–NBS1 complex results in the generation of a single-stranded region with a 3′ overhang. RAD51 polymerizes onto the single-stranded DNA to form a nucleoprotein filament, with the aid of the single-stranded DNA-binding protein, replication protein A (RPA), and RAD52. Other proteins involved in the proper RAD51 response include BRCA1, BRCA2, and the RAD51 paralogues XRCC2, XRCC3, RAD51B, RAD51C, and RAD51D. The rad 51 nucleoprotein filament searches for the homologous duplex DNA and when successful DNA strand exchange generates a joint molecule between the homologous damaged and undamaged duplex DNAs in a reaction stimulated by the RAD52 and RAD54 proteins. DNA synthesis fills in the break in the strand, and ligation and resolution of recombination intermediates results in accurate repair of the DSB.

determine whether the DSB is repaired via the HR or the NHEJ pathway. RAD52 interacts with RAD51 and with replication protein A (RPA) to stimulate the DNA strand-exchange activity of RAD51. RAD51 catalyzes strand-exchange events with the complementary strand in which the

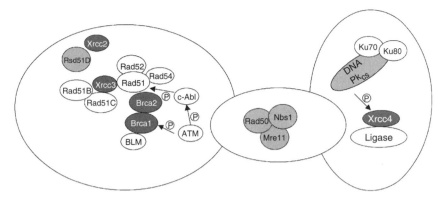

FIGURE 7.6
Components of DSB repair pathways. Nonhomologous end joining (NHEJ): Ku binds a DSB followed by recruitment and activation of DNA-PK$_{cs}$. XRCC4 and ligase IV are recruited directly or indirectly by the DNA-PK holoenzyme or are activated by DNA-PK-mediated phosphorylation. HR: The strand exchange reaction is mediated by RAD51 and facilitated by RAD52. RAD51 also interacts with BRCA2 and indirectly with BRCA1. The c-Abl tyrosine kinase modulates RAD51 strand exchange activity through phosphorylation. The MRE11/RAD50/NBS1 complex participates in both NHEJ and HR.

damaged DNA molecule invades the undamaged DNA duplex, displacing one strand as D-loop. In yeast, RAD54 displays double-strand DNA-dependent ATPase activity and uses the energy for unwinding of the dsDNA, thus stimulating DNA strand exchange. The assembly of the RAD51 nucleoprotein filament is facilitated by five different paralogues of RAD51 (RAD51B, C, and D; and XRCC2 and XRCC3), which play a role during presynapsis. After DSB recognition and strand exchange performed by RAD proteins, DNA repair is completed by DNA polymerases, ligases, and Holliday junction resolvases (Figure 7.5).

7.3.1.1   Role of BRCA1 and BRCA2

The role of BRCA1 and BRCA2 has been reviewed by Powell et al. [26,27]. The products of the human breast cancer susceptibility genes *BRCA1* and *BRCA2* are involved in DNA repair, and in fact, the carboxy-terminal region (BRCT domain) of these proteins is representative of a common motif among DNA-repair proteins. Loss of functional BRCA1 results in mild sensitivity to radiation and DNA-damaging chemicals. A fraction of human RAD51 co-localizes with BRCA1 and BRCA2 in mitotic cells and, following DNA damage, the three proteins are relocated to a structure, which also contains PCNA (proliferating cell nuclear antigen) and may represent a site of active repair. $BRCA1^{-/-}$ and $BRCA2^{-/-}$ mouse fibroblasts develop spontaneous chromosome aberrations consistent with the participation of these proteins in the repair of spontaneous DSBs [28].

BRCA1 is also implicated in the removal of oxidized bases from DNA. Damage is selectively removed from the transcribed DNA strand by the process of transcription-coupled repair (TCR). BRCA1-null cells are deficient in the removal by TCR of thymine glycol produced by ionizing radiation or hydrogen peroxide but perform TCR of UV-induced DNA damage normally. It is unclear whether BRCA1 participates directly in TCR or indirectly via a signaling role.

The involvement of the BRCA2 cancer susceptibility protein in homologous repair of DNA is suggested by its association with RAD51 and supported by the observation that there is relatively more DSB repair by nonhomologous pathways in $Brca2^{-/-}$ cells. Both BRCA1 and BRCA2 form discrete nuclear foci during S phase following DNA damage. BRC1 and BRC2 co-localize at subnuclear sites with Rad51 but though only a small percentage of BRCA1 is involved, BRCA2 contains 30–40 repeats that are major sites for Rad51 binding involving a substantial fraction of the BRCA2 pool. One model which has been proposed is that the Rad51–BRCA2 complex assembles DSB repair proteins required to offset breaks that accumulate during DNA replication. BRCA2-deficient cells have 10-fold lower levels of HR compared to proficient cells. Cells defective in the expression of the BRCA1 and BRCA2 proteins show decreased survival following radiation [29].

7.3.1.2 Single-Strand Annealing

Like HR, single-strand annealing (SSA) is a process for rejoining DSBs using homology between the ends of the joined sequences (Figure 7.4), and in fact, it can be considered a sub-routine of HR. SSA relies on regions of homology with which to align the strands of DNA to be rejoined. DNA strands are first resected to generate SBB tails and when this process has proceeded far enough to reveal complementary sequences, the two DNAs are annealed and then ligated. The genes which define SSA belong to the RAD52 epistasis group of HR.

7.3.1.3 Role of Homologous Recombination in DNA Repair in Mammalian Cells

Although mammalian cells are believed to rely less on HR, they do perform mitotic recombination and preferentially repair DSBs by HR in late S and G, phases of the cell cycle when an undamaged sister chromatid is available. This is illustrated by the fact that the DSB repair defect of murine SCID (severe combined immunodeficiency) cells, which occurs in the nonhomologous pathway, only influences survival if cells are exposed to DNA-damaging agents in G_1/early S phases [30]. The breast cancer susceptibility proteins BRCA1 and BRCA2 are only involved in HR in mammalian cells and have no obvious homologues in yeast indicating that, from an evolutionary point of view, they are a recent addition.

7.3.2 Nonhomologous End Joining

7.3.2.1 Demonstration of NHEJ in Mammalian Cells Using Repair-Deficient Mutants

Using physical methods that measure DNA size, it had been demonstrated that DSBs could be rejoined in mammalian cells some years before the mechanism of NHEJ, many aspects of which are still unresolved, became known. The study of DNA repair in microbial and lower eukaryote systems had been greatly facilitated by the use of repair deficient mutant strains which showed sensitivity to DNA-damaging agents. Applying an analogous approach in mammalian cells, Jeggo and coworkers used a standard procedure of microbial genetics involving transfer of cells from single colonies by means of sterile toothpicks to isolate x-ray sensitive mutants of Chinese hamster ovary (CHO) cells. From 9000 colonies tested, six mutants were found to be significantly radiosensitive while a further six were found showing slight sensitivity [31].

Radiation survival curves for these mutant cell lines, initially designated, *xrs-1* to *xrs-7* are shown in Figure 7.7. Cell fusion studies established that all except *xrs-3* belonged to a single complementation group, subsequently designated IR group 5 [32]). Soon after their discovery it was shown using

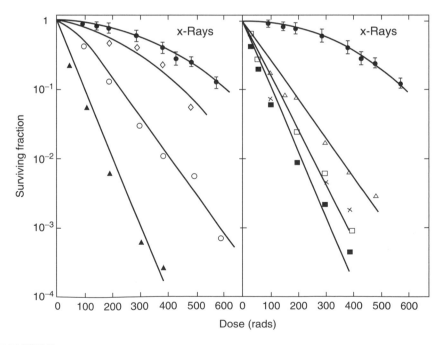

FIGURE 7.7

Radiation survival curves for CHO-K1 and mutant strains. ● CHO-K1, □ *xrs-1*, ○ *xrs-2*, ◇ *xrs-3*, △ *xrs-4*, ▲ *xrs-5*, ■ *xrs-6*, × *xrs-7*. (From Jeggo, P.A. and Kemp, L.M., *Mutat. Res.*, 112, 313, 1983. With permission.)

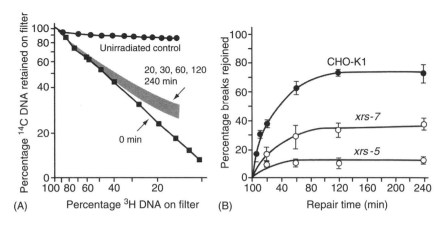

FIGURE 7.8
(A) Neutral elution of ^{14}C-labeled *xrs-5* cells irradiated with 90 Gy γ-rays and allowed various times to rejoin resultant double-strand breaks. The shaded area indicates the range of points for different incubation times. Very little repair was seen even at 240 min incubation. (B) Rate of rejoining of DSBs after 90 Gy γ rays. (From Kemp, L.M., Sedgwick, S.G., and Jeggo, P.A., *Mutat. Res.*, 132, 189, 1984. With permission.)

the neutral elution technique that the *xrs* mutants were defective in the ability to rejoin DSBs (Figure 7.8) [33]. Subsequently, this defect was characterized using a number of other techniques, including pulse-field gel electrophoresis (PFGE), nucleoid sedimentation, DNA unwinding, and the comet assay.

These cell lines were the first mammalian mutants to show a defect in DNA DSB rejoining and have been used extensively in many laboratories (especially *xrs-5* and *xrs-6*) to validate and optimize techniques for measuring DNA DSB rejoining, to examine the consequences of defective rejoining, and to investigate the mechanism of the rejoining process in mammalian cells. The defect in all the mutants represented by these complementation groups is manifest as a higher percentage of un-rejoined DSBs, with the magnitude of the defect increasing with dose. The defect is not large (two- to fourfold) and is of a similar magnitude for all the mutants examined. All the mutant cell lines show some residual DSB rejoining suggesting that more than one mechanism for handling DSBs is operative.

Later results of cell fusion studies showed that these cells belonged in fact to three distinct complementation groups, termed IR4, IR5, and IR7, and the human genes complementing them were reassigned to the XRCC nomenclature (x-ray cross-complementing); *XRCC4* is the gene that complements cells of IR4, *XRCC5* complements IR5, and *XRCC7* complements IR7 [32,34]. Cell lines from these complementation groups are extremely sensitive to ionizing radiation, anti-tumor drugs, such as bleomycin and other sources of DSBs, and all the cell lines show increased incidence of chromosomal aberrations like translocations and inversions after exposure to x-rays.

TABLE 7.2

Mutants Belonging to IR Complementation Groups 4–7

Complementation Group	Mutant	Gene	DNA End Binding Activity	DNA-PK Activity
4	XR-1	XRCC4	Wild type	Wild type
5	*xrs 1–6*	XRCC5	None	None
	XRV-17B	(Ku80)		
	XRV9-B			
	sxi-3			
6	Not identified	XRCC6 (Ku70)	Unknown	Unknown
7	V-3	XRCC7	Wild type	None
	SCID	SCID		
		(DNA-PK$_{cs}$)		

Mutants representing four x-ray sensitive complementation groups defective in DSB rejoining are listed in Table 7.2. The majority of the cell lines were isolated from mutagenized cultures of established rodent cell lines on the basis of their sensitivity to ionizing radiation as already described. Severe combined immunodeficiency (SCID) cell lines, in contrast, were derived from the SCID mouse, identified by its SCID phenotype. Cell lines established from the SCID mouse were subsequently shown to be radiosensitive, and to belong to the same complementation group as the radiosensitive hamster V-3 cells.

In addition to radiation sensitivity, group 4–7 mutants are sensitive to DSBs generated by restriction endonucleases introduced into cells by permeabilization and to bleomycin, a radiomimetic agent, but they show little cross-sensitivity to UV radiation or alkylating agents [31]. There is also some sensitivity to DNA crosslinking agents [35]. These results are consistent with a sensitivity to agents introducing DNA DSBs, either directly as in the case of radiation and bleomycin, or indirectly, for example, during the repair of other DNA damage. Group 4 and 7 mutants also display extreme sensitivity to a range of topoisomerase II inhibiting drugs, including the anticancer drugs VP-16, VM-26, and mAMSA [36]. These drugs inhibit the ligation step of topoisomerase II, with resulting accumulation of protein-bridged DSBs in DNA (called "cleavable complexes," since they are only revealed as DSBs following treatment with protein denaturants). Although cleavable complexes are lost at the same rate and to the same extent in normal and group 7 mutants after removal of the drug, a fraction of the breaks remain unrejoined in the mutant cells [35].

7.3.3 Genes and Proteins Involved in NHEJ

7.3.3.1 DNA-PK

The characterization of gene products defined by IR complementation groups 4–7 and consideration of the combined defects of all the mutants in DSB rejoining and in V(D)J recombination suggested that the gene products might operate in a single pathway and possibly within the same complex. This was identified as DNA-activated protein kinase (DNA-PK), which has specificity for double-stranded DNA ends and correlates with the specific and pronounced sensitivity of the group 5–7 mutants to agents that induce DSBs. The genetics and biochemistry of components of the DNA-PK complex have been extensively investigated and it has been established that DNA-PK consists of a regulatory subunit, the Ku protein, and a catalytic subunit DNA-PK$_{cs}$.

The Ku protein binds to double-stranded DNA ends before being complexed to DNA-PK$_{cs}$, and upon DNA-binding, it binds DNA-PK$_{cs}$ and activates its kinase function. Thus the formation of the DNA-PK complex specifically results in the activation of a kinase following the introduction of DNA DSBs. Upon activation, DNA-PK$_{cs}$ is able to phosphorylate DNA binding proteins including many transcription factors. Phosphorylation is most effective when these substrates are bound to the same DNA molecule as DNA-PK.

7.3.3.2   Ku Proteins

Genetic and biochemical studies led to the identification of *XRCC5*, the gene deficient in *xrs* mutants, which was found to encode the 80 kDa subunit of the Ku protein. (Ku80 is also called Ku86 and is actually 83 kDa [37].) Biochemical studies on the group 5 mutants identified a defect in a DNA end-binding activity with properties similar to the previously characterized Ku protein while a genetic study localized a complementing human gene to a 3 Mbp fragment mapping to the same chromosomal locus as the gene encoding the Ku80 subunit of Ku. Finally, Ku80 cDNA was shown to complement the radiosensitivity and V(D)J recombination defects of group 7 mutants.

The Ku protein had earlier been identified as an antigen present in the sera of certain autoimmune patients and was later shown to be a hetero-dimer comprising subunits of 70 and 80 kDa. *XRCC5* and *XRCC7* encode Ku80 and DNA-PK$_{cs}$, respectively, and mutations in Ku80 or DNA-PK$_{cs}$ lead to radiosensitivity. None of the original rodent cell lines was defective in Ku70, but it was assumed that cells lacking it would have a similar phenotype to others in IR4–7 since the DNA-PK complex is a crucial component of the mammalian DSB repair apparatus. This has been confirmed by targeted disruption of the gene for Ku70 in mouse cells, allowing such cells to be designated IR6 and the gene for Ku70 to be designated *XRCC6* [38].

Ku is a highly abundant nuclear protein in human cells and has the characteristic property of binding to double-strand DNA ends. Ku was identified as the DNA-binding component of DNA-PK with a characteristic specificity for its DNA substrates. It does not bind to single-stranded DNA ends, but binds tightly to double-stranded ends, having equal affinity for 5′-protruding, 3′-protruding, and blunt ends. Ku also binds to DNA nicks, gaps, bubbles, and DNA ending in stem-loop structures. This spectrum of DNA binding activity has been incorporated into a model in which Ku recognizes the transition between a region of double-stranded DNA and two non-annealed single strands. Once Ku binds to DNA ends, it is capable of translocating along the DNA, so that three or more molecules of Ku can be bound to a single DNA fragment.

7.3.3.3 DNA-PK$_{cs}$

The discovery that Ku80 is the product of IR group 5 implicated the other members of the DNA-PK complex Ku70 and DNA-PK$_{cs}$ in DNA repair and suggested them as potential candidates for the proteins defective in the remaining complementation groups. DNA-PK$_{cs}$ is a very large (467 kDa) nuclear protein, the kinase domain of which shares strong homology to the phosphatidylinositol-3 kinase superfamily. A number of lines of evidence now indicate that DNA-PK$_{cs}$ is defective in *XRCC7* (group 7) mutant cells. V-3 hamster cells and SCID mouse cells, which are deficient in DNA-PK activity, have intact Ku end-binding activity. DNA-PK activity is restored when mutant cell extracts are supplemented with purified DNA-PK$_{cs}$. V-3 and SCID cells are missing DNA-PK activity, and contain little or no protein cross-reacting with DNA-PK$_{cs}$ antibody. Yeast artificial chromosomes (YACs) encoding DNA-PK$_{cs}$ are able to complement hamster V-3 mutants, and the human *DNA-PK$_{cs}$* gene (now known as *PRKDC*) maps to a region of chromosome 8, which complements SCID cells.

In addition to V-3 and SCID cells a human glioma cell line (MO79J) has been found to be highly radiosensitive and defective in DNA-PK$_{cs}$ expression and DNA-PK activity (Figure 7.9) [39].

7.3.3.4 XRCC4 Gene Product

XR-1 is the only mutant cell line so far assigned to x-ray-sensitive complementation group 4. Like *XRCC7* mutants, XR-1 is defective for DSB rejoining and in both coding and signal joint formation during V(D)J recombination but XR-1 cell extracts contain normal levels of DNA end-binding activity. The x-ray sensitivity of XR-1 varies dramatically during the cell cycle, with

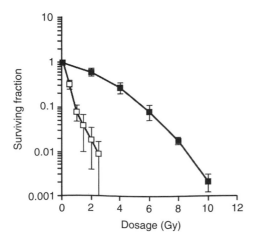

FIGURE 7.9
Radiation survival curves for human glioma cell lines M059 K and the radiosensitive variant MO59J that is defective in DNA-PK$_{cs}$ expression. (From Allalunis-Turner, M.J., Zia, P.K., Barron, G.M., Mirzayans, R., and Day, R.S. 3rd., *Radiat. Res.*, 144, 288, 1995. With permission.)

extreme sensitivity during G_1 and near-normal resistance during late S and G_2. The *XRCC4* gene was cloned by complementation of the V(D)J recombination deficiency and was shown to be deleted in IR4 cells. It encodes a protein that has been found to form a tight and specific association with DNA ligase, an enzyme which has been shown to be involved in the final stages DNA NHEJ.

7.3.3.5   NHEJ and Severe Combined Immunodeficiency

Much information regarding the proteins involved in NEHJ and their mechanisms of action has come from the understanding of the process of V(D)J rejoining. Defects in V(D)J rejoining result in a condition called severe combined immunodeficiency or SCID. The SCID mouse line was detected in 1983 as four littermates having a SCID phenotype [40], which results from defects in the process of V(D)J recombination in which the variable (V), diversity (D), and joining (J) segments of the immunoglobulin and T cell receptor genes are rearranged during development of the immune response.

Defects in many genes are now known to cause a SCID phenotype, and the defect most commonly found in human SCID patients is distinct from the defect that was identified in the SCID mouse. The mouse *SCID* gene, *XRCC7*, and the *DNA-PK$_{cs}$* gene are now known to be the same.

7.3.3.6   Role of NHEJ Proteins in V(D)J Recombination

Genes that encode active immunoglobulins (Ig) and T cell receptors (Tcr) are not present in the germline but occur as separate gene fragments that are assembled into active genes in specialized cells by V(D)J recombination. During lymphocyte development V(D)J recombination cuts, rearranges, and again pastes together the separate V, D, and J gene segments to form new genes. This process generates a large repertoire of different coding sequences and has a major role in enhancing the diversity of the immune system.

V(D)J recombination is initiated when the RAG1 and RAG2 proteins bind to specific recognition sequences that flank the Ig or Tcr gene segments (Figure 7.10). The RAG proteins act in concert to cleave the DNA precisely between the specific recognition sequences and the coding segments, producing blunt signal ends and covalently sealed (hairpinned) coding ends. The latter are then opened by endonucleolytic cleavage before forming the coding joint (i.e., the rearranged antigen–receptor gene). RAG1/RAG2 and NHEJ factors are required for forming both coding and signal joints. The generation of coding junctions necessitates initial cleavage of the hairpin, followed frequently by addition or deletion of nucleotides before rejoining. Hairpin opening requires several factors, including Artemis, DNA-PK$_{cs}$, and possibly the RAG proteins.

Cells lacking XRCC4, DNA ligase IV, or Ku are dramatically impaired in rejoining both signal and coding junctions while cells lacking DNA-PK$_{cs}$ show little coding joint formation but only a modest defect in signal junction

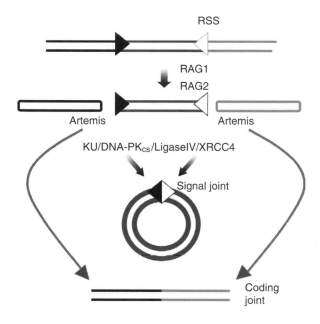

FIGURE 7.10

The basic mechanism of V(D)J recombination. The RAG I recombination-activating gene (I) and RAG2 proteins initiate V(D)J recombination by inducing a DSB at the border of the recombination signal sequence (RSS) of the gene segment. This produces a blunt end on the signal-sequence side and a DNA hairpin. The hairpins are opened by Artemis and the DSBs are repaired by the NHEJ mechanism (Ku70/Ku80, DNA-PK$_{cs}$, DNA ligase IV and XRCC4).

rejoining. Artemis-defective cells fail to carry out coding joint formation but are completely proficient in rejoining signal junctions. Consistent with this, mutations in Artemis can also contribute to human immunodeficiency, being responsible for RS-SCID, a human SCID condition associated with radiosensitivity. Unphosphorylated Artemis has 5′–3′ exonuclease activity but upon phosphorylation by DNA-PK Artemis gains hairpin cleavage activity and the ability to cleave 5′ and 3′ single-strand overhangs. In fact, DNA-PK$_{cs}$ has two roles in V(D)J recombination; in vitro it has been shown to phosphorylate and thus activate Artemis hairpin cleavage activity, which is required for coding joint formation, and it then facilitates the rejoining step via its role in NHEJ.

Another type of human immunodeficiency, LIG4 syndrome, has been attributed to mutations in DNA ligase IV.

7.3.3.7 *Artemis*

We return to the catalog of NHEJ proteins to describe Artemis, which has already been mentioned in the context of V(D)J recombination. A novel DNA repair gene was revealed as a result of a study of young patients exhibiting radiosensitivity and inherited SCID of B and T lymphocytes [41]. It was named after the Greek goddess Artemis, a guardian of young children and small animals.

Artemis is defective in a subpopulation of human radio-sensitive SCID (RS-SCID). The *Artemis* gene product is found in a stable protein complex with DNA-PK in human cells. Artemis is a single-strand-specific nuclease with 5′–3′ exonuclease activity, which, upon binding with DNA-PK acquires endonucleolytic activity on 5′ and 3′ ends as well as hairpin-opening activity. Artemis-defective cells are radiosensitive and possess a defective V(D)J recombination phenotype consistent with a role for Artemis in opening hairpin coding joints in V(D)J recombination.

Recently, Artemis has been designated a sixth component of NHEJ [42]. Artemis-defective cells display normal kinetics of DSB repair at early times post IR, but at longer times, a small repair defect is seen in Artemis-deficient cells. In fact, Artemis is one of a range of repair proteins which processes ends before DNA ligase IV and XRCC4 during NHEJ. This is necessary because DSBs generated by radiation rarely have 5′-P and 3′-OH termini, the prerequisite for ligation (Chapter 6).

Artemis is phosphorylated and potentially activated by ATM. The role of ATM in interacting with Artemis is described in Chapter 8. A core set of NHEJ proteins (rows 1–4 in Table 7.3) is conserved in all eukaryotic cells while the two additional proteins are present only in vertebrates. These proteins presumably evolved to accomplish features of NHEJ that are required for V(D)J recombination, a process found only in vertebrates. Nevertheless, DNA-PK$_{cs}$ and Artemis play an important role in general DSB repair indicated by the hyper-radiosensitivity of cells lacking these proteins.

7.3.3.8   Nuclear Foci: Mre11–Rad50–Nbs1 and Others

Joining of DSBs by NHEJ requires the MRE11–RAD50–NBS1 (MRN) complex. MRE11 is a nuclease with exonucleolytic activity similar to Artemis, which processes complex DNA ends before ligation. It has also been shown to interact with Ku70 following DNA damage. MRE11 and RAD50 congregate

TABLE 7.3

Summary of Genes and Proteins Important for NHEJ

Mammalian Gene Name	Protein
LIG4	DNA ligase IV; cooperates with XRCC4 to ligate broken DNA molecules after their ends have been properly processed
XRCC4	XRCC4; cooperates with DNA ligase IV to ligate broken DNA molecules after their ends have been properly processed
XRCC5	Ku80; cooperates with Ku70 to bind DNA ends and recruit other proteins
XRCC6	Ku70; cooperates with Ku80 to bind DNA ends and recruit other proteins
PRKDC (formerly *XRCC7*)	DNA-PK$_{cs}$; protein kinase activates Artemis
ARTEMIS	Artemis; nuclease regulated by DNA-PK$_{cs}$ important for preparing DNA-PK ends to make them ligatable

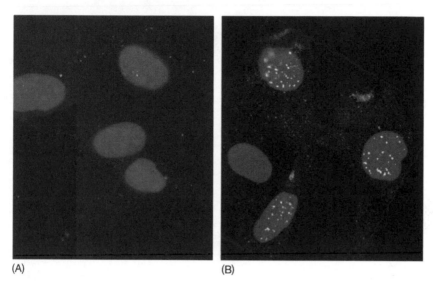

(A)   (B)

FIGURE 7.11

Foci induced by DNA damage. Exposing cells to ionizing radiation causes several proteins that are involved in double-stranded break repair to relocalize to nuclear foci. (A) Primary human fibroblasts that have not been irradiated. (B) Primary human fibroblasts after exposure to 12 gray (Gy) of ionizing radiation. The resulting nuclear foci have been detected using a fluorescently tagged antibody against the RAD51 protein. (From van Gent, D.C., Hoeijmakers, J.H., and Kanaar, R., *Nat. Rev. Genet.*, 2, 196, 2001. With permission.)

as discrete ionizing radiation-induced foci in a dose-dependent manner (Figure 7.11). Foci are not seen in irradiated fibroblasts that lack NBS1, since the NBS protein is needed to chaperone the complex to nuclear sites of damage. NBS-defective cells show a radiosensitive phenotype attributable to defective DSB repair but are proficient in V(D)J recombination.

Nuclear foci containing RAD51, RAD52, BRCA1, and BRCA2 are also seen following irradiation and during replication [43]. It has been postulated that there is a super complex of BRCA1-associated proteins consisting of MRE11/RAD50/NBS1, MMR-related proteins, BS-helicase, 53BP1, and BRCA1/2, which forms in the nucleus after DNA damage and acts as a damage surveillance "repairosome."

Mammalian cells also respond to radiation by phosphorylating serine 139 on histone H2AX [44]. Phosphorylated H2AX (γH2AX) is formed in thousands of copies per DNA DSB as an immediate response to damage [45]. γH2AX, which co-localizes with BRCA1, 53BP1, RAD50, and RAD51 over a period of hours, is phosphorylated by ATM and DNA-PK$_{cs}$. The detection and function of γH2AX is described in greater detail in Chapters 4 and 10.

7.3.3.9   Mechanism of NHEJ

The NHEJ system ligates the two ends of a DSB without the requirement of sequence homology between the DNA ends (Figure 7.12). Compared with

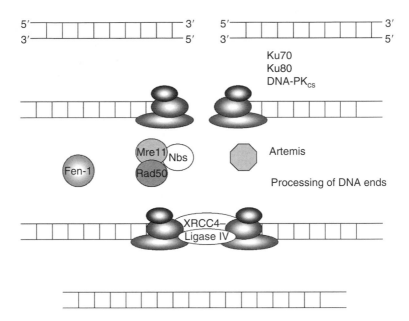

FIGURE 7.12
Mechanism of NHEJ. Following recognition and binding to damaged DNA by the Ku70/Ku80 complex the Ku heterodimer binds to DNA-PK$_{cs}$, forming the DNA-PK holoenzyme. DNA-PK activates XRCC4–ligase IV, which links the broken DNA ends together. Before re-ligation by XRCC4–ligase IV, the DNA ends are processed possibly by the MRE11–Rad50–NBS1 complex, with the involvement of FEN1 and Artemis.

the error-free precision of HR it is a rough and ready "duct tape" type solution to DNA damage. The process can be described as a series of steps:

- The initial step in NHEJ is the binding of a heterodimeric complex consisting of the proteins Ku70 and Ku80 (alias XRCC5) to the damaged DNA, thus protecting the DNA from exonuclease digestion. Ku binds to the DNA in such a way that Ku80 is distal to and Ku70 is proximal to the DNA break. The Ku70/Ku80 heterodimer has the ability to translocate from the DNA end in an ATP-independent manner.

- Following DNA binding, the Ku heterodimer associates with the catalytic subunit of DNA-PK (XRCC7, DNA-PK$_{cs}$) forming the active DNA-PK holoenzyme. DNA-PK$_{cs}$ is activated by interaction with a single-strand DNA at the site of DSB and displays Ser/Thr kinase activity.

- XRCC4 forms a stable complex with DNA ligase IV, and the XRCC4–ligase IV complex binds to the ends of DNA molecules and links together duplex DNA molecules with complementary but non-ligatable ends.

- The XRCC4–ligase IV complex cannot directly re-ligate most DSBs generated by mutagenic agents—they have to be processed first. In yeast, the processing of DSBs is performed by the MRE11–RAD50–NBS1 (MRN) complex which displays exonuclease, endonuclease, and helicase activity and removes excess DNA at 3′ flaps. One candidate responsible for removal of 5′ flaps is the flap endonuclease 1 (FEN1). Deficiency of this protein leads to a strong reduction in the usage of the NHEJ pathway. The involvement of the MRN complex in NHEJ of mammalian cells has not been definitively shown. A protein that is involved in processing overhangs during NHEJ in mammalian cells is Artemis. Artemis initially displays single-strand-specific exonuclease activity, but upon complexing with and being phosphorylated by DNA-PK$_{cs}$, Artemis acquires endonuclease activity, degrading single-strand overhangs and hairpins.

- In addition to frank DSBs, radiation causes clustered damage consisting of altered bases, AP sites, and SSBs which must also be repaired (more or less successfully) at this time.

- The NHEJ machinery must be removed from the DNA before the re-ligation of the DSB. Auto-phosphorylation of DNA-PK$_{cs}$ and/or DNA-PK mediated phosphorylation of accessory factors could play a role in the release of DNA-PK$_{cs}$ and Ku from the DSB before endjoining.

- Finally, DSB repair is completed although loss of nucleotides often occurs resulting in inaccurate repair.

Other roles have been suggested for Ku or the DNA-PK complex in addition to protecting the DNA ends from nucleolytic degradation. One possibility is that Ku/DNA-PK$_{cs}$ could potentiate ligation by tethering two DNA ends together. Ku is able to promote interactions between two DNA termini and can enhance end ligation by eukaryotic DNA ligases in vitro. It has been speculated that the weak helicase functions of Ku could play a role in dissociating the two strands of the DNA ends to allow microhomology alignments to be produced. However microhomology directed repair does occur in Ku80 knockout cells indicating that if Ku does play a role in this pathway, it cannot be essential.

Other possibilities are that once positioned at the DNA DSB, Ku and DNA-PK$_{cs}$ might provide a framework for recruiting and binding other NHEJ factors, Ku and DNA-PK$_{cs}$ might function to dissociate repair factors from the DNA after their job is complete, or they help to remove from the DNA other proteins, such as recombination factors that might block the repair process.

7.3.3.10 An Alternative, Microhomology-Dependent NHEJ Pathway

In some cases, NHEJ takes advantage of short stretches (1–4 nucleotides) of nucleic acid sequence complementarity near the ends of broken molecules.

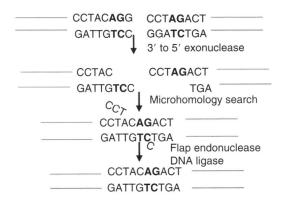

FIGURE 7.13
Alternative NHEJ pathway using micro-homology.

An example of microhomology usage is shown in Figure 7.13. Genetic evidence in mammalian systems suggests that a clean ligatable DSB as in line 1 of the figure would preferentially rejoin by the major NHEJ pathway, dependent on Ku, XRCC4, and ligase IV. However, the same studies suggest the presence of a second minor pathway in mammalian cells independent of normal NHEJ proteins, which requires microhomologies at DNA ends.

7.3.4   Telomere-Bound Proteins and DNA Repair

Telomeres are the network of proteins which bind to chromosomal ends and protect them from being inappropriately recognized as DNA damage. A number of double-stranded DNA-binding proteins have been characterized. One of these, TRF, negatively regulates telomerase access to the telomere and its DNA binding capacity is regulated by one of the poly-ADP ribosylation (PARP) enzymes. Another telomerase-binding protein TRF2 forms a complex with NHEJ proteins Ku70, Ku80, NBS1, MRE11, and RAD50. Ku70-deficient cells show increased rates of telomere–telomere fusion, and this and other lines of evidence suggest that repair proteins can act to protect chromosome ends from inappropriate fusion [46,47]. Removal of TRF telomeres through telomere uncapping or shortening leads to immediate activation of a DNA damage checkpoint, and critically damaged or shortened telomeres recruit DNA repair proteins as has been described for other types of DNA damage.

There is a clear association between DNA repair and telomerase activity and altered telomerase activity is frequently associated with the malignant phenotype. As described, telomeres interact with the repair proteins Mre11 and Ku70 to affect DNA repair and chromosomal stability following irradiation. Both telomere length and telomere function have been suggested to be possible determinants of chromosomal radiosensitivity. For example, mice that are deficient in telomerase are relatively radiosensitive secondary to an increased rate of apoptosis and an increase in terminal growth arrest in irradiated intestinal and stromal tissues, respectively. A relationship between telomere length and chromosomal radiosensitivity has been observed in lymphocytes from breast cancer patients and normal individuals [48].

7.4 Human Syndromes Involving DNA Repair Deficiency

Defects in DNA-repair pathways can result in cancer predisposition syndromes, one of the best documented of which is the defect in NER that causes xeroderma pigmentosum characterized by sensitivity to UV radiation (Table 7.4). However, human syndromes that involve proteins directly implicated in DSB repair are rare. Only a single leukemia patient has been identified as carrying a DNA ligase IV mutation resulting in severely increased sensitivity toward ionizing radiation [49].

A human disorder characterized by radiosensitivity that has been extensively studied is ataxia telangiectasia (A-T). Patients with A-T are radiosensitive and prone to cancer, with a particular predisposition to lymphoid malignancies. Cells from these patients show spontaneous chromosomal instability and fail to suppress DNA synthesis in response to ionizing radiation (radioresistant DNA synthesis). The gene that is mutated in these patients (ATM) encodes a member of the PI3K family,

TABLE 7.4

Human Repair Deficiency Disorders

Disease	Characteristics	Deficiency
Xeroderma pigmentosum	UV sensitive Sensitive to alkylating and cross-linking agents, defective NER	XPA-XPG
Ataxia telangiectasia	Radiosensitive, sensitive DNA breaking agents, defective DNA damage signaling	ATM
Trichothiodystrophy	Excision repair deficiency	
Fanconi's anemia	Deficient DNA cross-link repair	FANCA-FANCG
Bloom's syndrome	Deficient double-strand break repair High-level sister chromatid exchange	BLM (RecQ helicase)
Nijmegen breakage syndrome	DSB repair, radiosensitive	NBS
Werner Syndrome	DSB repair, radiosensitive	WRN
ATLD A-T like disease	DSB repair, radiosensitive	MRE-11

which also includes DNA-PK [50]. Cells from A-T patients have defective cell-cycle checkpoints and are defective in DNA repair (see Chapter 8).

The cellular phenotypes of A-T cells are very similar to those of cells from patients who are defective in the NBS 1 protein (also called Nibrin), a protein whose deficiency leads to the rare human genetic disorder called Nijmegen breakage syndrome (NBS). NBS is characterized by chromosomal instability, developmental abnormalities, and cancer predisposition (reviewed by Featherstone and Jackson [51]). NBS1 cells are radiosensitive and defects in NBS1 are reported to lead to impaired induction of p53 in response to radiation, thus providing a potential linkage between DNA DSB rejoining and DNA damage signaling. An explanation for the similarity in phenotypes of A-T and NBS cells is that the ATM kinase phosphorylates the polypeptide derived from the NBSl gene that is mutated in these patients. Phosphorylation of NBS1 catalyzed by ATM will result in regulation of the activity of the RAD50–MRE11–NBSl complex implying that ATM and the RAD50–MRE11–NBS1 complex function in the same pathway and in fact there is a human cancer predisposition syndrome with features very similar to A-T, known as ataxia telangiectasia-like disorder (ATLD), which is caused by mutations in the *MRE11* gene [52].

Several other chromosomal instability and cancer predisposition syndromes have links to DSB-repair pathways. Examples include Bloom's syndrome and Werner syndrome, which are caused by mutations in members of a class of DNA-unwinding enzymes from the so-called RecQ helicase family. Cells from Bloom's syndrome patients are defective in the *BLM* gene and show abnormally high levels of sister chromatid exchanges through the HR pathway [53]. The protein responsible for Werner syndrome (WRN) interacts with the Ku heterodimer resulting in increased WRN exonuclease activity and indicating a possible link between the WRN protein and DNA end-joining [54,55]. Finally, the breast-cancer-susceptibility gene products BRCAl and BRCA2 and their connection with HR were mentioned in the preceding section. Similar to NBSl, BRCAl is phosphorylated by the ATM kinase.

7.5   Relationship between DNA Repair and Cell Survival

No simple relationship can be discerned between expression of DNA repair genes or proteins and the relative radiosensitivity among unselected normal or tumor cells [56]. In some well-defined cell models, DNA repair capacity clearly does influence cellular radio-sensitivity, the extreme radiosensitivity of cells from some patients with DNA repair deficiency syndromes such as ataxia telangiectasia and the NBJ being obvious examples. Cells defective in the expression of the BRCA1 and BRCA2 proteins can have decreased HR-related repair of DNA-DSBs and decreased radiation cell survival [29]. A reduced capacity for repair of DNA DSBs is also seen for x-ray-sensitive

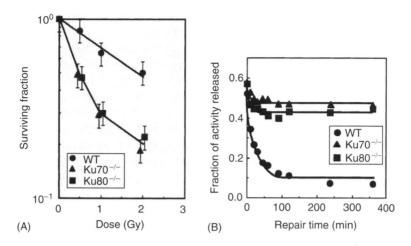

FIGURE 7.14

Disruption of Ku70 confers radiation hypersensitivity and a deficiency in DNA DSB repair. (A) Radiation survival curves for the CFU-GM in the bone marrow of wild-type (*WT*), *Ku70$^{-/-}$*, and *Ku80$^{-/-}$* mice. (B) Deficiency in the repair of radiation-induced DSB in *Ku70$^{-/-}$* and *Ku80$^{-/-}$* cells. Rejoining of DNA DSB produced by 40 Gy x-ray. (From Ouyang, H., Nussenzweig, A., Kurimasa, A., Soares, V., Li, X., Cordon-Cardo, C., Li, W.G., Cheong, N., Nussenzweig, M., Liakis, G., Chen, D.J., and Li, G.C., *J. Exp. Med.* 186, 921, 1997. With permission.)

mutant CHO cells and among radiosensitive fibroblasts derived from SCID mice in which deficient NHEJ was correlated to a lack of DNA-PK$_{cs}$ kinase expression [57]. Mouse cells made deficient for NHEJ proteins, cells which are knockouts for DNA-PK$_{cs}$ or Ku70 genes, show extreme radiosensitivity and defective rejoining of DNA-DSBs (Figure 7.14) [58].

A number of strategies have been designed to radiosensitize human tumor cell by targeting DSB repair. In human fibroblasts, small silencing RNAs (siRNA) have been used to decrease endogenous DNA-PK$_{cs}$ or ATM expression resulting in a reduced capacity for repair of radiation-induced chromosome breaks and an increased yield of acentric chromosome fragments. These chromosomal rearrangements are associated with increased radiation cell killing. Antisense RNA or specific pharmacological approaches targeting DNA repair factors such as Rad 51, DNA-PK$_{cs}$, ATR, or ATM have been shown to radiosensitize mammalian cells [59].

Molecular radiosensitization using inhibitors of DNA repair could be a promising area for clinical development if the repair of DNA-DSBs following irradiation in tumor tissues can be selectively targeted compared to that in normal tissues.

7.6   Summary

DNA repair mechanisms have evolved to remove or counter potentially cytotoxic, mutagenic, or clastogenic lesions. In the case of IR, the important mechanisms are BER, which repairs damaged bases and single-strand breaks, and HR and NHEJ, which repairs DSBs.

BER is responsible for removal of most of the base and sugar phosphate lesions produced by oxidative damaging agents such as IR and also repairs AP sites and SSBs. BER proceeds by sequential action of a DNA glycosylase, a 5′ AP endonuclease, and a DNA deoxyribophosphodiesterase to generate a single nucleotide gap in the DNA duplex which is filled in by DNA polymerase β (Polβ) using the opposing strand as the template for gap-filling synthesis. The final ligation step is performed by DNA ligases I and III. Members of the poly(ADP-ribose) polymerase (PARP) family play an important role in the regulation of DNA repair, particularly BER. BER processing of multiply damaged sites (MDS) may convert otherwise nonlethal lesions into lethal DSBs or other types of MDS.

There are two main pathways of DSB repair, HR and NHEJ. HR, which is error free, is mediated through the so-called RAD52 epistasis group of proteins and is initiated by a nucleolytic resection of the DSB in the 5′–3′ direction by the MRE11–RAD50–NBS1 (MRN) complex. Subsequently, RAD51 catalyzes strand-exchange events with the complementary strand during which the damaged DNA molecule invades the undamaged DNA duplex. DNA repair is completed by DNA polymerases, ligases, and Holliday junction resolvases. The BRCA2 cancer susceptibility protein is believed to be involved in homologous repair of DNA. The BRCA1 protein is phosphorylated by the ATM protein and appears to play a signaling role in response to ionizing radiation. Although mammalian cells rely less on HR than do lower eukaryotes, they may preferentially repair DSBs by HR in late S and G_2 phases of the cell cycle when an undamaged sister chromatid is available.

The mechanisms of NHEJ were elaborated using a group of x-ray-sensitive mutants of CHO cells that were found to be deficient in DSB rejoining. These cells belonged to three distinct complementation groups, termed IR4, IR5, and IR7, and the human genes complementing them were reassigned to the XRCC (x-ray cross-complementing) nomenclature. The gene products defined by IR complementation groups 5 and 7 operate within the same complex, identified as DNA-activated protein kinase (DNA-PK) that has specificity for double-stranded DNA ends. DNA-PK consists of a regulatory subunit, the Ku protein, and a catalytic subunit DNA-PK$_{cs}$. The Ku protein is a heterodimer comprising subunits of 70 and 80 kDa. The initial step in NHEJ is the binding of the heterodimeric complex consisting of the proteins Ku70 and Ku80 to DNA after which the Ku heterodimer associates with the catalytic subunit of DNA-PK, PK$_{cs}$ forming the active DNA-PK holoenzyme. Another target of DNA-PK$_{cs}$ is XRCC4, which forms a stable complex with DNA ligase IV and the XRCC4–ligase IV. The XRCC4–ligase IV complex cannot directly re-ligate most DSBs until they have been processed by the MRN complex.

The NHEJ proteins are also involved in the process of V(D)J rejoining, which creates new genes during lymphocyte development. Cells lacking Ku, XRCC4, or DNA ligase IV are defective in V(D)J rejoining. The *Artemis* gene product has been designated as the sixth component of NHEJ. Artemis is a single-strand-specific nuclease which, upon binding with DNA-PK, acquires endonucleolytic activity on 5' and 3' ends. Artemis-defective cells are radiosensitive and possess a defective V(D)J recombination phenotype.

There is a clear association between DNA repair and telomerase activity, and altered telomerase activity is frequently associated with the malignant phenotype. Telomeres interact with the repair proteins MRE11 and Ku70 to affect DNA repair and chromosomal stability following irradiation. Defects in DNA-repair pathways can result in cancer predisposition syndromes including a human disorder, ataxia telangiectasia (A-T), characterized by radiosensitivity. The phenotypes of A-T cells are very similar to those of cells from patients defective in the NBS1 protein and to those from another human cancer predisposition syndrome with features very similar to A-T, the ataxia telangiectasia-like disorder (ATLD) which is caused by mutations in the *MRE11* gene. Cellular radio-sensitivity is related to DNA repair capacity in the cases where DNA repair genes are mutated, including cells from patients with DNA repair deficiency syndromes, the x-ray-sensitive mutant CHO cells, and radiosensitive fibroblasts derived from SCID mice, which are deficient in NHEJ.

References

1. Ronen A and Glickman BW. Human DNA repair genes. *Environ. Mol. Mutagen.* 37: 241–283, 2001.
2. Wood RD, Mitchell M, Sgouros J, and Lindahl T. Human DNA repair genes. *Science* 291: 1284–1289, 2001.
3. Wallace SS. Enzymatic processing of radiation-induced free radical damage in DNA. *Radiat. Res.* 150: S60–S79, 1998.
4. Friedberg EC. How nucleotide excision repair protects against cancer. *Nat. Rev. Cancer* 1: 22–33, 2001.
5. Christmann M, Tomicic MT, Roos WP, and Kaina B. Mechanisms of human DNA repair: An update. *Toxicology* 193: 3–34, 2003.
6. Lindahl T, Satoh MS, Poirier GG, and Klungland A. Post-translational modification of poly(ADP-ribose) polymerase induced by DNA strand breaks. *Trends Biochem. Sci.* 20: 405–411, 1995.
7. D'Amours D, Desnoyers S, D'Silva I, and Poirier GG. Poly(ADP-ribosyl)ation reactions in the regulation of nuclear functions. *Biochem. J.* 342: 249–268, 1999.
8. Vidakovic M, Poznanovic G, and Bode J. DNA break repair: Refined rules of an already complicated game. *Biochem. Cell Biol.* 83: 365–373, 2005.
9. Burkle A. Poly(ADP-ribosyl)ation: A posttranslational protein modification linked with genome protection and mammalian longevity. *Biogerontology* 1: 41–46, 2000.

10. Veuger SJ, Curtin NJ, Smith GC, and Durkacz BW. Effects of novel inhibitors of poly(ADP-ribose) polymerase-1 and the DNA-dependent protein kinase on enzyme activities and DNA repair. *Oncogene* 23: 7322, 2004.
11. Haince JF, Rouleau M, Hendzel MJ, Masson JY, and Poirier GG. Targeting poly (ADP-ribosyl)ation: A promising approach in cancer therapy. *Trends Mol. Med.* 11: 456–463, 2005.
12. Chaudhry MA and Weinfeld M. Induction of double-strand breaks by S1 nuclease, mung bean nuclease and nuclease P1 in DNA containing abasic sites and nicks. *Nucleic Acids Res.* 23: 3805–3809, 1995.
13. Chaudhry MA and Weinfeld M. Reactivity of human apurinic/apyrimidinic endonuclease and *Escherichia coli* exonuclease III with bistranded abasic sites in DNA. *J. Biol. Chem.* 272: 15650–15655, 1997.
14. Weinfeld M. Complexities of BER. *Prog. Nucleic Acid Res. Mol. Biol.* 68: 125–127, 2001.
15. Yang N, Galick H, and Wallace SS. Attempted base excision repair of ionizing radiation damage in human lymphoblastoid cells produces lethal and mutagenic double strand breaks. *DNA Repair* 3: 1323–1334, 2004.
16. Jeggo PA. DNA breakage and repair. *Adv. Genet.* 38: 185–218, 1998.
17. Kanaar R, Hoeijmakers JH, and van Gent DC. Molecular mechanisms of DNA double strand break repair. *Trends Cell Biol.* 483–489, 1998.
18. Karran P. DNA double strand break repair in mammalian cells. *Curr. Opin. Genet. Dev.* 10: 144–150, 2000.
19. van Gent DC, Hoeijmakers JH, and Kanaar R. Chromosomal stability and the DNA double-stranded break connection. *Nat. Rev. Genet.* 2: 196–206, 2001.
20. Khanna KK and Jackson SP. DNA double-strand breaks: Signaling, repair and the cancer connection. *Nat. Genet.* 27: 247–254, 2001.
21. Jackson SP. Sensing and repairing DNA double-strand breaks. *Carcinogenesis* 23: 687–696, 2002.
22. Sancar A, Lindsey-Boltz LA, Unsal-Kacmaz K, and Linn S. Molecular mechanisms of mammalian DNA repair and the DNA damage checkpoints. *Annu. Rev. Biochem.* 73: 39–85, 2004.
23. Kurz EU L-MS. DNA damage-induced activation of ATM and ATM-dependent signaling pathways. *DNA Repair* 3: 889–900, 2004.
24. Collis SJ, DeWeese TL, Jeggo PA, and Parker AR. The life and death of DNA-PK. *Oncogene* 24: 949–961, 2005.
25. Kurz EU and Lees-Miller SP. DNA damage-induced activation of ATM and ATM-dependent signaling pathways. *DNA Repair* 3: 889–900, 2004.
26. Zhang J and Powell SN. The role of the BRCA1 tumor suppressor in DNA double-strand break repair. *Mol. Cancer Res.* 3: 531–539, 2005.
27. Powell SN. The roles of BRCA1 and BRCA2 in the cellular response to ionizing radiation. *Radiat. Res.* 163: 699–700, 2005.
28. Patel KJ, Yu VP, Lee H, Corcoran A, Thistlethwaite FC, Evans MJ, Colledge WH, Friedman LS, Ponder BA, and Venkitaraman AR. Involvement of Brca2 in DNA repair. *Mol. Cell.* 1: 347–357, 1998.
29. Powell SN and Kachnic LA. Roles of BRCA1 and BRCA2 in homologous recombination, DNA replication fidelity and the cellular response to ionizing radiation. *Oncogene* 22: 5784–5791, 2003.
30. Lee SE, Mitchell RA, Cheng A, and Hendrickson EA. Evidence for DNA-PK-dependent and -independent DNA double-strand break repair pathways in mammalian cells as a function of the cell cycle. *Mol. Cell Biol.* 17: 1425–1433, 1997.

31. Jeggo PA and Kemp LM. X-ray-sensitive mutants of Chinese hamster ovary cell line. Isolation and cross-sensitivity to other DNA-damaging agents. *Mutat. Res.* 313–327, 1983.

32. Thompson LH and Jeggo PA. Nomenclature of human genes involved in ionizing radiation sensitivity. *Mutat. Res.* 337: 131–134, 1995.

33. Kemp LM, Sedgwick SG, and Jeggo PA. X-ray sensitive mutants of Chinese hamster ovary cells defective in double-strand break rejoining. *Mutat. Res.* 132: 189–196, 1984.

34. Zdzienicka MZ. Molecular processes and radiosensitivity. *Strahlenther. Onkol.* 173: 457–461, 1997.

35. Caldecott K, Banks G, and Jeggo P. DNA double-strand break repair pathways and cellular tolerance to inhibitors of topoisomerase II. *Cancer Res.* 50: 5778–5783, 1990.

36. Jeggo PA, Caldecott K, Pidsley S, and Banks GR. Sensitivity of Chinese hamster ovary mutants defective in DNA double strand break repair to topoisomerase II inhibitors. *Cancer Res.* 49: 7057–7063, 1989.

37. Lieber MR. The biochemistry and biological significance of nonhomologous DNA end joining: An essential repair process in multicellular eukaryotes. *Genes Cells* 4: 77–85, 1999.

38. Gu Y, Jin S, Gao Y, Weaver DT, and Alt FW. Ku70-deficient embryonic stem cells have increased ionizing radiosensitivity, defective DNA end-binding activity, and inability to support V(D)J recombination. *Proc. Natl. Acad. Sci. USA* 94: 8076–8081, 1997.

39. Lees-Miller SP, Godbout R, Chan DW, Weinfeld M, Day RS, Barron GM, and Allalunis-Turner J. Absence of p350 subunit of DNA-activated protein kinase from a radiosensitive human cell line. *Science* 267: 1183–1185, 1995.

40. Bosma GC, Custer RP, and Bosma MJ. A severe combined immunodeficiency mutation in the mouse. *Nature* 301: 527–530, 1983.

41. Moshous D, Callebaut I, de Chasseval R, Corneo B, Cavazzana-Calvo M, Le Deist F, Tezcan I, Sanal O, Bertrand Y, Philippe N, Fischer A, and de Villartay JP. Artemis, a novel DNA double-strand break repair/V(D)J recombination protein, is mutated in human severe combined immune deficiency. *Cell* 105: 177–186, 2001.

42. Jeggo PA and Lobrich M. Artemis links ATM to double strand break rejoining. *Cell Cycle* 4: 359–362, 2005.

43. Paull TT, Rogakou EP, Yamazaki V, Kirchgessner CU, Gellert M, and Bonner WM. A critical role for histone H2AX in recruitment of repair factors to nuclear foci after DNA damage. *Curr. Biol.* 10: 886–895, 2000.

44. Rogakou EP, Boon C, Redon C, and Bonner WM. Megabase chromatin domains involved in DNA double-strand breaks in vivo. *J. Cell Biol.* 146: 905–916, 1999.

45. Modesti M and Kanaar R. DNA repair: Spot(light)s on chromatin. *Curr. Biol.* 11: R229–R232, 2001.

46. Song K, Jung D, Jung Y, Lee SG, and Lee I. Interaction of human Ku70 with TRF2. *FEBS Lett.* 481: 81–85, 2000.

47. Zhu X, Kuster B, Mann M, Petrini JH, and de Lange T. Cell-cycle-regulated association of RAD50/MRE11/NBS1 with TRF2 and human telomeres. *Nat. Genet.* 25: 347–352, 2000.

48. McIlrath J, Bouffler SD, Samper E, Cuthbert A, Wojcik A, Szumiel I, Bryant PE, Riches AC, Thompson A, Blasco MA, Newbold RF, and Slijepcevic P. Telomere length abnormalities in mammalian radiosensitive cells. *Cancer Res.* 61: 912–915, 2001.

49. Riballo E, Critchlow SE, Teo SH, Doherty AJ, Priestley A, Broughton B, Kysela B, Beamish H, Plowman N, Arlett CF, Lehmann AR, Jackson SP, and Jeggo PA. Identification of a defect in DNA ligase IV in a radiosensitive leukaemia patient. *Curr. Biol.* 9: 699–702, 1999.

50. Rotman G and Shiloh Y. ATM: From gene to function. *Hum. Mol. Genet.* 7: 1555–1563, 1998.

51. Featherstone C and Jackson SP. DNA repair: The Nijmegen breakage syndrome protein. *Curr. Biol.* 8: R622–R625, 1998.

52. Stewart GS, Maser RS, Stankovic T, Bressan DA, Kaplan MI, Jaspers NG, Raams A, Byrd PJ, Petrini JH, and Taylor AM. The DNA double-strand break repair gene hMRE11 is mutated in individuals with an ataxia-telangiectasia-like disorder. *Cell* 99: 577–587, 1999.

53. Karow JK, Newman RH, Freemont PS, and Hickson ID. Oligomeric ring structure of the Bloom's syndrome helicase. *Curr. Biol.* 9: 597–600, 1999.

54. Cooper MP, Machwe A, Orren DK, Brosh RM, Ramsden D, and Bohr VA. Ku complex interacts with and stimulates the Werner protein. *Genes Dev.* 14: 907–912, 2000.

55. Li B and Comai L. Displacement of DNA-PKcs from DNA ends by the Werner syndrome protein. *Nucleic Acids Res.* 30: 3653–3661, 2002.

56. Tenzer A and Pruschy M. Potentiation of DNA-damage-induced cytotoxicity by G2 checkpoint abrogators. *Curr. Med. Chem. Anticancer Agents* 3: 35–46, 2003.

57. Jeggo PA. The fidelity of repair of radiation damage. *Radiat. Prot. Dosimetry* 99: 117–122, 2002.

58. Ouyang H, Nussenzweig A, Kurimasa A, Soares V, Li X, Cordon-Cardo C, Li WG, Cheong N, Nussenzweig M, Liakis G, Chen DJ, and Li GC. Ku70 is required for DNA repair but not for T cell antigen receptor gene recombination in vivo. *J. Exp. Med.* 186: 921–929, 1997.

59. Collis SJ, Swartz MJ, Nelson WG, DeWeese TL, and Earle JD. Enhanced radiation and chemotherapy-mediated cell killing of human cancer cells by small inhibitory RNA silencing of DNA repair factors. *Cancer Res.* 63: 1550–1554, 2003.

8

Cellular Response to DNA Damage

8.1   Passing on the Message that DNA Has Been Damaged

As described in the preceding chapter, the cell can mobilize extensive resources to repair DNA damage. For the effective exploitation of these resources, the mechanism by which the damage is detected must be rapid, sensitive, and discriminate between multiple types of damage. Endogenous DNA damage occurs at high frequency with the loss of bases due to the spontaneous hydrolysis of DNA glycosyl bonds being of the order of 10^4 events per day per cell. At the same time, an array of physically dissimilar DNA lesions produced by exogenous insults must be distinguished from each other. Ultraviolet (UV) light induces dimerization of adjacent pyrimidines, photoproducts, and bulky DNA adducts creating obstacles to DNA transcription and replication, which are ultimately repaired by nucleotide excision repair (NER) proteins (Chapter 7). Ionizing radiation or the reactive oxygen intermediates produced as a consequence of oxidative metabolism cause DNA double-strand breaks (DSBs), which are repaired by homologous recombination or nonhomologous end-joining (NHEJ). Efficient detection of DNA damage is particularly important for dividing cells where replication or segregation of chromosomes bearing unrepaired lesions could seriously compromise genome integrity.

The spectrum of DNA lesions is recognized by DNA-damage-response pathways that transduce a signal from a damage sensor or detector to a series of downstream effector molecules via a sequence of events that is referred to as a signal-transduction cascade or pathway. This is initiated when the signal (DNA damage) is detected by a sensor (DNA-damage binding protein) triggering the activation of a transducer system (protein kinase cascade) that amplifies and diversifies the signal by targeting a series of downstream effectors of the DNA-damage response (shown schematically in Figure 8.1). This system is extremely sensitive and selective; it is triggered rapidly and efficiently by one or a few chromosomal DNA DSBs, while remaining inactive under other conditions.

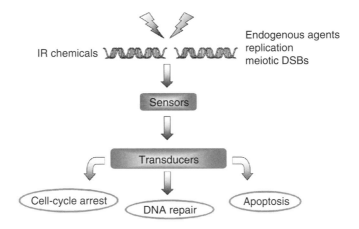

FIGURE 8.1
DNA-damage-response pathway. DSBs are recognized by sensor, which transmits the signal through transduction cascade to a series of downstream effector molecules, to activate signaling mechanisms for cell-cycle arrest, DNA repair, or cell death.

8.1.1   Signal Transduction

Much recent research in molecular biology has focused on the elucidation of signal-transduction pathways that are involved in all aspects of intracellular metabolism. In general, signal-transduction cascades act through a sequence of phosphorylation–dephosphorylation reactions catalyzed by specific enzymes (kinases) that transfer phosphate groups from ATP to the hydroxyl groups on amino acids and by other enzymes (phosphatases) that remove the phosphate groups (Figure 8.2). The protein kinases and phosphatases are divided into three groups based on the amino acids, which they phosphorylate and dephosphorylate. One class recognizes serine and threonine residues, another recognizes tyrosine residues, and a third, small group recognizes both serine–threonine and tyrosine residues. The kinases and phosphatases can have one or several substrates. The final stage is phosphorylation or dephosphorylation of an amino acid on the effector molecule,

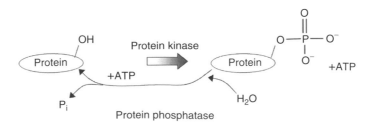

FIGURE 8.2
Phosphorylation catalyzed by protein kinase is the basic reaction of signal transduction. The protein is dephosphorylated by phosphatase.

which can cause stimulation or inhibition of the protein's enzymatic activity or its ability to bind to certain other proteins or to specific DNA sequences. Multiple signal-transduction cascades can operate in tandem, with variable amounts of connection or "cross-talk" between them.

8.1.2 Signal-Transduction Cascade Initiated by Radiation-Induced DNA Damage

In the course of the DNA damage–induced signal-transduction cascade, specific proteins bind preferentially to certain classes of DNA lesion. For example, the MSH2/3/6 proteins bind to mismatched bases, the Ku hetero-dimer binds to DSBs, and the Xeroderma pigmentosum (XP) group C protein (XPC) selectively recognizes UV-induced DNA photoproducts. In addition to detecting different types of DNA lesions, the cell must also be able to recognize very low levels of DNA damage anywhere in the genome. The rapidity and potency of the DNA-damage response indicates that the signaling proteins involved are very sensitive and have the capacity to greatly amplify the initial stimulus. In mammalian cells, this is the role of the ATR/ATM network, a highly conserved protein kinase cascade that is critical for cellular responses to many types of DNA damage.

8.2 ATM Protein

The material in this part of Chapter 8 is based on reviews by Barzilai et al. [1], Khanna et al. [2], Fei and El-Deiry [3], Canman and Lim [4], Shiloh [5], Abraham [6], Lobrich and Jeggo [7], and the references cited therein.

The research that led to the identification of the ATM (mutated in ataxia telangiectasia) protein began with the recognition of patients with the heritable syndrome ataxia telangiectasia (A-T). A-T is a highly pleiotropic disease that includes progressive degeneration of the cerebellar cortex leading initially to ataxia (lack of balance) and to severe neuromotor dysfunction when other parts of the central nervous system are affected in the later stages; telangiectases or dilated blood vessels that typically appear in the eyes and sometimes on the facial skin; primary immunodeficiency involving the cellular and humoral arms of the immune system; degeneration of the thymus and gonads; retarded somatic growth; varying signs of premature aging; and acute predisposition to lymphoreticular malignancies. These patients were also found to be extremely sensitive to ionizing radiation, which became apparent when they were occasionally treated with radio-therapy for cancer.

Cells derived from A-T patients, which were induced to grow in tissue culture, were found to exhibit chromosomal instability and extreme sensitivity to DSB-inducing agents, including ionizing radiation and radiomimetic

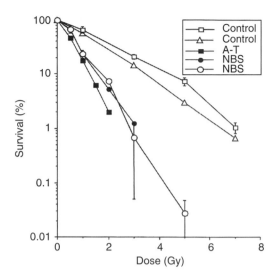

FIGURE 8.3
Radiation survival curve for normal human fibroblasts control, fibroblasts from a patient with A-T and fibroblasts from a patient with NBS. (From Girard, P.M., Foray, N., Stumm, M., Waugh, A., Riballo, E., Maser, R.S., Phillips, W.P., Petrini, J., Arlett, C.F., and Jeggo, P.A., *Cancer Res.*, 60, 4881, 2000. With permission.)

chemicals such as bleomycin and neocarzinostatin, but little or no sensitivity to other forms of DNA damage. A radiation survival curve for cells derived from an A-T patient is shown in Figure 8.3. A-T cells are also deficient in radiation-induced G_1-S, intra-S, and G_2-M cell-cycle checkpoints, but they are not as profoundly deficient in DSB repair as are cells of the XRCC cell lines described in Chapter 7. Further studies revealed the cellular basis for these defects and led eventually to the cloning of the ATM gene and characterization of the ATM protein [8].

The ATM gene encodes a 370 kDa protein characterized by a carboxy-terminal region containing the signature motif of phosphatidylinositol 3-kinases (PI 3-kinases) although ATM differs from other PI 3-kinases in that it phosphorylates proteins rather than lipids. The PI 3-kinases, most of which are serine/threonine kinases, function in eukaryotes ranging from yeast to mammals in various pathways associated with DNA-damage responses and cell-cycle control. Another mammalian member of the PI 3-kinase family is DNA-PK$_{cs}$, the catalytic subunit of the DNA-dependent protein kinase described in Chapter 7, while the family member most closely related to ATM is the ATR protein that mediates cellular responses to unreplicated or damaged DNA.

8.2.1   Functions of ATM

The radiosensitivity of cells derived from A-T patients reflects a broad defect in the cellular response to DSBs spanning the activation of all the signaling pathways that respond to the DSB lesion. ATM mediates this response by phosphorylating numerous substrates, which are involved in specific signaling pathways (Figure 8.4). Many signaling pathways are activated by DNA damage. Some of them respond specifically to DSBs,

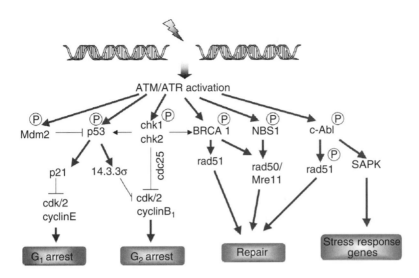

FIGURE 8.4
ATM is activated in response to DSBs and signals the presence of DNA damage by phosphorylating downstream targets including p53. Downstream effectors of p53 are p21/Cip1 and 14-3-3σ. p21 inhibits the activity of cdk2/cyclinE and 14-3-3σ inhibits the activity of cdc2/cyclin B causing cell-cycle arrest, which is also mediated by activation of Chk1 and Chk2. c-Abl activates stress-activate protein kinase (SAPK) for transcriptional regulation of stress-response genes. Other proteins (BRCA1, NBS1) are involved in DNA repair.

the major damage caused by ionizing radiation while others are activated by base modifications, such as those created by alkylating agents or UV light. The induction of a cellular response to DSBs depends on a functional ATM protein. However, even in ATM-deficient cells, the responses are attenuated or delayed rather than completely abolished. This indicates that the ATM kinase is necessary for the immediate, rapid response to damage, but other enzymes can kick in to induce the same pathways at a later stage. There is a rapid enhancement of ATM kinase activity, which follows treatment with agents that generate DSBs, such as radiation, radiomimetic agents, and topoisomerase inhibitors. ATM appears to be neither activated by nor involved in the cellular response to other types of DNA damaging agents to which A-T cells are insensitive such as alkylating agents or UV. The ATM-related protein ATR responds to UV-induced damage and stalled replication forks, but ATR is also involved in maintaining the later stages of the response to DSBs that was initiated by ATM. ATM and ATR share many targets that are initially phosphorylated by ATM immediately after DSB induction, but are later targeted by ATR, which maintains these phosphorylations for extended periods of time. Thus the downstream targets for ATM, which include p53, Mdm2, CHK1, CHK2, BRCA1, and NBS, are still phosphorylated but at a slower rate in ATM deficient A-T cells, and this activity is maintained primarily by ATR. ATR and ATM have overlapping substrate specificities, however phosphorylation and stabilization of p53 after radiation are largely ATM-dependent, whereas ATR mediates the UV-induced rapid phosphorylation of p53.

8.2.2 How Does ATM Respond to Radiation-Induced DNA Damage?

The mechanisms by which eukaryotic cells sense DNA strand breaks are not completely explained, but the rapid induction of ATM kinase activity following radiation suggests that it acts at an early stage of signal transduction in mammalian cells and raises the possibility that ATM kinase activity is modulated by post-translational events.

It was proposed by Bakkenist and Kastan [9] that ATM is sequestered in unperturbed cells as a dimer or a higher-order multimer with its kinase domain bound to an internal domain of a neighboring ATM molecule containing serine 1981. This interdomain interaction must occur for ATM to fold correctly and to remain stable in the cell. While contained in this complex, ATM is unable to phosphorylate other cellular substrates. After DNA damage, the kinase domain of one ATM molecule phosphorylates serine 1981 of an interacting ATM molecule. In effect, ATM can phosphorylate itself in a similar manner to that by which it phosphorylates the target substrates. The phosphorylation event does not directly regulate the activity of the kinase, but instead disrupts ATM oligomers allowing accessibility of substrates to the ATM kinase domain. Phosphorylated ATM dissociates from the complex and is free to phosphorylate other substrates in the cell. The rapidity and stoichiometry of the reaction indicate that ATM is not activated by binding directly to DNA strand breaks, and the same authors presented data to support a model in which DNA damage rapidly causes changes in higher-order chromatin structures that initiates ATM activation.

Experimental evidence supporting the autophosphorylation model includes the fact that exposure of cells to IR significantly increases the incorporation of ^{32}P-orthophosphate into both transfected ATM and endogenous ATM, but no such increase is observed after the irradiation of cells that had been transfected with kinase-inactive ATM. In addition, incorporation of radioactive phosphate into ATM occurs in in vitro assays of ATM kinase activity and this in vitro phosphorylation of ATM, which is inhibited by exposure to kinase inhibitors, is dependent on the presence of cofactors and is not dependent on the addition of exogenous DNA. Identical properties are characteristic of phosphorylation of target substrates by ATM. These observations are all consistent with a model in which ATM phosphorylates itself.

A high fraction of cellular ATM is rapidly phosphorylated in the presence of a small number of DNA strand breaks, in fact, over 50% of the ATM molecules in the cell become phosphorylated within a 5 min exposure to 0.5 Gy x-rays, a dose that would be expected to induce only about 18 DSBs in the genome of a mammalian cell. This is not consistent with the idea that ATM has to bind directly to DNA DSBs to become activated, but rather suggests that the introduction of DNA breaks must signal to ATM molecules at a distance in the cell. It is known that DNA strand breaks introduced by exposure to radiation can rapidly alter topological constraints on DNA (discussed in Chapter 9), and alterations in some aspect of chromatin structure could meet the criteria of a rapid change that can occur at a distance in the nucleus. This interpretation is supported by results of experiments in which changes in chromatin structure induced hypotonic conditions, or by chromatin modifying drugs activated ATM in the absence of DNA strand breaks. In these conditions, ATM substrates (for example, H2AX) that would be phosphorylated at the site of breaks if a DSB were the signal fail to become phosphorylated, whereas substrates present elsewhere in the nucleus (for example p53) can still be phosphorylated. It is suggested that the amount of ATM phosphorylation becomes maximal at a dose of 0.5 Gy because the number of breaks produced by this dose is sufficient to relax enough higher-order loops to signal to all of the ATM contained in a cell.

8.2.3   Role of ATM in DNA Repair

ATM deficient cells show cell-cycle checkpoint defects and profound radiosensitivity. Checkpoint responses allow more time for damage repair and may also lead to permanent cell-cycle arrest, while ATM also controls the onset of apoptosis in response to radiation. However, these factors alone do not confer radiosensitivity, and until recently the basis of A-T radiosensitivity has remained unclear. There is now substantial evidence that A-T cells harbor a DNA repair defect. Support for this comes from several sources. Prolonged holding of normal G_0/G_1 cells after ionizing radiation leads to an enhanced survival, a process called repair of potentially lethal damage (PLDR). A-T cells do not show PLDR. Analysis of cells held under these conditions for chromosome breaks visualized by premature chromosome condensation (PCC) (see Chapter 4 for methods) shows A-T cells to have elevated chromosome breakage compared with normal cells. These cells have not traversed any checkpoints, so this strongly suggests deficient repair. A-T cells also show a small DNA repair defect detectable by pulsed field electrophoresis (PFGE) (see Chapter 4).

The analysis of DSB repair relied for many years on PGFE and neutral elution, techniques which require the use of high radiation doses. Recently, it has been demonstrated that the formation and loss of γ-H2AX foci is a highly sensitive technique to monitor DSB formation and repair. Analysis of DSB repair by this technique has confirmed that A-T cells have a small DSB repair defect detectable at physiologically relevant doses manifest as a failure to repair approximately 10% of radiation-induced breaks (Figure 8.5). The initial kinetics of repair are normal in A-T cells with the defect only being observed at longer times (>4 h) after radiation. Cells lacking the nuclease Artemis show an identical defect. DSBs generated by radiation rarely have 5'-P and 3'-OH termini, which are a prerequisite for ligation and Artemis is one of a range of repair proteins that process broken ends before

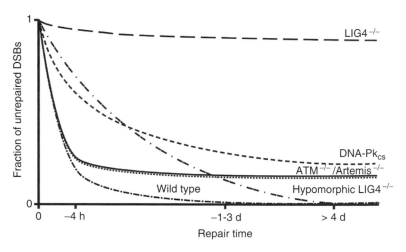

FIGURE 8.5

Kinetics of DSB rejoining in cell lines lacking damage-response proteins. Cells with hypomorphic DNA ligase IV mutation can slowly repair all breaks and can reach a level of un-rejoined DSBs similar to wild-type cells. In contrast, ATM$^{-/-}$ and Artemis$^{-/-}$ cells show proficient DSB repair at early times post-radiation but fail to rejoin a subset of DSBs (10%) normally repaired with slow kinetics in wild-type cells. DNA-PK$_{cs}$-deficient cells show slightly impaired repair kinetics at early times post-radiation, a more pronounced defect at intermediate times (4–24 h), and a dramatic defect at longer repair times. (From Lobrich, M. and Jeggo, P.A., *DNA Repair (Amst)*, 12, 749, 2005. With permission.)

the action of DNA ligase IV and XRCC4 during NHEJ. The role of ATM appears to be to phosphorylate and potentially activate Artemis. ATM-dependent DSB rejoining also requires H2AX, 53BP3, and the Mre11-Rad50-nibrin (MRN) complex. DNA damage and repair are discussed in greater detail in Chapters 6 and 7, respectively.

8.2.4 ATM and the MRN Complex

Among the many downstream ATM targets is the MRN complex. Both nibrin and Mre11 are phosphorylated by ATM in response to ionizing radiation. In vitro studies have shown Mre11-Rad50 can bind to DNA and that Mre11 possesses a nuclease activity that can process broken DNA ends. Nibrin is responsible for translocating the MRN complex to the nucleus and relocalizing the complex to the sites of DSBs following irradiation. The connection between ATM and the MRN complex in the DSB response is underscored by the existence of radiosensitivity disorders caused by mutations in the genes encoding these proteins. Mutations in the *ATM* gene result in A-T, *MRE11* gene mutations cause A-T-like disorder (ATLD), and mutations in the *NBS1*gene are responsible for Nijmegen breakage syndrome (NBS) (see Table 7.4). These autosomal recessive disorders share some common features including immunological defects and a predisposition to lymphoid cancers. Cells derived from A-T, ATLD, and NBS patients are hypersensitive to ionizing radiation and display spontaneous chromosomal instability, frequently involving chromosomes 7 and 14, as well as induced chromosomal instability. In response to ionizing radiation, A-T, ATLD, and NBS cells have cell-cycle checkpoint defects, notably in the S-phase checkpoint as measured by radioresistant DNA synthesis [10]. The clinical similarities between these syndromes in addition to the genome instability and checkpoint insufficiencies observed in cell lines derived from the patients suggest that the MRN complex plays an important role in ATM-dependent DNA-damage signaling in addition to its essential roles in DNA repair. Recent evidence suggests that the MRN complex is important for the activation of ATM by DSBs in cells, and in vitro studies of the effects of MRN on ATM indicate that MRN stimulates ATM through multiple protein–protein contacts, and that this interaction increases the affinity of ATM for its substrates [11].

8.3 Tumor Suppressor Gene p53

ATM regulates the G_1/S checkpoint after IR exposure, at least in part by controlling the activation and stabilization of the tumor suppressor protein p53. The unmutated form (wild type) of the p53 gene has become one of the most extensively studied genes in both normal and tumor cells. (This part of Chapter 8 is based on reviews by Cuddihy and Bristow [12], Giaccia and Kastan [13], Amundson et al. [14], Freedman et al. [15], and Fei and El-Deiry [3]. Mutations in the p53 gene have been found in the majority of human cancers including brain, lung, breast, prostate, thyroid, bladder, colon, liver, musculoskeletal, lymphoma, leukemia, head and neck, and gynecologic cancers. Because of its role in eliminating cells containing damaged DNA, p53 has been described as the "guardian of the genome" [16].

The product of the p53 tumor suppressor gene is a 53 kDa nuclear phosphoprotein that contains well-defined domains involved in gene transactivation, protein oligomerization, specific and nonspecific DNA binding, binding of cellular and viral oncoproteins, and nuclear localization (Figure 8.6). The central core region of the p53 protein (amino acids 100–293), which is essential for its site-specific double-stranded DNA binding, folds in such a way as to form a domain that interacts with DNA in a sequence-specific manner. The majority of missense mutations seen in human (and animal) tumors occur in regions of the gene encoding this domain and involve either alteration of critical residues involved in DNA contact or disruption of the conformation of the core domain of the protein. The consequence of either of these events is the loss of the ability of p53 to bind to DNA in a sequence-specific manner. The consensus p53 DNA binding site is composed of two copies of a 10 bp sequence separated by 0–13 bp. The binding to DNA is optimal when the protein is in a tetrameric state and there is interaction of

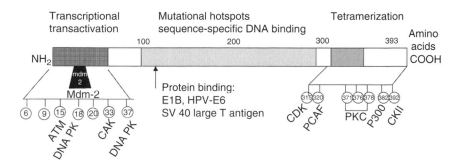

FIGURE 8.6

Structure of the p53 protein and location of stress-induced modifications. The transactivation domain is important for transactivation of a number of transcription factors. The central domain (amino acids 100–300) is involved in sequence-specific DNA binding and is evolutionarily conserved. The final 100 amino acids are concerned with nonspecific DNA binding and damage recognition. Sites at the N-terminus include those known to be phosphorylated in vitro by ATM, DNA-PK, or cyclin activating kinase complex (CAK). C-terminal phosphorylation sites include those for CDK, PK-C, and CKI1 protein kinases and acetylation sites PCAF and p300 for acetyl transferases.

four separate p53 molecules via the tetramerization or oligomerization (amino acid residues 324–355). Mutation may result in a dominant-negative form of the protein that can heterodimerize with wild-type p53 protein and disrupt its DNA binding, inhibiting downstream gene activation.

The function of p53 is to integrate signals emanating from a variety of cellular stresses and mediate response to these insults by activating a set of genes whose products facilitate adaptive and protective activities including apoptosis and growth arrest. p53 effector genes with roles in growth arrest and apoptosis are listed in Table 8.1.

The ability of p53 to either inhibit proliferation or to kill a cell must be restrained under normal circumstance; and the potentially dangerous p53

TABLE 8.1

Partial List of p53 Effector Genes

Cell-Cycle Control	Apoptosis	DNA Repair
CIP1/WAF1	BAX	XPC
ClnG	BCL-X	DDB2
ClnD1	PAG608	GADD45
WIP1	FAS/APO1	PCNA
EGF-R	TRUNDD	
Rb	TRID	
PCNA	IGF-BP3	
GADD45		
14-3-3σ		
IGF-BP3		

TABLE 8.2

Regulation p53—Proteins that Positively Regulate p53

Positive Regulators of p53	Binding/Modification of p53	Effects on p53
ATM	Phosphorylation (ser 15)	Stabilization, activation
DNA-PK	Phosphorylation (ser 15, 3)	Stabilization, activation
ATR	Phosphorylation (ser 15)	Stabilization, activation
JNK	Phosphorylation (ser 33)	Stabilization(stressed cells)
CDK7/cycII/p36	Phosphorylation (ser 33)	Activation
Cdks	Phosphorylation (ser 315)	Activation
p38	Phosphorylation (ser 392)	Activation
Casein kinase	Phosphorylation (ser 392)	Activation
Protein kinase C	Phosphorylation (ser 371, 376, 378)	Activation
P300	Acetylation(lys382)	Stabilization, activation
PCAF	Acetylation(lys320)	Activation
c-Abl	Binding to p53, Antagonizing Mdm2	Stabilization, activation
p19^{ARF}	Antagonizing Mdm2	Stabilization, activation
E2F-1	Induction of p19^{ARF}	Stabilization, activation
c-myc	Induction of p19^{ARF}	Stabilization, activation
Rb	Partially antagonizing mdm2	Stabilization
Ref-1	Regulation of the redox state	Activation
PARP	Binding to p53	Stabilization, activation
BRCA1	Binding to p53	Activation
P33^{ING}	Binding to p53	Activation

protein is controlled by a variety of regulatory mechanisms that act at two levels, that of p53 protein stability and that of biochemical activity, whereby p53 is converted from a latent protein into an active protein. Both the expression and activity of p53 are tightly regulated by multiple positive and negative feedback loops. Tables 8.2 and 8.3 give a partial list of proteins that regulate p53 by binding to the protein, by post-translational activation or by indirect means.

TABLE 8.3

Regulation p53—Proteins that Negatively Regulate p53

Negative Regulators of p53	Binding/Modification of p35	Effects on p53
Mdm-2	Binding to p53, nuclear cytoplasmic export, ubiquitination	Inhibition
JNK	ubiquitination	Destabilization (in unstressed cells)
Bcl-2	Blocking p53 nuclear export	Inhibition
BRCA2	Binding to p53	Inhibition
MDMX	Binding to p53	Inhibition
IGF-1, bFGF, T3R	Induction of Mdm-2 expression	Antagonizing p53 through Mdm2

8.3.1 Turnover of p53: Mdm2

Wild type p53 protein has a very short half-life under normal conditions but exposure of cells to DNA damage, such as that caused by ionizing radiation, UV light, or anticancer drugs, can lead to a rapid accumulation of p53 protein. A key player in this regulation is the Murine Double Minute-2 (Mdm2) proto-oncoprotein. Mdm2 binds p53 within its transactivation domain, blocks its transcriptional activity, and consequently abrogates the ability of p53 to induce growth arrest and apoptosis. Since Mdm2 itself is a direct target of p53, a negative autoregulatory feedback loop exists between these two proteins (shown schematically in Figure 8.7). Activated p53 increases transcription from the Mdm promotor, and Mdm in turn binds to p53 inhibiting its transcription promoting activity and targeting it for degradation by the ubiquitin–proteasome system and also facilitating the nuclear-cytoplasmic export of p53. The human papillomavirus (HPV) E6 protein also promotes p53 degradation via the ubiquitin–proteasome system and here too nuclear export is required for p53 degradation suggesting some similarities in the mechanisms by which E6 and Mdm2 promote p53 degradation.

8.3.1.1 Protein Degradation by the Ubiquitin–Proteasome System

The proteasome is an intracellular complex whose function is to degrade unneeded proteins. Protein degradation is as essential to the cell as is protein synthesis since it supplies amino acids for fresh protein synthesis, removes excess enzymes, and removes transcription factors that are no longer needed.

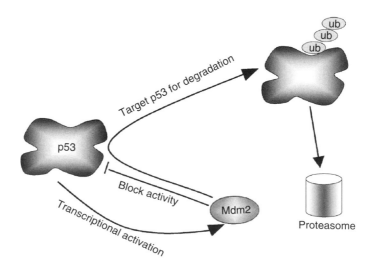

FIGURE 8.7
Negative autoregulatory feedback loop between p53 and Mdm2. The basal level of p53 activates the basal level of downstream genes including Mdm2. Mdm2 protein binds p53 inactivating its activity as a transcription factor and targeting p53 for ubiquitin-mediated degradation.

The two major intracellular programs by which damaged or unneeded proteins are broken down are lysosomes and proteasomes. Lysosomes deal primarily with extracellular proteins such as plasma proteins that are taken into the cell by endocytosis, or cell-surface membrane proteins that are used in receptor-mediated endocytosis. Proteasomes, in contrast deal primarily with endogenous proteins, i.e., with proteins that are synthesized within the cell. These include

- transcription factors, for instance p53,
- cyclins (which must be destroyed to prepare for the next step in the cell cycle),
- proteins encoded by viruses and other intracellular parasites, and
- proteins that are folded incorrectly because of translation errors or are encoded by faulty genes or have been damaged by other molecules in the cytosol.

The proteasome consists of a core particle (CP), a regulatory particle (RP) and the small protein ubiquitin. The CP is made of 2 copies of each of 14 different proteins, which are assembled in groups of 7 forming a ring with the 4 rings stacked on top of each other like a pile of doughnuts (Figure 8.8).

There are two identical regulatory particles, one at each end of the CP. Each consists of 14 different proteins (none of which are the same as those in the CP), 6 of which are ATPases. Some of the subunits have sites that recognize ubiquitin, a small protein (76 amino acids) that has been conserved throughout evolution and has virtually the same sequence whether

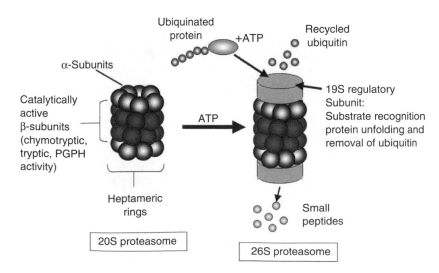

FIGURE 8.8
Schematic representation of a proteasome.

in bacteria, yeast, or mammals. The function of ubiquitin in all these organisms is to target proteins for destruction.

Proteins that are destined for destruction are conjugated to a molecule of ubiquitin that binds to the terminal amino group of a lysine residue. Additional molecules of ubiquitin bind to the first forming a chain, and the complex binds to ubiquitin-recognizing sites on the regulatory particle. The protein is unfolded by the ATPases of the regulatory molecule using the energy generated by the breakdown of ATP. The unfolded protein is then translocated into the central cavity of the CP where several active sites on the inner surface of the two middle molecules break specific bonds of the peptide chain. This produces a set of smaller peptides averaging about eight amino acids. These leave the CP and may be further broken down into individual amino acids by peptidases in the cytosol or in mammals, incorporated in a class I histocompatibility molecule to be presented on the cell surface to the immune system as a potential antigen. The regulatory particle releases the ubiquitins for reuse.

8.3.2   Modulation of p53 Stability and Activity

In unstressed cells, p53 has a short half-life and is strictly maintained at low levels by ubiquitin-mediated proteolytic degradation. Nevertheless, several mechanisms exist, which can regulate the stability and activity of p53. These act by modulation of the inhibitory effect of Mdm2, through subcellular localization of the p53 protein and through post-translational modifications that modulate both its stability and activity.

8.3.2.1   Bypassing Mdm2

The major cause of p53 turnover and short half-life is its interaction with Mdm2; however, there are several mechanisms through which the inhibitory effect of Mdm2 can be bypassed. These include down-regulation of Mdm2 expression, prevention of the Mdm2–p53 interaction, and inhibition of Mdm2-mediated degradation of p53. The p53 protein is stabilized after DNA damage, even in tumor cells over-expressing Mdm2, if the Mdm2 effect is neutralized. This regulation occurs when the p53 molecule is phosphorylated or acetylated at key sites in such a way as to impair its interaction with Mdm2. For example p53 is stabilized when it is phosphorylated on serine 15 by DNA-dependent protein kinase (DNA-PK) by ATM or by ATR. Similarly, phosphorylation of p53 on Ser-20 and possibly other sites is mediated by Chk1 and Chk2. These phosphorylations impair the interaction between p53 and Mdm2 and account for the stabilization of p53 after activation by these kinases.

The human tumor suppressor p14^{ARF} stabilizes p53 by binding and antagonizing Mdm2 and has also been shown to affect p53 directly with modulation of p53 trans-activation activity. Over-expression of p14^{ARF} activates p53-dependent cell-cycle arrest in both G_1 and G_2/M phases.

Other regulators of p53 stability include the Jun-NH$_2$-terminal kinase-1 (JNK1) (Chapter 11) and the c-Abl nonreceptor tyrosine kinase, which neutralizes the effect of Mdm2 on p53 degradation by blocking the ability of Mdm2 to promote p53 degradation.

8.3.2.2 Localization of p53

For p53 to bind DNA and act as a transcription factor, the stabilized protein must be localized to the nucleus, thus activation of p53 can be regulated through its subcellular localization. The carboxy-terminal domain of the protein contains a nuclear localization and nuclear export sequences which regulate the subcellular localization of p53. Tetramerization of p53 masks the nuclear export signal, thus retaining active p53 in the nucleus and preventing it from being exported to the cytoplasm where it is subject to degradation.

8.3.2.3 Post-Translational Modification

The carboxy terminus of the p53 protein is the site of phosphorylation and acetylation, and these post-translational modifications may regulate the ability of p53 to function as a transcription factor. p53 is activated by several stress conditions in addition to DNA damage, including hypoxia, changes in the redox potential, a reduction in the ribonucleoside triphosphate pool, adhesion, and the expression of several oncogenes (Figure 8.9). In response to these signals, p53 is subjected to extensive post-translational modifications,

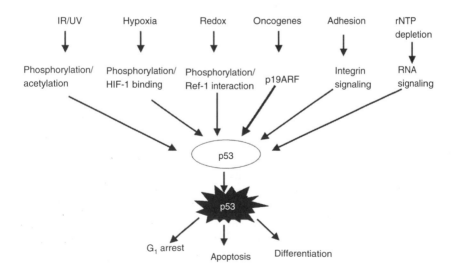

FIGURE 8.9
A number of stimuli, mediated by post-translational modifications and protein–protein interactions bring about inactivation of p53.

including phosphorylation and acetylation, which modulate both its stability (as described above) and its activity. p53 is phosphorylated on numerous serines in the carboxy and amino terminal regions of the protein with the central core remaining relatively free of post-translational modification. Phosphorylation of sites in the amino terminal region which influence binding of Mdm2 and p53 stability have already been described. In the carboxy-terminal region of the protein which is implicated as the regulatory region of the protein, a number of kinases have been identified as phosphorylating specific serine sites including cdk2 and cdc2 (serine 315), protein kinase C (serine 378), and casein kinase II (serine 292). p53 is also the substrate for several histone acetylases and acetyl transferases.

8.4 Radiation-Induced Growth Arrest

Radiation-induced delays in the G_1-S and G_2 phases of the cell cycle of human and rodent cell lines were first described by radiobiologists many years ago [17]. They were interpreted as being periods of time during which the cells could survey and repair DNA damage. A role for the p53 gene in the radiation-induced G_1 delay was first reported by Kastan et al. [18] who observed that increased expression of the p53 protein correlated with the onset of a G_1 arrest.

Before discussing the mechanisms of radiation-induced growth arrest, it is necessary to describe in more detail how the cell cycle works.

8.4.1 Cell Cycle: Cyclins and Cyclin Dependent Kinases

The cell cycle is governed by a family of cyclin-dependent kinases (cdks), which regulate a series of biochemical pathways, or checkpoints, that integrate mitogenic and growth-inhibitory signals, monitor chromosome integrity and coordinate the orderly sequence of cell-cycle transitions (Figure 8.10). This section of Chapter 8 is largely based on reviews by Hartwell [19], Sherr [20], Morgan [21], Solomon [22], King et al. [23], Slingerland and Tannock [24], and the references contained therein.

The activity of the cdks is regulated, as the name implies, by the binding of each cdk to its specific positive effector protein, a cyclin. The oscillation of cyclin abundance is an important mechanism by which these enzymes control phosphorylation of key substrates to ensure that cell-cycle events occur at the correct time and sequence (Figure 8.11). While cyclin binding to its cdk is required for kinase activation, an additional level of control is provided by the activities of two different families of cdk inhibitory molecules.

In mammalian cells, the family of cdks, cdk 1–7 (the nomenclature of the cyclins and cdks is largely derived from the order in which they were identified), are conserved in size, ranging from 32 to 40 kDa, and share sequence homology (>40% identity). They are small serine/threonine

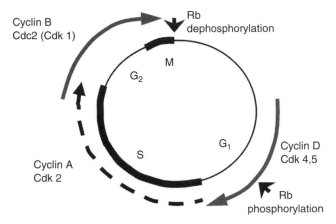

FIGURE 8.10
Overview of the regulation cell cycle by protein kinases (cdks) activated by cyclins.

kinases, expressed at constant levels throughout the cell cycle and are cata-
lytically inactive unless they are bound to cyclins. The interaction of cyclins
and cdks is shown schematically in Figure 8.11. Phosphorylation of a conserved threonine, located in the catalytic cleft of the kinase (Thr 161 for cdkl
and Thr 160 for cdk2) is required for full activation, and this is catalyzed
by the cdk-activating kinase, or CAK. Cdk activation can be inhibited by
phosphorylation of conserved inhibitory sites at Thr 14, and Tyr 15 by a
group of enzymes, called the wee-l kinases. Complete activation of cdk
requires dephosphorylation of these inhibitory sites by phosphatases of the
cdc25 family.

While cdks remain constant, cyclin levels oscillate during the cell cycle
and cyclin mRNA and protein expression peak at the time of maximum

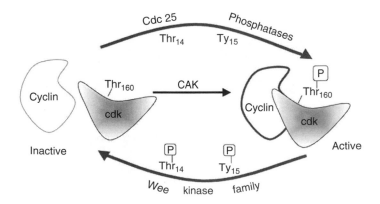

FIGURE 8.11
Schematic diagram of cyclin–cdk interaction. Phosphorylation of threonine [Thr] in the catalytic
cleft is required for full activity. Cdk activity is inhibited by phosphorylation of threonine and
tyrosine [Tyr] at positions 14 and 15.

kinase activation, contributing to discrete bursts of kinase activity at specific cell-cycle transitions. Degradation of cyclin proteins (by the ubiquitin–proteasome mechanism) is tightly regulated and contributes to the precise regulation of cdk scheduling. The family of mammalian cyclins includes cyclins A to H, which have shared a conserved sequence of about 100 amino acids, mutation of which can disrupt both kinase binding and activation. Cyclins come in different subfamilies; there are two B-type cyclins and three D-type cyclins. Usually more than one D-type cyclin is expressed in any mammalian cell, but the combinations differ in different cell types. The interactions of cyclins, cdks, and activating and inhibitory proteins are shown schematically in Figure 8.12 and can be summarized as follows:

1. $G_0 \rightarrow G_1$: Early cell-cycle progression in G_1 is under the control of the D-type cyclins and of Cdks 4 and 6. The formation of cyclin D–Cdk4/6 complexes is promoted by two proteins, p21$^{Cip1/Waf1}$ and p27^{kip1} and the cdks are activated by phosphorylation by CAK. Their activity can also be inhibited by the binding of several small cdk-inhibitory proteins (CKIs).

2. $G_1 \rightarrow S$: Passage through G_1 and into S phase is regulated by the activities of cyclin D-, cyclin E-, and cyclin A-associated kinases. The role of p21$^{Cip1/Waf1}$, which earlier in G_1 promoted formation of cyclin/cdk complexes, is reversed and binding of p21$^{Cip1/Waf1}$ and p27^{kip1} is inhibitory of the formation of cyclin E–Cdk2 complexes, which control the transition from G_1 into S-phase.

3. As the cyclin E:Cdk2 complexes are formed, Cdk2 is maintained in an inactive state by phosphorylation by the Wee1 and Myt1 kinases.

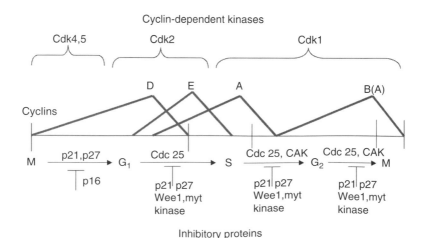

FIGURE 8.12
Variations in cell-cycle regulatory proteins.

Dephosphorylation by Cdc25A activates Cdk2, and is coordinated with the cells reaching the proper size, and with the DNA synthesis machinery being in place.

4. The retinoblastoma (pRb) protein is the most important target of the cdks. Hypophosphorylated pRb is complexed with the transcription factor E2F preventing E2F from switching on genes essential for passage into the DNA synthetic phase. Phosphorylation of pRb late in G_1 releases E2F from the complex and transcription is activated resulting in transactivation of important genes for later cell-cycle events. Both cyclin E and cyclin A activated kinases play a role in modulating E2F transcriptional activity. These cyclin/cdks are found in complexes that contain E2F and the Rb related protein p107 in late G_1 and early S phase, respectively.

5. **S→G$_2$**: During S phase, B-type cyclins increase but phosphorylation of the inhibitory Thr14 and Tyr15 sites on cdk1 keeps the kinase inactive until the G_2/M transition.

6. **G$_2$→M**: Mitotic or B-type cyclins associate with cdk1 to control entry and exit from mitosis. Dephosphorylation of the inhibitory sites by cdc25 phosphatase and activation of cyclin B triggers cdk1 activation essential for entry into mitosis.

The abrupt and controlled degradation of cyclin B by the ubiquitin–proteasome pathway allows exit from mitosis. The retinoblastoma protein pRb is also dephosphorylated at this time.

In summary, as the cell-cycle proceeds, a tightly regulated sequence of events unfolds around each cell-cycle transition. For each of these scenarios the script is essentially the same, but the identities and roles of the players change. Radiation-induced damage to DNA in normal nonmutant cells instigates an unscheduled break in the action during which damage control can occur.

8.4.2 Radiation-Induced Cell-Cycle Arrest

Delays at cell-cycle checkpoints are believed to prevent the replication of damaged DNA (G_1/S and intra-S checkpoints) or segregation of damaged chromosomes (the G_2/M checkpoint). Indirect evidence for the importance of such controls in the maintenance of genome stability is provided by genetic disorders such as A-T, ATLD, and NBS, which are characterized by checkpoint deficiencies, are radiosensitive and associated with a high incidence of tumorigenesis. However, as described, there is evidence that the radiosensitivity associated with syndromes such as A-T may also correlate with a defect in DNA repair as well as with a cell-cycle checkpoint deficiency. A number of DNA repair proteins have been found to be regulated either transcriptionally or post-translationally after DNA damage in a manner that requires "checkpoint" genes [25] while several repair proteins have been established as substrates for checkpoint kinases.

8.4.2.1 G_1 Arrest

Tumor cells that are devoid of endogenous p53 protein or express a mutant p53 or HPV16-E6 protein have an abrogated G_1 arrest response indicating a requirement for a functional p53 to maintain an intact G_1 checkpoint. Transcriptional activation of a p53-dependent pathway leads to increased expression of the p21[Waf1/Cip1] protein, which, as has been described, inhibits the activity of the cyclin D- and cyclin E-associated kinases that would otherwise phosphorylate the retinoblastoma gene product (pRb). In its hypophosphorylated form, pRb sequesters the E2F transcription factor, thereby preventing transition from G_1 to S phase. The action of radiation in causing G_1 arrest is shown in Figure 8.13. pRb also acts by recruiting histone deacetylase (HDAC1), which blocks transcription by promoting nucleosome compaction. p21 has another role in promoting growth arrest by preventing proliferating cell nuclear antigen (PCNA) from activating DNA polymerase, which is essential for DNA replication. It has been shown that the G_1 arrest in normal human fibroblasts can be permanent with the cells acquiring a morphology similar to that observed in senescent fibroblasts. This permanent G_1 arrest is radiation-dose dependent and may be associated with ongoing expression of p21[Waf1/Cip1] protein.

8.4.2.2   S-Phase Checkpoint

The DNA-damage checkpoint activated within S phase is manifest as a transient reduction in the rate of DNA synthesis. This reflects regulation at

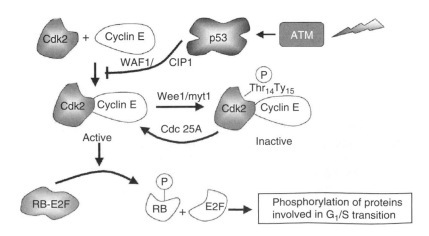

FIGURE 8.13
Radiation-induced G_1 delay is mediated by p53. Combination of cyclin E/cdk2 is inhibited by WAF1/CIP1 with the result that phosphorylation of Rb and release of E2F does not take place.

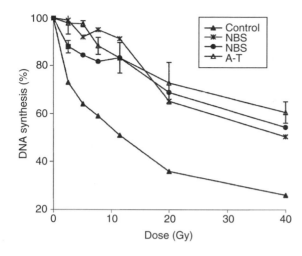

FIGURE 8.14
RDS in A-T and NBS cells. Cells were prelabeled with [14C] thymidine, treated with γ-rays at the doses indicated, and incubated for 4 h in medium containing [3H] thymidine. The results shown represent the percentage of DNA synthesis relative to unirradiated cells. (From Girard, P.M., Foray, N., Stumm, M., Waugh, A., Riballo, E., Maser, R.S., Phillips, W.P., Petrini, J., Arlett, C.F., and Jeggo, P.A., *Cancer Res.* 60, 4881, 2000. With permission.)

the level of replicon initiation and allows repair of the damage before the lesions are permanently fixed by DNA replication into irreparable chromosomal breaks. The ATM protein again plays an important role in this pathway and Λ-T cells display relative resistance to radiation-induced inhibition of DNA replication, a phenomenon termed radioresistant DNA synthesis (RDS) (Figure 8.14).

Of all the cellular checkpoints, the S-phase checkpoint is the most complex and elaborate. This is because it needs to respond to endogenous and exogenous signals in a manner that ensures stability and fidelity of replication [26]. The current view of the S-phase checkpoint envisions at least two parallel pathways. The first represents a reversible, fast response to DNA damage and involves Chk kinase-dependent Cdc25 degradation. Chemical or genetic ablation of human Chk1 was found to trigger a supraphysiological accumulation of the S phase-promoting Cdc25A phosphatase, to prevent IR-induced degradation of Cdc25A, and to cause radioresistant DNA synthesis. Turnover of Cdc25A required Chk1-dependent phosphorylation on four serines and IR-induced acceleration of Cdc25A proteolysis corresponded with increased phosphate uptake by these residues [27]. Phosphorylation of Chk1 by ATM was required for effective IR-induced degradation of Cdc25A. The irreversible slower response to DNA damage requires p53 stabilization [26].

The role of the BRCA1 and BRCA2 proteins in DNA repair was described in the previous chapter (Section 7.3.1.1). In fact, BRCA1 is a large protein with multiple functional domains, which interacts with numerous proteins that are involved in many important biological processes. Mounting evidence indicates that BRCA1 is involved in all phases of the cell cycle and regulates orderly events during cell-cycle progression. BRCA1 deficiency has been shown to cause abnormalities in the S-phase checkpoint, the G_2/M checkpoint, the spindle checkpoint, and centrosome duplication [28].

8.4.2.3 G₂/M Arrest

Downstream targets of ATM which mediate G_2/M checkpoint are again the protein kinases CHK1 and CHK2. Phosphorylation of Chk2 and its activation in response to radiation are both ATM-dependent, with ATM phosphorylating Chk2 directly on Thr-68. Cells expressing a kinase-defective mutant ATR are compromised for the radiation-induced G_2 arrest indicating that Chk1 is also an effector of ATR. Activated CHK1 and CHK2 phosphorylate the protein phosphatase Cdc25C with the result that it is inactivated and bound by the 14-3-3σ protein. Thus disabled, Cdc25C is unable to remove an inhibitory phosphate group on Tyr-15 of Cdc2 and entry into mitosis is prevented. This sequence of events is summarized in Figure 8.15. Additional controls probably exist, and it seems likely that the activities of other Cdc2 regulators are modulated in response to DNA damage. In addition, p53-dependent transcriptional repression of the Cdc2 and cyclin B promoters may contribute to the maintenance of the G_2/M checkpoint in mammalian cells.

8.4.2.4 Modifying the G₂/M Delay: Caffeine and Related Drugs

Low levels of caffeine and other methyl xanthines added to cells can increase mRNA levels of cyclin B1, the coenzyme regulating cdk1 activity. Caffeine has been shown to affect the activation of Chk-l, the kinase

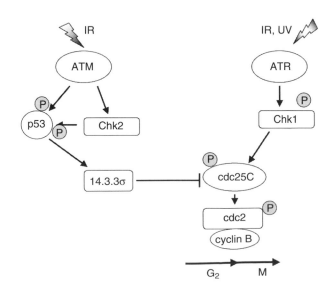

FIGURE 8.15
G_2/M checkpoint. In response to radiation, Cdc25 is phosphorylated by Chk1 or Chk2. This is the signal for the degradation of Cdc25.

that phosphorylates Cdc25C [29] that is required for the activation of p34 Cdc2. At the cellular level, caffeine results in radiosensitization of tumor cells and also abrogates or shortens the radiation-induced G_2 delay. Caffeine can also alter the cellular response to damage into an apoptotic pathway. Hela cells, which in general are quite resistant to induction of apoptosis by a variety of means, are radiosensitized and undergo apoptosis within 24 h of exposure to caffeine [30]. Cells with wild type p53 are resistant to this action of caffeine, but cells that lack functional p53 and are treated with caffeine are presumably sensitized because they bypass both G_1 and G_2 checkpoints and have no opportunity to repair DNA damage. Drugs similar to caffeine, such as pentoxyphylline, produce similar effects [31] but the clinical applicability of these drugs is limited since the doses of caffeine and of pentoxyphylline required to produce radiosensitization are far above the amounts that can be safely administered in vivo. The possibility remains that a safer drug with a similar action could selectively sensitize tumor cells with inactive p53 leaving the normal cells with wild type p53 unaffected.

8.4.3 Oncogenes and Cell-Cycle Checkpoints

Oncogenic transformation can result in continuous cell proliferation due to abrogation of cell-cycle checkpoints. The oncogene *ras* links extracellular signals and the cell-cycle regulatory machinery. Activation of *ras* leads to induction of cyclin D1 and to down-regulation of the cdk inhibitor p27^{KIP-1}. Using rat embryo fibroblasts transfected with *myc* or oncogenic H-*ras* + *myc*, it was shown by McKenna and collaborators [32,33] that *ras* transfection led to increased radiation resistance, reduction in radiation-induced apoptosis, and significantly prolonged G_2 delay. The duration of the G_2 delay correlated with the delay in cyclin mRNA expression. The significance of cell-cycle delay in this process is indicated by the fact that radioresistance is not associated with modification of DNA damage or DNA damage repair. The most recent experiments by this group used synthetic interfering RNA (siRNA) to selectively block specific isoforms of *ras*. Inhibition of oncogenic ras by this method reduced clonogenic survival of T24 human bladder carcinoma cells [34] (Figure 8.16).

As just described, *ras* genes with oncogenic mutations confer radioresistance both on human tumor cells and on transformed rat fibroblasts. Since the *ras* gene family are frequently mutated in human cancers, it has been proposed that targeting *ras* could allow radiosensitization of tumor cells without affecting normal tissues [33,35]. One method to block ras activity depends on the fact that oncogenic ras is constitutively bound to guanosine triphosphate (GTP) but requires membrane attachment to be active in transformation. Membrane binding of ras is mediated by post-translational prenylation of the ras protein. This understanding has led to the development of inhibitors of ras prenylation with potential for use as clinical radiosensitizers. This is discussed in more detail in Chapter 11.

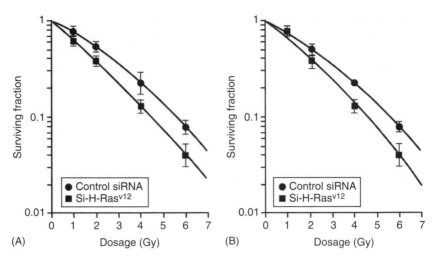

FIGURE 8.16

Effects of inhibiting ras isoforms and tumor cell radiation survival. (A) T24 cells were treated with siRNA to H-*ras*V12 or treated with nonspecific control siRNA. (B) SW480 cells were treated as in (A), except that an siRNA specific to K-*ras*V12 was used. (From Kim, I.-A., Bae, S.-S., Fernandes, A.-M., Wu, J.-M., Muschel, R.J., McKenna, W.G., Birnbaum, M.J., and Bernhard, E.J., *Cancer Res.*, 65, 7902, 2005. With permission.)

8.4.4   Variation in Radiosensitivity through the Cell Cycle

Variation in radiosensitivity through the cell cycle has been reviewed by Wilson [36]. One of the long-standing dogmas of cellular radiobiology concerns the differential radiosensitivity associated with different phases of the cell cycle that arose from the original work of Terashima and Tolmach [37]. This was summarized by Sinclair in 1968 [38] as follows:

- Mitotic cells are generally the most radiosensitive.
- If G1 has an appreciable length, there is usually a resistant period which declines toward S phase.
- In most cell lines, resistance increases during S-phase with a maximum increase in the latter part of the phase.
- In most cell lines, G_2 phase is almost as sensitive as mitosis.

The first three of these conclusions have been corroborated in many experimental studies, but the last one was not supported by later experimental findings. Using mitotic shake-off to select G_2 cells, Schneiderman et al. demonstrated that the D_0 for CHO cells in G_2 was 2.45 Gy [39], indicating a relatively resistant population. More recently, Biade et al. [40] used mitotic shake-off to study the cell-cycle variation in radiosensitivity of some commonly used human tumor cell lines that showed a wide variation in their linear quadratic characteristics. The conclusions based on these studies were in agreement with those of Sinclair, with the exception that all cell lines became increasingly radioresistant as they progressed through S-phase and remained relatively radioresistant in G_2, especially HT29 colon carcinoma cells. Similar conclusions were reached by Hill et al. [41] using A549, HT29, and U1 (melanoma) cell lines. These findings, and subsequent improved understanding of the molecular biology of the cell cycle, prompted the suggestion that some of the radiosensitivity originally attributed to G_2 cells might have been caused by imperfect synchronization, which resulted in the presence of significant number of exquisitely sensitive mitotic cells [42] in the selected population [36].

DNA repair occurs throughout the cell cycle [43] including in G_2 cells [44], and in fact, both NHEJ and HR DNA repair modes are available in G_2 and the cells are further protected by the presence of checkpoints for G_2 arrest [45]. The two major repair pathways associated with repair of DNA DSBs show very different cell-cycle periodicity. It is generally accepted that NHEJ is active throughout the cell cycle [43,46,47] whereas repair by HR predominates during S and G_2 phases of the cell cycle when sister chromatids are present [43,47–49]. It is not clear whether these two pathways cooperate or compete to repair DSBs [46,50]. It would seem reasonable that the DNA-damage response is an integrated defense against damaged chromosomes. However, this is not supported by evidence that defects or manipulation of checkpoint activity do necessarily translate into differences in radiosensitivity. It has been shown that neither S-phase checkpoint defects, G_2 arrest abnormalities, nor abrogation of the G_1 checkpoint alone confer radiosensitivity [45,51], and in fact, defects in more than one checkpoint may be required to alter radiosensitivity. The dual role of p53 in both G_1 and G_2 checkpoints may explain some of the data in the literature claiming increased resistance associated with mutant p53 [3,52], although this has not been a universal finding [53,54] and the role of p53 is still the subject of much debate [3,55]. There is evidence of DNA repair activity during G_2 delay [44]. As already described, abrogation of the G_2 checkpoint using caffeine in cells that were competent or deficient in G_1 checkpoint control (by manipulation of p53 using E6) further sensitized wild type p53 cells or corrected the radioresistance in E6-inactivated cells [30,31] and there are a number of reports of radiosensitization by manipulation of the G_2 checkpoint in a background of p53 deficiency.

8.5 p53-Mediated Apoptosis

The guardian of the genome, p53, regulates normal responses to DNA damage and other forms of genotoxic stress and is a key element in maintaining genomic stability and in protecting the cell from acquiring and

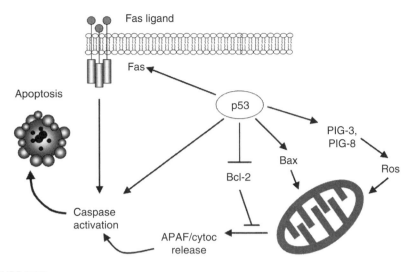

FIGURE 8.17
p53-mediated apoptotic signaling. Several p53 target genes can promote apoptosis including
the pro-apoptotic Bax and the Fas death receptor. p53 inhibits the expression of the anti-
apoptotic Bcl-2 and induces IGF-BP3 which induces apoptosis by blocking IGF-IR survival
signaling. p53 can mediate mitochondrial signaling by elevating ROS through PIG3 and PIG8
induction.

propagating carcinogenic change. p53-dependent mechanisms of cell-cycle
arrest in response to radiation insult have already been described. The
alternate route of DNA-damage response is the initiation of a chain of
reactions leading to apoptosis. A variety of stimuli, both physiological and
pathological, have been shown to induce apoptosis, including extracellular
stress (ionizing and ultraviolet radiation, heat shock, and oxidative and
osmotic stress), receptor-mediated processes (death receptors, such as
tumor necrosis factor [TNF], CD95, and the hormone receptors), growth
factor withdrawal, loss of cell adhesion (anoikis), cytotoxic lymphocytes,
and many chemotherapeutic drugs.

Radiation-induced apoptosis can take place through several routes, some
of which are p53-mediated and are summarized in Figure 8.17. The reader
should fast-forward to Chapter 9, which is devoted to apoptosis exclusively,
to read details of the downstream effects of DNA damage on apoptotic
processes.

8.6   Summary

A spectrum of DNA lesions is recognized by DNA-damage-response pathways that transduce a signal from a damage sensor or detector to a series
of downstream effector molecules via the ATR/ATM network, a highly

conserved protein kinase cascade. The identification of the ATM protein began with the recognition of patients with the heritable syndrome ataxia telangiectasia (A-T). Cells derived from these patients were found to exhibit chromosomal instability and extreme sensitivity to ionizing radiation and to radiomimetic chemicals and to be deficient in IR-induced G_1-S, intra-S, and G_2-M cell-cycle checkpoint regulation but not deficient in DSB repair. The induction of ATM kinase activity following radiation is very rapid, suggesting that it acts at an early stage of signal transduction and is modulated by post-translational events.

ATM regulates the G_1-S checkpoint after IR exposure, at least in part by controlling the activation and stabilization of the tumor suppressor protein p53.

The function of p53 is to integrate signals from a variety of cellular stresses and to mediate responses to these insults by activating a set of genes whose products facilitate adaptive and protective responses. Wild type p53 protein has a very short half-life under normal conditions. It is regulated by the Mdm2 proto-oncoprotein, which itself is a direct target of p53, and a negative autoregulatory feedback-loop exists between the two proteins. The inhibitory effect of Mdm2 on p53 can be modulated by impairing the interaction of p53 with Mdm2, by post-translational modification of p53 to change the DNA binding capacity or by control of the sub-cellular localization of the p53 protein.

Radiation-induced delays in the G_1-S and G_2 phases of the cell cycle allow periods of time during which cells could survey and repair DNA damage. The cell cycle is governed by a family of cdks that regulate a series of biochemical pathways, or checkpoints. The activity of the cdks is regulated by the binding of each cdk to its specific positive effector protein, a cyclin; and the oscillation of cyclin abundance controls phosphorylation of key substrates to ensure that cell-cycle events occur at the correct time and sequence. Translational activation of a p53-dependent pathway leads to increased expression of the p21$^{Waf1/Cip1}$ protein which inhibits the activity of the cyclin D- and cyclin E-associated kinases, which would otherwise phosphorylate the retinoblastoma gene product (pRb) and precipitate entry into S phase. The DNA-damage checkpoint activated within S phase is manifested as a transient reduction in the rate of DNA synthesis. The ATM protein plays an important role in this pathway and A-T cells display relative resistance to radiation-induced inhibition of DNA replication, a phenomenon termed radioresistant DNA synthesis (RDS).

Protein kinases Chk1 and Chk2 mediate the G_2/M checkpoint. Phosphorylation of Chk2 and its activation in response to IR are both ATM-dependent, with ATM phosphorylating Chk2 directly. Caffeine radiosensitizes tumor cells lacking p53 function and abrogates or shortens the radiation-induced G_2 delay. Activation of the oncogene *ras* leads to induction of cyclin D1 and to down-regulation of the cdk inhibitor p27^{KIP-1} causing increased radiation resistance and prolongation of G_2 delay.

p53 has a dual role in the response to DNA damage acting both as a regulator of cell-cycle arrest and a positive regulator of programmed cell death or apoptosis. Radiation-induced apoptosis can take place through several routes, three of which are p53-mediated.

References

1. Barzilai A, Rotman G, and Shiloh J. ATM deficiency and oxidative stress: A new dimension of defective response to DNA damage. _DNA Repair_ 1: 3–25, 2002.
2. Khanna KK, Gatti R, Concannon P, Weemaes C, Hoekstra MF, and Lavin M. Cellular responses to DNA damage and human chromosome instability syndromes. In: Nickloff JA, Hoekstra MF (Eds.) _DNA Damage and Repair_, Humana Press, Totowa, New Jersey, 1998, pp. 395–442.
3. Fei P and El-Deiry WS. P53 and radiation responses. _Oncogene_ 22: 5774–5783, 2003.
4. Canman C and Lim D-S. The role of ATM in DNA damage responses and cancer. _Oncogene_ 17: 3301–3308, 1998.
5. Shiloh Y. ATM and ATR: Networking cellular responses to DNA damage. _Curr. Opin. Genet. Dev._ 11: 71–77, 2001.
6. Abraham RT. Cell cycle checkpoint signaling through the ATM and ATR kinases. _Genes Dev._ 15: 2177–2196, 2001.
7. Lobrich M and Jeggo PA. Harmonising the response to DSBs: A new string in the ATM bow. _DNA Repair_ 4: 749–759, 2005.
8. Savitsky K, Bar-Shira A, Gilad S, Rotman G, Ziv Y, Vanagaite L, Tagle DA, Smith SJ, Uziel T, and Sfez S, et al. A single ataxia telangiectasia gene with a product similar to PI-3 kinase. _Science_ 268: 1749–1753, 1995.
9. Bakkenist CJ and Kastan MB. DNA activates ATM through intermolecular autophosphorylation and dimer dissociation. _Nature_ 421: 499–506, 2003.
10. Cerosaletti K and Concannon P. Independent roles for Nibrin and Mre11 in the activation and function of ATM. _J. Biol. Chem._ 279: 38813–38819, 2004.
11. Paull TT and Lee J-H. The Mre11/rad50/Nbs1 complex and its role as a double-strand break sensor for ATM. _Cell Cycle_ 4: 737–740, 2005.
12. Cuddihy AR and Bristow RG. The p53 protein family and radiation sensitivity: Yes or no? _Cancer Metastasis Rev._ 23: 237–257, 2004.
13. Giaccia AJ and Kastan MB. The complexity of p53 modulation: Emerging patterns from divergent signals. _Genes Dev._ 19: 2973–2983, 1998.
14. Amundson SA, Myers TG, and Fornace AJ Jr. Roles for p53 in growth arrest and apoptosis: Putting on the brakes after genotoxic stress. _Oncogene_ 17: 3287–3299, 1998.
15. Freedman DA, Wu L, and Levine AJ. Functions of the MDM2 oncoprotein. _Cell Mol. Life Sci._ 55: 96–107, 1999.
16. Lane DP. p53, guardian of the genome. _Nature_ 358: 15–16, 1992.
17. Elkind M, Han A, and Volz K. Radiation response of mammalian cells grown in culture: IV Dose dependence of division delay and post-irradiation growth of surviving and non-surviving Chinese hamster cells. _J. Natl. Cancer Inst._ 30: 705–721, 1964.

18. Kastan MB, Onyekwere O, Sidransky D, and Vogelstein B, RW. C. Participation of p53 protein in the cellular response to DNA damage. *Cancer Res.* 51: 6304–6311, 1991.
19. Hartwell L. Defects in a cell cycle checkpoint may be responsible for the genomic instability of cancer cells. *Cell* 71: 543–546, 1992.
20. Sherr CJ. Cancer cell cycles. *Science* 274: 1672–1677, 1996.
21. Morgan DO. Principles of Cdk regulation. *Nature* 374: 131–134, 1995.
22. Solomon MJ. Activation of various cyclin/cdc 2 protein kinases. *Curr. Opin. Cell Biol.* 5: 180–186, 1993.
23. King RW, DesRaies RJ, Peters J-M, and Kirschner MW. How proteolysis drives the cell cycle. *Science* 274: 1652–1659, 1996.
24. Slingerland JM and Tannock IF. Cell Proliferation and cell death. In: Tannock IF, Hill RP (Eds.) *The Basic Science of Oncology*, McGraw Hill, 1998, pp. 134–165.
25. Bashkirov VI, King JS, Bashkirova EV, Schmuckli-Maurer J, and Heyer WD. DNA repair protein Rad55 is a terminal substrate of the DNA damage checkpoints. *Mol. Cell Biol.* 20: 4393–4404, 2000.
26. Gottifredi V and Prives C. The S phase checkpoint: When the crowd meets at the fork. *Semin. Cell Dev. Biol.* 16: 355–368, 2005.
27. Sørensen CS, Syljuåsen RG, Falck J, Schroeder T, Rönnstrand L, Khanna KK, Zhou B-B, Bartek J, and Lukas J. Chk1 regulates the S phase checkpoint by coupling the physiological turnover and ionizing radiation-induced accelerated proteolysis of Cdc25A. *Cancer Cells* 3: 247–258, 2003.
28. Deng C-X. BRCA1: Cell cycle checkpoint, genetic instability, DNA damage response and cancer evolution. *Nucleic Acids Res.* 34: 1416–1426, 2006.
29. Kumagai A, Guo Z, Emami KH, Wang SX, and Dunphy WG. The Xenopus Chk1 protein kinase mediates a caffeine-sensitive pathway of checkpoint control in cell-free extracts. *J. Cell Biol.* 142: 1559–1569, 1998.
30. Bernhard EJ, Muschel RJ, Bakanauskas VJ, and McKenna WG. Reducing the radiation-induced G2 delay causes HeLa cells to undergo apoptosis instead of mitotic death. *Int. J. Radiat. Biol.* 69: 575–584, 1996.
31. Powell SN, DeFrank JS, Connell P, Eogan M, Preffer F, Dombkowski D, Tang W, and Friend S. Differential sensitivity of p53(−) and p53(+) cells to caffeine-induced radiosensitization and override of G2 delay. *Cancer Res.* 1643–1648, 1995.
32. McKenna WG, Bernhard EJ, Markiewicz DA, Rudoltz MS, Maity A, and Muschel RJ. Regulation of radiation-induced apoptosis in oncogene-transfected fibroblasts: Influence of H-ras on the G2 delay. *Oncogene* 12: 237–245, 1996.
33. Bernhard EJ, Kao G, Cox AD, Sebti SM, Hamilton AD, Muschel RJ, and McKenna WG. The farnesyltransferase inhibitor FTI-277 radiosensitizes H-ras-transformed rat embryo fibroblasts. *Cancer Res.* 56: 1727–1730, 1996.
34. Kim I-A, Bae S-S, Fernandes A-M, Wu J-M, Muschel RJ, McKenna WG, Birnbaum MJ, and Bernhard EJ. Selective inhibition of Ras, phosphoinositide 3 kinase, and akt isoforms increases the radiosensitivity of human carcinoma cell lines. *Cancer Res.* 65: 7902–7910, 2005.
35. Bernhard EJ, McKenna WG, Hamilton AD, Sebti SM, Qian Y, Wu JM, and Muschel RJ. Inhibiting Ras prenylation increases the radiosensitivity of human tumor cell lines with activating mutations of ras oncogenes. *Cancer Res.* 58: 1754–1761, 1998.
36. Wilson GD. Radiation and the cell cycle revisited. *Cancer Metastasis Rev.* 23: 209–225, 2004.

37. Terasima T and Tolmach LJ. Variations in several response of HeLa cells to x-irradiation during the division cycle. *Biophys. J.* 3: 11–33, 1963.
38. Sinclair WK. Cyclic x-ray responses in mammalian cells in vitro. *Radiat. Res.* 33: 620–643, 1968.
39. Schneiderman MH, Schneiderman GS, and Rusk CM. A cell kinetic method for the mitotic selection of treated G2 cells. *Cell Tissue Kinet.* 16: 41–49, 1983.
40. Biade S, Stobbe CC, and Chapman JD. The intrinsic radiosensitivity of some human tumor cells throughout their cell cycles. *Radiat. Res.* 147: 416–421, 1997.
41. Hill AA, Wan F, Acheson DK, and Skarsgard LD. Lack of correlation between G1 arrest and radiation age-response in three synchronized human tumour cell lines. *Int. J. Radiat. Biol.* 75: 1395–1408, 1999.
42. Stobbe CC, Park SJ, and Chapman JD. The radiation hypersensitivity of cells at mitosis. *Int. J. Radiat. Biol.* 78: 1149–1157, 2002.
43. Rothkamm K, Kruger I, Thompson LH, and Lobrich M. Pathways of DNA double-strand break repair during the mammalian cell cycle. *Mol. Cell. Biol.* 23: 5706–5715, 2003.
44. Kao GD, McKenna WG, and Yen TJ. Detection of repair activity during the DNA damage-induced G2 delay in human cancer cells. *Oncogene* 20: 3486–3496, 2001.
45. Xu B, Kim ST, Lim DS, and Kastan MB. Two molecularly distinct G(2)/M checkpoints are induced by ionizing irradiation. *Mol. Cell Biol.* 22: 1049–1059, 2002.
46. Wang H, Perrault AR, Takeda Y, Qin W, Wang H, and Iliakis G. Biochemical evidence for Ku-independent backup pathways of NHEJ. *Nucleic Acids Res.* 31: 5377–5388, 2003.
47. Yoshida M, Hosoi Y, Miyachi H, Ishii N, Matsumoto Y, Enomoto A, Nakagawa K, Yamada S, Suzuki N, and Ono T. Roles of DNA-dependent protein kinase and ATM in cell-cycle-dependent radiation sensitivity in human cells. *Int. J. Radiat. Biol.* 78: 503–512, 2002.
48. Valerie K and Povirk LF. Regulation and mechanisms of mammalian double-strand break repair. *Oncogene* 22: 5792–5812, 2003.
49. Henning W and Sturzbecher HW. Homologous recombination and cell cycle checkpoints: Rad51 in tumour progression and therapy resistance. *Toxicology* 193: 91–109, 2003.
50. Allen C, Halbrook J, and Nickoloff JA. Interactive competition between homologous recombination and non-homologous end joining. *Mol. Cancer Res.* 1: 913–920, 2003.
51. Kastan MB and Lim DS. The many substrates and functions of ATM. *Nat. Rev. Mol. Cell Biol.* 1: 179–186, 2000.
52. Lee J and Bernstein A. p53 mutations increase resistance to ionizing radiation. *Proc. Natl. Acad. Sci. USA* 90: 5742, 1993.
53. Slichenmyer WJ, Nelson WG, Slebos RJ, and Kastan MB. Loss of a p53-associated G1 checkpoint does not decrease cell survival following DNA damage. *Cancer Res.* 53: 4164–4168, 1993.
54. Brachman DG, Beckett M, Graves D, Haraf D, Vokes E, and Weichselbaum RR. p53 mutation does not correlate with radiosensitivity in 24 head and neck cancer lines. *Cancer Res.* 53: 3667, 1993.
55. Bristow RG. The p53 gene as a modifier of intrinsic radiosensitivity: Implications for radiotherapy. *Radiother. Oncol.* 40: 197–223, 1996.
56. Girard PM, Foray N, Stumm M, Waugh A, Riballo E, Maser RS, Phillips WP, Petrini J, Arlett CF, and Jeggo PA. Radiosensitivity in Nijmegen breakage syndrome cells is attributable to a repair defect and not cell cycle checkpoint defects. *Cancer Res.* 60: 4881–4888, 2000.

9

Chromatin Structure and
Radiation Sensitivity

9.1 Cell Nucleus

The early conception of the cell nucleus was that it was a passive compartment, the main role of which was to keep genetic information separate from
the surrounding cytoplasm. The DNA, organized in chromatin fibers, was
envisioned as floating like spaghetti in a bowl of fluid nucleoplasm in which
the only other structures were ribonucleoprotein particles and the nucleolus.

The presence of a more complex substructure was revealed by microscopy and confirmed by the subsequent isolation of the nuclear matrix (NM)
[1,2]. The NM is defined as the residual framework structure of the nucleus
that maintains many of the overall architectural features of the cell nucleus
including the nuclear lamina with complex pore structures, residual nucleoli, and an extensive fibrogranular structure in the nuclear interior. It is
operationally defined as a subnuclear structure, which is insoluble following
several different extraction procedures (Figure 9.1). It has been identified in
a wide range of eukaryotes from yeast to man.

9.1.1 Hierarchical Structure of Chromatin

DNA in the cell nucleus needs to be tightly packaged to fit in the nucleus.
Four meters of DNA are wrapped up into an approximately 5 μm diameter
nucleus in the typical diploid mammalian cell requiring it to be compacted
approximately 4000-fold. At the first level, the DNA is wrapped around core
octameric histone complexes (nucleosomes) forming a "beads-on-a-string"
array (Figure 9.2C). In the process, the packing ratio of DNA in the nucleus
(the ratio of the DNA double helix length to the actual length of the chain of
nucleosomes) increases by nearly sevenfold. The nucleosomal array is then
folded into a 30 nm chromatin fiber which further increases the packing
ratio to approximately 40-fold (Figure 9.2D). The chromatin fibers are then
folded into repeating loop domains which are anchored to components of
the NM. Higher levels of chromatin organization than the loop domains

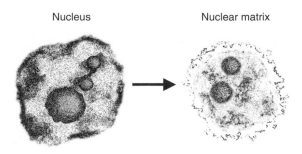

Nucleus   Nuclear matrix

FIGURE 9.1
Isolation of the nuclear matrix
(NM). Sequential extraction with
nonionic detergent, brief DNAse
digestion, and a hypertonic salt
buffer results in a residual structure which is essentially devoid
of histones and lipids.

exist in mitotic chromosomes, where the highly condensed chromatin is
packaged into precise chromosome structures by radial arrangement of
repeating loops continuously wound and stacked along a central axis to
form a chromatid (Figure 9.2) [3,4].

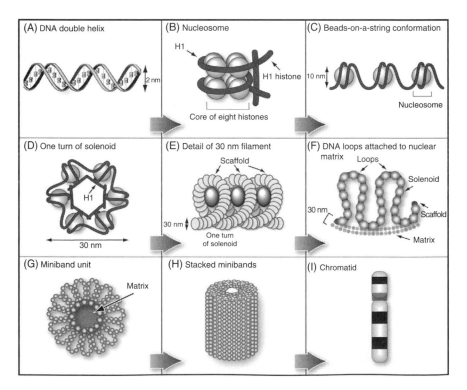

FIGURE 9.2
Structure of chromatin. The DNA double helix (A) is wrapped around a core of histone proteins
to form the nucleosome (B). Nucleosomes form a ''beads-on-a-string'' fiber (C), which winds in
a solenoid fashion, six nucleosomes per turn, to form a 30 nm chromatin filament (D). Each
nucleosome is associated with an H1 histone that stabilizes the solenoid. The chromatin
filament (E) is attached at intervals to the nuclear matrix at matrix attachment regions
(MARs) forming looped domains (F). This extended form of the chromosome (G) is condensed
by radial arrangement around a central axis (H) to form a chromatid (I).

9.1.2 Structure and Function: Chromatin and the Nuclear Matrix

In eukaryotic cells, compaction and organization of nuclear DNA is mediated, in part, by the NM, which imposes a higher order organization on nuclear DNA by periodic attachments that define loop domains ranging from 5 to 200 kb in length [5]. DNA sequences known as matrix attachment regions (MARs) or scaffold attachment regions (SARs) are AT (adenine, thymine)-rich [6], often contain topoisomerase II consensus sequences and are prone to unwinding. The loop domains are not only important for the organization and packaging of chromatin in the interphase nucleus and mitotic chromosomes, but also represent fundamental units for the genomic functions of DNA replication and transcription. The three-dimensional organization of the chromatin loop domains is the basis for the regulation of replicational and transcriptional programming in the cell while the NM provides the underlying structure for DNA replication, transcription, and some repair complexes and contributes to the control of these processes (reviewed by Jackson and Cook [7]).

A schematic of the functional nucleus is shown in Figure 9.3. Discrete domains have been identified for DNA replication sites, transcription sites, domains where splicing factors are highly concentrated, and other nuclear functions [8,9]. In addition, chromatin in the interphase nucleus is arranged in territories specific for each chromosome. It has been shown that all the functional domains of intact cells persist following NM extraction,

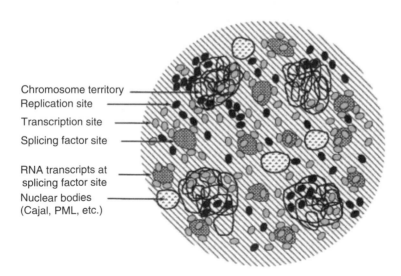

Chromosome territory
Replication site
Transcription site
Splicing factor site

RNA transcripts at splicing factor site
Nuclear bodies (Cajal, PML, etc.)

FIGURE 9.3

Schematic of the functional mammalian nucleus. The known structures of the nucleus are omitted to highlight the presence of discrete functional domains. (From Berezney, R., *Adv. Enzyme Regul.*, 42, 39, 2002. With permission.)

including the chromosome territories. Using specific chromosome paints, the territorial organization of the chromosomes in human diploid fibroblasts has been shown to be maintained despite the extraction of 90% of proteins [10].

The average amount of DNA in the mouse fibroblast replication site has been calculated to be 1 mbp [11] suggesting that each site of replication is composed of multiple (6–12) replication units or chromatin loop domains. These higher order chromatin domains each consisting of several loops are called multiple loop chromatin (MLC) domains. During replication, the replicational machinery assembles along the base of the loops of the MLC domains and replication occurs across the cluster of loops (each loop represents one bidirectional replication unit or replicon) in a parasynchronous wave. At the completion of replication, the replicational components are disassembled but the MLC domains persist as a fundamental higher order structure of chromatin in the cell nucleus [9]. Studies of individual genes have demonstrated that actively transcribed genes are preferentially replicated in early S phase while genes which are not going to be transcribed are replicated later in S phase, implying that the timeline of replication is highly coordinated.

In spite of the need for extensive compaction of DNA, nuclear DNA must remain accessible for replication, transcription, and repair. Simultaneous labeling for replication and transcription sites revealed that the individual sites were segregated into separate zones of replication and transcription and are part of spatially separate networks of replication and transcription, which extended throughout the nuclear volume. These higher order zones form three-dimensional networks in the cell nucleus with the replication network completely separated spatially from the transcription network. As noted, entire chromosomes also maintain their characteristic territorial organization following the extraction of cells for NM. This separation is maintained as cells progress through the cell cycle [12]. Some DNA repair pathways are coupled to transcription while other repair pathways are independent of transcription and replication [7,13].

9.2   Protection of DNA from Radiation Damage by Nuclear Proteins

Using a variety of methods, it has been found that the efficiency with which radiation induces DNA strand breaks depends on the extent to which DNA is associated with nuclear proteins. Histones and other DNA-bound proteins and soluble proteins are effective protectors against ionizing radiation. The histones act as scavengers for radiation-induced radicals and they expel water from the vicinity of DNA, while the compaction of the chromatin fiber reduces the access of radiation-induced free radicals to DNA. The effect of

TABLE 9.1

Yield of DNA DSBs for Different Chromatin Structures after High and Low
LET Radiation

Protein (μg/10^5 cells)[a]	Chromatin Structure	Yield of DSBs[b]		RBE[b]
		Low LET Radiation	High LET Radiation	
22.4	Intact Cell	1.0	1.0	1.6
7.4	Condensed DNA	1.2–5.8	1.5	0.42
7.4	Decondensed DNA	4–18	3–4	0.25
3.6	Nucleoids	22–93	12–18	0.42
0	Deproteinized DNA	83–195	25.4	0.45

Sources: [a] From Warters, R.L., Newton, G.L., Olive, P.L., and Fahey, R.C., *Radiat. Res.*, 151, 354, 1999. [b] From Radulescu, I., Elmrothe, K., and Stenerlow, B., *Radiat. Res.*, 161, 1, 2004. (Combined with values from other sources quoted in the same publication.)

Note: The chromatin structures listed in were isolated as follows: condensed chromatin by permeabilization of intact cells and removal of soluble proteins in buffer containing physiological concentrations of Mg^{2+} and NA/K^+; decondensed chromatin by similar treatment using hypertonic buffer (condensed and decondensed chromatin are found within the permeabilized nucleus and are attached to the nuclear matrix); nucleoid chromatin by removal of all soluble proteins and histones leading to a structure consisting of chromatin loops attached to the nuclear matrix; naked genomic DNA devoid of proteins by standard methods following lysis of intact cells.

sequential stripping of proteins and relaxation of chromatin structure on the yields of DNA DSBs is shown in Table 9.1.

Decondensation of the chromatin with hypotonic buffers markedly sensitizes the DNA to radiation whereas treatment of nuclei with hypertonic buffers strips the DNA of histones and other nuclear proteins and enhances the radiosensitivity of the DNA with respect to DSB formation. Condensed chromatin will be protected from radiation damage because it has less available water for the formation of hydroxyl radical and is less accessible to attack by •OH formed in the bound water external to the condensed structure. In addition, hydroxyl and other radiation-induced radicals may be sequestered by the histones and DNA-bound nonhistone proteins preventing their interaction with DNA or with low molecular weight scavengers. This is indicated by the finding that addition of the radical scavenger DMSO reduces the yield of strand breaks in dehistonized chromatin but gives no additional protection native chromatin at any concentration (Table 9.2).

Radiation damage to chromatin may also be mediated by binding sites in the NM for metal catalysis of Fenton reactions generating •OH radicals. The NM is known to bind metal ions in association with MARs sequences consisting of 300 + bp of AT-rich DNA. Chui and Oleinick [14] compared the action of hydroxyl radical generated by Cu^{2+} with that of Fe^{2+}-EDTA catalyzed Fenton reactions for the induction of DSB and DNA protein cross-links (DPC). The size of the DNA fragments produced, the effect of

TABLE 9.2

Effect of Radical Scavenging

DNA Lesion	Source	Chromatin	DMSO Concentration		
			1 mM	0.1 M	1 M
SSB	Human fibroblasts	Native	1	1	1
		Dehistonized	5	2	1
DSB	V79 cells	Native	1	1	1
		Dehistonized	9	2.5	1

Source: From Oleinick, N.L., Balasubramaniam, U., Xue, L., and Chiu, S., *Int. J. Radiat. Biol.*, 66, 523, 1994.

expanding or dehistonizing chromatin and the effects of radical scavengers suggested that γ-radiation and Fe^{2+}-catalyzed Fenton reactions produce DSB at open chromatin sites, whereas Cu^{2+}-generated DSB are similar to radiation-induced DPC in being located at the NM. Both metal ions produce damage by site-specific generation of hydroxyl radicals. The interaction of cloned MARs with isolated NMs has been found to be hypersensitive to cross-linking upon γ-irradiation in comparison with associations formed by similarly sized DNA fragments lacking MARs sequences. The binding of MARs to the NM at specific sites may thus create structures with high susceptibility for nuclear damage.

9.2.1   DNA-Protein Cross-Link Formation

As already described, DNA DSBs are not the only form of damage to result from irradiation of chromatin. DPC were first reported following extremely high/supra-lethal doses of radiation but more recently have been detected by sensitive methods following doses of ~4 Gy (Chapter 6). As described above, DPC were found to form at or near the NM at sites where it is stabilized and radiosensitized by Cu^{2+} and are relatively unaffected by the removal of the majority of histones from chromatin [14].

9.2.2   DSB Yields and RBE

With the exception of intact cells, the DSB yields after irradiation of different chromatin structures with γ-rays were higher than those after high LET radiation (Table 9.1). DSB yield is partially dependent on the quality of radiation and partially on the relative input of direct effects and of indirect effects mediated by free radicals. As chromatin structure becomes more open, DSB yield increases because of the greater accessibility of the DNA to free radical attack and there is an increase in DNA DSBs for each chromatin decondensation step. For low LET radiation, there is an increase in the free-radical-mediated component of damage after the first three chromatin modifications which do not occur for high LET radiation with the result that the RBE decreases from 1.6 for intact cells to 0.45 for naked DNA.

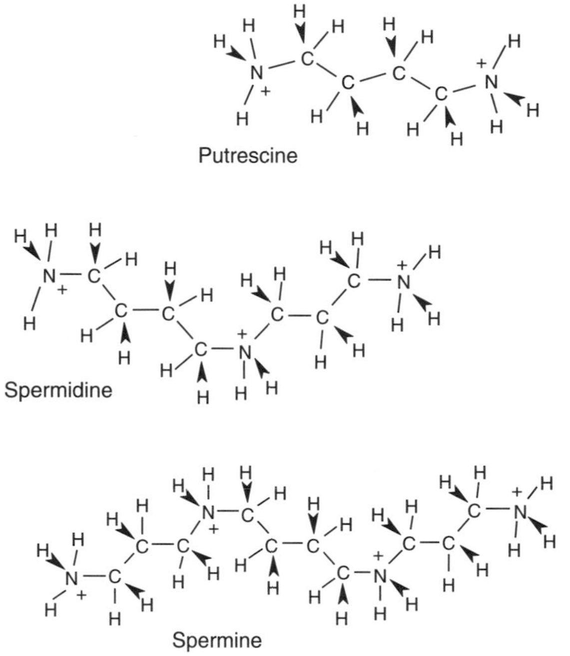

Putrescine

Spermidine

Spermine

FIGURE 9.4
Structure of three common polyamines.

9.2.3 Role of Polyamines

Polyamines are organic compounds, including putrescine (PUT), spermine
(SPM), and spermidine, which act as growth factors in pro- and eukaryotes.
Polyamines as cations bind to DNA by means of positive charges distri-
buted at regularly spaced intervals throughout the molecule (Figure 9.4).
SPM is found tightly bound to chromatin in intact nuclei and provides about
twofold protection of DNA against radiation-induced DSBs [15].

At concentrations below 1 mM, PUT or SPM provided equivalent levels of
protection to deproteinized nuclear DNA consistent with their capacity to
scavenge radiation-induced radicals. A sharp increase in protection of deproteinized and nucleoid DNA which occurs for concentrations above 1 mM has
been shown to result from the capacity of the polycationic polyamine to bind
to and condense DNA, supporting the hypothesis that polyamines protect
against radiation by compaction and aggregation of chromatin [16]. Condensation of DNA by SPM reduces the availability to radiation-induced radicals.

9.3 Radiation Sensitivity and the Stability of the DNA–Nuclear Matrix

As described in Section 9.1, DNA compaction within the nucleus is maintained in part via negative supercoiling within separate domains maintained

by periodic DNA MARs. Changes in DNA supercoiling status and the stability of DNA anchoring have been studied using the nucleoid halo assay which was introduced by Vogelstein et al. [17] and developed further by Roti Roti et al. [18]. In this assay, salt-extracted (1–2 M) nuclei are exposed to varying concentrations of the DNA-intercalating dye propidium iodide (PI). DNA is visualized as a fluorescent halo extending beyond the NM which changes in diameter at different PI concentrations. Starting from low PI concentrations, the supercoils relax until the optimum PI concentration is reached, after which the supercoils rewind in the opposite sense. The end point is the diameter of the fluorescent halo at different dye concentrations measured using an inverted fluorescence microscope. Representative nucleoids are shown in Figure 9.5A and the change in nucleoid diameter with increasing PI concentration in Figure 9.5B.

In irradiated cells, the ability to rewind DNA supercoils is inhibited in a dose-dependent manner presumably because of the presence of

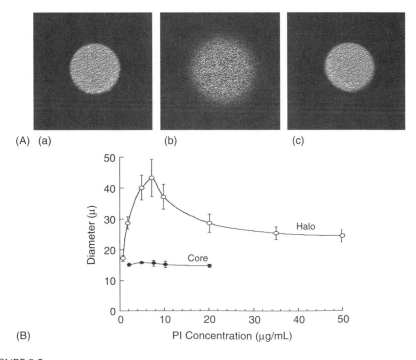

FIGURE 9.5

(A) Interphase nuclei depleted of histones and other nuclear proteins by extraction in nonionic detergent and high salt concentration were incubated in various concentrations of ethidium bromide (EB). (a) Permeabilized cell, not treated with high salt, 4 μg/mL EB; (b) high salt, 4 μg/mL EB; and (c) high salt 100 μg/mL EB. (From Vogelstein, B., Pardoll, D.M., and Coffey, D.S., *Cell*, 22, 79, 1980. With permission.) (B) Effects of increasing DNA concentration on DNA loop diameter. HeLa cells were lysed in the presence of propidium iodide (PI). (From Roti Roti, J.L. and Wright, W.D., *Cytometry*, 8, 461, 1987. With permission.)

radiation-induced DNA strand breaks that remove the topological constraints on the DNA loops. The ability to rewind DNA loops in irradiated nuclei is restored with time after irradiation, presumably as the lesion is repaired. Considerable evidence has accumulated that DNA MARs are different in radiosensitive cells [19]. Specifically, the radiosensitive cells show a greater inhibition of DNA supercoil rewinding for a given radiation dose and level of DNA damage. This effect has been observed in radiosensitive mutants of CHO cells, in spontaneous variants of L5178Y cells, and in rat embryonic cells transfected with the *RAS* oncogene (reviewed by Roti Roti et al. [20]). Nucleoids prepared from cells of the more radiosensitive cell lines are also more sensitive to protease and DNAase treatment, and results of DNA filter elution studies suggest that there are differences between resistant and sensitive cell lines in DNA nuclear protein interactions [21]. Roti Roti and colleagues have identified specific protein alterations associated with changes in nucleoids [19]. DNA NM attachments appear to be loosened when nucleoids are isolated in the presence of the reducing agent dithiothreitol (DTT) as shown by enhanced unwinding of the DNA loops and inhibition of loop rewinding. When nucleoids were prepared from irradiated cells, the effects of radiation and DTT were additive, implying that the two agents act on separate parts of the DNA-protein complex. This suggests that a redox sensitive component of the DNA nuclear MAR is compromised in radiosensitive cells.

A different but not exclusive model that has been proposed by Schwartz [21,22] is partially based on the observation that DNA–NM interactions near actively transcribing regions of the genome are more resistant to endonuclease digestion. It has been shown by fluorescent in situ hybridization (FISH) analysis of nucleoid preparations that most active genes remain tightly associated with the NM while inactive genes reside in the DNA loops. These tightly bound regions could insulate transcribing regions of the genome from adjacent non-transcribing regions by preventing the unwinding of adjacent loops [23]. On this basis, the differences in nucleoid response associated with variations in radiation sensitivity would reflect variations in the number or distribution of actively transcribing regions of the genome. This effect could influence radiosensitivity if changes in DNA supercoiling were to reorient a strand break and make it a poor substrate for repair.

9.4   Radiosensitivity of Condensed Chromatin

Measurements of cellular radiosensitivity linked to position in the cell cycle invariably demonstrate that mitotic cells are hypersensitive to ionizing radiation exhibiting single-hit inactivation coefficients similar to those of repair deficient cell lines. Other observations have led to the suggestion that cellular radiosensitivity is correlated with the extent of chromatin condensation in the irradiated cell [24]. Cells at mitosis are characterized by the dissolution of the nuclear membrane while the functional organization of mitotic spindle proteins and chromatin is in a maximally compacted state. The term chromatin condensation describes the process that occurs in the chromosomes observed at mitosis when an additional level of organization and packing is imposed on the compacted chromatin present in the interphase nucleus.

In addition to the hypersensitivity of mitotic cells, there are several lines of evidence implicating condensed chromatin as a radiosensitive entity. Firstly, while different mammalian cell lines show different intrinsic radiosensitivities, the sensitivity of the mitotic cells is the same for all cell lines and does not correlate with the sensitivity of interphase cells. Secondly, chemicals that cause chromatin condensation are radiosensitizing to an extent that correlates with the degree of chromatin condensation produced as measured by electron microscopy. Thirdly, the cross-sectional area of condensed chromatin in the nuclei of interphase cells measured by EM correlates with the intrinsic radiosensitivity of these cell lines (Figure 9.6).

The molecular basis for the hypersensitivity of chromatin in the compact form is not known. Possibilities include an increased cross section for multiple lethal damages produced by electron track ends, the inability of bulk-repair complexes to associate with radiation-induced lesions (steric hindrance) before their fixation, the interaction of sublethal lesions in DNA—which are required for chromatin compaction with radiation-induced lesions, and an increased sensitivity of DNA in compacted structures to radiation damage. The data suggest that the intrinsic radiosensitivity of the clonogens in individual tumors might be predicted from measurements of their chromatin in a compacted state [25].

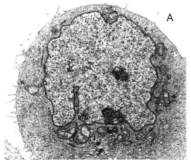
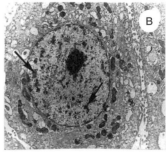

FIGURE 9.6
Electron micrographs of cross-sections of (A) radioresistant HT-29 human colon cancer cells and (B) radiosensitive PC-3 human prostate cancer cells. Arrows indicate sites of compacted chromatin. (From Chapman, J.D., Stobbe, C.C., and Matsumoto, Y., *Am. J. Clin. Oncol.*, 24, 509, 2001. With permission.)

9.5 Role of Chromatin in DNA DSB Recognition and Repair

Genetic and biochemical studies on transcription have identified two classes of enzymes that modify chromatin structure. The first class functions through covalent modifications of histone tails, including posttranslational changes. The second class consists of large multi-protein complexes that use the energy from ATP hydrolysis to alter the position or composition of nucleosomes within chromatin.

The posttranslational modification of histone tails by phosphorylation, acetylation, poly(ADP) ribosylation, and ubiquitination provides a vast number of possible signals for translation, replication, and repair. This signaling resource together with changes in nucleosome conformation has been called the "histone code." However, since these changes only make sense in the context of the chromatin fiber where their physicochemical message can be translated into regulatory programs at the genome level, some authorities have suggested the "chromatin code" to be a more appropriate name. One crucial chromatin modification, the phosphorylation of histone H2A, links the recruitment of histone modifiers and ATP-dependent chromatin-remodeling complexes to sites of DNA damage [26].

9.5.1 Histone 2AX

As has already been described, the fundamental structural and functional unit of chromatin is the nucleosome, composed of 145 base pair (bp) of DNA wrapped around eight histones (small basic proteins), two from each of the four core histone families, H4, H3, H2B, and H2A. A minimum of another 20 bp of DNA complexed with the linker histone H1 extends between the nucleosomes (Figure 9.2C).

The nucleosome plays a role in the control of information flow from the DNA involving not only the regulation of transcription but also the regulation of the condensation state of chromatin for meiosis and mitosis, and the maintenance of genomic integrity. These roles are mediated by modification of specific amino acid residues, located primarily but not exclusively in the N-terminal tails of the histones and by the presence of specialized histone species, which confer particular properties on the chromatin. Generally, the conserved core regions of the histone proteins are involved in histone–histone interaction while the N-terminal, and sometimes the C-terminal tails interact with DNA. Unique among members of the histone H2A family, the conserved C-terminal tail region of H2AX is concerned with localizing and repairing DNA DSBs [27]. When a DSB occurs, many molecules of histone H2AX in the chromatin adjacent to the break become phosphorylated on the serine residues (S136 and S139 in the mouse) located in the C-terminal tail, and H2AX phosphorylated on these serine residues is called γ-H2AX. H2AX is phosphorylated to γ-H2AX whenever a DSB is

0 Gy   0.6 Gy, 3 min   0.6 Gy, 15 min

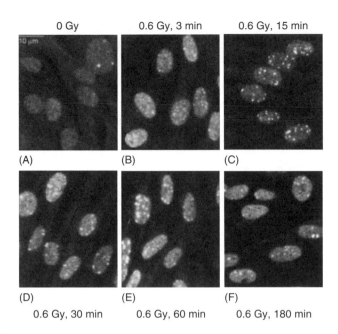

(A)   (B)   (C)

(D)   (E)   (F)
0.6 Gy, 30 min   0.6 Gy, 60 min   0.6 Gy, 180 min

FIGURE 9.7
H2AX foci formation in human cells after irradiation. IMR90 normal human fibroblasts
were exposed to 0.6 Gy ionizing radiation and permitted to recover for various lengths of
time (0–180 min). Cells were processed for laser scanning confocal microscopy. (From Rogakou,
E.P., Boon, C., Redon, C., and Bonner, W.M., *J. Cell Biol.*, 146, 905, 1999. With permission.)

formed regardless of the origin of the break and is detectable within 3 min of
radiation insult (Figure 9.7). The phosphorylated fraction of γ-H2AX
increases until a plateau is reached at 10–30 min after radiation, the level
of the plateau being proportional to the radiation dose. γ-H2AX molecules
are grouped in foci adjacent to the break sites with the number of foci
per cell being proportional to the number of DSBs. About 2000 H2AX are
phosphorylated per break site indicating that the DSB signal is highly
amplified and further amplification of the signal results from the fact that
each focus is distributed over a large area of chromatin. Foci are initially
visible as small compact structures close to the site of the break, which then
grow along the protein strand close to the site of the break finally stabilizing
as thick discs traversing the chromosome arm.

γ-H2AX is phosphorylated by the PI-3 kinase family members ATM,
ATR, and DNA-PK (Chapter 8), which are central transducers of DNA
damage response [27]. In addition to damage control, phosphorylation
by the PI-3 kinases signals nuclear foci formation. Many components
of the DNA damage response, including ATM, P53, MDC1, RAD51, and
the MRE11/RAD50/NBS1 complex form foci, which colocalize with the
γ-H2AX foci. One function of γ-H2AX as an active participant of the DNA
damage response process could be that it is involved in the recruitment of

repair factors on DSBs, but in fact, while γ-H2AX is essential for the formation of foci, it does not appear to regulate recruitment of repair factors to DSBs. Nevertheless, the biological function of the foci must be related to the localization of repair factors close to DSBs, and possibly the localized concentration of proteins generates signal amplification, which is important at threshold levels of DNA damage.

Other possible functions of γ-H2AX might include the possibility that the foci help in keeping broken DNA ends together over a period. γ-H2AX foci increase in size over about 30 min during which a substantial number of DSBs have been rejoined although some still remain open. By keeping broken DNA ends together γ-H2AX might make successful and faithful repair more likely. If DNA ends drift apart, inappropriate chromatin fragments could be joined resulting in translocations and other aberrations. A related role for γ-H2AX may be to differentiate broken DNA ends from telomeres.

9.5.2 ATM Signaling from Chromatin

In Chapter 8, it was noted that one of the explanations for the rapid propagation of the signal from a DNA break to induction of a nuclear response is that a highly localized DNA lesion could act by initiating widespread perturbation of chromatin structure. If this is the case, then yet another role can be ascribed to chromatin restructuring, i.e., as a trigger for the conversion of inactive ATM dimers into auto-phosphorylated active monomers. This idea is supported by results of experiments in which cells are treated with agents such as histone deacetylase inhibitors that alter the interaction between histones and DNA. No DSBs are formed but these drugs induce the phosphorylation of ATM on Ser-1981 and the concomitant phosphorylation of p53, a known ATM target protein. This supports the idea that ATM monitors the genome for DNA DSBs by detecting changes in chromatin structure. To date, the nature of these changes is unknown.

9.5.3 Modulation of Chromatin Structure and Function by Acetylation

As already noted, chromatin structure can be modified as a result of various posttranslational processes, including acetylation, phosphorylation, methylation, and ribosylation. Acetylation and deacetylation of nucleosome core histones play important roles in the modulation of chromatin structure and the regulation of gene transcription. Transcriptionally active genes are associated with highly acetylated core histones, whereas transcriptional repression is associated with low levels of histone acetylation. The acetylation status of histones is controlled by the activities of two families of enzymes, the histone acetyltransferases (HATS) and histone deacetylases (HDACs). In the nucleosome, hypoacetylated histones bind tightly to the phosphate backbone of DNA and maintain chromatin in a transcriptionally silent state by inhibiting access of transcription factors, transcription regulatory complexes, and RNA polymerases to DNA. Acetylation neutralizes the positive charge on histones to allow a more open conformation of chromatin, thereby enhancing access of transcription factors and the transcription apparatus to promotor regions of DNA. Conversely, histone deacetylation restores a positive charge to lysine residues, condensing the chromatin into a tightly coiled conformation.

Nuclear receptor coactivators have intrinsic HAT activity whereas corepressor proteins exist in large complexes with HDAC enzymes. Direct alterations in HDAC genes have not been demonstrated in human cancers, but HDACs are associated with several well-characterized oncogenes that stimulate cell proliferation including E2F. Rb/E2F-mediated cell-cycle control is disrupted in almost every human tumor and Rb is known to interact with class I HDACs through several complex pathways.

The "access, repair, restore" model (ARR) that was proposed to explain how the conserved process of nucleotide excision repair operates within chromatin has been extended to other DNA repair processes [28]. In the ARR model, initial repair steps include reactions that permit access of the repair machinery to DNA damage and later steps to restore the nucleosomal organization of DNA. Histone modifications, including acetylation and deacetylation are important factors in these processes. ATM, ATR, and Rad9 have been shown to interact with deacetylases, and deacetylation activity is necessary for the restoration event. HDAC4 is recruited to double-strand breaks in vivo and HDAC4 inhibition by RNAi results in increased radiosensitivity and deficient G_2-damage checkpoint maintenance. Modifications of the H2A N and C tails have been shown to contribute to DSB repair in yeast. Interestingly certain H2A modifications appear to be specific for homologous recombination and not NHEJ pathways, supporting the hypothesis that particular patterns of modification have a role in the choice or commitment to a given pathway of DSB repair [29].

9.5.4   Radiosensitization by Histone Deacetylase Inhibitors

A number of structurally dissimilar inhibitors of HDACs have been identified, many of which are natural products. As described above, the histone acetylases acetylate lysine groups in nuclear histones resulting in the neutralization of the charges on the histones giving a more open and transcriptionally active chromatin structure while HDACs function to deacetylate and suppress transcription. Inhibition of HDACs results in arrest of cell growth, cell-cycle delay, differentiation, disruption of transcription, and apoptosis in transformed cells. Investigation of HDAC inhibitors for the treatment of cancer has intensified as their versatility and low toxicity have become apparent, and a number of these inhibitors have been tested on animals and in clinical trials. Radiosensitization by HDAC inhibitors has been reported but the mechanisms have not been defined. Results of a study of human squamous carcinoma cells implicated G_1 arrest and inhibition of DNA synthesis as mechanisms underlying radiosensitization by trichostatin

A [30]. Recent studies have shown little toxicity of HDAC inhibitors in normal cells [31] supporting the use of HDAC inhibitors for targeting radio-resistant cancers.

9.6 Summary

In eukaryotic cells, DNA is organized in a hierarchically structured nucleo-protein complex. At the first level, the DNA is wrapped around core octameric histone complexes (nucleosomes) forming a beads-on-a-string array; the nucleosomal array is folded into 30 nm chromatin fibers which are then folded into repeating loop domains, which are anchored to components of the NM. The loop domains ranging from 5 to 200 kb in length are defined by periodic attachments to the nuclear matrix (NM) containing defined DNA sequences known as matrix attachment regions (MARs) or scaffold attachment regions (SARs). The loop domains are not only important for the organization and packaging of chromatin in the interphase nucleus and mitotic chromosomes but also represent fundamental units for the genomic functions of DNA, replication, transcription, the regulation of the condensation state of chromatin for meiosis and mitosis, and the maintenance of genomic integrity.

The efficiency with which radiation induces DNA strand breaks depends on the extent to which DNA is associated with nuclear proteins. Histones and other DNA-bound proteins and soluble proteins are effective protectors against ionizing radiation while polyamines bind to DNA and protect against radiation by free-radical scavenging and by compaction and aggregation of chromatin.

Changes in DNA supercoiling status and the stability of DNA anchoring have been studied using the nucleoid halo assay. In irradiated cells, the ability to rewind DNA supercoils is inhibited in a dose-dependent manner presumably because of the presence of radiation-induced DNA strand breaks which remove the topological constraints on the DNA loops. Radio-sensitive cells show a greater inhibition of DNA supercoil rewinding for a given radiation dose and level of DNA damage than do resistant cells.

Mitotic cells are hypersensitive to ionizing radiation exhibiting single-hit inactivation coefficients similar to those of repair deficient cell lines. In addition to mitotic cells, other lines of evidence have also led to the suggestion that cellular radiosensitivity is correlated with the extent of chromatin condensation in the irradiated cell.

In addition to a structural role, the nucleosome is involved in the control of information flow from the DNA, including the intracellular signaling initiated by DNA damage. One histone, H2AX, is concerned with localizing and repairing DNA DSBs. When a DSB occurs, many molecules of histone H2AX in the chromatin adjacent to the break become phosphorylated and other components of the DNA damage response colocalize with the γ-H2AX foci. Another function which has been ascribed to chromatin restructuring is to act as a trigger for the conversion of inactive ATM dimers into auto-phosphorylated active monomers. HDACs function to deacetylate and suppress transcription. Inhibition of HDACs results in arrest of cell growth, cell-cycle delay, differentiation, disruption of transcription, and apoptosis in transformed cells. Radiosensitization by HDAC inhibitors has been reported but the mechanisms have not been defined.

References

1. Berezney R and Coffey DS. Identification of a nuclear protein matrix. *Biochem. Biophys. Res. Commun.* 60: 1410–1417, 1974.
2. Berezney R and Coffey DS. Nuclear matrix. Isolation and characterization of a framework structure from rat liver nuclei. *J. Cell Biol.* 73: 616–637, 1977.
3. Pienta KJ and Coffey DS. A structural analysis of the role of nuclear matrix and DNA loops in the organization of the nucleus and chromosome. *J. Cell. Sci. Suppl.* 1: 123–135, 1984.
4. Pienta KJ, Murphy BC, Getzenberg RH, and Coffey DS. The effect of extracellular matrix interactions on morphologic transformation in vitro. *Biochem. Biophys. Res. Commun.* 179: 333–339, 1991.
5. Razin SV, Gromova II, and Iarovaia OV. Specificity and functional significance of DNA interaction with the nuclear matrix: New approaches to clarify the old questions. *Int. Rev. Cytol.* 162B: 405–448, 1995.
6. Gasser SM and Laemmli UK. The organisation of chromatin loops: Characterization of a scaffold attachment site. *EMBO J.* 5: 511–518, 1986.
7. Jackson DA and Cook PR. The structural basis of nuclear function. *Int. Rev. Cytol.* 162A: 125–149, 1995.
8. Berezney R, Mortillaro M, Ma H, Wei X, and Samarabandu J. The nuclear matrix: A structural milieu for genomic function. *Int. Rev. Cytol.* 162A: 1–65, 1995.
9. Berezney R. Regulating the mammalian genome: The role of nuclear architecture. *Adv. Enzyme. Regul.* 42: 39–52, 2002.
10. Ma H, Siegel AJ, and Berezney R. Association of chromosome territories with the nuclear matrix. Disruption of human chromosome territories correlates with the release of a subset of nuclear matrix proteins. *J. Cell Biol.* 146: 531–542, 1999.
11. Ma H, Samarabandu J, Devdhar RS, Acharya R, Cheng PC, Meng C, and Berezney R. Spatial and temporal dynamics of DNA replication sites in mammalian cells. *J. Cell Biol.* 143: 1415–1425, 1998.
12. Wei Y, Mizzen CA, Cook RG, Gorovsky MA, and Allis CD. Phosphorylation of histone H3 at serine 10 is correlated with chromosome condensation during mitosis and meiosis in tetrahymena. *Proc. Natl. Acad. Sci. U.S.A.* 95: 7480–7484, 1998.
13. Tornaletti S and Hanawalt PC. Effect of DNA lesions on transcription elongation. *Biochimie* 81: 139–146, 1999.
14. Chiu S and Oleinick NL. Radioprotection against the formation of DNA double-strand breaks in cellular DNA but not native cellular chromatin by the polyamine spermine. *Radiat. Res.* 148: 188–192, 1997.

15. Warters RL, Newton GL, Olive PL, and Fahey RC. Radioprotection of human cell nuclear DNA by polyamines: Radiosensitivity of chromatin is influenced by tightly bound spermine. *Radiat. Res.* 151: 354–362, 1999.
16. Newton GL, Aguilera JA, Ward JF, and Fahey RC. Polyamine-induced compaction and aggregation: A major factor in radioprotection of chromatin under physiologic conditions. *Radiat. Res.* 145: 776–780, 1996.
17. Vogelstein B, Pardoll DM, and Coffey DS. Supercoiled loops and eucaryotic DNA replicaton. *Cell* 22: 79–85, 1980.
18. Roti Roti JL and Wright WD. Visualization of DNA loops in nucleoids from HeLa cells: Assays for DNA damage and repair. *Cytometry* 8: 461–467, 1987.
19. Malyapa RS, Wright WD, and Roti Roti JL. DNA supercoiling changes and nucleoid protein composition in a group of L5178Y cells of varying radiosensitivity. *Radiat. Res.* 145: 239–242, 1996.
20. Roti Roti JL, Wright WD, and VanderWaal R. The nuclear matrix: A target for heat shock effects and a determinant for stress response. *Crit. Rev. Eukaryot. Gene Expr.* 7: 343–360, 1997.
21. Schwartz JL, Mustafi R, Beckett MA, Czyzewski EA, Farhangi E, Grdina DJ, Rotmensch J, and Weichselbaum RR. Radiation-induced DNA double-strand break frequencies in human squamous cell carcinoma cell lines of different radiation sensitivities. *Int. J. Radiat. Biol.* 59: 1341–1352, 1991.
22. Schwartz JL. Alterations in chromosome structure and variations in the inherent radiation sensitivity of human cells. *Radiat. Res.* 149: 319–324, 1998.
23. Bode J, Kohwi Y, Dickinson L, Joh T, Klehr D, Mielke C, and Kohwi-Shigematsu T. Biological significance of unwinding capability of nuclear matrix-associating DNAs. *Science* 255: 195–197, 1992.
24. Chapman JD, Stobbe CC, and Matsumoto Y. Chromatin compaction and tumor cell radiosensitivity at 2 gray. *Am. J. Clin. Oncol.* 24: 509–515, 2001.
25. Chapman JD. Single-hit mechanism of tumour cell killing by radiation. *Int. J. Radiat. Biol.* 79: 71–81, 2003.
26. van Attikum H and Gasser SM. The histone code at DNA breaks: A guide to repair? *Nat. Rev. Mol. Cell Biol.* 6: 757–765, 2005.
27. Redon C, Pilch D, Rogakou E, Sedelnikova O, and Newrock KWB. Histone H2A variants H2AX and H2AZ. *Curr. Opin. Genet. Dev.* 12: 162–169, 2002.
28. Green CM and Almouzni G. When repair meets chromatin. First in series on chromatin dynamics. *EMBO Rep.* 3: 28–33, 2002.
29. Moore JD and Krebs JE. Histone modifications and DNA double strand break repair. *Biochem. Cell Biol.* 82: 446–452, 2004.
30. Zhang Y, Adachi M, Zhao X, Kawamura R, and Imai K. Histone deacetylase inhibitors FK228, N-(2-aminophenyl)-4-[N-(pyridin-3-yl-methoxycarbonyl)amino-methyl]benzamide and *m*-carboxycinnamic acid bis-hydroxamide augment radiation-induced cell death in gastrointestinal adenocarcinoma cells. *Int. J. Cancer* 110: 301–308, 2004.
31. Vigushin DM and Coombes RC. Histone deacetylase inhibitors in cancer treatment. *Anticancer Drugs* 13: 1–13, 2002.
32. Radulescu I, Elmrothe K, and Stenerlow B. Chromatin organization contributes to non-randomly distributed double strand breaks after exposure to high LET radiation. *Radiat. Res.* 161: 1–8, 2004.
33. Oleinick NL, Balasubramaniam U, Xue L, and Chiu S. Nuclear structure and the microdistribution of radiation damage in DNA. *Int. J. Radiat. Biol.* 66: 523–529, 1994.

10

Radiation-Induced Chromosome Damage

10.1 DNA, Chromosomes, and the Cell Cycle

10.1.1 Organization of DNA into Chromatin and Chromosomes

Most of the DNA of mammalian cells is sequestered in the cell nucleus, a structure which is a defining feature of eukaryotes, distinguishing them from prokaryotes such as bacteria. In the cell nucleus, a vast amount of DNA is arranged in a spatially organized fashion defining the domains of individual chromosomes and providing a framework for replication, transcription, and repair. Just before cell division, DNA and the associated proteins which are together called chromatin, become "condensed" and can be stained and viewed under the light microscope as rod-like structures called chromosomes. The human genome consists of 23 pairs of chromosomes (including one pair of sex chromosomes) giving a total of 46 for every diploid cell.

The hierarchical structure of chromatin has been discussed in Chapter 9. The condensed form of the chromatid in the metaphase chromosome represents the ultimate level of chromatin organization and, in fact, exists in this state for only a very small proportion of the life of a dividing cell. For most of the time the nucleus is in the interphase mode, with the 30 nm chromatin filaments which form approximately 50,000 loop domains attached at their bases to the inner portions of the nuclear matrix, in a relaxed conformation (Figure 9.1). The transition to the metaphase chromatid requires that the 60 kbp loops wind into 18 radial loops to form miniband units that in turn are continuously wound and stacked along a central axis to form each chromatid (Figure 10.1).

10.1.2 Cell Cycle

Cell division is a cyclic phenomenon, repeated for each generation of cells and is usually represented as a circle (Figure 10.2). A detailed description of the cell cycle in terms of the cyclic expression of the regulatory proteins involved is given in Chapter 8. On an operational level it can be divided into

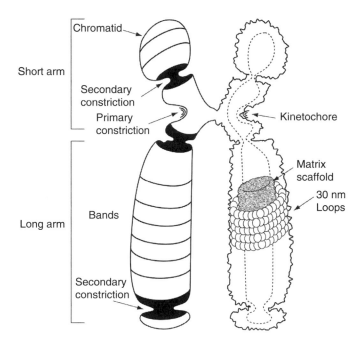

FIGURE 10.1
A metaphase chromosome.

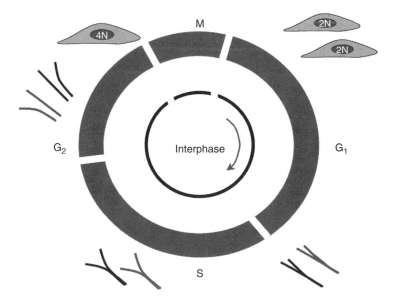

FIGURE 10.2
The cell cycle. After cell division the cell passes through the G_1 phase (variable duration). DNA synthesis takes place during the S phase when the amount of DNA in the nucleus doubles. G_2 is a relatively short period ending when the cell enters mitosis and divides. At the end of mitosis the DNA content has been halved and the two daughter cells reenter G_1.

four stages, starting at the mitotic (M) phase, which consists of nuclear division (karyokinesis) and cytoplasmic division (cytokinesis) to produce two daughter cells that begin the intermitotic period (interphase) of a new cycle. Interphase starts with G_1 phase during which the biosynthetic processes had slowed down during mitosis return to full operation. The S phase starts with the onset of DNA synthesis and is completed when the DNA content of the nucleus has doubled (from 2n to 4n) and the chromosomes have replicated to form two identical sister chromatids. The cell then enters G_2 phase, which ends when mitosis begins. During most of the interphase period chromosomes are extended and individual chromosomes cannot be detected. In the latter part of G_2, chromosome condensation starts and is completed during the first phase of prophase.

10.1.3 Mitosis

The need for complex mitotic machinery arose from the evolution of cells having greatly increased amounts of DNA packaged in a number of discrete chromosomes. The function of the machinery is to ensure that the genetic material of the replicated chromosomes is precisely divided between the two daughter cells at the time of division.

During interphase, when the cell is engaged in metabolic activity, chromosomes are not clearly discernible in the nucleus. A pair of centrioles that are microtubule organizing centers are apparent in the cytoplasm during interphase.

Mitosis is divided into stages that can be distinguished by light microscopy and are shown schematically in Figure 10.3:

1. In the first stage of mitosis, prophase, the chromatin, which is diffuse in interphase, condenses into chromosomes. Since each chromosome has duplicated during the preceding S phase, it now consists of two sister chromatids joined at a specific point along their length by a region known as the centromere. At the beginning of prophase, cytoplasmic microtubules that are part of the cytoskeleton disassemble, forming a pool of tubulin molecules that are then reused to construct the main component of the mitotic apparatus, the mitotic spindle that initially assembles outside the nucleus. By late prophase, bundles of polar microtubules that interact between the two asters appear to elongate in such a way as to push the two centrioles apart along the outside of the nucleus to form a bipolar mitotic spindle. At the end of prophase the nuclear membrane dissolves.

2. During metaphase, the chromosomes are arranged with their centromeres aligned in one plane, sometimes referred to as the metaphase plate. This is brought about by proteins that attach to the centromeres forming kinetochore fibers and are responsible for aligning the chromosomes halfway between the spindle poles

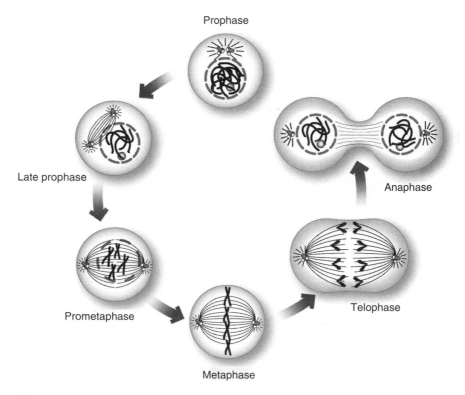

FIGURE 10.3
The stages of mitosis.

and for orienting them with their long axes at right angles to the
spindle axis. Each chromosome is held in tension at the metaphase
plate by the paired kinetochores and their associated fibers.

3. During anaphase, paired kinetochores on each chromosome
   detach and the pairs of chromatids separate, each moving toward
   an opposite pole of the cell at about 1 $\mu m/m$. As they move, the
   kinetochore fibers shorten (depolymerize) as the chromosomes
   approach the poles while the spindle fibers elongate (polymerize)
   and the two poles of the polar spindle move further apart.

4. Mitosis ends with telophase, during which the separated daughter
   chromatids arrive at the poles and the kinetochore fibers disap-
   pear and the polar fibers elongate still further. The interphase
   characteristics of the nucleus reappear as a new nuclear envelope
   forms around each group of new daughter chromosomes. The
   condensed chromatin expands and the nucleoli reassemble.

The division of the cytoplasm or cytokinesis results from cleavage occurring
during late anaphase or telophase. The membrane around the middle of the

cell, perpendicular to the spindle axis and between the daughter nuclei (the contractile ring), is drawn inward to form a cleavage furrow, which gradually deepens until it encounters the narrow remains of the mitotic spindle between the two nuclei. This narrow bridge may persist for some time until it finally breaks at each end leaving two complete, separate daughter cells.

It is clear from this description that damage caused to the chromosome by ionizing radiation or other insult will throw a spanner in the complex works of mitotic division and may result in mechanical disruption of the mitotic process, failure to correctly segregate genetic material between daughter cells and ultimately to cell death.

10.2 Radiation-Induced Chromosome Aberrations

Chromosome damage is one of the most obvious cytological effects of radiation. Experimental work on radiation-induced chromosome aberrations has been ongoing for almost 70 years since the early studies of Sax [1] with *Tradescantia*. Techniques developed over this period have expanded the scope and power of cytogenetic analysis and yielded new insights into the mechanisms of chromosome aberration formation.

Initially, the microscopic analysis of chromosomes was possible only during mitosis when condensed chromosomes could be visualized and cytogenetic analysis was confined to cells in the first mitosis following exposure to radiation or other clastogenic agent. In fact, aberrations seen at metaphase are an historical record of damage to the cell which was incurred during the preceding interphase and the analysis of metaphase chromosomes is a measure of the residual damage which remains after repair mechanisms have processed the initial lesions.

Aberrations produced by ionizing radiation can be classified as chromatid or chromosome-type aberrations. When normal cells are exposed to radiation in early interphase (i.e., in G_1), the first postirradiation mitoses yield exclusively chromosome-type aberrations, which result from damage to the G_1 chromosome. When this chromosome replicates during S phase, the lesion is replicated and affects both sister chromatids at the same site. In contrast, in cells that are irradiated in S or G_2 phase, after DNA has replicated, aberrations are produced which show defects in only one chromatid of a chromosome and are referred to as chromatid-type aberrations. This relationship between the type of damage and the point in the cell cycle at which the cell was exposed to the damaging agent is characteristic of radiation and a small number of chemical agents that produce double-strand DNA breaks. For UV radiation and the majority of chemical clastogens, which produce DNA single-strand breaks and base damage, chromatid-type aberrations are produced in cells exposed during G_1 [2] (Figure 10.4). In the light of what we now know of chromosome structure, it is almost

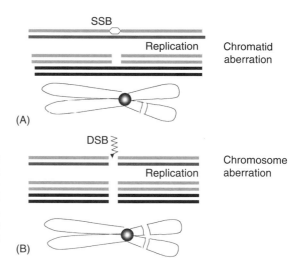

FIGURE 10.4
(A) Chemical damage to one strand of DNA results in a single-strand break which replicates to affect one double strand and only one chromatid. (B) A double-strand break replicates to form a chromosome aberration affecting both chromatids.

intuitive to relate this observation to the fact that ionizing radiation is one of very few cytotoxic modalities that produce double-strand DNA breaks. However, the nature of the initial lesion underlying chromosome aberrations was not always obvious and even today its identification relies on circumstantial evidence.

10.2.1 Nature of the Initial Lesion

Based on their observation of metaphase chromosomes, the early investigators developed the concept that radiation produced a frank "chromosome break" as the initial lesion. The broken ends of chromosomes or chromatids, depending on when during the cell cycle radiation occurred, could rejoin with those close to them (the broken ends were frequently referred to as being "sticky") on the same or an adjacent chromosome. Since two breaks have to be open at the same time in order to join, they would have to be temporally as well as spatially coincident. This scenario does not presuppose the nature of the initial lesion; however, the accepted model of the chromosome involves a single double-stranded DNA molecule passing through the length of the chromosome, so a break in the chromosome must involve the equivalent of at least one DSB. The idea that the DSB is the fundamental lesion in chromosome breakage is supported by considerable indirect evidence including the following:

1. Natarajan and Zwanenburg [3] showed that x-irradiated mammalian cells treated with *Neurospora* endonuclease had increased numbers of chromosome aberrations compared to cells exposed to x-rays alone. They argued that this increase was due to the conversion of x-ray-induced DNA single-strand damage into DSBs by the endonuclease.

2. When restriction enzymes (nucleases that have a specificity for particular DNA sequences and cause DSBs) are introduced into the cell nucleus they cleave DNA in vivo as they do in vitro and cause chromosome aberrations [4,5]. In human cells, restriction enzymes introduced into cells cause frank chromosome breaks which can be detected by the technique of premature chromosome condensation (PCC, described below). These enzymes which can produce only DNA DSBs mimic the effects of ionizing radiation both in terms of the types of aberrations produced with respect to the cell cycle, and their distribution with respect to chromosomal substructure.

3. When mammalian cells are exposed to hydrogen peroxide under conditions of reduced temperature, extensive DNA single-strand breaks (SSBs) and various types of base damages are produced but few if any DNA DSBs, even for doses of hydrogen peroxide which produce a very high yield of SSBs [6]. This is consistent with the belief that the majority of radiation-induced cell killing is the result of DNA DSBs and subsequent formation of chromosome aberrations [7].

4. Most importantly, and as described above, few chemical agents are capable of causing the pattern of chromosome damage characteristic of ionizing radiation, i.e., chromosome-type aberrations after G_1 exposure and chromatid-type aberrations following exposure of S and G_2 cells. These chemicals include bleomycin and neocarzinostatin, which produce damage to DNA via attack by radical species which mimic the indirect effects of ionizing radiation [8]. These agents are also known for their ability to produce DNA DSBs in large numbers.

10.2.2 Partial Catalog of Chromosome and Chromatid Aberrations

When chromosome breaks are produced, they can behave in different ways. Firstly, the breaks may restitute, that is, rejoin in their original conformation; in this case, nothing is apparent at the next mitosis. Another possibility is that the breaks do not rejoin and that an aberration that is seen as a chromosome deletion and associated acentric fragment is scored at the next mitosis. The third option is that breaks re-assort and rejoin to other broken ends. This gives rise to many forms of chromosome damage, including such aberrations as chromosome translocations, inversions, and insertions.

10.2.2.1 *Chromosome Aberrations Produced by Irradiation in G_1 or Early S Phase*

Radiation-induced G_1 or early S phase (pre- or early in DNA replication) chromosome aberrations can be classified as either symmetrical or asymmetrical. Symmetrical rearrangements include reciprocal translocations

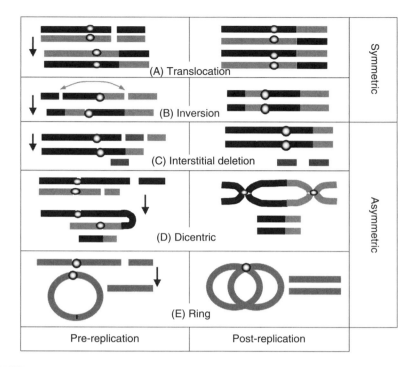

FIGURE 10.5
Chromosome aberrations.

(Figure 10.5A) in which breakage of two pre-replication chromosomes is followed by exchange of the broken fragments, and inversions (Figure 10.5B) that can arise when two breaks occur within the same chromosome. The inversion can be either paracentric if both breaks occur within the same chromosome arm, or pericentric (encompassing the centromere) if the breaks occur in both chromosome arms. These symmetrical aberrations do not result in gross distortions (gain or loss) of chromosome structure and are probably compatible with cell survival.

In contrast, failure to apportion genetic material between daughter cells at mitosis due to the formation of asymmetrical chromosome aberrations is more likely to be a lethal event. For example, two breaks on the same arm of a pre-replicated chromosome can rejoin in such a way as to eliminate an interstitial fragment causing an interstitial deletion (Figure 10.5C) and loss of genetic material. Formation of a dicentric chromosome involves an interchange between two separate chromosomes. When the breaks occur in early interphase and the ends are close to each other, they can rejoin as shown in Figure 10.5D. When this structure replicates during S phase the result is a distorted chromosome with two centromeres, i.e., a dicentric, and a fragment with no centromeres (acentric fragment). The formation of a ring is shown in Figure 10.5E. In this case, breaks occur in both arms of one chromosome and the ends rejoin to form a ring and a fragment. When the chromosome replicates two overlapping rings with one centromere (like Siamese-twinned doughnuts) are formed together with two acentric fragments.

10.2.2.2 Chromatid Aberrations Produced in Mid to Late S and in G_2

Chromatid aberrations are produced after DNA has replicated and each chromosome consists of a pair of chromatids joined at the centromere. Aberrations can result from interaction between chromatid arms of different chromosomes or between the two chromatid arms of the same chromosome. Shown are a chromatid deletion when a terminal fragment of one chromatid arm is deleted (Figure 10.6A); a triradial that involves deletions and rejoining of two terminal fragments from one chromosome and one from another (Figure 10.6C); a symmetrical interchange between terminal fragments of one chromatid arm from each of two chromosomes (Figure 10.6E); an asymmetrical interchange in which two chromosomes become joined as a result of a deletion of a fragment from one chromatid arm of each resulting in a dicentric structure and an acentric fragment (Figure 10.6D). One of the most destructive lesions occurs when breaks occur in each of the two chromatid arms of one chromosome and the pairs of broken ends join to form a sister union and an acentric fragment. When the chromatid pairs attempt to separate at anaphase, the acentric fragment is lost and the two centromeres

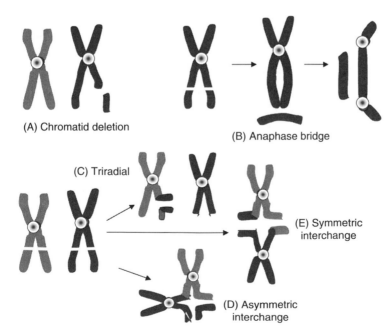

(A) Chromatid deletion

(B) Anaphase bridge

(C) Triradial

(E) Symmetric interchange

(D) Asymmetric interchange

FIGURE 10.6
Chromatid aberrations.

of the dicentric head in opposite directions stretching the chromatid between opposite poles of the cell and preventing complete separation of the daughter cells. This is called an anaphase bridge (Figure 10.6B).

10.3   Visualization of Chromosome Breaks during Interphase: Premature Chromosome Condensation

The limitations of metaphase analysis for the study of chromosome repair kinetics were overcome when Johnson and Rao [9] discovered that the chromatin of interphase cells became visibly condensed as "interphase chromosomes" when such an interphase cell was fused with a mitotic cell.

The morphology of prematurely condensed chromosomes (PCC) reflects the position of the interphase cell in the cell cycle at the time of fusion. PCC from G_1 and G_2 cells exhibit one and two chromatids per chromosome, respectively while the S phase PCC have a fragmented appearance (Figure 10.7). The PCC technique allows direct visualization of breaks in

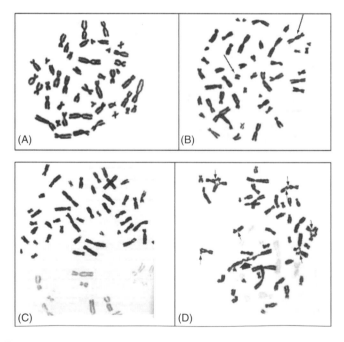

FIGURE 10.7
Ionizing radiation-induced chromosomal aberrations seen at metaphase and in prematurely condensed chromosomes. (A) Metaphase, control, (B) Metaphase, 2 Gy (arrows indicate dicentrics and fragments), (C) G_2 premature condensed chromosomes, control, and (D) G_2 premature condensed chromosomes following exposure to 2Gy (arrows indicate gaps, breaks and exchanges). (Courtesy of Dr Tej Pandita.)

interphase chromatin within minutes following exposure to IR, providing information about the extent of initial chromosome breakage and the characteristics of break repair as a function of time following exposure [10,11].

Dose–response curves for the induction of initial PCC breaks (i.e., immediately after exposure) are linear for virtually all types of radiation, with slopes that increase with increasing ionization density. These yields account for only a fraction (about 10%–15%) of the total DNA DSBs initially produced in such cells. The data provided by PCC analysis suggests that a subset of DSBs is effectively converted to chromosome breaks, while the remainder is rejoined. In most cases, the kinetics of PCC break rejoining can be adequately described as a first-order process, with half times ranging from about 0.5 [12,13] to 2 h depending on cell type.

Rejoining of breaks as demonstrated by PCC can explain the small number of aberrations scored at metaphase and has been claimed to parallel the increase in cell survival that accompanies the recovery from potentially lethal and sublethal damage will be discussed later in this chapter. PCC break rejoining is also in reasonable agreement with the repair kinetics of radiation-induced DSBs, as measured by neutral velocity sedimentation [14].

10.3.1 FISH, mFISH, SKY, mBAND FISH, and Chromosome Painting

Fluorescence in situ hybridization (FISH) is a process whereby a fluorescently labeled DNA probe binds to complementary DNA segment in chromosomes of metaphase spreads. Using chromosome specific probes, the technique can identify many types of chromosome aberrations and is particularly effective for translocations. Technical details are described in Chapter 4. In the first FISH studies, the probes were directed against a single homologous chromosome pair or a few chromosomes with the rest of the genome being distinguished by a nonspecific DNA counterstain. Interchanges were seen as color switches at the break junctions (also known as color junctions). Second generation painting schemes involved labeling individual subsets of chromosomes with two or three distinct colors. These limitations were overcome by multifluor FISH (mFISH), spectral karyotyping (SKY), and mBAND FISH, which recognizes unique combinations of different fluors which are directed against specific chromosomes and chromosome bands, respectively [15,16]. If a total of five spectrally distinct hybridization fluors are used and two or three of these fluors are directed against a given chromosome, each of 24 different chromosome types in the human karyotype is represented by a unique combinatorial signature. Computer software then recognizes these signatures and arbitrarily assigns convenient pseudocolors on a pixel by pixel basis. Combinatorial painting can produce a 24 color karyotype that is capable of revealing multiple radiation-induced interchromosomal rearrangements simultaneously (Figure 10.8).

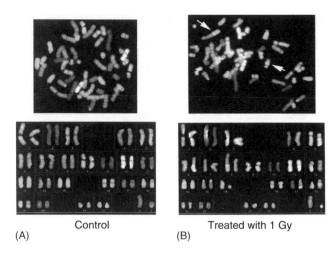

Control	Treated with 1 Gy
(A)	(B)

FIGURE 10.8 (See color insert following page 288.)
Spectral karyotype of human chromosomes. (A) Control. (B) Treated with 1 Gy. (Courtesy of Dr Tej Pandita.)

10.3.2   Results of Whole Chromosome Painting

One of the more striking results coming from whole-chromosome painting is the detection of "complex exchanges" that result from three (or more) illegitimate breakpoints among two (or more) chromosomes. This was already detectable to some extent using simple chromosome painting methods. Using a two-color painting scheme to identify exchanges simultaneously involving chromosomes 1 and 2 in human lymphocytes following a 2.3 Gy dose of ^{137}Cs γ-rays, it was observed that about 7% of the exchanges were actually "three-way" complex exchanges that also involved unidentified (counterstained) chromosomes [17]. After 6 Gy of x-rays, Brown and Kovacs [18] observed that 26% of the complete exchanges involving chromosome 1, 4, or 8 of primary human fibroblasts were visibly complex and estimated that up to half of the total exchanges formed may be of this type. These results were later confirmed and extended for exchanges among chromosomes 1, 5, and 7 of human fibroblasts that were exposed to either 4 or 6 Gy of x-rays [19].

The full range of these effects was only appreciated when the technique of mFISH was used. Loucas and Cornforth [20] used mFISH to study the production of complex chromosome exchanges in human lymphocytes and found that these aberrations are quite common even at doses of 2 or 4 Gy. At 4 Gy, 44% of cells contained at least one complex exchange. Complex exchanges require a minimum of at least 3 breaks and as many as 11 were observed, and at this dose, more than 40% of gross cytogenetic damage as measured by total number of exchange breakpoints was complex in origin. Clearly the list of chromosome aberrations in

Section 10.2.2 is by no means complete because only those interchanges involving two breaks in one or two chromosomes or chromatids are included. The number of possible rearrangements involving three or more breaks is enormous.

10.4 Mechanisms of Aberration Formation

The mechanisms of aberration formation has been reviewed by Savage [21], Bryant [22], and Hlatky et al. [23]. This chapter is not a suitable forum to recapitulate in detail the various hypotheses of chromosome aberration formation. Briefly, the debate has centered about the three pathways shown schematically in Figure 10.9. These pathways, which are not the only ones to have been proposed, are not necessarily exclusive and could exist in parallel.

1. The breakage and reunion mechanism (Figure 10.9A) corresponds at the molecular level to a process of double-strand breakage followed by joining of the broken ends by a process of nonhomologous end joining (NHEJ, described in Chapter 7).

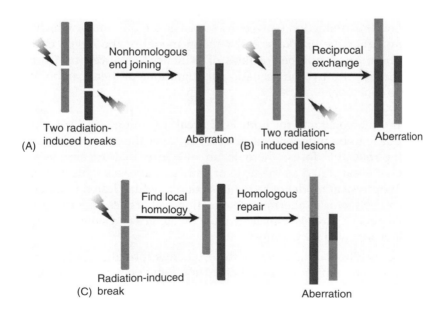

FIGURE 10.9
Models for exchange formation. (A) Breakage-and-reunion, (B) exchange theory, and (C) recombinational misrepair (one hit).

2. The one-hit model (Figure 10.9C) envisages that a single radiation-induced DSB is sufficient to initiate an exchange with an otherwise undamaged portion of the genome [24]. This could involve repair by a process of homologous recombination (HR) (Chapter 7) with interaction between limited sequence homologies occurring at different sites such as the abundant repetitive DNA elements which are known to occur. This process, which involves only one initial lesion has also been called recombinational misrepair.

3. The third pathway (Figure 10.9B), the exchange hypothesis introduced by Revell in 1959 [25], departs from the idea that exchanges form from frank radiation-induced chromosome breaks. This model postulates that the primary lesions destined for pairwise interaction are not true breaks in the chromosome but some other form of damage that does not immediately compromise the integrity of the chromosome. Only after such primary lesions attempt the process of exchange is gross chromosomal structure disrupted so as to produce the aberrations visible at mitosis. An important consequence of this is that most terminal deletions would be the result of failed exchange events rather than unrejoined breaks.

Much of what will be discussed later in this chapter concerning the relationship of chromosome aberrations to genetic effects, carcinogenesis and cell survival depends on the conventional interpretation of chromosome aberration data, i.e., that the formation of a lethal chromosome aberration depends on the interaction between two breaks, a requirement which could be met by either the breakage and reunion model or the Revell exchange theory. There are a number of lines of evidence (reviewed by Cornforth [26,27]) that favor a two-hit scenario although none of them can be considered to be conclusive.

1. The rate constants for both PCC break rejoining and DSB repair appear to be independent of dose and over the range of biologically relevant doses; there is no evidence that repair systems become saturated as the level of damage increases. This supports a biophysical model of radiation action in which lesion interaction is a key factor in aberration generation and argues against one-hit cytogenetic models, which attribute exchanges to a single damaged site (reviewed by Cornforth [26]).

2. The available data suggests that exchange formation is driven by recombinational processes that are nonhomologous in nature. Consequently, one-hit models of aberration formation which depend on extensive sequence homologies, including homologies provided by repetitive elements must be rejected. Elevated frequencies of chromosome aberrations occur to various extents in cell lines mutated in NHEJ and HR pathways of DSB end joining.

3. Aberration formation, proximity effects and other aspects of repair/misrepair mechanisms have been quantified and complicated data sets clarified by computer modeling [28]. Models of chromatin structure are integrated with simple models of DSB formation and mis-rejoining as well as with previously developed, radiation track codes describing physical and radiochemical aspects of radiation action [29,30]. Based on such biophysical modeling, it has been suggested that breakage-and-reunion is the dominant, but not the sole pathway for aberrations produced in mammalian cells following exposure to ionizing radiation during G_0/G_1. In support of this, simulations based on other mechanisms are unable to reproduce the full range of aberration spectra that are observed experimentally, especially the frequency and extent of complex aberrations. However, the results of modeling do not preclude some involvement of the other pathways. While there is balance of opinion that chromosome aberrations are formed by a breakage first, two-hit scenario, it has been questioned whether chromatid aberrations are formed by the same mechanisms since there is little correspondence between chromosome aberrations and chromatid aberrations [22].

10.4.1 Chromosome Localization and Proximity Effects

An early controversy in the field of cytogenetics was how close together do chromosomes have to be for radiation-induced rearrangements to occur? It was not clear whether interaction could occur when broken ends came in contact after radiation damage (breakage first hypothesis) or that exchange occurred only where close contact already existed (contact first hypothesis). Breakage first was the dominant idea because it was felt to be more consistent with quantitative data on chromosome aberrations [31,32].

Over the last decade it has been shown, that at any one time, a chromosome is predominantly localized to a territory whose volume is only a few percent of the volume of the whole nucleus (reviewed by Cremer and Cremer [33]). Indirect evidence of the juxtaposition of interphase chromosomes, sometimes referred to as proximity effects, include a bias in interchange frequency to favor chromosomes or segments of chromosomes that are close together. Thus, for intrachromosomal rearrangements there is a statistical bias for small intrachanges over large intrachanges, compared to what might be expected on the basis on genomic content [34], for all types of intrachange over two-chromosome aberrations [35] and for two-chromosome aberrations over three-chromosome aberrations. Direct evidence indicates little overlap of chromosome territories [33] and recent results for the dependence of aberration frequency on chromosome DNA content support a picture where interchromosomal interactions occur mainly near the surface of territories [36,37]. However, the formation of multi-chromosome aberrations suggests that territorial overlap must occur [28] and this might be explained if there was preferential induction of aberrations in special locations outside the main territories, where loops from many different chromosomes come close to each other [38].

In cancer cells, exchanges are frequently observed to occur at precisely the same location suggesting that the functional association between apparently remote DNA sequences is a regular feature of the interphase nucleus. An example of this is seen in many radiation-induced thyroid tumors where there is an intrachromosomal inversion resulting in the fusion of the tyrosine kinase domain of the *RET* gene with a section of the *H4* gene. The regions containing the gene encoding the RET receptor tyrosine kinase and that containing the *H4* gene are separated by 30 Mb. It was shown that in 35% of normal thyroid cells the *RET* and *H4* genes were actually in close proximity in the interphase nucleus as judged by resolution of fluorescent probes by three-dimensional microscopy [39]. The authors postulate that such a preformed molecular association would favor the intrachange between the regions containing these genes, which is seen in many radiation-induced thyroid tumors. The proximity of the chromosomal regions involved in the *RET-H4* inversion suggests that the association of these regions may be important in thyroid cell differentiation. The existence in cancer cells of many similar recurrent aberrations with precise exchange point positioning suggests that the functional association between remote sequences is a regular feature of the interphase nucleus even in normal cells. In actively dividing cells many of these associations will be transitory and not easily seen but in differentiated and differentiating cells they may be much longer-lived and readily detectable. The sites of chromosomal association in normal cells may be also highly favored locations where the exchange process can take place, possibly in the DNA loops in the regions between contiguous domains.

10.5   Implications of Chromosome Damage

Many of the effects of exposure to ionizing radiation can be interpreted in terms of the response to radiation-induced chromosome damage.

10.5.1   Genetics

It was proposed some years ago on theoretical grounds that most radiation-induced germline mutations in the mouse are the result of large-scale genetic alterations exemplified by gross structural chromosomal changes [40] and this has been confirmed for several mammalian genetic loci for ionizing radiations of various qualities [41].

Large-scale deletions and rearrangements often dominate radiation-induced mutational spectra of mammalian cells and overwhelm the contributions of point mutations and small-scale intragenic changes [42]. Thus, although radiation can produce in mammalian cells small-scale genetic changes such as nucleotide base transition, transversion, and frame-shift mutations, its most important effect is to cause large-scale genetic alterations, most notably deletions. These often involve the excision of an entire locus under study and can actually be seen as chromosome aberrations.

Radiation mutation spectra of one particular gene, the *HPRT* gene, have been the subject of extensive experimentation [43,44]. The *HPRT* gene, which encodes the hypoxanthine-guanine phosphoribosyl transferase enzyme protein, is relatively large (~45 kb) and has been completely sequenced. The *HPRT* protein is nonessential in normal medium but mutants can be selected on the basis of their resistance to 6-thioguanine added to the medium. A high resolution cytogenetic analysis of seven independent x-ray-induced *HPRT* mutants of primary human fibroblasts [45] showed five of the mutants to have total gene deletions while two had partial deletions having a breakpoint contained within the remaining *HPRT* sequence. The size and position of partial deletions and rearrangements of the hamster *HPRT* gene were studied in a series of spontaneous, γ-ray-induced or α particle-induced mutants. The deletions produced by α particles were larger on average than those produced by gamma rays and ranged in size from relatively small to those which involved much of the gene as well as 3′ or 5′ flanking sequences.

10.5.2 Carcinogenesis

The molecular consequences of chromosome aberrations influence the functioning of two distinct classes of genes in human tumors: dominantly acting oncogenes and tumor suppressor genes. Fusion of an oncogene with a second gene at a site of a translocation or an inversion generating a chimeric gene and a new protein are seen primarily in leukemias, lymphomas, and sarcomas. The best known example of fusion is the activation of the *abl1* (*formerly abl*) oncogene by translocation with *bcr* generating the Philadelphia chromosome as the result of a reciprocal translocation between chromosomes 9 and 22. Of particular interest in terms of radiation carcinogenesis is the fusion of the *RET* and *H4* genes, described above, which occurs in radiation-induced thyroid cancer [39].

The function of tumor suppressor genes is to restrain tumor growth and the loss or inactivation of one of these genes removes the normal constraints on tumor growth contributing to the malignant phenotype. The two-hit model of tumorigenesis [46] predicts that certain genes (now designated as tumor suppressor genes) will be tumorigenic when both alleles are inactivated by mutation or chromosome deletion. The first allele can be inactivated either in the germ cells resulting in a familial susceptibility or predisposition to cancer or in the somatic cells while the second allele is inactivated in the somatic cells. Mutation or deletion of the allele on both maternal and paternal chromosomes is referred to as a loss of heterozygosity (LOH) at the locus of the tumor suppressor gene and is a precondition of malignant transformation. LOH can come about by several mechanisms including chromosomal nondisjunction and mitotic recombination or gene conversion in which a wild-type allele is replaced by a duplicated copy of the homologous chromosome region that carries the mutant allele. LOH may also occur by deletion of the normal allele or by complete or partial chromosome loss in which case the cell contains one copy of the mutated allele and no wild-type allele.

10.5.3   Cell Survival, Dose Rate, and Fractionation Response

Evidence of a cause and effect relationship between chromosome aberrations and cell killing has been frequently documented although in many cases these reports present evidence which is largely circumstantial. Close correlations have been drawn between mean lethal dose and the dose required to produce an average of one visible aberration per cell in the first mitosis after irradiation; however, the existence of such a correlation would not establish a cause and effect relationship between chromosomal aberrations and cell killing since there is the possibility that damage resulting in chromosome aberrations and damage resulting in cell death are separate but show the same radiosensitivities. Ingenious experiments by Joshi et al. [47] showed that there was in fact a one-to-one correspondence between the two endpoints. Cells were irradiated during G_1 where only chromosome aberrations would be produced and the anticipated lethal aberrations would be asymmetrical resulting in formation of acentric fragments. Acentric fragments were visualized as extra-nuclear bodies or micronuclei apparent at the first mitosis after radiation in at least one daughter cell. By scoring individual live cells the investigators were able to show that nearly all irradiated cells generating a micronucleus failed to continue to proliferate whereas those without a micronucleus continued to proliferate thus establishing a one-to-one relationship between chromosome aberrations and cell death.

The formation of asymmetric exchanges requires that two (or more) lesions are close enough together in time and space to interact. Analysis of experimental data and theoretical modeling (reviewed by Hlatky et al. [23]) both support a consistent model which calls for two types of radiation action one-track action and two-track action (Figure 10.10). Two-track action requires the interaction of uncorrelated damage between two different primary radiation tracks, such as two different x-ray photons. The relationships between radiation damage and cell killing associated with these two modes of energy deposition will have certain characteristics:

1. Two-track action will produce an approximately quadratic yield of lesions (proportional to dose squared) since if each break is produced independently, the number of individual breaks capable of forming an exchange will increase as a linear function of dose, while the number of coincident pairs of breaks that interact

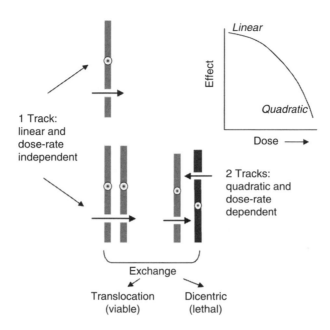

FIGURE 10.10
Dose and dose-rate dependence of radiation-induced chromosome aberrations. Damage can be induced in one or two chromosomes by a single radiation track. The effect will be linear and dose-rate independent. Two chromosome breaks can also be produced by the action of two radiation tracks. The effect will be quadratic and dose-rate dependent. These relationships are reflected in the shape of the dose response curve; linear at low doses, quadratic at high doses.

together will be proportional to the square of the radiation dose. Two-track action will have a reduced effect when the dose rate is decreased since there will be time available between interactions produced by separate tracks for repair to take place.

2. One-track action in which the two interactive lesions are both produced by the same radiation track or x-ray photon will produce a dose linearly proportional to dose and independent of dose rate.

It is believed that most lethal lesions result from a pair of interphase chromosome breaks occurring sufficiently close together in time and space to interact and form a ring or dicentric as a result of misrepair. Requirement for two different radiation-induced lesions to initiate an exchange will lead to a mixture of linear and quadratic dose dependence. The linear component of dose response which results from a single track producing two or more DSBs will dominate at low doses, while the quadratic term predominates at higher doses [29] since, with increasing dose there is increased probability that two breaks produced by separate tracks will interact (Figure 10.11). A second possible scenario does not involve an interaction but a single chromosome break, that remains unrepaired and results in a terminal deletion (Figure 10.10 upper chromosome) leading to cell death as a result of loss

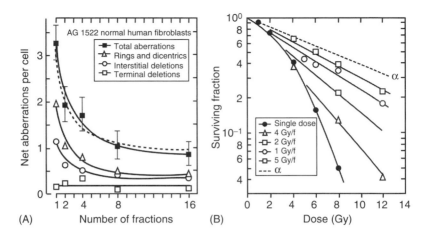

FIGURE 10.11

(A) Decrease in yield of chromosome type aberrations (rings dicentrics and interstitial deletions) at the first mitosis following radiation of non-cycling normal human fibroblasts to a total dose of 8 Gy with the dose given as 1, 2, 4, 8, or 16 fractions. (with 6 h between fractions). (B) Survival curves for the same cells under the same conditions. (From Bedford, J.S., *Int. J. Radiat. Oncol. Biol. Phys.*, 21, 1457, 1991. With permission.)

of genetic material. This could be the result of a single track, producing one DSB, or from two tracks producing two lesions close together. The shape of the typical mammalian cell survival curve reflects the fact that lethal lesions can be produced by either one or two-track action by having an initial linear portion at low dose and moving at higher doses to a curvilinear quadratic response.

The two-track action model will result in a smaller effect when a given dose is prolonged, since if the dose is spread out over time, repair may take place during the time between the first lesion being formed and the production of the second. One-track action on the other hand usually produces a yield that is linearly proportional to dose and independent of dose rate since two lesions are produced almost simultaneously.

Chromosome break rejoining kinetics as revealed by PCC can explain both the decrease in aberrations seen at metaphase (and associated increase in cell survival) that accompanies the repair of potentially lethal damage (PLD) as shown by delayed plating experiments and the repair of sublethal damage which occurs during fractionated exposure. Changes in the frequency of chromosome-type deletions and asymmetrical exchange aberrations measured in the first postirradiation mitosis corresponded closely with cell killing when dose was fractionated supporting the hypothesis that sublethal damage repair results from rejoining of breaks in interphase chromatin so they are no longer capable of reacting with the second dose [48] (Figure 10.11 and 10.12B). Repair of PLD has also been examined in relation to PCC rejoining kinetics. Figure 10.12A shows survival as a

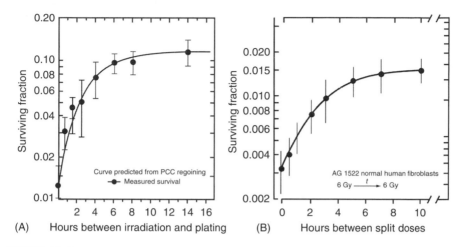

FIGURE 10.12
(A) Repair of potentially lethal damage (PLD) rate of rejoining of interphase chromosome breaks (solid line) is in good agreement with rate of increase in survival with delayed plating (data points). (B) Repair of sublethal damage. Rate of rejoining of interphase chromosome breaks (solid line) is in good agreement with survival (data points) with increasing time between doses of 6 Gy. (From Bedford, J.S., *Int. J. Radiat. Oncol. Biol. Phys.*, 21, 1457, 1991. With permission.)

function of postirradiation incubation time for AG1522 normal human fibroblasts. There was a close correspondence between the rates of cell survival in terms of colony formation and the increase in survival predicted on the basis of PCC rejoining kinetics [49].

10.5.4 Genomic Instability

Genomic instability is a term used to describe damage observed in cells at delayed times after radiation and may be seen in the progeny of exposed cells at multiple generations after the initial radiation insult (reviewed by Morgan [50,51]). It is manifested as chromosome aberrations, changes in ploidy, micronucleus formation, gene mutations and amplifications, microsatellite instability, and reduced plating efficiency. Chromosome changes are, in fact, the best described of the endpoints associated with genomic instability. In colonies of surviving cells exposed to relatively low doses of radiation, karyotypic abnormalities have been reported in 40%–60% of murine stem cells exposed to doses of α particles which would produce approximately one hit per cell [52]. The prevailing hypothesis is that radiation destabilizes the genome and initiates a cascade of genomic events that increases the rate of mutation and chromosomal change in the progeny of the irradiated cell. It is unlikely that directly induced DNA damage such as DSBs are the causative agent [53], instead deficient response to DNA damage, changes in gene expression [54] and perturbation of cellular homeostasis [55] have been invoked. While the nucleus may be the ultimate target of instability there is evidence of a role for cytoplasmic persistent oxidative stress (i.e., an increase in ROS) [56]. There is evidence implicating extranuclear or extracellular events in initiating and perpetuating chromosomal instability and several investigators have reported that in a population where not all cells are irradiated it is nonirradiated cells which manifest chromosomal instability [52]. This will be discussed further in Chapter 15 which deals with bystander effects.

10.5.5 Biodosimetry and Risk Estimation

In biodosimetry, the goal is usually to infer dose retrospectively from the level of chromosome aberrations in an individual's peripheral blood lymphocytes. This procedure requires knowing how aberration frequencies vary with dose for the particular radiation type and exposure conditions involved. Under the breakage-and-reunion scenario described above, radiation-induced dicentric or translocation frequency is expected to have a linear quadratic dependence on dose for sparsely ionizing radiation in the relevant dose range. Densely ionizing radiation, in contrast, operates almost exclusively via intra-track action over the relevant dose range, so near-linearity in dose is expected and has in fact been observed [29,57]. In addition, because of the interplay between chromatin geometry and radiation track structure, the spectrum of aberration types is expected to be different for densely ionizing radiation [58]. A different spectrum is observed in vitro with higher frequencies of aberrations involving several exchange breakpoints within the same chromosome arm, compared to interchromosomal interactions. At low doses, there is a higher frequency of complex aberrations compared to simple ones [59,60]. On the basis of such differences, retrospective biodosimetry should eventually be able to identify the type of radiation as well as the dose received.

A second potential application of biodosimetry involves estimating cancer risk from radiation exposure. A major problem has always been that the doses of interest are too small to produce quantifiable or even detectable biological effects, either experimentally or epidemiologically but at the same time, it is the effects of such low doses that are of the most serious concerns when very large populations are involved. Uncertainties in low-dose and low dose-rate experimental estimates, together with the major health and economic issues involved, have made this area controversial. Carcinogenic changes are likely to result from translocations and inversions while the aberrations being scored are the lethal dicentrics and rings, a mode of risk estimation that is valid only if the dose and dose-rate dependence of radiation carcinogenesis parallels that of aberration formation. The majority of cytogenetic data used to derive dose response relationships comes from the scoring of the mainly lethal asymmetric exchanges such as dicentrics on the assumption that their yields parallel those of their non-lethal counterparts such as reciprocal translocations. Some observations have indicated that there can be unequal frequencies between symmetrical and asymmetrical exchanges recovered in cells at their first postirradiation mitosis [61,62] and this casts some doubt on the application of this approach to risk estimation. Currently risk estimate extrapolations from higher doses are based on the linear-quadratic model with consideration given to effects of dose rate [63,64]. While there is good evidence for a causal link between translocations and certain cancers, especially leukemias, radiogenesis of some solid tumors may be more closely related to other forms of radiation damage, having different dose dependencies and dose-rate dependencies [64].

10.6 Summary

In early studies, microscopic analysis of chromosomes was possible only when condensed chromosomes could be visualized during mitosis, and the type of aberration seen was dependent on the phase of the cell cycle at which the cell was exposed to radiation. Cells exposed in G_1 replicated damage during S phase and showed only chromosome aberrations involving both sister chromatids at the same site, whereas cells irradiated in late S or G_2 after DNA had replicated showed chromatid aberrations affecting only one chromatid arm of the chromosome.

Indirect evidence strongly supports the view that the initial lesion leading to a chromosome break is a DNA double-strand break. The breakage–reunion hypothesis that corresponds at the chromosome level to the molecular process of double-strand DNA breakage with joining of the broken ends by Non-Homologous End Joining (NHEJ) is the mechanism of aberration formation best supported by the evidence but is not necessarily exclusive. Development of premature chromosome condensation (PCC), which allowed direct visualization of chromosome damage within minutes of irradiation, facilitated demonstration of the kinetics of chromosome break rejoining. A more recent technical advance is the fluorescence *in situ* hybridization (FISH) technique for specific labeling of gene segments of condensed chromosomes. Multifluor FISH (mFISH), mBAND FISH, and SKY are combinatorial chromosome painting techniques which reveal multiple radiation-induced chromosomal rearrangements simultaneously. An important result of the application of FISH techniques was the detection of "complex exchanges" which result from three or more illegitimate break points among two or more chromosomes.

Symmetrical exchanges such as reciprocal translocations and inversions that occur without acentric fragment formation and do not result in gross distortion of chromosome structure are probably compatible with cell survival, but may however be associated with malignant transformation. Asymmetric exchanges in pre-replication chromosomes resulting in the formation of dicentrics, rings, or frank chromosomal deletions cause loss of genetic material and a distorted chromosome following replication, which is probably not compatible with continued cell survival. Lethal lesions can also result from chromatid aberrations that are produced after

DNA is replicated, such as the formation of an anaphase bridge. Genetic changes produced by radiation are mostly large-scale genetic alterations, notably deletions that may involve the excision of an entire locus under study and override smaller radiation-induced changes.

Lethal lesions result from a pair of interphase chromosome breaks occurring sufficiently close together in time and space to interact and form a ring or dicentric as a result of misrepair or as the result of an unrepaired single chromosome break leading to loss of genetic material. This results in a mixture of linear and quadratic dose dependence in terms of cell survival with the linear component dominating at low doses whereas the quadratic term is predominant at higher doses with the increasing probability that breaks produced by separate radiation tracks will interact. It is hypothesized that the linear quadratic survival curve typical of mammalian cells reflects the linear quadratic relationship of complex chromosome aberrations to dose.

The interphase chromosome is predominantly localized to a territory occupying a volume that is only a small proportion of the whole nucleus. Consequences of what are termed proximity effects include a bias in the frequency of interchanges in favor of those between chromosomes or segments of chromosomes that are close together, the so-called proximity effect. The reproducible occurrence of the same fusion genes in cancer cells or following radiation exposure suggests that certain loci may be positioned in a way that favors interaction.

In biodosimetry, the goal is to infer dose retrospectively from the level of chromosome aberrations in peripheral blood lymphocytes. The spectrum of aberration types differs for radiations of different quality and retrospective biodosimetry should eventually be able to identify the type of radiation as well as the dose received. However, estimation of cancer risks from radiation exposure uses easily scored aberrations, dicentric, and rings to reflect the number of symmetric aberrations which are formed. This is based on the assumption that there is a fixed ratio between the two types of aberration which may not always be the case.

References

1. Sax K. Chromosome aberrations induced by X-rays. *Genetics* 23: 494–516, 1938.
2. Bender MA, Griggs HG, and Bedford JS. Mechanisms of chromsomal aberration production III chemicals and ionizing radiation. *Mutat. Res.* 23: 197–212, 1974.
3. Natarajan AT and Zwanenburg TSB. Mechanisms for chromosomal aberrations in mammalian cells. *Mutat. Res.* 95: 1–6, 1982.
4. Ager D, Phillips JW, Columna EA, Winegar RA, and Morgan WF. Analysis of restriction enzyme-induced DNA double strand breaks in Chinese hamster ovary cells by pulsed field gel electrophoresis: Implications for chromosome damage. *Radiat. Res.* 128: 150–156, 1991.

5. Bryant PE. Enzymatic restriction of mammalian DNA using Pvu II and Bam H1: Evidence for the double strand break origin of chromosomal break aberrations. *Int. J. Radiat. Biol.* 46: 57–65, 1984.

6. Ward JF, Blakely WF, and Joiner EI. Mammalian cells are not killed by DNA single strand breaks caused by hydroxy radicals from hydrogen peroxide. *Radiat. Res.* 103: 383–392, 1985.

7. Dewey WC, Miller HH, and Leeper DB. Chromosomal aberrations and mortality of x-irradiated mammalian cells: Emphasis on repair. *Proc. Natl. Acad. Sci. U.S.A.* 68: 667–671, 1971.

8. Vig KB and Lewis R. Genetic toxicology of bleomycin. *Mutat. Res.* 55: 121–145, 1978.

9. Johnson RT and Rao PN. Mammalian cell fusion: II. Induction of a premature chromosome condensation in interphase nuclei. *Nature* 226: 717–722, 1970.

10. Cornforth MN and Bedford JS. Ionizing radiation damage and its early development in chromosomes. In: Lett JT, Sinclair WK (Eds.) *Advances in Radiation Biology*, Vol. 17, Academic, San Diego, 1993, pp. 423–496.

11. Hittelman WN and Rao PN. Premature chromosome condensation. I. Visualization of X-ray-induced chromosome damage in interphase cells. *Mutat. Res.* 23: 251–258, 1974.

12. Iliakis G and Pantelias GE. Production and repair of chromosome damage in an X-ray sensitive CHO mutant visualized and analyzed in interphase using the technique of premature chromosome condensation. *Int. J. Radiat. Biol.* 57: 1213–1223, 1990.

13. Cornforth MN and Bedford JS. On the nature of a defect in cells from individuals with ataxia telangiectasia. *Science* 227: 1589–1591, 1985.

14. Blocher D. DNA double strand breaks in Ehrlich ascites tumor cells at low doses of X-rays. I. Determination of induced breaks by centrifugation at reduced speed. *Int. J. Radiat. Biol.* 42: 317–328, 1982.

15. Szeles A, Joussineau S, Lewensohn R, Lagercrantz S, and Larsson C. Evaluation of sectral kayotyping (SKY) in biodosimetry for the triage situation following gamma radiation. *Int. J. Radiat. Biol.* 82: 87–96, 2006.

16. Mitchell CR, Azizova TV, Hande MP, Burak LE, Tsakok JM, Khokhryakov VF, Geard CR, and Brenner DJ. Stable intrachromosomal biomarkers of past exposure to densely ionizing radiation in several chromosomes of exposed individuals. *Radiat. Res.* 162: 257–263, 2004.

17. Lucas JN and Sachs RK. Using three color chromosome painting to test chromosome aberration models. *Proc. Natl. Acad. Sci. U.S.A.* 90: 1484–1487, 1993.

18. Brown JM and Kovacs MS. Visualization of non-reciprocal chromosome exchanges in irradiated human fibroblasts by fluorescent in situ hybridization. *Radiat. Res.* 136: 71–136, 1993.

19. Simpson PJ and Savage JRK. Estimating the true frequency of X-ray-induced complex chromosome exchanges using fluorescence in situ hybridization. *Int. J. Radiat. Biol.* 67: 37–45, 1995.

20. Loucas BD and Cornforth MN. Complex chromosome exchanges induced by gamma rays in human lymphocytes: An mFISH study. *Radiat. Res.* 155: 660–671, 2001.

21. Savage JR. Insight into sites. *Mutat. Res.* 366: 81–95, 1996.

22. Bryant PE. Repair and chromosomal damage. *Radiother. Oncol.* 72: 251–256, 2004.

23. Hlatky LR, Sachs RK, Vazquez M, and Cornforth MN. Radiation-induced chromosome aberrations: insights gained from biophysical modeling. *Bioessays* 24: 714–723, 2002.

24. Cucinotta FA, Nikjoo H, O'Neill P, and Goodhead DT. Kinetics of DSB rejoining and formation of simple chromosome exchange aberrations. *Int. J. Radiat. Biol. Relat. Stud. Phys. Chem. Med.* 76: 1463–1474, 2000.
25. Revell SH. The accurate estimation of chromatid breakage and its relevance to a new interpretation of chromatid aberrations induced by ionizing radiation. *Proc. Roy. Soc. Lond. B Biol. Sci.* 150: 563–589, 1959.
26. Cornforth MN. Radiation-induced damage and the formation of chromosome aberrations. In: Nickoloff JA and Hoekstra MF (Eds.) *DNA Damage and Repair*, Vol. 2, Humana Press Inc., Totowa, New Jersey, 1999, pp. 559–584.
27. Cornforth MN. Perspectives on the formation of radiation-induced exchange aberrations. *DNA Repair* 5: 1182–1191, 2006.
28. Sachs RK, Hlatky LR, and Trask BJ. Radiation-induced chromosome aberrations: Colourful clues. *Trends Genet.* 16: 143–146, 2000.
29. Edwards AA. Modelling radiation-induced chromosome aberrations. *Int. J. Radiat. Biol.* 78: 551–558, 2002.
30. Chatterjee A, Holley W, Rydberg B, Mian S, and Alexander RD. Theoretical modeling of DNA damages and cellular responses. *Physica Medica*, XVII, Supplement 1: 59–61, 2001.
31. Serebrovsky AS. A general scheme for the origin of mutations. *Am. Nat.* 53: 374–378, 1929.
32. Muller HJ. Further studies on the nature and causes of gene mutation. In: *6th Int Cong Genet*, Vol. 1, Brooklyn Botanical Gardens, New York, 1932, pp. 213–255.
33. Cremer T and Cremer C. Chromosome territories, nuclear architecture and gene regulation in mammalian cells. *Nat. Rev. Genet.* 2: 292–301, 2001.
34. Sachs RK, Hahnfeldt P, and Brenner DJ. The link between low-LET dose-response relations and the underlying kinetics of damage production/repair/misrepair. *Int. J. Radiat. Biol. Relat. Stud. Phys. Chem. Med.* 72: 351–374, 1997.
35. Savage JR and Papworth DG. Frequency and distribution studies of asymmetrical versus symmetrical chromosome aberrations. *Mutat. Res.* 95: 7–18, 1982.
36. Ostashevsky JY. Higher-order structure of interphase chromosomes and radiation-induced chromosomal exchange aberrations. *Int. J. Radiat. Biol.* 76: 1179–1187, 2000.
37. Wu H, Durante M, and Lucas JN. Relationship between radiation-induced aberrations in individual chromosomes and their DNA content: Effects of interaction distance. *Int. J. Radiat. Biol.* 77: 781–786, 2001.
38. Savage JRK. Proximity matters. *Science* 290: 62–63, 2000.
39. Nikiforova MN, Stringer JR, Blough R, Medvedovic M, Fagin JA, and Nikiforov YE. Proximity of chromosomal loci that participate in radiation-induced rearrangements in human cells. *Science* 290: 138–141, 2000.
40. Abrahamson S and Wolff S. Re-analysis of radiation-induced specific locus mutations in the mouse. *Nature* 264: 715–719, 1976.
41. Thacker J. Radiation-induced mutation in mammalian cells at low doses and dose rates. In: Nygaard OF, Sinclair WT, and Lett JT (Eds.) *Advances in Radiation Biology*, Vol. 16, Academic, San Diego, 1992, pp. 77–124.
42. Evans HH, Nielsen M, Mencl J, Horng MF, and Ricanati M. The effect of dose rate on X-radiation-induced mutant frequency and the nature of DNA lesions in mouse lymphoma L5178Y cells. *Radiat. Res.* 122: 316–325, 1990.
43. Thacker J. The nature of mutants induced by ionising radiation in cultured hamster cells. III. Molecular characterization of HPRT-deficient mutants induced

by gamma-rays or alpha-particles showing that the majority have deletions of all or part of the hprt gene. *Mutat. Res.* 160: 267–275, 1986.

44. Thacker J, Fleck EW, Morris T, Rossiter BJF, and Morgan TL. Localization of deletion points in radiation-induced mutants of the *hprt* gene in hamster cells. *Mutat. Res.* 232: 163–170, 1990.

45. Simpson P, Morris T, Savage JRK, and Thacker J. High-resolution cytogenetic analysis of X-ray induced mutations of the HPRT gene of primary human fibroblasts. *Cytogenet. Cell Genet.* 64: 39–45, 1993.

46. Knudson AG. Two genetic hits (more or less) to cancer. *Nat. Rev. Cancer* 1: 157–162, 2001.

47. Joshi GP, Nelson WJ, Revell SH, and Shaw CA. X-ray-induced chromosome damage in live mammalian cells, and improved measurements of its effects on their colony-forming ability. *Int. J. Radiat. Biol. Relat. Stud. Phys. Chem. Med.* 41: 161–181, 1982.

48. Bedford JS and Cornforth MN. Relationship between the recovery from sublethal X-ray damage and the rejoining of chromosome breaks in normal human fibroblasts. *Radiat. Res.* 111: 406–423, 1987.

49. Cornforth MN and Bedford JS. A quantitatve comparision of potentially lethal damage repair and the rejoining of interphase chromosomes in low passage normal human fibroblasts. *Radiat. Res.* 111: 385–405, 1987.

50. Morgan WF, Day JP, Kaplan MI, McGhee EM, and Limoli CL. Genomic instability induced by ionizing radiation. *Radiat. Res.* 146: 247–258, 1996.

51. Morgan WF. Non-targeted and delayed effects of exposure to ionizing radiation: I. radiation-induced genomic instability and bystander effects *In Vitro. Radiat. Res.* 159: 567–580, 2003.

52. Kadhim MA, Macdonald DA, Goodhead DT, Lorimore SA, Marsden SJ, and Wright EG. Transmission of chromosomal instability after plutonium alpha-particle irradiation. *Nature* 355: 738–740, 1992.

53. Morgan WF, Corcoran JJ, Hartmann A, Kaplan MI, Limoli CL, and Ponnaiya B. DNA double-strand breaks, chromosomal rearrangements, and genomic instability. *Mutat. Res.* 404: 125–128, 1998.

54. Baverstock K. Radiation-induced genomic instability: a paradigm-breaking phenomenon and its relevance to environmentally induced cancer. *Mutat. Res.* 454: 89–109, 2000.

55. Barcellos-Hoff MH and Brooks AL. Extracellular signaling through the microenvironment: a hypothesis relating carcinogenesis, bystander effects, and genomic instability. *Radiat. Res.* 156: 618–627, 2001.

56. Limoli CL, Corcoran JJ, Jordan R, Morgan WF, and Schwartz JL. A role for chromosomal instability in the development of and selection for radioresistant cell variants. *Br. J. Cancer* 84: 489–492, 2001.

57. Bauchinger M. Retrospective dose reconstruction of human radiation exposure by FISH/chromosome painting. *Mutat. Res.* 404: 89–96, 1998.

58. Brenner DJ and Sachs RK. Chromosomal "fingerprints" of prior exposure to densely ionizing radiation. *Radiat. Res.* 140: 134–142, 1994.

59. Boei JJWA, Vermeulen S, Mullenders LHF, and Natarajan AT. Impact of radiation quality on the spectrum of induced chromosome exchange aberrations. *Int. J. Radiat. Biol. Relat. Stud. Phys. Chem. Med.* 77: 847–857, 2001.

60. Anderson RM, Marsden SJ, Wright EG, Kadhim MA, Goodhead DT, and Griffin CS. Complex chromosome aberrations in peripheral blood lymphocytes as a

potential biomarker of exposure to high-LET alpha-particles. *Int. J. Radiat. Biol. Relat. Stud. Phys. Chem. Med.* 76: 31–42, 2000.

61. Bauchinger MSE, Zitzelsberger H, Braselmann H, and Nahrstedt U. Radiation-induced chromosome aberrations analysed by two-color fluorescence in situ hybridization with composite whole chromosome-specific DNA probes and a pancentromeric DNA probe. *Int. J. Radiat. Biol.* 64: 179–184, 1993.

62. Griffin CS, Marsden SJ, Stevens DL, Simpson P, and Savage JR. Frequencies of complex chromosome exchange aberrations induced by 238Pu alpha-particles and detected by fluorescence in situ hybridization using single chromosome-specific probes. *Int. J. Radiat. Biol.* 67: 431–439, 1995.

63. NCRP. *Evaluation of the Linear-Nonthreshold Dose–Response Model for Ionizing Radiation: Report No. 136,* 2001.

64. Little MP. A comparison of the degree of curvature in the cancer incidence dose-response in Japanese atomic bomb survivors with that in chromosome aberrations measured in vitro. *Int. J. Radiat. Biol. Relat. Stud. Phys. Chem. Med.* 76: 1365–1375, 2000.

11

Modulation of Radiation Response via Signal Transduction Pathways

11.1 Intracellular Signaling

Exposure of cells to stress activates multiple intracellular signaling or signal transduction pathways along which signals move from the outside of the cell to the inside and between different intracellular locations. The result is an alteration in cellular activity and changes in the program of genes expressed in the responding cells. Ionizing radiation is one of the stresses that can initiate intracellular signaling and the response to that signaling can in turn modulate radiation response. In this chapter, the key players in the signal transduction process are briefly described and the dual effects of radiation-induced signaling and signal-mediated modulation of radiation response are discussed.

Much of the material in this chapter is based on reviews by Schmidt-Ullrich et al. [1], Dent et al. [2], and the references listed therein.

11.2 Transmembrane Receptors

Signaling is initiated by cytokines, low molecular weight hormones, growth factors, and other proteins that arrive at the plasma membrane and precipitate intracellular events by interacting with cell surface receptors that span the plasma membrane. The outer (extracellular) domain of these receptors is activated by selectively binding a receptor-specific protein or ligand and signals are transferred to the intracellular domain, setting off an intracellular signal transduction cascade.

In the case of growth factor (GF) receptors, transmembrane signal transduction is initiated at the outer surface of the cell membrane by the binding of a GF to the extracellular domain of its cognate receptor. Outside the cell, GF receptor proteins have a ligand binding N-terminal ectodomain, followed by a membrane-spanning transmembrane domain. At their C-termini

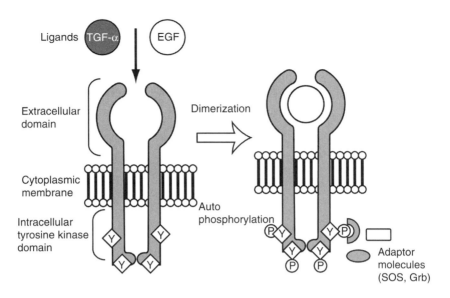

FIGURE 11.1

Schematic representation of the EGFR in the transmission of signals regulating gene activation, cell-cycle progression, growth arrest, apoptosis, and proliferation. Ligands such as EGF and TGF-α bind to the extracellular domain activating the receptor tyrosine kinase and initiating signal transduction pathways.

in the cytoplasm, they have a specialized enzyme domain that becomes activated whenever the extracellular domain of the receptor encounters and binds a GF ligand (Figure 11.1). For GF receptors, the cytoplasmic enzyme domain is a protein kinase, i.e., an enzyme that catalyzes protein phosphorylation. Consequently, these proteins are frequently referred to as receptor tyrosine kinases or RTKs. Binding of a GF ligand to a single receptor molecule facilitates the dimerization of the receptor with another one nearby in the plasma membrane. Often the GF ligand itself has two receptor-binding ends, enabling it to serve as a bridge between the two receptors encouraging their dimerization, and stabilizing the resulting receptor dimer pair. The dimerization of the extracellular domain in turn drags the cytoplasmic domains of the two receptor molecules into closer proximity. The tyrosine kinase (TK) of one receptor molecule can then phosphorylate the kinase domain of the second receptor molecule with which it is in close contact resulting in its functional activation. In effect, the two kinase domains, once they are juxtaposed, phosphorylate and activate each other.

Examples of signal transduction receptors that penetrate the plasma membrane and are capable of autophosphorylation are the tyrosine kinase GF receptors platelet-derived growth factor (PDGF), insulin-like growth factor 1 (IGR-1), epidermal growth factor (EGF), and fibroblast growth factor (FGF) receptors). These growth factors are polypeptides, usually 50–100 amino

TABLE 11.1

Properties of Selected Growth Factors and Growth Factor Receptors

Receptor	Ligand	Receptor Description
Epidermal growth factor receptors	Epidermal growth factor	Transmembrane tyrosine kinases
Epidermal growth factor (EGFR/ErbB1/HER1)	Epidermal growth factor	
EGFR/ErbB1/HER1	Transforming growth factor α (TGF-α)	
EGFR/ErbB1/HER1	Amphiregulin	
EGFR/ErbB1/HER1	Epigen	
EGFR/ErbB1/HER1; ErbB4/HER4	Heparin binding (HB)-EGF	
ErbB4/HER4	Betacellulin	
ErbB4/HER4	Epiregulin	
EGFR2/ErbB2/HER3 EGFR3/ErbB2/HER2	No natural ligand identified	
PDGFR	Platelet-derived growth factors (PDGF)	Homo or heterodimer of transmembrane tyrosine kinases
PDGFRα and PDGFRβ	PDGFA, PDGFB, PDGFC, PDGFD	
Fibroblast growth factor receptors (FGFR) 1,2,3,4	Fibroblast growth factor 1–20	Transmembrane tyrosine kinases
TGF-β	Transforming growth factor β family (TGF-β)	Complex of one Type 1 and one Type 5 receptors with ser/thr kinase activity
IGFR	Insulin-like growth factor IGF	Dimer of transmembrane tyrosine kinases
IGF-1R	IGF-1	
IGF-1R, IGF-2R	IGF-2	
c-Met	Hepatocyte growth factor–scatter factor (HGF/SF)	Tyrosine kinase

acids in length, which are sometimes termed mitogens because they induce the cell to grow and divide (Table 11.1).

11.2.1   ErbB Family of Receptor Kinases

Based on the nature of their extracellular domains, RTKs have been classified into a number of different families. The ErbB family of RTks comprises ErbB1–ErbB4 (also called HER1-4). ErbB1 is more commonly known as the epidermal growth factor (EGF) receptor, and these molecules are also referred to as EGFR and ErbB2-4. When ErbB1 binds to one of its ligands (EGF or TGFα) homo- and heterodimerization with other ErbB family molecules can occur enabling the tyrosine kinase domain of each ErbB1 molecule to trans-phosphorylate its partner. In this way ErbB1 can mediate the activation not only of ErbB1 but of ErbB2-4. Currently, no ligand which binds to ErbB2 has been described, but ErbB2 is thought to have a role in the activation of all the ErbB family members via heterodimerization. ErbB2 is over-expressed in 15%–25% of solid tumors (including mammary carcinoma), and is believed, together with ErbB1, to play a protective role against cyto-toxic therapeutic agents such as IR and chemotherapy.

11.2.2 Cytoplasmic Signaling

Signaling pathways downstream of activated receptors act through inter-actions of proteins that create signaling networks. Cytoplasmic signaling proteins are characterized by certain common features in the form of one or more noncatalytic domains that mediate sequence-specific protein–protein interactions. These domains are, in fact, modular components that occur in different signaling molecules to facilitate recognition and interaction. Pro-teins that contain SH2 (Src homology 2) or PTB (phosphotyrosine binding) domains that recognize tyrosine phosphorylated sequence motifs are the keys to the formation of signaling complexes following activation of growth factor RTKs. Two additional protein interaction domains frequently found in signaling proteins are the SH3 (Src homology 3) and the PH (pleckstrin homology) domain.

Most cytoplasmic signaling molecules that are targets of tyrosine kinases contain one or more SH2 domains and may also contain SH3 domains. In fact, some signaling proteins have no catalytic function (e.g., nck, crk, grb-2) and act solely as adaptors, coupling proteins that lack an SH2 domain to a tyrosine kinase signaling complex. Activation of GF recep-tors results in the autophosphorylation of the receptor at multiple tyrosine residues creating docking sites for cytoplasmic proteins containing the adaptor motifs.

The SH2 and SH3 domain adaptor protein, grb-2 (GF receptor bound-2), plays a critical role in the activation of the small GTPase protein Ras which is a central transducer of growth factor signals. The SH2 domain of grb-2 associates with activated GF receptors while its SH3 domain binds to proline-based motifs in SOS, a guanine nucleotide exchange protein that activates Ras (Figure 11.2).

11.2.3 *Ras* Proto-Oncogene Family

Proteins of the Ras family belong to a superfamily of small GTP-binding switch proteins that bind guanine nucleotides (GTP and GDP) and also have GTPase activity. The three distinct Ras protein isoforms that have been identified in mammals, H-ras, N-ras, and K-ras, are part of a large family of low molecular weight, guanine nucleotide triphosphate (GTP) binding proteins. Ras is a regulator of cell growth in eukaryotic cells and 30% of human tumors contain *Ras* mutations, the frequency being dependent on tumor type. The activation of Ras proteins is dependent on posttranslational modification via prenylation, a critical step mediated by farnesyl transferase

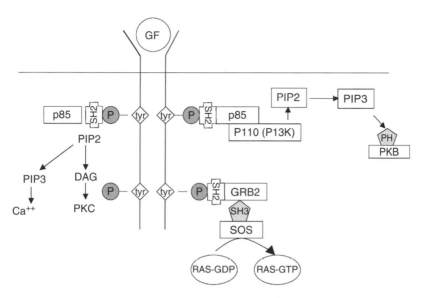

FIGURE 11.2

Recruitment of cytoplasmic signaling molecules by receptor protein tyrosine kinases. Binding of a receptor to a growth factor leads to phosphorylation of the intracellular domain on tyrosine residues allowing the SH2 mediated association of enzymes such as phospholipase C. Adaptor molecules such as Grb and p85 also bind to activated receptors by their SH2 domains. PI3K is a heterodimer of a catalytic subunit and an adaptor or regulatory subunit, p85 that containing 2 SH2 domains.

(Ftase), which allows Ras protein to dock onto the plasma membrane. Once prenylated, Ras proteins behave as a molecular switch that oscillates between active (guanosine triphosphate GTP-bound state) and inactive (guanosine diphosphate, GDP-bound state) in response to upstream growth factor signals.

Ras is activated by a family of guanine nucleotide exchange factors (GEFs) (e.g., SOS described above) that release Ras-bound GDP and allow GTP to be bound. Termination of Ras activity occurs through hydrolysis of GTP converting it to GDP through the activation of GTPase-activating proteins or GAPs which promote intrinsic GTPase activity of the Ras proteins themselves (Figure 11.3). Mutated Ras is permanently activated in the GTP-bound signaling state and provides proliferative signals in the absence of growth factor ligands, which lead to permanent growth and transformation (Figure 11.3).

11.2.4 Signal Transduction Cascades

Signal transduction cascades act through a sequence of phosphorylation–dephosphorylation reactions catalyzed by specific enzymes (kinases) that transfer phosphate (Pi) groups from ATP to the hydroxyl groups on amino acids and by other enzymes (phosphatases) that catalyze removal of the Pi

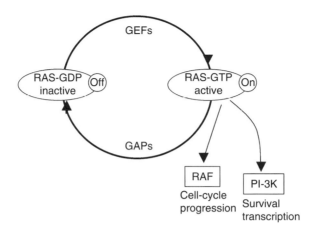

FIGURE 11.3
Ras protein activation and downstream signaling. Ras cycles between inactive GDP-bound state and active GTP-bound state. The exchange of GDP for GTP is regulated by guanine nucleotide exchange factors (GEFs). GTP hydrolysis requires GTPase activating proteins (GAPs), which enhance the weak intrinsic GTPase activity of ras proteins. In its active form, Ras interacts with different families of effector proteins including RAF protein kinases and PI3K.

by hydrolysis (described in Chapter 8 for the signal transduction pathway initiated by damage to DNA [Figure 8.2]). The protein kinases and phosphatases are divided into three groups based on the amino acids they phosphorylate and dephosphorylate. One class recognizes serine and threonine residues, another recognizes tyrosine residues, and a smaller group recognizes both serine–threonine and tyrosine residues. The kinases and phosphatases may have one or several substrates. The final stage of the signal transduction cascade is phosphorylation or dephosphorylation of a target amino acid on an effector molecule stimulating or inhibiting the protein's enzymatic activity or its ability to bind to certain proteins or specific DNA sequences.

11.2.4.1 MAPK Superfamily of Cascades

Mitogen activated protein kinases (MAPK) are a family of serine/threonine protein kinases widely conserved among eukaryotes and involved in many cellular programs such as cell proliferation, cell differentiation, cell movement, and cell death. The acronym MAPK stands for mitogen activated protein (MAP) kinase since many growth factors and mitogens have been shown to activate this pathway. MAPK signaling cascades are organized hierarchically into three-tiered modules. MAPKs are phosphorylated and activated by MAPK-kinases (MAPKKs), which in turn are phosphorylated and activated by MAPKK-kinases (MAPKKKs) (Figure 11.4). The MAPKKKs are activated by interaction with the family of small GTPases described above, which connect the MAPK module to surface receptors and

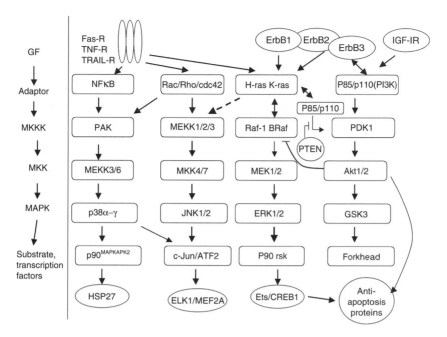

FIGURE 11.4

Some signal transduction pathways in mammalian cells. Growth factors, for instance the ErbB family, signal through GTP-binding proteins into multiple signal transduction pathways. The prototype of the MAPK three-tiered hierarchical signaling cascade is shown on the left. Three of the MAP kinase superfamilies (ERK1/2, JNK, and p38) are shown in the middle. On the right is the PI3K pathway. Some of the connections within the MAPK families and with the PI3K pathway are shown.

external stimuli. Stimulation through external receptors of functionally distinct but structurally related MAPK pathways results in physiologic consequences, i.e., mitogenesis or stress response.

In mammals, there are three well-characterized subfamilies of the MAPK signal transduction cascade. The three pathways are the extracellular signal-regulated kinase 1 and 2 (ERK1/2); the c-Jun NH_2 terminal kinases or stress-activated kinase (JNK/SAPK), and the p38 MAP kinases p38α, p38β, p38γ, and p38δ. Additional protein kinases and MAPK-related pathways are continually being described.

Activation of MAPK pathways brings about changes in gene expression, which drive the cellular responses to external stimuli. Cell proliferation, differentiation, and survival can all be regulated through MAPK signaling. Nuclear transcription factors are key MAPK targets that directly affect gene transcription. ERKs phosphorylate and activate the ELK-1 transcription factor whereas JNK/SAPK activates c-Jun, p38 phosphorylates MEF2A, and most MAPKs activate the Ets family of transcription factors. Activated MAPKs can also act on cytoplasmic targets such as effector kinases, MAP-KAP2, and MNK1, which regulate protein translation.

Thus the MAPK super-cascade can be summarized as

Stimulus > MAPKKK > MAPKK > MAPK > Effector

with the effector end of the chain usually involving transcription factors, promoters, mitogenic agents, apoptotic or anti-apoptotic factors that influence the proliferation, and survival of the cell and its response to toxic agents and endogenous stressors.

11.2.4.2   MAPK Pathway

The protein kinase MAPK was originally described as a 42-kDa insulin-stimulated protein kinase activity whose tyrosine phosphorylation was increased after insulin exposure. At about the same time, a 44 kDa isoform of MAPK, ERK1 (extracellular signal regulated kinase) was described. Activation of Ras proteins causes the activation of Raf-1, a MAP3K upstream of ERK1/2. ERK kinase activation is part of a common pathway used by GF receptors such as those for EGF, PDGF, and FGF and for more diverse stimuli from antigen receptors and cytokine receptors. RAF-1 directly activates MEK-1/2 by phosphorylating it on serine residues, which enhances the availability of the catalytic site to potential substrates. Activated MEK1/2 is a dual specificity kinase that phosphorylates the ERK kinases. MEK-induced phosphorylation of ERK occurs on the threonine and tyrosine of a threonine-glutamic acid-tyrosine motif, which induces both catalytic activation of ERK and its translocation to the nucleus. Nuclear ERK interacts with transcription factors such as ELK-1, which contain an ERK docking site leading to their phosphorylation and activation of specific transcriptional targets.

There are multiple components at each level including three Rafs, two MKKs, and two ERKS. Many different stimuli including GFs, cytokines, viral infection, ligands for G-protein-coupled receptors, transforming agents, and carcinogens can activate the MAPK pathway. Mutations which convert *Ras* to an activated oncogene are common in human tumors and activated Ras persistently activates the MAPK pathway contributing to the increased proliferation of tumor cells.

11.2.4.3   c-JUN NH₂ Terminal Kinase Pathway

The c-JUN NH_2 terminal kinase (JNK) and p38 pathways mediate responses to cellular stresses such as heat, UV and ionizing radiation, anticancer drugs, and exposure to potentially damaging biological agents such as the cytokines IL-1 and tumor necrosis factor (TNF). In recognition of the fact that multiple stresses increase JNK1/2 activity, the pathway is sometimes called the stress-activated protein kinase (SAPK) pathway. The core components of the JNK/SAPK and p38 pathways parallel those of the ERK pathway although the upstream activation steps are less well defined.

JNK/SAPK is phosphorylated by the dual specificity kinase MKK4/7, which in turn is phosphorylated on serine by protein kinases at the MKKK level. In keeping with the heterogeneity of stimuli that activate the JNK/SAPK and p38 pathways, there are at least 13 MKKKs that regulate the JNKs, and this diversity allows a wide range of stimuli to activate the JNK pathway. This is in contrast to the MAPK pathway, which depends primarily on the three protein kinases of the Raf family to activate MKK1/2. Upstream of the MAPKKK enzymes are low-molecular-weight GTP-binding proteins of the Rho family, the mechanism of activation of which is not completely understood. One proposed mechanism is by activation by the *Ras* proto-oncogene, while involvement of phosphatidyl inositol 3-kinase (PI3K) and/or protein kinase C isoforms has also been suggested.

The JNK1/2 stress-induced protein kinases phosphorylate the NH_2-terminus of the DNA-binding protein c-Jun and increase its transcriptional activity. c-Jun is a component of the AP-1 transcription complex and an important regulator of gene expression contributing to the control of many cytokine genes (Chapter 13). AP-1 is activated in response to growth factors, environmental stress, UV and ionizing radiation, cytotoxic agents, and reactive oxygen species (ROS)—all stimuli that activate JNKs. It has also been suggested that JNK can phosphorylate the NH_2-terminus of c-Myc, giving it a potential role in both proliferative and apoptotic signaling.

11.2.4.4 p38 MAPK Pathway

There are four p38 kinases α, β, γ, and δ. The p38α enzyme, which is expressed in most cell types, is the best characterized. p38 MAPKs are activated by many stimuli including inflammatory cytokines, hormones, ligands for the G-protein coupled receptors, and stresses such as osmotic and heat shock. Rho family GTPases have an important role as upstream activators of the p38 MAPK pathway acting through MAPKKK enzymes to regulate the MAPKK enzymes MKK3 and MKK6. There are several protein kinases downstream of the p38 MAPK enzymes with differing functions including activation of heat shock protein (HSP) 27 and the phosphorylation and activation of transcription factors such as CREB.

11.2.4.5 Phosphatidyl Inositol 3-Kinase Pathway

Activation of GF receptors results not only in activation of the MAPK cascade but also of another important signal transduction pathway which is not represented by the MAPK scheme of hierarchical kinases, the phosphatidyl inositol 3-kinase (PI3K) pathway. Phosphoinositides are phospholipids of cell membranes that respond to growth factor signaling and contribute to signal propagation by two main mechanisms: by serving as precursors of second messengers IP_3 and DAG or by binding to signaling proteins that contain specific phosphoinositide binding modules. Inositol phospholipids were among the first molecules to be recognized as second messengers. It was found that when phospholipase Cγ was activated by mitogens such as EGF and TGFα, inositol phospholipids were cleaved into diacyl glycerol (DAG) and inositol triphosphate (IP$_3$) with the release of IP$_3$ into the cytoplasm. IP$_3$ interacts with a receptor in the endoplasmic reticulum leading to the release of Ca^{2+} into the cytosol, and Ca^{2+}, together with diacyl-glycerol, can cause activation of protein kinase C (PKC). The release of Ca^{2+} also activates Ras-GTP to recruit RAF-1 to the cytoplasm, initiating signaling through the MAPK cascade, which has already been described.

The PI3K enzymes consist of two subunits: a catalytic p110 subunit and a regulatory and localizing subunit, p85. Binding of p85 to an activated ErbB receptor (predominantly ErbB3) results in PI3K activation. The major catalytic function is in the p110 subunit and the role of this enzyme is to phosphorylate inositol phospholipids in the plasma membrane at the 3 position in the inositol sugar ring. When other positions within the inositol ring are phosphorylated by additional PI kinases (e.g., PI 4 kinase, PI 5 kinase), the inositol 3, 4, 5 trisphosphate molecule becomes an acceptor site in the plasma membrane for molecules that contain a pleckstrin binding domain (PH domain), in particular, the protein kinases PDK1 and Akt (also called protein kinase B, PKB). PDK1 is believed to phosphorylate and activate Akt. The PI3K-dependent phosphorylation of the inositol sugar ring can be reversed by the tumor suppressor lipid phosphatase PTEN. Loss of PTEN expression occurs frequently in some types of tumor, e.g., glioblastoma multiforme, resulting in an apparent constitutive activation of PDK1 and Akt (Figure 11.5).

In effect, the lipid products of PI3K activity provide an anchor for assembling signaling proteins at specific locations in the membrane in response to cell stimulation. Signaling by PDK1 to Akt and by PDK1 and Akt

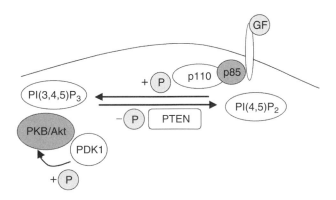

FIGURE 11.5
Phosphatidyl inositol 3' signaling. Following interaction of growth factors (GFs) with receptor tyrosine kinase (RTK), the regulatory subunit of PI3K (p85) is recruited to the receptor resulting in the activation of the catalytic subunit (p110) and increase in levels of PI(3,4,5)P$_3$. Interaction of proteins PKB1 and PKB/Akt with PI(3,4,5)P$_3$ result in activation of PKB/Akt and effect numerous cellular processes. PTEN acts as a negative regulator of these pathways.

downstream to other protein kinases such as PKC isoforms, GSK3, mTOR, p70S6K, and p90rsk has been shown to play a key role in mitogenic and metabolic responses of cells, particularly the inhibition of apoptosis.

11.3   Modulation of Radiation Response by Interaction of Signal Transduction Pathways

Ionizing radiation has been shown to initiate signaling through different signal transduction pathways usually by activation of specific transmembrane receptors.

11.3.1   Activation of ErbB Receptors by Ionizing Radiation

The epidermal growth factor receptor (EGFR), also called ErbB1 or HER1, has been shown to be activated in response to irradiation in several carcinoma cell lines (reviewed by Schmidt-Ullrich et al. [1] and Dent et al. [2]). Radiation exposure in the range of 1–2 Gy can activate the MAPK pathway through activation of the EGFR to a level similar to that observed by physiological, growth-stimulatory EGF concentrations [3]. The threshold dose at which radiation could induce ErbB1 phosphorylation in breast cancer cell lines MCF7, A431, and MDA-MB-231 was found to be 0.5 Gy [4].

It has also been shown that radiation can activate the other ErbB family members ErbB2, ErbB3, and ErbB4. In this case, radiation-induced activation of ErbB2 is independent of ErbB1, suggesting that radiation can cause indiscriminate activation of multiple plasma membrane RTKs. Following activation of the receptors by radiation, signaling can proceed by several routes including radiation-induced anti-apoptotic signaling, which has been demonstrated to be mediated by PI3K [5].

11.3.2   Mechanism of Receptor Activation by Ionizing Radiation

Stimulation of metabolic generation of ROS and reactive nitrogen species (RNS) by ionizing radiation was described in Chapter 5. These processes have been suggested as mechanisms by which cells sense and amplify cellular ionization events [6,7]. Radiation-stimulated ROS and RNS have been shown to be essential activators of signal transduction pathways with the role of ROS being that of the initial reactants produced by an ionization event and that of RNS being that of the actual effectors/activators of the cellular signal transduction pathway.

11.3.3   Role of Other Growth Factors

In addition to the ErbB family, other growth factors and cytokine receptors are believed to play a role in cellular radiation responses. Cytokines such as

TGFα, IL6, urokinase-type plasminogen activator (uPA), and TGFβ have all been proposed to control cell-survival responses after irradiation via interaction with their cognate receptors. The reactions and interactions are complex and the ultimate responses represent the sum of often-conflicting cell signaling. It has been demonstrated that radiation-stimulated signaling through the MAPK and p38 pathways can cause release of the TGFα, a ligand of EGFR [8]. TGFα signaling after irradiation may lead in one case to the activation of pro-caspase enzymes and in another to the generation of the cytoprotective transcription factor nuclear factor kappa B (NFκβ) [9]. The overall effect of radiation-induced TGFα-receptor signaling is thus the composite result of opposing cellular signals.

IL6 is a cytokine that regulates immune cell function as well as the ability of epithelial cells to proliferate [10]. It has been shown that IL6 can generate anti-apoptosis signaling in cells that is protective against the toxic effects of radiation [11]. In some cell types, the protective effect of IL6 has been proposed to be mediated by the PI3K pathway [12] while in others the radiation-induced expression of IL6 is dependent upon prior activation of NFκB [13].

Another cytokine, TGFβ, can cause growth arrest and differentiation in nontransformed cells. TGFβ has been reported to confer a protective effect via MAPK signaling and the downstream expression of anti-apoptotic molecules in some tumor cells [14] while in other cell types TGFβ can protect cells in a Ras- and MAPK-independent manner that is dependent on PI3K signaling [15].

11.3.4 Effects of Activation of GF Receptors on Cell Survival

Initiation of signaling through the transmembrane GF receptors has a number of downstream effects that influence radiation response. These are mediated mainly through interaction with apoptosis regulatory proteins or by modulation of cell-cycle regulation.

11.3.4.1 Anti-Apoptotic Signaling

Signaling by the ErbB family of receptors in response to growth factors is believed to play an important anti-apoptotic role in both normal and tumor cells. In addition to PI3K and MAPK and other pathways and molecules downstream of ErbB signaling including K and H-Ras molecules, JAK/STAT molecules and the JNK pathway have been shown to mediate ErbB receptor anti-apoptotic signaling in a cell type and stress-specific manner [16].

The intrinsic and extrinsic pathways that lead to the activation of effector caspases and apoptosis have been described elsewhere (Chapters 7 and 12). A large and expanding group of pro- and anti-apoptotic Bcl-2 family proteins has been described, which may act directly or by modulating Bax/Bak interactions and mitochondrial pore function. There is accumulating evidence that diverse signaling pathways downstream of ErbB receptors regulate cell survival and response to radiation and chemotherapy by modulating the apoptotic threshold.

The anti-apoptosis role of the PI3K/Akt pathway in response to toxic stimuli has been well documented [17]. ErbB signaling to PI3K/Akt has been proposed to enhance the expression of the mitochondrial anti-apoptosis proteins Bcl-XL and Mcl-l and of caspase inhibitor proteins [18]. Bcl-XL and Mcl-l protect cells from apoptosis via the intrinsic mitochondrial pathway whereas caspase inhibitors act by blocking the extrinsic pathway through the death receptors. In addition, Akt has been shown to phosphorylate pro-apoptotic BAD and human pro-caspase 9, rendering these proteins inactive in the processes that lead to apoptosis [19]. In view of these anti-apoptotic effects, it is not surprising that inhibitors of ErbB signaling, which have been shown to decrease the activity of the PI3K/Akt pathway in a variety of cell types, are reported to increase the sensitivity of cells to a wide range of toxic stresses, including cytotoxic drugs and radiation [20].

Two of the major isozymes of the small GTPase Ras, Ha-Ras, and Ki-Ras, have been shown to play opposing roles in the modulation of cell sensitivity to ionizing radiation. Over-expression of the active isoform of Ha-Ras increases resistance to radiation through PI3K signaling and Akt is detected specifically on Ha-Ras-over-expressing cells. Over-expression of Ki-Ras, on the other hand, can increase radiation sensitivity, and P38 MAPK activity is selectively enhanced by ionizing radiation in cells over-expressing Ki-Ras. An experimental demonstration of this effect is shown in Figure 11.6. The mechanism that underlies potentiation of cell death in cells over-expressing Ki-Ras involves Bax translocation to the mitochondrial membrane. These findings help to explain the opposite effect of Ha-Ras and Ki-Ras on modulation of radiosensitivity and suggest differential activation of PI3K/Akt and p38MAPK signaling. The ability of H-Ras to influence radiation response is partially mimicked by downstream intermediates in different pathways, Raf and PI3K, but neither of these effectors can substitute completely for Ras signaling (Figure 11.7).

In contrast to the MAPK and JNK pathways, where radiation-induced activation has been consistently observed in diverse cell types, the ability of ionizing radiation to regulate p38 MAPK activity appears to be highly variable, with different groups reporting no activation [21], weak activation [22], or strong activation [23].

11.3.4.2 Cell-Cycle Regulation

The ability of MEK 1/2 inhibitors to enhance cell killing by radiation has been linked in part to a derangement of radiation-induced G_2/M-phase growth arrest as well as to enhanced apoptosis [24]. In contrast, activation of the MAPK pathway after irradiation has been found to promote radiosensitivity in some cell types by abrogating the G_2/M-phase checkpoint [23]. The dual nature of MAPK signaling in the control of cell survival has also

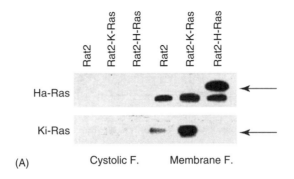

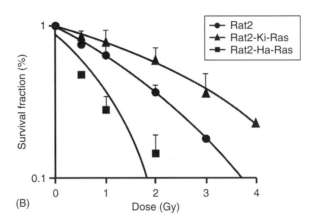

FIGURE 11.6
Differential modulation of clonogenic survival after radiation by activated Ha-Ras and Ki-Ras. Exponentially growing Rat2 cells were stably transfected with activated Ha-Ras or Ki-Ras. (A) Expression levels of Ha-Ras and Ki-Ras were determined by Western blot analysis using antibody specific for Ha-Ras or Ki-Ras, respectively. (B) The Rat2 cells expressing activated forms of Ha-Ras or Ki-Ras were plated and treated with indicated doses of radiation. After 10–15 days, the plates were scored for colony formation. (From Choi, J.A., Park, M.T., Kang, C.M., Um, H.D., Bae, S., Lee, K.H., Kim, T.H., Kim, J.H., Cho, C.K., Lee, Y.S., Chung, H.Y., and Lee, S.J., *Oncogene*, 23, 9, 2004. With permission.)

been observed for DNA-damaging agents other than radiation including Adriamycin.

The ability of MAPK signaling to impact both differentiation and proliferative responses in cells appears to depend on the amplitude and duration of MAPK activation and on the type of cell. A short activation of the MAPK cascade by growth factors has been correlated with increased proliferation via both increased cyclin D1 expression and increased ability to progress through the G_2-M transition. In contrast, prolonged elevation of MAPK activity has been shown to inhibit DNA synthesis via induction of the cyclin-dependent kinase inhibitor protein p21Cipl/WAF1. Thus MAPK signaling may potentially affect both G_1/S and G_2/M transitions and have a

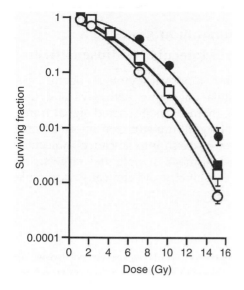

FIGURE 11.7
Constitutively activated PI3K and Raf are each partially sufficient to confer radioresistance compared with constitutively activated Ras. Radiation survival curves were made using stable RIE-1 cell lines expressing empty vector or constitutively activated H-Ras (12V), PI3K (p110-CAAX), or Raf-1 (Raf22W). The ability of both PI3K and Raf to induce similar clonogenic survival intermediate between that of Ras and vector indicates that neither of these Ras effectors can substitute completely for Ras signaling. O, RIE-Vector; ●, RIE-Ras; □, RIE-Raf; ■, RIE-PI3-K. (From Grana, T.M., Rusyn, E.V., Zhou, H., Sartor, C.I., and Cox, A.D. *Cancer Res.*, 62, 4142, 2002. With permission.)

dual positive and negative role in the regulation of cell-cycle progression after irradiation, which is dependent on cell type and duration of signal. In other studies where p38 MAPK activation was observed after irradiation, the p38γ isoform has been proposed to signal G_2/M-phase arrest [25] dependent on expression of a functional ATM protein.

11.3.5 Autocrine Signaling

In addition to causing ligand-independent activation of ErbB receptors, ionizing radiation and other stresses also cause the synthesis and release from tumor cells of autocrine growth factors such as TGFα, which can reactivate the ErbB receptor system hours after the initial exposure to the stress [26,27]. Autocrine loops are established when soluble factors secreted by cells bind to and stimulate receptors on their own surfaces. In the EGFR system, ligands such as TGFα are shed in the form of membrane bound precursors. Ligand-releasing proteases (also known as 'sheddases') process the membrane-associated precursors into their active soluble form. The characteristically high levels of cognate receptors expressed by autocrine cells make them very efficient at recapturing endogenous ligands. Response to a primary stimulus such as an exogenous growth factor, a component of the extracellular matrix (ECM), or ionizing radiation can lead to ligand release and recapture and stimulation of intracellular signaling. In nonautocrine cells soluble growth factors are shown to activate sheddases through the Ras-activated MAPK pathway. This suggests that in an autocrine EGFR system, a ligand, the receptor, the sheddase, and the intracellular signaling network can form a positive feedback loop [27].

11.4   Radiosensitization by Modulation of Signal Transduction Intermediates: Molecular Radiosensitizers

The understanding that cell survival in response to cytotoxic agents was influenced by perturbation of intracellular signaling pathways and that cytotoxins, including ionizing radiation, themselves impacted signal transduction pathways and the availability of signal transduction intermediates has stimulated a very productive area of research into so-called molecular radiosensitizers. In addition to the vast number of potential targets, the appeal of this strategy lies in their target specificity and clinically acceptable toxicity.

11.4.1   ErbB Family Signal Inhibitors

Signaling by the ErbB family of receptors, with some exceptions, is pro-proliferative and cytoprotective. Because both receptor expression and autocrine growth factor levels are often increased in carcinoma cells compared to normal tissue, blockade of signaling is being intensively studied as a chemo/radiosensitizing stratagem.

Radiosensitivity of cells may be influenced by exogenous growth factors or hormones before or after radiation. For example, insulin-like growth factor-1 receptor (IGF-1R) is a cell surface receptor with tyrosine kinase activity, which has been linked to increased radioresistance. IGF-1R is expressed at low levels in A-T cells, possibly contributing to their radiosensitivity. In another case, tyrosine kinase activity of the EGFR is increased following radiation exposure, and addition of exogenous EGF to cells in culture makes them more radioresistant [28]. EGFR and the ErbB2 receptor are over-expressed in a wide variety of human tumors and this has been linked to poor clinical outcome following radiotherapy [29] while targeting EGF and ErbB2 receptor signaling using monoclonal antibody or specific inhibitors of EGFR or ErbB2 leads to radiosensitization in vitro and in vivo.

Signaling from ErbB family receptors can be blocked by use of inhibitory antibodies such as C225 or Herceptin, which bind to the extracellular portions of EGFR and ErbB2, respectively [24,30]. In the case of ErbB1, C225 binds to the portion of the molecule that associates with growth factor ligands including EGF and TGFα [31], abolishing the ability of growth factors, to stimulate receptor function. In agreement with the ligand-independent nature of the process, C225 does not block the primary activation of the receptor or MAPK after irradiation.

The anti-proliferative and anti-survival mechanisms of action of Herceptin appear to be more complex. Herceptin binds to ErbB2, but this receptor has no known ligand and herceptin appears to act by causing the internalization and degradation of ErbB2, as well as by blocking ErbB2 heterodimerization with other ErbB family members [32]. Both C225 and Herceptin have been shown individually to kill cells and to interact in a synergistic fashion in combination with standard therapeutic regimens such as ionizing radiation, cisplatin, and Taxol to reduce tumor cell survival both in vitro and in vivo [33].

Recent studies have used monoclonal antibodies to target truncated forms of ErbB1, e.g., EGFR VIII [34]. In these studies, a novel monoclonal antibody, 806, was found to potently inhibit truncated forms of EGFR and inhibit full-length receptors more weakly. The inhibition of receptor function correlated with reduced tumor cell growth in vitro and in vivo. Some of the antitumor effects of anti-ErbB receptor antibodies may be attributable, however, to enhanced immunological reactivity of the antibody.

Small molecule inhibitors of the tyrosine kinase domains of the ErbB family of receptors have been used with some success in blocking tumor cell growth and survival both in vitro and in vivo. The inhibitors include the tyrphostin AGl478 and ZDl83 (Iressa) that binds to the catalytic kinase domain of EGFR and inhibits tyrosine/kinase activity [35]. Inhibition of EGFR kinase activity not only blocks phosphorylation of EGFR itself in response to the growth factors to which it binds but also inhibits the trans-phosphorylation of other ErbB family members by EGFR. In addition to inhibiting EGFR, the tyrphostin AGl478 has been shown to inhibit ErbB4 [36] and these drugs have the potential to impact not only on EGF/TGFα, signaling through EGFR, but also on signaling through ErbB4 and ErbB3 [37].

The ErbB family of receptors can be inhibited by the use of dominant negative and anti-sense approaches. In particular, expression of truncated forms of ErbB1 (EGFR–CD533), ErbB2, and ErbB3 in a variety of cell types has been shown to reduce proliferation and survival of both normal and tumor cells in vitro and in vivo [38,39]. The dominant negative approaches are believed to act by blocking homo- and heterodimerization of ErbB family members, reducing receptor transphosphorylation and down-stream signaling by the receptors. Investigations using dominant negative EGFR–CD533 demonstrated that it could block radiation-induced phosphorylation of the EGFR [5,40]. Transfection with EGFR–CD533 by means of intratumoral injection of a recombinant adenovirus was shown to increase the radiosensitivity of glioblastoma and mammary carcinoma cells in vitro and in vivo [38,40]. Some of the results indicating radiosensitization by inhibition of ErbB family signaling are summarized in Table 11.2.

11.4.2 Clinical Applications of EGFR Signal Inhibitors

EGFR inhibitors from both the monoclonal antibody (mAb) and tyrosine kinase inhibitor (TKI) class have shown clinical activity in the treatment of several human cancers (reviewed by Harari and Huang [29]). Three EGFR inhibitors have been approved for cancer therapy by the Food and Drug Administration in the United States, the mAB cetuximab (Erbitux) and the small molecule TKIs gefitinib (Iressa) and erlotinib (Tarceva). Results from a phase-III head-and-neck cancer trial of EGFR inhibitor/radiation combination identified a survival advantage for the EGFR inhibitor cohort [41].

TABLE 11.2

Interaction of Ionizing Radiation with Inhibitors of Growth Factor Receptors and Signal Transduction Intermediates

Agent	Experimental Model	Effects	References
EGFR Inhibitors			
C225 (cetuximab) (antibody)	TC: HNSCC (SCC13-Y)	Enhance G_1 arrest, apoptosis	[56,57]
	BC (MCF-7)	Growth inhibition	[58]
	XG: HNSCC (A431)	Complete response, repeated C225 dose better than single	[59]
	GBM (U251)	Improved host survival	[60]
Trastuzumab (antibody)	TC: BC HER+	Inhibit DNA repair	[61]
	XG: BC HER+	Abrogate G_1 arrest	
ZD1839 (TK/EGFR inhibitor)	TC: NSCLC (four cell lines)	Synergistic/additive to RT	[62]
	TC: Pancreatic Ca	Radiosensitize	[63]
	XG: Colon Ca. (LoVo)	Enhance tumor growth delay	[61]
CI-1033 (truncated ErbB1)	TC: Breast cancer	Reduced survival	[64]
Farnesyl Transferase Inhibitors			
FTI-277	TC: *H-Ras+* bladder Ca, BC.*K-Ras+* colon, lung Ca	Synergism with acute dose RT, tumors with mutated *Ras* only	[52]
	HeLa, wild-type *Ras*	Reduce survival, abrogate G_2 arrest	[65]
L74832	TC: *H-Ras+* bladder, *K-Ras+* colon, lung Ca	Synergistic to RT in tumors expressing mutated *Ras*	[51,52]
	XG: Wt Ras *colon*		
BIM-46228	TC: HeLa-3A	Additive to RT	[66]
Inhibitors of Ras-mediated Downstream Pathways			
AS-ON Raf	TC: Rt: Laryngeal ca	Enhance cytotoxicity of single RT	[67]
LY294002 (PI3K inhibitor)	TC: Mutated *Ras*: bladder,colon, Wt *Ras* bladder, colon	Radiosensitize cells expressing mutated *Ras* only	[49]
Wortmannin (PI3K inhibitor)	TC: Colon (HCT-116), BC(MCF-7), Cervical (HeLa)	Radiosensitize	[68]
AS-ON p21	TC: Colon (HCT-116)	Enhance apoptosis	[69]
MEK inhibitor	HNSCC(A431), BC (MDA-MB-231)	Radiosensitize to single RT	[26]

Note: TC: tissue culture; XG: xenograft; HNSCC: head and neck squamous cell carcinoma; BC: breast cancer; NSCLC: non-small cell lung cancer.

This was the first phase-III trial to show a survival benefit attributable to a molecular targeted agent administered in conjunction with high-dose irradiation. A number of other clinical trials are ongoing. Retrospective studies have identified an inverse relationship between tumor EGFR expression and clinical outcome after radiotherapy [42–44]. One of these reports suggests that tumor EGFR expression may serve as a predictive factor for those patients who would benefit from more aggressive treatment schedules aimed at limiting the possibility for tumor cell repopulation during therapy [44]. Apart from the head-and-neck patients, the overall clinical gains realized to date with EGFR inhibitors are modest for the total cancer population and it appears that much remains to be learned about the integration of EGFR inhibitors into cancer treatment and the selection of patients who will benefit from combined treatment.

11.4.3   Inhibition of the Ras-Mediated Signaling Pathway

The ability of the *Ras* oncogene to contribute to radio-resistance has been demonstrated by several independent lines of enquiry. Over-expression or transformation of rodent and human cell lines by *Ras* has been shown in many cases to result in cell lines that are resistant to radiation [45–47], whereas inhibition of Ras activation has resulted in radiosensitization in rodent cells transfected with *Ras* and of human tumor cell lines bearing endogenous mutations in *Ras* [47,48]. As described earlier, when the *Ras* oncogene undergoes mutation, it is permanently activated in the GTP-bound signaling state, providing proliferative signals in the absence of growth factor ligands leading to altered cell growth, transformation, and occasionally radioresistance. Increased radioresistance is more commonly observed, however, in cells transfected with the *Ras* gene in combination with a nuclear cooperating oncogene such as *c-myc* or mutant *p53*.

Because Ras activation holds such a key role in the etiology of human cancers, Ras was one of the first oncoproteins to be targeted for anticancer therapy. Ras activation and localization to cellular membranes is essential for function and dependent on a posttranslational modification called prenylation. Farnesyltransferase (Ftase) was identified as the enzyme that catalyzes the addition of a farnesyl lipid moiety to the CAAX motif at the carboxyl end of H-Ras protein leading to the development of Ftase inhibitors (FTIs) as Ras antagonists.

FTIs have been reported to enhance radiation-induced cytotoxicity in human cancer cells expressing *H*- or *K-Ras* genes [49]. Extensive experimentation has been done in vivo and in vitro with a number of FTIs (reviewed by Brunner et al. [50]) and the results indicate that Ras activation by mutation or upstream signaling can increase radioresistance of tumor cells by activation of PI3K. Inhibitors of the Ras signaling pathway can reduce tumor radioresistance when Ras is directly or indirectly activated and inhibition of Ras inactivation can enhance both apoptosis and oxygenation in tumors. Figure 11.8 shows the effect on cell survival of treatment with

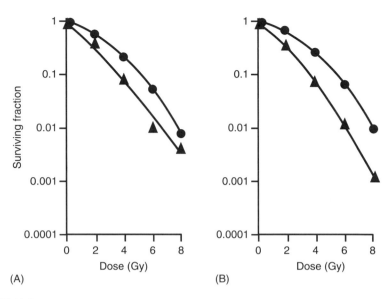

(A)   (B)

FIGURE 11.8

Clonogenic survival in T24 cells after treatment with inhibitors of the Ras signal transduction pathway. Survival after treatment with (A) FTI L744,832: ● control; ▲ L744,832 5 μM. FTI. (B) LY294002: ● control; ▲ LY294002 10 μM. (From McKenna, W.G., Muschel, R.J., Gupta, A.K., Hahn, S.M., and Bernhard, E.J., *Oncogene*, 22, 5866, 2003. With permission.)

drugs with FTI action while Figure 11.9 shows the combined effect of radiation and injected FTI on tumor growth in vivo.

Inhibition of Ras by treating cells with FTIs before irradiation has a synergistic effect on radiation-induced cell killing in human and rodent cells with Ras mutations [51,52]. H-Ras requires only farnesylation for activation and so blocking Ftase alone can sensitize cells with activated H-Ras. However, K-Ras

FIGURE 11.9

T24 tumor survival after irradiation and FTI treatment in vivo. Regrowth of T24 tumor xenografts after treatment with L744,832 and radiation. Tumor-bearing mice were treated with L744,832 (40 kg/day × 7 days). (From McKenna, W.G., Muschel, R.J., Gupta, A.K., Hahn, S.M., and Bernhard, E.J., *Oncogene*, 22, 5866, 2003. With permission.)

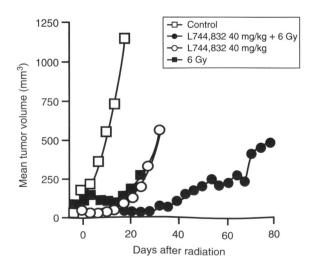

is the Ras family member that is primarily activated in human cancer and is prenylated not by farnesyl tranferase action but by geranylgeranyltransferase (GGTase). Combined blocking of Ftase and GGTase is required to sensitize cells with K-Ras. GGT inhibitors (GGTIs) are under development. Findings demonstrating the effects of FTIs on radiosensitization of cell lines and preclinical animal models are summarized in Table 11.2.

Downstream of *Ras*, the RAF-MEK-ERK and phosphatidylinositol-3 kinase PI3K/AKT/PKB pathways are two separate signaling pathways that have been linked to radioresistance. Using antisense nucleotides against *Raf* increased radiosensitivity has been observed in a human squamous cell SQ-20B. Inhibitors of PI3K signaling such as wortmannin and LY294002 enhanced the response of radiation in lung, bladder, colon, breast, HNSCC, and cervical cancer cells. There are limited reports of anticancer activities of this drug although they can also target other important PI3K related proteins.

FTIs that block processing of Ras result in radio-sensitization and a number of FTIs have been developed specifically to inhibit the activity of oncogenic *Ras* in tumor cells.

11.4.4 Clinical Application of Farnesyl Transferase Inhibitors

A dual farnesyl and geranylgeranyl transferase inhibitor Merck L-77,123 is the only FTI to be tested clinically so far.

A multicenter phase I trial of L778,123 plus radiotherapy in patients with head and neck cancer (HNC) and nonsmall cell lung cancer (NSCLC) has been completed. Among nine patients enrolled, five (2HNC and 3 NSCLC) had a complete response and one NSCLC patient had a partial response. No dose-limiting toxicities were observed. A separate cohort of patients with locally advanced pancreatic cancer was enrolled in the same trial. Combination of radiation and L778,123 at the lowest dose level tested showed acceptable toxicity. Radiosensitization of a patient-derived pancreatic cell line was observed [53].

11.4.5 Clinical Implications of Radiation-Induced Cell Signaling: Accelerated Cell Proliferation

Accelerated proliferation or repopulation after repeated exposure of cells to ionizing radiation is a response of acute reacting normal tissues and tumors. The recognition that accelerated proliferation may counteract the effectiveness of fractionated radiation is based on retrospective analyses of clinical data. Tumor control data for squamous cell carcinoma (SCC) of the head and neck treated with different fractionation schedules and treatment times showed an inverse relationship between overall treatment time and tumor control probability. Although there was a close linear relationship between treatment time and tumor control probability for the first 4 weeks of treatment, at 4 weeks a break point occurred to a steeper response curve indicating increasing treatment failures as treatment time became longer. This finding was interpreted as evidence that an accelerated proliferation response of tumor clonogens occurred during the latter part of treatment [54].

It has been suggested that a radiation-induced cytoprotective mechanism underlies the accelerated proliferation response to fractionated radiation [55]. Ionizing radiation has been shown to exert GF-like effects in autocrine regulated human carcinoma cells through activation of signal transduction pathways. Immediate activation of ErbB RTKs in the plasma membrane and the possible secondary release of TGFα at later times could cause the activation of a cytoprotective signaling cascade involving MAPK, PI3K, and transcriptional activation events resulting in anti-apoptotic and pro-proliferative cellular responses. In certain tumors these events would be initiated and maintained by exposure to fractionated radiation and will become apparent as accelerated proliferation and treatment failure if treatment time is prolonged.

11.5   Summary

Exposure of cells to stress activates multiple intracellular signaling or signal transduction pathways by which signals move from the outside of the cell to the inside and between different intracellular locations. Signaling is initiated by cytokines, hormones, growth factors, and other proteins, which arrive at the plasma membrane and precipitate intracellular events by interacting with cell surface receptors that span the plasma membrane.

Signal transduction cascades act through a sequence of phosphorylation–dephosphorylation reactions catalyzed by specific enzymes (kinases). There are three, well-characterized subfamilies of a signal transduction cascade called MAPK—the extracellular signal-regulated kinases ERKl and ERK2, the c-Jun NH 2 terminal kinases JNK1–3, and the p38 MAP kinases. The effector end of the chain involves transcription factors, promoters, mitogenic agents, and apoptotic or anti-apoptotic factors, which influence the proliferation and survival of the cell and its response to toxic agents and endogenous stressors.

A signal transduction pathway not represented by the MAPK scheme of hierarchical kinases is the PI3K pathway. Activation of the PI3K enzyme causes phosphorylation of inositol phospholipids in the plasma membrane and the lipid products of PI3K activity provide an anchor for assembling signaling proteins at specific locations in the membrane. Signaling by these proteins downstream to other protein kinases plays a key role in mitogenic and metabolic responses of cells, particularly the inhibition of apoptosis.

Radiation can cause indiscriminate activation of plasma membrane RTKs EGFR and ErbB2-4. Inhibitors of ErbB signaling have been shown to decrease the activity of the PI3K/Akt pathway in a variety of cell types and to increase the sensitivity of cells to a wide range of toxic stresses, including cytotoxic drugs and radiation. In part, this is attributable to the fact that signaling by the ErbB family of receptors in response to growth factors plays an important anti-apoptosis role in normal and tumor cells.

Receptor expression and autocrine growth factor levels are often increased in carcinoma cells compared to normal tissue, suggesting block-ade of signaling as a selective stratagem. Signaling from ErbB family of receptors can be blocked by use of inhibitory antibodies such as C225 or Herceptin, which bind to the extracellular portions of ErbB1 and ErbB2. Both C225 and Herceptin have been shown to interact in a synergistic manner with standard therapeutic regimens including radiation. Small molecule inhibitors of the tyrosine kinase domains of the ErbB family of receptors have been used with some success to block tumor cell growth and survival both in vitro and in vivo. FTIs which block processing of the Ras protein result in radio-sensitization and a number of FTIs have been developed specifically to inhibit the activity of oncogenic Ras in tumor cells. Inhibition of Ras by treating cells with FTIs before irradiation has a synergistic effect on radiation-induced cell killing in human and rodent cells with *Ras* mutations. A phase I trial of one FTI with radiation has shown encouraging results consistent with preclinical findings.

References

1. Schmidt-Ullrich RK, Dent P, and Grant S. Signal transduction and cellular radiation responses. *Radiat. Res.* 153: 245–257, 2000.
2. Dent P, Yacoub A, Contessa JN, Caron R, Amorino G, Valerie K, Hagan MP, Grant S, and Schmidt-Ullrich R. Stress and radiation-induced activation of multiple intracellular signaling pathways. *Radiat Res.* 159: 283–300, 2003.
3. Schmidt-Ullrich RK, Mikkelsen RB, Dent P, Todd DG, Valerie K, Kavanagh BD, Contessa JN, Rorrer WK, and Chen PB. Radiation-induced proliferation of the human A431 squamous carcinoma cells is dependent on EGFR tyrosine phosphorylation. *Oncogene* 15: 1191–1197, 1997.
4. Carter S, Auer KL, Birrer M, Fisher PB, Schmidt-Ullrich RK, Valerie K, Mikkelsen R, and Dent P. Inhibition of mitogen activated protein kinase cascade potentiates cell killing by low dose ionizing radiation in A431 human squamous carcinoma cells. *Oncogene* 16: 2787–2796, 1998.
5. Contessa JN, Hampton J, Lammering G, Mikkelsen RB, Dent P, Valerie K, and Schmidt-Ullrich RK. Ionizing radiation activates Erb-B receptor depen-dent Akt and p70 S6 kinase signaling in carcinoma cells. *Oncogene* 21: 4032–4041, 2002.
6. Leach JK, Black SM, Schmidt-Ullrich RK, and Mikkelsen RB. Activation of constitutive nitric-oxide synthase activity is an early signaling event induced by ionizing radiation. *J. Biol. Chem.* 277: 15400–15406, 2002.
7. Mikkelsen RB and Wardman P. Biological chemistry of reactive oxygen and nitrogen and radiation-induced signal transduction mechanisms. *Oncogene* 22: 5734–5754, 2003.
8. Rutault K, Hazzalin CA, and Mahadevan LC. Combinations of ERK and p38 MAPK inhibitors ablate tumor necrosis factor-alpha (TNF-α) mRNA induction.

Evidence for selective destabilization of TNF-a transcripts. *J. Biol. Chem.* 276: 6666–6674, 2001.

9. Basu S, Rosenzweig KR, Youmell M, and Price BD. The DNA-dependent protein kinase participates in the activation of NFkB following DNA damage. *Biochem. Biophys. Res. Commun.* 247: 79–83, 1998.

10. Legue F, Guitton N, Brouazin-Jousseaume V, Colleu-Durel S, Nourgalieva K, and Chenal C. IL-6 a key cytokine in in vitro and in vivo response of Sertoli cells to external gamma irradiation. *Cytokine* 16: 232–238, 2001.

11. Miyamoto Y, Hosotani R, Doi R, Wada M, Ida J, Tsuji S, Kawaguchi M, Nakajima S, Kobayashi H, Masui T, and Imamura M. Interleukin-6 inhibits radiation induced apoptosis in pancreatic cancer cells. *Anticancer Res.* 21: 2449–2456, 2001.

12. Chung TD, Yu JJ, Kong TA, Spiotto MT, and Lin JM. Interleukin-6 activates phosphatidylinositol-3 kinase, which inhibits apoptosis in human prostate cancer cell lines. *Prostate* 42: 1–7, 2000.

13. Zhou D, Yu T, Chen G, Brown SA, Yu Z, Mattson MP, and Thompson JS. Effects of NF-κB1 (p50) targeted gene disruption on ionizing radiation-induced NF-κB activation and TNFα, IL-1a, IL-1b and IL-6 mRNA expression in vivo. *Int. J. Radiat. Biol.* 77: 763–772, 2001.

14. Saile B, Matthes N, Armouche E, Neubauer K, and Ramadori G. The bcl, NFκB and p53/p21WAF1 systems are involved in spontaneous apoptosis and in the anti-apoptotic effect of TGF-β or TNF-α on activated hepatic stellate cells. *Eur. J. Cell Biol.* 80: 554–561, 2001.

15. Chen RH, Su YH, Chuang RL, and Chang TY. Suppression of transforming growth factor-beta-induced apoptosis through a phosphatidylinositol 3-kinase/Akt-dependent pathway. *Oncogene* 17: 1959–1968, 1998.

16. Liu B, Fang M, Lu Y, Mills GB, and Fan Z. Involvement of JNK-mediated pathway in EGF-mediated protection against paclitaxel-induced apoptosis in SiHa human cervical cancer cells. *Br. J. Cancer.* 85: 303–311, 2001.

17. Daly JM, Olayioye MA, Wong AM, Neve R, Lane HA, Maurer FG, and Hynes NE. NDF/heregulin-induced cell cycle changes and apoptosis in breast tumour cells: Role of PI3 kinase and p38 MAP kinase pathways. *Oncogene* 18: 3440–3451, 1999.

18. Kuo ML, Chuang SE, Lin MT, and Yang SY. The involvement of PI3-K/Akt-dependent up-regulation of Mcl-1 in the prevention of apoptosis of Hep3B cells by interleukin-6. *Oncogene* 20: 677–685, 2001.

19. Li Y, Tennekoon GI, Birnbaum MJ, Marchionni MA, and Rutkowski JL. Neuregulin signaling through a PI3K/Akt/Bad pathway in Schwann cell survival. *Mol. Cell. Neurosci.* 17: 761–767, 2001.

20. Pianetti S, Arsura M, Romieu-Mourez R, Coffey RJ, and Sonenshein GE. Her-2/neu overexpression induces NF-κB via a PI3-kinase/Akt pathway involving calpain-mediated degradation of IkB-alpha that can be inhibited by the tumor suppressor PTEN. *Oncogene* 20: 1287–1299, 2001.

21. Kim SJ, Ju JW, Oh CD, Yoon YM, Song WK, Kim JH, Yoo YJ, Bang OS, Kang SS, and Chun JS. ERK-1/2 and p38 kinase oppositely regulate nitric oxide-induced apoptosis of chondrocytes in association with p53, caspase-3, and differentiation status. *J. Biol. Chem.* 277: 1332–1339, 2002.

22. Taher MM, Hershey CM, Oakley JD, and Valerie K. Role of the p38 and MEK-1/2/p42/44 MAP kinase pathways in the differential activation of human

immunodeficiency virus gene expression by ultraviolet and ionizing radiation. *Photochem. Photobiol.* 71: 455–459, 2000.

23. Lee YJ, Soh JW, Dean NM, Cho CK, Kim TH, Lee SJ, and Lee YS. Protein kinase CD overexpression enhances radiation sensitivity via extracellular regulated protein kinase 1/2 activation, abolishing the radiation-induced G2-M arrest. *Cell Growth Differ.* 13: 237–246, 2002.

24. Mendelsohn J and Baselga J. The EGF receptor family as targets for cancer therapy. *Oncogene* 19: 6550–6565, 2000.

25. Wang X, McGowan CH, Zhao M, He L, Downey JS, Fearns C, Wang Y, Huang S, and Han J. Involvement of the MKK6-p38g cascade in gamma-radiation-induced cell cycle arrest. *Mol. Cell. Biol.* 20: 4543–4552, 2000.

26. Dent P, Reardon DB, Park JS, Bowers G, Logsdon C, Valerie K, and Schmidt-Ullrich R. Radiation-induced release of transforming growth factor alpha activates the epidermal growth factor receptor and mitogen-activated protein kinase pathway in carcinoma cells, leading to increased proliferation and protection from radiation-induced cell death. *Mol. Biol. Cell.* 10: 2493–2506, 1999.

27. Zhou BB. Targeting ligand cleavage to inhibit the ErbB pathway in cancer. *Ann. NY Acad. Sci.* 1059: 56–60, 2005.

28. Kwok TT and Sutherland RM. The influence cell-cell contact on radiosensitivity of human squamous carcinoma cells. *Radiat. Res.* 126: 52–57. 1991.

29. Harari PM and Huang S. Radiation combined with EGFR signal inhibitors: Head and neck cancer focus. *Semin. Radiat. Oncol.* 16: 38–44, 2006.

30. Yarden Y and Sliwkowski MX. Untangling the ErbB signalling network. *Nat. Rev. Mol. Cell. Biol.* 2: 127–137, 2001.

31. Herbst RS and Langer CJ. Epidermal growth factor receptors as a target for cancer treatment: The emerging role of IMC-C225 in the treatment of lung and head and neck cancers. *Semin. Oncol.* 29: 27–36, 2002.

32. Baselga J and Albanell J. Mechanism of action of anti-HER2 monoclonal antibodies. *Ann. Oncol.* 12: 35–41, 2001.

33. Pegram MD, Lopez A, Konecny G, and Slamon DJ. Trastuzumab and chemotherapeutics: Drug interactions and synergies. *Semin. Oncol.* 27: 21–25, 2000,

34. Luwor RB, Johns TG, Murone C, Huang HJ, Cavenee WK, Ritter G, Old LJ, Burgess AW, and Scott AM. Monoclonal antibody 806 inhibits the growth of tumor xenografts expressing either the de2-7 or amplified epidermal growth factor receptor (EGFR) but not wild-type EGFR. *Cancer Res.* 61: 5355–5361, 2001.

35. Barker AJ, Gibson KH, Grundy W, Godfrey AA, Barlow JJ, Healy MP, Woodburn JR, Ashton SE, Curry BJ, and Richards L. Studies leading to the identification of ZD1839 (IRESSA): An orally active, selective epidermal growth factor receptor tyrosine kinase inhibitor targeted to the treatment of cancer. *Bioorg. Med. Chem. Lett.* 11: 1911–1914, 2001.

36. Bowers G, Reardon D, Hewitt T, Dent P, Mikkelson RB, Valerie K, Lammering G, Amir C, and Schmidt-Ullrich RK. The relative role of ErbB1-4 receptor kinases in radiation signal transduction responses of human carcinoma cells. *Oncogene* 20: 1388–1397, 2001.

37. Pinkas-Kramarski AEL R, Bacus SS, Lyass L, van de Poll ML, Klapper LN, Tzahar E, Sela M, van Zoelen EJ, and Yarden Y. The oncogenic ErbB-2/ErbB-3 heterodimer is a surrogate receptor of the epidermal growth factor and betacellulin. *Oncogene* 16: 1249–1258, 1998.

38. Reardon D, Contessa JN, and Mikkelson RB. Dominant negative EGFR-CD533 and inhibition of MAPK modify JNK1 activation and enhance radiation toxicity of human mammary carcinoma cells. *Oncogene* 18: 4756–4766, 1999.

39. Schmidt-Ullrich RK, Valerie K, Fogleman PB, and Walters J. Radiation-induced autophosphorylation of epidermal growth factor receptor in human malignant mammary and squamous epithelial cells. *Radiat. Res.* 145: 79–83, 1996.

40. Lammering G, Valerie K, Lin PS, Mikkelsen RB, Contessa JN, Feden JP, Farnsworth J, Dent P, and Schmidt-Ullrich RK. Radiosensitization of malignant glioma cells through overexpression of dominant-negative epidermal growth factor receptor. *Clin. Cancer Res.* 7: 682–690, 2001.

41. Bonner JA, Buchsbaum DJ, Russo SM, Fiveash JB, Trummell HQ, Curiel DT, and Raisch KP. Anti-EGFR-mediated radiosensitization as a result of augmented EGFR expression. *Int. J. Radiat. Oncol. Biol. Phys.* 59: 2–10, 2004.

42. Ang KK, Andratschke NH, and Milas L. Epidermal growth factor receptor and response of head-and-neck carcinoma to therapy. *Int. J. Radiat. Oncol. Biol. Phys.* 58: 959–965, 2004.

43. Chua DT, Nicholls JM, Sham JS, and Au GK. Prognostic value of epidermal growth factor receptor expression in patients with advanced stage nasopharyngeal carcinoma treated with induction chemotherapy and radiotherapy. *Int. J. Radiat. Oncol. Biol. Phys.* 59: 11–20, 2004.

44. Bentzen SM, Atasoy BM, Daley FM, Dische S, Richman PI, Saunders MI, Trott KR, and Wilson GD. Epidermal growth factor receptor expression in pretreatment biopsies from head and neck squamous cell carcinoma as a predictive factor for a benefit from accelerated radiation therapy in a randomized controlled trial. *J. Clin. Oncol.* 23: 5560–5567, 2005.

45. Ling CC and Endlich B. Radioresistance induced by oncogenic transformation. *Radiat Res.* 120: 267–279, 1989.

46. McKenna WG, Weiss MC, Bakanauskas VJ, Sandler H, Kelsten ML, Biaglow J, Tuttle SW, Endlich B, Ling CC, and Muschel RJ. The role of the H-ras oncogene in radiation resistance and metastasis. *Int. J. Radiat. Oncol. Biol. Phys.* 18: 849–859, 1990.

47. Sklar MD. The ras oncogenes increase the intrinsic resistance of NIH 3T3 cells to ionizing radiation. *Science* 239: 645–647, 1988.

48. Pirollo KF, Tong YA, Villegas Z, Chen Y, and Chang EH. Oncogene-transformed NIH 3T3 cells display radiation resistance levels indicative of a signal transduction pathway leading to the radiation-resistant phenotype. *Radiat. Res.* 135: 234–243, 1993.

49. Gupta AK, Bakanauskas VJ, Cerniglia GJ, Cheng Y, Bernhard EJ, Muschel RJ, and McKenna WG. The Ras radiation resistance pathway. *Cancer Res.* 61: 4278–4282, 2001.

50. Brunner TB, Hahn SM, Gupta AK, Muschel RJ, McKenna WG, and Bernhard EJ. Farnesyltransferase inhibitors: An overview of the results of preclinical and clinical investigations. *Cancer Res.* 63: 5656–5668, 2003.

51. Cohen-Jonathan E, Muschel RJ, McKenna GW, Evans SM, Cerniglia GJ, Mick R, Kusewitt D, Sebti SM, Hamilton AD, Oliff A, Kohl N, Gibbs JB, and Bernhard EJ. Farnesyltransferase inhibitors potentiate the antitumor effect of radiation on a human tumor xenograft expressing activated HRAS. *Radiat. Res.* 154: 125–132, 2000.

52. Bernhard EJ, McKenna WG, Hamilton AD, Sebti SM, Qian Y, Wu JM, and Muschel RJ. Inhibiting Ras prenylation increases the radiosensitivity of human

tumor cell lines with activating mutations of ras oncogenes. *Cancer Res.* 58: 1754–1761, 1998.

53. Martin NE, Brunner TB, Kiel KD, DeLaney TF, Regine WF, Mohiuddin M, Rosato EF, Haller DG, Stevenson JP, Smith DC, Pramanik B, Tepper J, Tanaka WK, Morrison B, Deutsch P, Gupta AK, Muschel RJ, McKenna WG, Bernhard EJ, and Hahn SM. A phase I trial of the dual farnesyltransferase and geranylgeranyltransferase inhibitor L-778,123 and radiotherapy for locally advanced pancreatic cancer. *Clin. Cancer Res.* 10: 5447–5454, 2004.

54. Wang JZ and Li XA. Impact of tumor repopulation on radiotherapy planning. *Int. J. Radiat. Oncol. Biol. Phys.* 61: 220–227, 2005.

55. Schmidt-Ullrich RK, Contessa JN, Dent P, Mikkelsen RB, Valerie K, Reardon DB, Bowers G, and Lin PS. Molecular mechanisms of radiation-accelerated repopulation. *Radiat. Oncol. Invest.* 7: 321–330, 1999.

56. Milas L, Mason K, and Hunter N. In vivo enhancement of tumor response by C225 antiepidermal growth factor receptor antibody. *Clin. Cancer Res.* 6: 701–708, 2000.

57. Saleh MN, Raisch KP, and Stackhouse MA. Combined modality therapy of A431 human epidermoid cancer using anti-EGFr antibody C225 and radiation. *Cancer Biother. Radiopharm.* 14: 451–463, 1999.

58. Wollman R, Yahalom J, Maxy R, Pinto J, and Fuks Z. Effect of epidermal growth factor on the growth and radiation sensitivity of human breast cancer cells in vitro. *Int. J. Rad. Oncol. Biol. Phys.* 35: 751–757, 1994.

59. Nasu S, Ang KK, and Fan Z. C225 antiepidermal growth factor receptor antibody enhances tumor radiocurability. *Int. J. Radiat. Oncol. Biol. Phys.* 51: 474–477, 2001.

60. Raben D, Buchsbaum DJ, and Gillespie Y. Treatment of human intracranial gliomas with chimeric monoclonal antibody against the epidermal growth factor receptor increases survival of nude mice when treated concurrently with irradiation. *Proc. Am. Assoc. Cancer Res.* 40: 184, 1999.

61. Pietras RJ, Poen JC, Gallardo D, Wongvipat PN, Lee HJ, and Slamon DJ. Monoclonal antibody to HER-2/neureceptor modulates repair of radiation-induced DNA damage and enhances radiosensitivity of human breast cancer cells overexpressing this oncogene. *Cancer Res.* 59: 1347–1355, 1999.

62. Raben D, Helfrich B, Chan D, Shogun T, Zundel W, Scafani R, and Bunn PJ. The effects of ZD1839 on cell signaling processes and its growth effects with radiation and chemotherapy in human non-small cell lung cancer cells in vitro. In: AACR-NCI-EORTC International Conference. Miami Beach, Florida, 2001.

63. Raben D, Phistry M, and Helfrich B. ZD1839 (Iressa), a selective epidermal growth factor receptor tyrosine kinase inhibitor (EGFR-TKI), enhances radiation induced cytotoxicity in human pancreatic and cholangiocarcinoma cell lines in vitro. In: *Gastrointestinal Cancer Research Conference*, Orlando, Florida, 2000.

64. Rao GS, Murray S, and Ethier SP. Radiosensitization of human breast cancer cells by a novel ErbB family receptor tyrosine kinase inhibitor. *Int. J. Radiat. Oncol. Biol. Phys.* 48: 1519–1528, 2000.

65. Cohen-Jonathan E, Toulas C, Ader I, Monteil S, Allal C, Bonnet J, Hamilton AD, Sebti SM, Daly-Schveitzer N, and Favre G. The farnesyltransferase inhibitor FTI-277 suppresses the 24-kDa FGF2-induced radioresistance in HeLa cells expressing wild-type RAS. *Radiat. Res.* 152: 404–411, 1999.

66. Prevost GP, Pradines A, Brezak MC, Lonchampt MO, Viossat I, Ader I, Toulas C, Kasprzyk P, Gordon T, Favre G, and Morgan B. Inhibition of human tumor cell

growth in vivo by an orally bioavailable inhibitor of human farnesyltransferase, BIM-46228. *Int. J. Cancer.* 91: 718–722, 2001.

67. Soldatenkov VA, Dritschilo A, Wang FH, Olah Z, Anderson WB, and Kasid U. Inhibition of Raf-1 protein kinase by antisense phosphorothioate oligodeoxyribonucleotide is associated with sensitization of human laryngeal squamous carcinoma cells to gamma radiation. *Cancer J. Sci. Am.* 3: 13–20, 1997.

68. Price BD and Youmell MB. The phosphatidylinositol 3-kinase inhibitor wortmannin sensitizes murine fibroblasts and human tumor cells to radiation and blocks induction of p53 following DNA damage. *Cancer Res.* 56: 246–250, 1996.

69. Tian H, Wittmack EK, and Jorgensen TJ. p21WAF1/CIP1 antisense therapy radiosensitizes human colon cancer by converting growth arrest to apoptosis. *Cancer Res.* 60: 679–684, 2000.

70. Choi JA, Park MT, Kang CM, Um HD, Bae S, Lee KH, Kim TH, Kim JH, Cho CK, Lee YS, Chung HY, and Lee SJ. Opposite effects of Ha-Ras and Ki-Ras on radiation-induced apoptosis via differential activation of PI3K/Akt and Rac/p38 mitogen-activated protein kinase signaling pathways. *Oncogene* 23: 9–20, 2004.

71. Grana TM, Rusyn EV, Zhou H, Sartor CI, and Cox AD. Ras mediates radioresistance through both Phosphatidylinositol 3-kinase-dependent and Raf-dependent-dependent but mitogen activated protein kinase/Extracellular Signal-regulated kinase-independent signaling pathways. *Cancer Res.* 62: 4142–4150, 2002.

72. McKenna WG, Muschel RJ, Gupta AK, Hahn SM, and Bernhard EJ. The RAS signal transduction pathway and its role in radiation sensitivity. *Oncogene* 22: 5866–5875, 2003.

12

Radiation-Induced Apoptosis

12.1 Apoptosis

Apoptosis is a cell suicide program that removes damaged, infected, and superfluous cells. The term apoptosis was coined in a classic paper by Kerr et al. in 1972 [1] to describe a distinct mode of cell destruction, which is the major mechanism for elimination of overabundant and unwanted cells during embryonic development, growth, differentiation, and normal cell turnover. It is a programmed sequence of biochemical events triggered by specific physiological and stress stimuli leading to the activation of a set of cysteine proteases known as caspases. These enzymes are the effectors for the programmed disintegration of subcellular structures leading ultimately to cell death.

In addition to the involvement of apoptosis in the programming of normal development, a number of pathological stimuli have been shown to induce apoptosis, including extracellular stress (ionizing and ultraviolet radiation, heat shock, and oxidative and osmotic stress), receptor-mediated processes, growth-factor withdrawal, loss of cell adhesion (anoikis), cytotoxic lymphocytes, and many chemotherapeutic drugs.

Radiation biologists were aware of the process before the use of the term apoptosis and before the phenomenon was widely studied. The rapid death of small lymphocytes which is observed following irradiation results from an apoptotic mechanism but was originally termed interphase death because cell death preceded cell division rather than following it as is the case with reproductive or mitotic death.

This chapter describes the basic mechanisms of apoptosis, the apoptotic cascades, and how they may be set in motion by ionizing radiation.

12.2 Mechanisms of Apoptosis

Parts of this chapter are based in part on reviews by Danial and Korsmeyer [2] and Zimmerman et al. [3] and the papers referenced therein.

In contrast to necrosis, which is a passive series of events resulting from acute injury to cells, apoptosis is a programmed sequence of reactions triggered by specific stimuli. Morphologically, apoptosis is characterized by nuclear chromatin condensation, cytoplasmic shrinking, dilated endoplasmic reticulum (ER), and membrane blebbing with the mitochondria remaining morphologically unchanged (Figure 12.1A). This type of cell death is often hard to observe in vivo because the dying cells are rapidly phagocytosed by tissue macrophages.

A number of biochemical events have been found to correlate with these morphological features of apoptosis. Among the first described and most striking is DNA fragmentation. When DNA from apoptotically dying cells was subjected to agarose gel electrophoresis, ladders with ~200 bp repeats were observed, corresponding to histone protection of the nucleosomes of native chromatin. Pulsed field gel techniques have shown these ~200 bp ladders to result from earlier DNA cleavage into larger fragments (Figure 12.1B). Since even a few double-stranded DNA breaks will render the cell unable to undergo mitosis successfully, such DNA fragmentation can be regarded as a biochemical signature of cell death.

The universal biochemical hallmark of apoptotic death is the activation of caspases, which are cysteine proteases characterized by their unusual ability to cleave proteins at aspartic acid residues. Different caspases have different specificities with respect to recognition of amino acids neighboring aspartic acid. Evidence that these proteases are involved in apoptotic cell death comes from the demonstration that specific caspase inhibitors can block cell death and that knockout mice lacking caspases-3, -8, and -9 fail to complete normal embryonic development for which apoptosis is required.

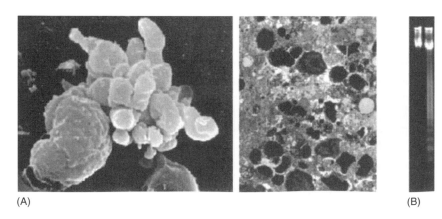

(A)   (B)

FIGURE 12.1
Apoptotic changes: (A) T cells undergoing apoptosis in vitro and in the thymus after activation, showing blebbing and nuclear condensation. (B) Right lane: Oligosomal DNA fragmentation in apoptotic T cells. Left lane: Untreated cells. (From Zimmermann, K.C., Bonzon, C., and Green, D.R., *Pharmacol. Ther.*, 92, 9257, 2001. With permission.)

12.2.1 Caspases

Caspases can be divided into initiator and effector caspases. Initiator caspases (e.g., caspase-9) activate the effector caspases (e.g., caspase-3) which are the executioners of the cell. The protein degradation by caspases is responsible for the characteristic morphological features of apoptosis; membrane blebbing, cytoplasmic and nuclear condensation, DNA fragmentation, and the formation of apoptotic bodies.

Active caspases are potentially lethal to the cell and must be effectively regulated, but at the same time they have to be deployed within a short-time frame when required. This is achieved by the regulation of caspases at the translational level. Caspases initially exist as precursors or procaspases with three basic domains. They are activated by cleavage of the N-terminal prodomain and the asymmetrical splitting of the remaining protein into large (17–22 kD) and small (10–12 kD) subunits (Figure 12.2A). The large and small subunits come together to form two heterodimers which combine to form the tetramer which is the active form of the enzyme. Activated initiator caspase-9 then participates in the mobilization of the effector caspases by a complex series of events which occur at the mitochondrial membrane and are mediated by Bcl-2.

The effector caspases target numerous specific cellular substrates. Some effector caspases cleave an inhibitor or an effector protein, one example

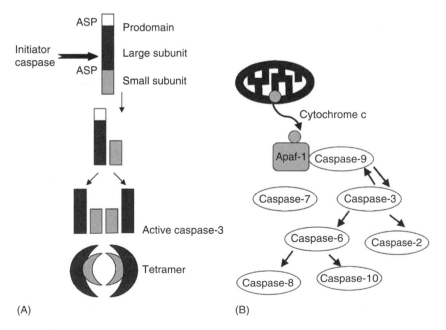

(A)

(B)

FIGURE 12.2

Activation of procaspase-3 by cleavage. (A) Cleavage of N-terminal pro-domain followed by splitting of the remaining protein into large and small subunits. The subunits form heterodimers which combine to form a tetramer, the active form of the enzyme (ASP: aspartic acid). (B) Caspase cascade.

being CAD (caspase-activated deoxyribonuclease) and ICAD (inhibitor of CAD). CAD is inactive as long as it remains bound to ICAD, but when the active effector caspases cleave ICAD, CAD is released and partially digests the DNA molecule into the fragments which form the characteristic DNA ladder of apoptotic cells. Other caspase substrates are structural proteins, the nuclear lamins, which maintain the integrity of the nucleus. When these proteins are cleaved by the effector caspases, the nucleus is condensed in the manner characteristic of apoptotic cells. Yet another target of effector caspases is p21-activated kinase 2 (PAK2). Activation of PAK2 by cleaving of the auto-inhibitory domain affects the cytoskeleton promoting the membrane blebbing that is characteristic of apoptotic cells.

12.3 Apoptotic Signaling Pathways

All apoptotic pathways culminate in the death-dealing caspase cascade (Figure 12.2B). Up to that point, apoptotic signaling can be initiated in different cellular compartments, including the nucleus, cytosol, mitochondria, and cell membrane and proceed by different routes. An overview of apoptotic pathways are listed below and their interaction with ionizing radiation is shown in Figure 12.3.

- A pathway initiated by extrinsic signaling results from the activation of plasma membrane death receptors (DRs).
- Another extrinsic pathway is receptor independent and is mediated by the sphingolipid ceramide acting as a second messenger in an apoptotic cascade initiated by damage to the cell membrane.
- Intrinsically activated pathways require signaling through the mitochondria. Various stimuli, including BAX, ROS, and caspases can directly trigger mitochondria to release caspase activating factors like cytochrome *c* [4].
- Signaling through the mitochondria mediated by p53-dependent regulation of apoptotic proteins.
- p53-mediated apoptosis also occurs via extrinsic signaling DRs following up-regulation of the *Fas* and *FasL* (*CD95* and *CD95L*) genes.

12.3.1 Intrinsic Apoptotic Signaling: The Mitochondrial Pathway

The intrinsic pathway is triggered by various extracellular and intracellular stresses such as growth factor withdrawal, hypoxia, DNA damage, and oncogene induction. Signals that are transduced in response to these stresses converge on the mitochondria, where a series of biochemical events are induced resulting in the permeabilization of the outer mitochondrial

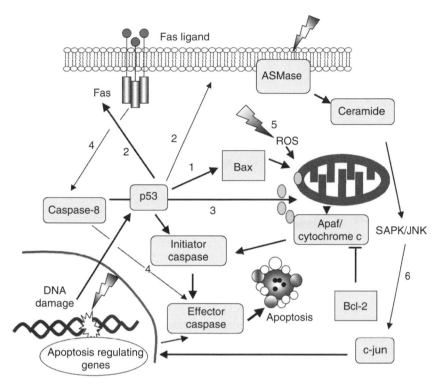

FIGURE 12.3

Overview of radiation-induced apoptotic pathways. The p53-dependent mechanisms are regulation of Bcl-2 family members (1) or up-regulation of Fas or Fas ligand (2). Downstream of the Fas death receptors activation of caspases can proceed through mitochondria-dependent (3) and independent mechanisms (4). Various stimuli (Bax, ROS, caspases) can trigger mitochondria to release caspase activating factors (5). Activation of the SAPK/JNK pathway can occur downstream of membrane-derived signals such as ceramide (6).

membrane and the release of cytochrome c and other pro-apoptotic molecules. In this process, the permeabilization of the mitochondrial membrane is regulated by proteins of the Bcl-2 family.

12.3.1.1 Bcl-2 Proteins

Bcl-2 family members can be pro-survival or pro-apoptotic. Bcl-2 family molecules are characterized by the presence of at least one of four conserved motifs, designated BH domains 1–4. Anti-apoptotic proteins are characterized by the domains BHl and BH2 while there are two pro-apoptotic groups: the Bax group and the BH3-only proteins. Members of the Bax group are structurally similar to Bcl-2, whereas the BH3-only proteins (including Bim, Bad, Bim, Bik, Noxa, and Puma) share only the BH3 interaction domain (Figure 12.4). The pro-survival Bcl-2 proteins can prevent the release of cytochrome c from the mitochondria by forming homodimers or

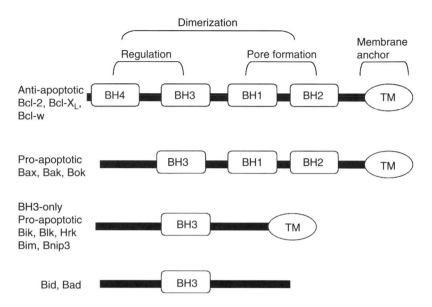

FIGURE 12.4
Members of the Bcl-2 family. Three sub-families are indicated. The anti-apoptotic Bcl-2 members promote survival while the pro-apoptotic and BH3-only members promote apoptosis. BH1–4 are the conserved sequence motifs. The membrane anchoring domain, TM, is not carried by all family members.

by neutralizing pro-apoptotic family members by combining with them to form heterodimers. An excess of Bax at the mitochondria allows the formation of more Bax–Bax homodimers (since there are not sufficient Bcl-2 proteins to titrate out Bax) and leads to the release of cytochrome *c*.

The ratio of pro-survival to pro-apoptotic Bcl-2 proteins which is a matter of life and death to the cell must be strictly regulated, and one of the multiple roles of p53 is in participation in this regulatory process. Transactivation by p53 induces expression of the pro-apoptotic Bax protein perturbing the balance of pro-survival to pro-apoptotic Bcl-2 proteins. As described, the presence of excess of Bax will favor the formation of Bax–Bax homodimers leading to the release of cytochrome *c* and initiation of the caspase cascade.

Subcellular localization studies have shown that pro-survival Bcl-2 and Bcl-X$_L$ reside in the mitochondrial outer membrane, while the pro-apoptotic family members are found either in the cytosol or the mitochondrial membrane.

12.3.1.2   *Precipitating the Caspase Cascade*

The Bcl-2 proteins act at the point of initiation of the caspase cascade by activation of the apoptosis activating factor-l (Apaf-l) protein. Apaf-l requires the addition of two cofactors, ATP, and cytochrome *c*, to become active. ATP is already present in the cell, but cytochrome *c* is normally found in the intermembrane space of the mitochondria where it participates in the electron transport chain and oxidative phosphorylation. Cytochrome *c* must be released into the cytoplasm in order to bind with and activate Apaf-l. The crystal structure of the Bcl-2 proteins is very similar to that of the pore-forming proteins of bacteria suggesting that Bcl-2 proteins can form pores in the outer membrane of the mitochondria and allow cytochrome *c* to escape into the cytoplasm.

The first stage of the activation process is that caspase-8 activates Bid by cleavage in two fragments of 15 and 11 kD. The 15 kD part (tBid) translocates to the mitochondria, inserts into the outer membrane, and allows the release of cytochrome *c*. Cytochrome *c* then associates with ATP, Apaf-1, and caspase-9, to form a complex, the apoptosome, which activates the effector caspases-3, -6, and -7 (Figure 12.5). This event marks the point of no return and leads to the irreversible phase of apoptosis. Release of cytochrome *c* from the mitochondria is blocked by the pro-survival Bcl-2 proteins but facilitated by pro-apoptotic Bcl-2 proteins (especially Bax).

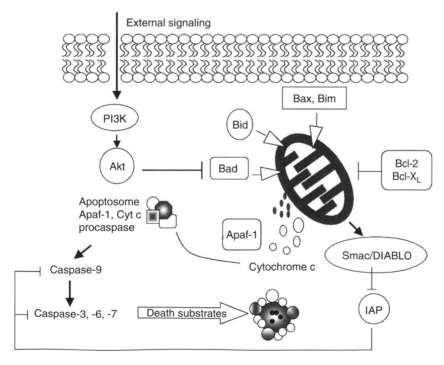

FIGURE 12.5

Apoptosis via the mitochondrial pathway. Bcl-2 family members translocate to the cytosol where they induce release of cytochrome *c*. Cytochrome *c* induces the oligomerization of Apaf-1 which recruits and promotes the activation of procaspase-9, which in turn activates procaspase-3 leading to apoptosis. Smac/DIABLO is another mitochondrial activator of caspases which acts by relieving the inhibition on caspases by binding to IAPs.

12.3.1.3 Activation of Mitochondrial Caspases by Inhibition of IAPs

A recent development is the identification of another protein with a double-barreled name, Smac/DIABLO, which is released from mitochondria in conjunction with cytochrome c during apoptosis. This protein promotes caspase activation by associating with the apoptosome and blocking naturally occurring inhibitors of apoptosis (IAPs). At least five different mammalian IAPs—X-linked inhibitor of apoptosis (XIAP), c-IAP1, c-IAP2, neuronal IAP, and survivin—exhibit anti-apoptotic activity in tissue culture. The apoptotic stimuli that are blocked by mammalian IAPs include ligands and transducers of the tumor necrosis factor (TNF) family of receptors, pro-apoptotic Bcl-2 family members, cytochrome c, and chemotherapeutic agents. The strongest anti-apoptotic activity is attributable to XIAP. XIAP, c-IAP1, and c-IAP2 are direct caspase inhibitors.

In some cellular systems, cytochrome c is necessary but not sufficient for cell death. In these cells Smac/DIABLO is required as backup to counteract the effect of IAPs.

12.3.1.4 p53-Regulated Apoptosis Initiated by DNA Damage

As described in Chapter 8, the tumor suppressor protein p53 plays a pivotal role in the cellular response to nuclear DNA damage. Irradiation-induced p53 activation causes a delay in cell-cycle progression, predominantly at the G_1–S transition, allowing the damaged DNA to be repaired before replication and mitosis occur. However, if repair fails, p53 may trigger the deletion of cells through apoptosis. One way by which this occurs is through the ability of p53 to regulate the expression of pro- and anti-apoptotic members of the Bcl-2 family. p53-mediated apoptosis can also occur through extrinsic signaling via DRs as a result of the up-regulation of the *Fas* and *FasL* genes (Figure 12.3).

In the unperturbed cell, Bcl-2 is bound to Apaf on the surface of the outer mitochondrial membrane. Internal damage to the cell from radiation initiates a signal transduction pathway through the ATM protein and p53 culminating in the up-regulation of the Bax protein, throwing off the ratio of pro-survival to pro-apoptotic Bcl-2 proteins, and causing the release of cytochrome c from the mitochondria.

As described in the previous section, cytochrome c, Apaf-1, and ATP bind to molecules of procaspase-9 resulting in the cleavage of procaspase-9 and triggering the sequential activation of one caspase by another to create an expanding cascade of proteolytic enzymes.

Another way of shifting the balance is by regulation at the posttranslational level. The BH3 subfamily of pro-apoptotic Bcl-2 proteins, of which one example is Bad, can dimerize with Bcl-2 via its BH3 domain preventing Bcl-2 from binding to Bax and favoring the formation of Bax–Bax homodimers. Bad is regulated by means of posttranslational phosphorylation by a pro-survival protein called Akt (sometimes called PKB) which is activated by PI3-K. The pro-survival function of the PI3K signal transduction cascade via the Akt protein was described in Chapter 11. Phosphorylated Bad is sequestered by the 14-3-3σ protein in the cytoplasm and thus prevented from binding to Bcl-2 and promoting apoptosis.

Several BH3-only proteins, which function upstream of Bax/Bak to induce apoptosis have been shown to be critical regulators of p53-dependent apoptosis. Puma and Noxa are activated in a p53-dependent manner following DNA damage [5]. Gene-targeting experiments showed that Puma mediates apoptosis induced by p53, hypoxia, DNA-damaging agents, and ER stress in human colorectal cancer cells [6]. Puma induces a very rapid apoptosis within hours of its expression.

12.3.2 Extrinsic Apoptotic Signaling

Apoptosis initiated by extrinsic signaling can result from the activation of plasma membrane DRs by binding to their ligands or following p53-mediated up-regulation of the *Fas* and *FasL* (*CD95* and *CD95L*) genes.

A receptor-independent extrinsic pathway is mediated by the sphingolipid ceramide acting as a second messenger in an apoptotic cascade initiated by damage to the cell membrane.

12.3.2.1 Mechanism of Apoptosis Mediated through the Death Receptors

Mechanism of apoptosis mediated through the DRs has been reviewed by Zimmerman et al. [3] and Almasan and Ashkenazi [7]. Cell surface DRs belong to the tumor necrosis factor receptor (TNFR) superfamily and transmit their apoptotic signals following binding of specific compounds termed death ligands. Following ligand binding, apoptotic cascades are initiated within seconds and apoptotic cell death can result within hours. The best-characterized family members are Fas (also known as Apol or CD95) and TNFRl.

DRs are transmembrane proteins characterized by extracellular cysteine-rich domains (CRD) and intracellular death domains (DDs). The extracellular domains are responsible for receptor–ligand interactions and for receptor self-association which is required before ligand binding. The cytoplasmic tail contains a DD consisting of approximately 80-amino acids. Following receptor–ligand binding, receptor DDs interact and recruit other DD-containing proteins to function as adaptor proteins in the signal transduction cascade. These proteins interact with a variety of other proteins downstream to complete the DR-signaling pathways. The Fas-signaling pathway is shown schematically in Figure 12.6.

The physiological ligand of Fas is Fas ligand (FasL). Stimulation of Fas by FasL induces trimerization of the receptor complex followed by the recruitment of the adapter molecule Fas-associated death domain (FADD)/ mediator of receptor-induced toxicity (MORT-l) to the cytoplasmic DD of Fas. Signal transduction through the TNFR runs follows a parallel route to

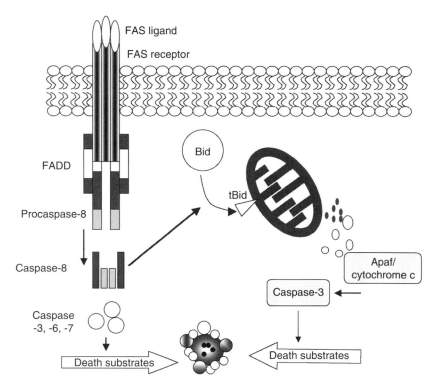

FIGURE 12.6

Signaling from a death receptor; the activation of caspase-8. FasL recruits FADD to the intra-cellular region which in turn recruits procaspase-8. The procaspase-8 is transactivated and the mature caspase can now cleave and activate procaspase-3 leading to apoptosis. Signaling from the Fas receptor to mitochondria involves cleavage of the BH3-only protein Bid by caspase-8. Bid subsequently induces cytochrome *c* release and downstream apoptotic events.

Fas-initiated signaling in that after trimerization of TNFRl, the DD (TNFR TRADD) is recruited.

Recruitment of procaspase-8 to the site of Fas–FADD interaction results in formation of the death-inducing signaling complex (DISC) which is com-posed of the Fas receptor, FADD/MORT-1 and procaspase-8. During com-plex formation, the procaspase-8 is autocatalytically activated and in turn activates the effector caspase-3, which cleaves vital cellular structures and enzymes and causes apoptotic cell death. Caspase-10 is also believed to transmit apoptotic signals mediated through Fas in some types of cells.

A less direct path to cell death exists in the form of an amplification loop through the mitochondria that releases cytochrome *c* and activates caspase-9. Signaling from the Fas receptor to mitochondria requires cleavage of the BH3-only protein Bid by caspase-8. Bid subsequently induces cytochrome *c* release and downstream apoptotic events.

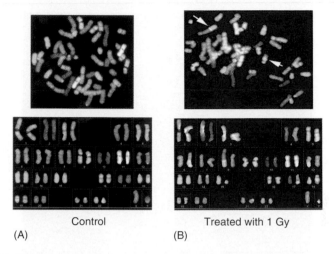

Control   Treated with 1 Gy
(A)   (B)

COLOR FIGURE 10.8
Spectral Karyotype of human chromosomes. (A) Control. (B) Treated with 1 Gy. (Courtesy of Dr Tej Pandita.)

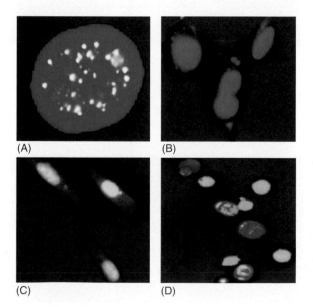

(A)   (B)

(C)   (D)

COLOR FIGURE 15.11
The proposed sequence by which the DIE kills cells. (A) Within 30 min of transferring unstable media, a significant number of γH2AX foci are seen in recipient cells indicative of induced DNA double-strand breaks. (B) After 24 h of growth in unstable media, many cells show induced micronuclei. (C) Evidence of apoptosis by Annexin V staining, note the micronuclei. (D) Evidence of apoptosis by TUNEL assay. (From Sowa Resat, M.B. and Morgan, W.F., *J. Cell Biochem.*, 1, 92, 1013, 2004. With permission.)

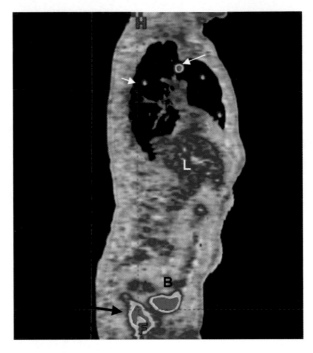

COLOR FIGURE 16.4
Positron-emission tomography imaging with 18-fluorodeoxyglucose (FdG) showing primary prostate tumor (black arrow) and metastases in the lung (white arrows). B—bladder, L—lung. (Courtesy of Dr Slobodan Devic. With permission.)

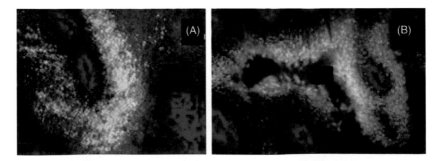

COLOR FIGURE 16.9
Antibody detection of chemical and endogenous hypoxia markers in frozen sections of tumors. (A) Expression of CA9 (red, membranous), HIF-1α (green, nuclear), and Hoechst 33342 (blue, nuclear) in a representative section from a SiHa cervical carcinoma xenograft. After Hoechst 33342 intravenous injection, frozen sections were stained sequentially with CA9 primary antibody followed by Alexa-594 conjugated antimouse immunoglobulin G, and then HIF-1α primary antibody followed by Alexa-488 conjugated antimouse immunoglobulin G. (B) Pimonidazole (green) and HIF-1α (red nuclei) staining are shown in a SiHa xenograft tumor. (From Olive, P.L. and Aquino-Parsons, C., *Semin. Radiat. Oncol.*, 14, 241, 2004. With permission.)

The Fas/FasL system is responsible for three types of cell killing: (1) activation-induced cell death (AICD) of T cells, (2) cytotoxic T lymphocyte-mediated killing of target cells, and (3) killing of inflammatory cells in immune privilege sites and killing of cytotoxic T lymphocytes by tumor cells. More recently, Fas/FasL has been demonstrated to be implicated in developmental and pathological apoptosis in neural tissue.

12.3.2.2 Fas-Mediated Cell Death Initiated by Damage to Genomic DNA

Damage to genomic DNA induces up-regulated expression of Fas and apoptotic cell death. The expression of Fas can be up-regulated following DNA damage by radiation by activation of the p53 tumor suppressor protein which transcriptionally regulates the *Fas* gene [8,9]. The *Fas* gene is transcriptionally activated when p53 consensus elements in the promotor and first intron of *Fas* become co-localized and cooperatively interacts with p53 [10]. Tumor cells that contain a mutation in any of the evolutionarily conserved sequences of the DNA binding region of the p53 protein fail to up-regulate their expression of *Fas* on the cell surface after γ-irradiation [11] and Fas up-regulation in response to a variety of chemotherapeutic agents has also been demonstrated to be dependent on the presence of wild-type p53.

12.3.2.3 Therapeutic Application of Tumor Necrosis-Factor Related Apoptosis-Inducing Ligand for Radiosensitization

As described above, cell death can result from activation of DR-mediated apoptosis. Tumor necrosis-factor related apoptosis-inducing ligand (TRAIL) is a recently identified member of the TNF superfamily which binds to its cognate DR4 and DR5. TRAIL is nontoxic to normal cells and preferentially kills tumor cells. Other DR ligands, TNF and FasL have apoptotic activity against cancer cells, but these agents also display significant toxicities. TNF activates the pro-inflammatory transcription factor NF-κB and was shown to cause a lethal septic shock-like state when infused into mice, while activation of the FasL pathway in vivo causes massive liver damage resulting from Fas-mediated hepatocyte apoptosis.

Interaction of TRAIL and radiation was studied in the MCF-7 human breast cancer cell line which is only marginally sensitive to TRAIL alone. In this system, TRAIL and radiation were synergistic in inducing apoptosis (Figure 12.7). A similar response was seen for other p53-wild-type breast carcinoma cells tested. p53-negative cell lines, on the other hand, failed to exhibit a combined effect as did an MCF-7 line which expresses a dominant negative version of p53. This could be explained on the basis of p53-mediated up-regulation of DR5 receptor mRNA following radiation.

Systemic TRAIL inhibits the growth of breast and colon cancer xenografts in mice and combined TRAIL and radiotherapy (TRAIL + RT) have been

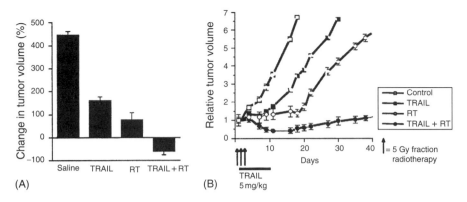

FIGURE 12.7

TRAIL and radiotherapy (RT), in combination, induce regression and slow the growth of breast cancer xenografts. NIH III nude mice harboring MCF7 xenograft tumors were treated with saline, TRAIL (5 mg/kg for 7 days), RT (three 5 Gy fractions), or a combination of TRAIL and RT (TRAIL + RT). Mean tumor volumes were calculated 15 days after the start of treatment and are expressed as a percent change in tumor volume (A). Mean tumor volumes were assessed 2–3 times a week for over 4 weeks and are expressed in relation to the starting tumor volume (B). Six to ten xenografts were used in each experiment, and the data shown is derived from the mean volume ± SEM. (From Chinnaiyan, A.M., Prasad, U., Shankar, S., Hamstra, D.A., Shanaiah, M., Chenevert, T.L., Ross, B.D., and Rehemtulla, A., *Proc. Natl. Acad. Sci. U.S.A.*, 97, 1754, 2000. With permission.)

shown to induce regression and slow the growth of breast cancer xeno-grafts. Combined therapy with TRAIL and chemo- or RT is undergoing clinical trials [12].

12.3.2.4 Suppression of Apoptosis by Anti-Apoptotic Signaling

Signaling culminating in anti-apoptotic activity was discussed in part in the preceding Chapter 11. Growth Factors such as EGF and PDGF, cyto-kines such as IL-2 and IL-3, and some hormones such as insulin act as survival factors by activating the PI3K pathway. Active PI3K generates the 3-phosphorylated lipid phospatidylinositol-3,4,6 triphosphate (PtdIns[3,4,5] P_3). This leads to recruitment of kinases PDK1, PDK2, and Akt (also known as protein kinase B or PKB) to the plasma membrane. A complex is formed in which PDK1 and PDK2 activate Akt by phosphorylation. Active Akt interferes with the apoptotic machinery by phosphorylating and thus inhi-biting the pro-apoptotic BCl-2 family member Bad. In addition, it influences gene expression by inactivating the forkhead family transcription factors which can induce pro-apoptotic genes including *FasL* and can also activate the transcription factor NFκB leading to expression of anti-apoptotic genes. Survival signaling by Akt is counteracted by the tumor suppressor PTEN, a lipid phosphatase that antagonizes the action of PI3K by dephosphorylating PtdIns $(3,4,5)P_3$ (Figure 11.5).

These survival mechanisms apply particularly in tumor cells, where the mechanisms which protect normal cells from perpetuating genomic damage are frequently overridden. Akt is over-expressed in some cancer cells, while PTEN, the PI3K antagonist, is frequently deleted in advanced-stage tumors and significant PTEN mutation can be found in various types of cancer (reviewed by Igney and Krammer [13]).

12.3.3 Extrinsic Apoptotic Signaling Initiated at the Plasma Membrane: The Ceramide Pathway

There is strong evidence in support of the existence of a second extrinsic apoptotic mechanism, a receptor-independent apoptotic cascade originating at the plasma membrane and mediated by the sphingomyelin ceramide. As described in Chapter 5, ionizing radiation interacts with the plasma membrane via free-radical species to inflict lipid oxidative damage. This damage can occur after very low doses of radiation, which is cumulative and is not repaired. In vitro studies have demonstrated that the damaged plasma membrane may function as a source of bioactive molecules, activating various signal transduction pathways including the alteration of Ca^{2+} and K^+ influx, the release of arachidonic acid, and the generation of ceramide.

12.3.3.1 *Ceramide as a Second Messenger Regulating Stress Responses*

The sphingomyelin pathway for induction of apoptosis is a ubiquitous, evolutionarily conserved signaling system initiated by hydrolysis of sphingomyelin by the action of sphingomyelin-specific forms of the enzyme phospholipase C, called sphingomyelinases (SMases), to generate ceramide [14,15]. Ceramide then acts as a second messenger to stimulate a cascade of kinases and transcription factors that are involved in a variety of cellular responses. The sphingomyelin pathway has been reported to be involved in the generation of apoptosis via a variety of stresses, including ionizing radiation, UV radiation, oxidizing agents, and heat.

Evidence supporting ceramide as a message for stress-induced apoptosis includes data from many cell systems that show an increase in ceramide levels preceding biochemical and morphologic manifestations of apoptosis. Addition of natural sphingomyelinase (SMase) or pharmacologic agents, which interfere with enzymes of ceramide metabolism, mimic the effects of stress on apoptosis induction (reviewed by Haimovitz-Friedman [16]).

Studies by Haimovitz-Friedman and coworkers [17] were the first to show that radiation targets the cell membrane of bovine aortic endothelial cells (BAEC) to induce, within seconds, activation of SMase—a sphingomyelin-specific form of phospholipase C, which generates ceramide by enzymatic hydrolysis of the phosphodiester bond of sphingomyelin. Much additional biochemical and genetic evidence has been accumulated to support the involvement of acid sphingomyelinase (ASMase) and ceramide in radiation-induced apoptosis.

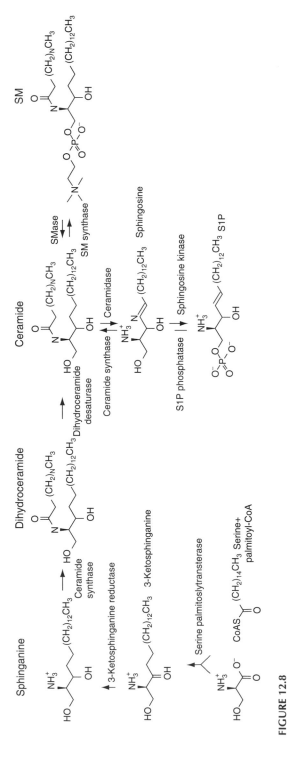

FIGURE 12.8

Schematic representation of intermediary metabolism of sphingolipids. (From Kolesnick, R. and Fuks, Z., *Oncogene*, 22, 5897, 2003. With permission.)

12.3.3.2 Synthesis and Metabolism of Ceramide

Sphingolipids consist of a backbone of a long-chain sphingoid base, an amide-linked long-chain fatty acid, and a polar head group, which is specific to the sphingolipid subtype (ceramide has a hydroxyl head group). The main features of sphingolipid intermediary metabolism are shown schematically in Figure 12.8. Ceramide is generated by enzymatic hydrolysis of sphingomyelin via activation of SMase or by de novo synthesis coordinated through the enzyme ceramide synthase (CS) [18]. Mammalian cells utilize three distinct forms of SMases, distinguished by their pH optimum—the acid, neutral, and alkaline SMases. ASMase and neutral sphingomyelinase (NSMase) are rapidly activated by stress stimuli, including ionizing radiation and are known to be involved in signal transduction in mammalian cells.

Once generated, ceramide may accumulate, may be converted into a variety of metabolites (Figure 12.8), or may be converted back to sphingomyelin by transfer of phosphorylcholine from phosphatidylcholine to ceramide via sphingomyelin synthase.

The metabolic products of ceramide may serve as effector molecules, often inducing proliferation, but, with the exception of sphingosine and GD3, they do not signal apoptosis.

12.3.3.3 Apoptotic Signaling via Ceramide

The mechanism for irradiation-induced apoptosis via the sphingomyelin pathway is shown schematically in Figure 12.9. One route for the transfer of signals from the irradiated cell membrane via ceramide leading to apoptosis goes through the stress-activated protein kinase (SAPK/JNK) cascade, involving sequential activation of MEKKl, MEKK4/7 and SAPK/JNK culminating in the transcription factor c-Jun (described in Chapter 11). Verheij and coworkers [19] showed that ionizing radiation and other stresses which induce rapid generation of ceramide also initiate activation of the SAPK/JNK pathway and apoptosis. Exposure of cell lines to ionizing radiation, hydrogen peroxide, UV radiation, heat shock, or TNF α rapidly induce significant elevations of ceramide and morphological and biochemical features of apoptosis concomitant with activation of SAPK/JNK. Once activated, SAPK/JNK binds to the amino-terminal transactivation domain of c-Jun thereby increasing the AP-1-dependent gene expression. In addition, the transcription factors ATF2 and ELK are involved in the transcription of apoptotic proteins. The key role of c-Jun is indicated by the finding that dominant-negative c-Jun mutant cells showed inhibition of stress- and ceramide-induced apoptosis while the processes of ceramide generation and activation of SAPK/JNK were not compromised [19].

In rat 1 Myc-ER cells, the ceramide-induced response to stress couples phosphoinositide 3'-kinase (PI3K) to caveolin 1, thereby inactivating PI3K. As described in Chapter 11, PI3K normally provides sustained anti-apoptotic signaling in mammalian cell lines, so its inactivation promotes an apoptotic response to stress. Exogenous ceramide mimicked the radiation

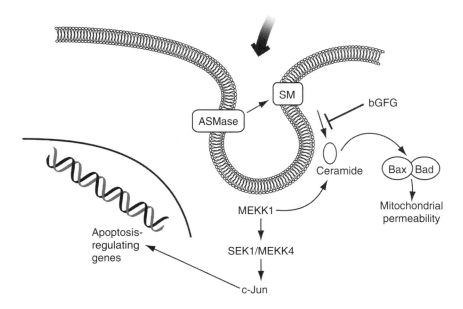

FIGURE 12.9

Acid sphingomyelinase (ASMase) mediates pro-apoptotic signaling induced by ionizing radiation, heat, UV, and H_2O_2. ASMase generates ceramide by hydrolysis of sphingomyelin. Ceramide then acts as a second messenger to stimulate a cascade of kinases and transcription factors culminating in an apoptotic response.

effect by inactivating PI3K and cells from patients deficient in radiation-induced ceramide generation fail to show PI3K inactivation and concomitant apoptosis [20].

12.3.3.4 Ceramide-Mediated Apoptosis Induced by DNA Damage: Role of Ceramide Synthase

Ceramide has been shown to mediate apoptosis that occurs in response to DNA damage but the generation of ceramide in this case does not involve activation of SMase. CS is part of an alternate pathway for the generation of ceramide via de novo synthesis in the ER and mitochondria. Synthesis occurs by the condensation of the sphingoid base sphinganine and fatty acyl-CoA to form dihydroceramide, which is oxidized to ceramide. In HAEC, it has been shown that low doses of radiation (2.5–5.0 Gy) can activate the enzyme CS and that this activation contributes to the apoptotic response.

Elevation of ceramide levels in response to radiation-induced activation of SMase occur within minutes of exposure but do not exceed a 1.5–2-fold increase even after high doses (10–20 Gy). In contrast, CS-mediated elevation of ceramide does not start until hours after irradiation but can persist over a long period and may gradually increase to greater than fourfold–sixfold of the baseline levels [16]. This biphasic response in terms of ceramide levels correlates with the two modes of apoptotic response, prompt response which occurs immediately after radiation insult and delayed response which results in post-mitotic death of cells which exit G_2 with unrepaired DNA damage.

12.3.3.5 Ceramide Rafts

Ceramide rafts have been reviewed by Gulbins and Kolesnick [21]. The classical fluid mosaic model of the plasma membrane [22] has been modified in recent years to accommodate a role for distinct micro domains in the cell membrane that may serve signaling functions. The cell membrane is mainly composed of glycerophospholipids, sphingolipids, and cholesterol. The head groups of sphingolipids trigger a lateral association of lipids of this class with one another, which is further enhanced by hydrophobic interactions between the saturated side chains. Cholesterol fills voids between the large glycerosphingolipids and tightly interacts with sphingolipids, in particular sphingomyelin. The tight interaction of sphingolipids with one another and with cholesterol results in the segregation of these lipids into discrete membrane structures characterized by a liquid-ordered or possibly gel-like phase, while glycerophospholipids in the bulk cell membrane reside in a more fluid liquid-disordered phase. These distinct sphingolipids and cholesterol-enriched membrane micro-domains (30–300 nm) are considered to be floating in a "sea" of phospholipids, and hence have been termed rafts. The generation of ceramide-enriched membrane platforms represents an evolutionarily conserved cellular function of ceramide which does not exclude the other functions of ceramide.

The ceramide-enriched membrane platforms serve to trap and cluster pro-apoptotic receptor molecules and pathogens to achieve a critical density required for biochemical transfer of the stress across the plasma membrane. Ionizing radiation and heat can induce the formation of ceramide-enriched platforms in the plasma membrane of cells destined to die by apoptosis. In effect, the response to external stress of ceramide-enriched membrane platforms is to trigger apoptosis after irradiation by recruitment and activation of pro-apoptotic molecules.

12.3.3.6 Upstream Pathways That Antagonize to Ceramide-Signaled Apoptosis

During radiation- and drug-induced apoptosis, the pro-apoptotic ceramide signal may be counter-balanced by anti-apoptotic signaling. Several pathways are involved, the most important being mediated via 1,2-diacylglycerol (DAG) and PKC (Chapter 11) [23]. Activation of PKC by phorbol esters blocks radiation-induced ceramide generation and apoptosis in several different cell types, while pharmacologic inhibition of PKC enhances radiation-induced cell death, and DAG or phorbol ester can also block the action of ceramide to initiate apoptosis. These protective effects of PKC may occur through inhibition of SMase [24] or in some cells by up-regulation of the apoptosis-regulating proteins Bcl-2 or Bcl-XL [25,26]. Basic fibroblast growth factor (bFGF) also protects endothelial cells against radiation-induced apoptosis, in part, via activation of the PKC and MAPK pathways [27].

The enzyme ceramidase has been found to have significant impact on apoptotic signaling [28]. Increasing the rate of metabolic removal of ceramide produces an anti-apoptotic phenotype, and ceramidase also catalyze the hydrolysis of ceramide to free fatty acid and sphingosine, thus attenuating ceramide and depleting the pro-apoptotic signal. Another outcome of ceramide hydrolysis by ceramidase is the generation of the sphingoid base sphingosine which in many cell types is rapidly phosphorylated to the anti-apoptotic SI-P via the enzyme sphingosine.

12.4 Why Do Some Cells Die as the Result of Apoptosis and Not Others?

The most important determinant of the mode and probability of apoptotic death is cell type. This can be illustrated by a comparison of T lymphocytes which usually respond to DNA damage by undergoing extensive apoptosis, whereas fibroblasts undergo prolonged cell-cycle arrest. The difference in the apoptotic threshold of the two cell types is a reflection of their different biology. The T lymphocytes have to respond rapidly to antigen stimulus, undergo clonal expansion, and later die in order to limit the immunological response. In these cells, death is highly dependent on the Fas–FasL death-signaling pathway. Typical examples of cells that display this type of acute apoptotic response are those from the lymphoid and myeloid lineage and epithelial cells located in the intestinal crypts and salivary glands [29].

For cells of these groups, there is no choice. There is no evidence that cells of hematopoietic lineage can respond to ionizing radiation in any way other than apoptosis, whereas fibroblasts never respond by apoptosis and are strongly committed to growth arrest. In other situations, either apoptosis or growth arrest could occur and the route taken depends on growth conditions and severity of damage [30] with the decision between apoptosis and growth arrest largely determined by the status of p53.

Several factors have been proposed to influence the decision making between death and growth arrest in response to activation of p53. The first is the amount of activated p53 and the duration of its activation. The stronger and longer the activation of p53, the greater is the chance of the outcome being apoptosis. In effect, p53 is an evaluator of the severity of damage; more severe damage results in more p53 and persistence of damage ensures continuing generation of the signal. Low levels of p53 expression have anti-apoptotic activity while high levels promote apoptosis. Mdm2 is the major regulator of p53 expression and factors affecting Mdm2 expression and activity have a major impact on the level of p53-response expression.

The second factor determining outcome is the spectrum of p53-responsive genes which are available. Different cell lineages have different sets of p53-responsive genes available predetermining a specific pattern of transactivation and transrepression that dictates the final outcome, and cells expressing more pro-apoptotic genes will be more likely to undergo apoptosis. The third factor governing the choice between growth arrest and apoptosis is the availability of p53 cofactors that differentially regulate the ability of p53 to bind to specific subsets of target genes (see Chapter 8).

The oncogenic composition of the cell is also important. The activation of p53 in normal cells results in cell-cycle arrest or senescence, whereas in transformed cells p53 may promote apoptosis. The p53-pRb signaling pathway (described in Chapter 8) is abrogated in cancer cells and the inactivation of Rb results in the loss of G_1 arrest and induction of apoptosis after DNA damage. In rat embryo cells or Rat-1 fibroblasts, radiation-induced apoptosis can be enhanced by over-expression of c-Myc, whereas it is suppressed by the over-expression of Ha-Ras.

Growth factors have been shown to affect both radiation-induced apoptosis and clonogenic cell survival. Mitogenic growth stimulatory signals, such as cytokines, growth, and survival factors, usually protect cells from apoptotic response to DNA damage by promoting growth arrest [31]. The addition of bFGF to cultures of BAEC within 24 h after irradiation increases clonogenic capacity in the low-dose range, producing a wider shoulder region of the clonogenic survival curve. It was later shown that bFGF rescues the cells by inhibiting the apoptotic pathway both in vitro and in vivo [32,33]. Interleukin 6 (IL-6), IL-3, and erythropoietin have a cell-specific role in protecting cells from apoptotic death. The anti-apoptotic effect of these survival factors may be explained by the cross talk between growth-signaling pathways and cell death-signaling involving the Bcl-2 family members, such as Bad [34].

12.5 Apoptotic Processes and the In Vivo Radiation Response

Most mammalian adult tissues are quite resistant to ionizing radiation and do not undergo morphological or physiological alteration even after treatment with high doses. Death induced by a lethal dose of radiation is caused by the failure of a few sensitive tissues, in some cases by apoptotic processes. Tumor systems, although displaying generally radiosensitive phenotype related to their rate of proliferation, are also variable in the extent to which cells are killed by apoptosis.

12.5.1 Normal Tissue

Normal tissue has been reviewed by Gudkov and Komarova [29]. Comparison of transgenic, p53-deficient mice with wild-type p53 mice shows that p53 is important for the rapid apoptosis which occurs in sensitive tissues after exposure to ionizing radiation. Analysis of the response of these mice indicated that the extent of apoptosis correlated with the amount of p53 the cell can accumulate as a result of stress and that radiosensitive tissues have a high basal levels of p53 mRNA. Most of the known p53-responsive genes are activated in a tissue-specific manner, the most vigorous responses being found in the highly radiosensitive hemopoietic and lymphoid tissues, spermatogonia, small intestine, hair follicles, and early embryos [35–37]. The importance of p53 as a determinant of radiosensitivity is indicated by the fact that p53-deficient mice can survive doses of radiation which cause lethal bone marrow depletion in wild-type animals.

Although the differences in the radiosensitivity of different normal tissues reside partly in the tissue-specific activation of the p53 pathway, there are also differences in the activity and composition of the apoptotic machinery in different cell types. Radiation-induced apoptosis in lymphoblasts requires ASMase, a mediator of ceramide-mediated apoptosis which, in some cases, seems essential for the execution of p53-mediated apoptosis [38,39].

p53 as a determinant of the apoptotic response is not the only factor which defines tissue-specific radiosensitivity. This is indicated by the fact that p53-deficient mice which do not respond to radiation with early apoptotic death in the small intestine manifest a later wave of apoptosis starting at 24 h after irradiation. The scale of the secondary p53-independent response, which is presumably secondary to mitotic catastrophe after DNA damage, increases with radiation dose. The exact mechanism is unknown but seems likely to involve the sphingomyelin–ceramide pathway [40]. A number of investigators have demonstrated that the survival of the clonogenic cells which are responsible for the recovery of epithelial tissues of the skin, intestine, colon, and testis, is independent of p53 status [41–44].

The radiation-generated gastrointestinal (GI) syndrome results from radiation-induced depletion of villi and crypts, causing potentially lethal loss of the barrier and resorptive functions of the GI tract. It has been attributed to direct damage to the intestinal epithelium which undergoes p53-dependent apoptosis. However, studies by Kolesnick and coworkers [18,45] indicated that the GI syndrome in mice treated with high doses of ionizing radiation was not caused by a direct effect on the epithelium but by apoptosis in the vascular endothelium triggered by a p53-independent ceramide-mediated mechanism. An analysis of the molecular mechanisms mediating the GI syndrome detected apoptosis of endothelial cells as early as 1 h and peaking at 4 h after irradiation, preceding the onset of apoptosis of epithelial cells in the crypts and villi observed at 8–10 h after irradiation. The extent of endothelial cell apoptosis correlated closely with the onset of the GI syndrome. Radiation-induced apoptosis of endothelial cells in the GI tract was prevented by ASM deficiency or intravenous injection of bFGF that blocked ASM-induced generation of ceramide and since endothelial cells but not crypt stem cells display FGF receptors, this indicates that the protection afforded must be via the microvasculature [18] (Figure 12.10). The clinical

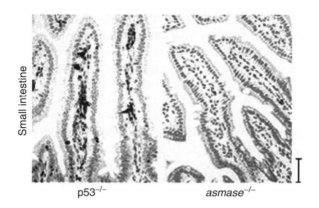

FIGURE 12.10
Acid sphingomyelinase (ASMase) deficiency but not p53 deficiency protects against radiation-induced microvascular apoptosis and death from the GI syndrome. Small intestine of p53$^{-/-}$/C57BL/6 mice or 16 Gy-irradiated *asmase*$^{-/-}$/SV129/C57BL/6 mice were stained by TUNEL. The scale bar represents 140 μm. (From Paris, F., Fuks, Z., Kang, A., Capodieci, P., Juan, G., Ehleiter, D., Haimovitz-Friedman, A., Cordon-Cardo, C., Kolesnick, R., *Science*, 293, 293, 2001. With permission.)

relevance of these observations remains unclear because the described effect occurs at doses which exceed those routinely used in treatment protocols.

The endothelium of the central nervous system (CNS) has also been identified as a primary target for radiation-induced apoptosis regulated by ASM-expression. Radiation to the CNS in two strains of mice resulted in dose-dependent apoptosis occurring within the first 12 h after radiation [46]. As many as 20% of all apoptotic cells were confirmed as endothelial cells. Oocytes are another cell type whose radiosensitivity is regulated by ASM. Irradiation of ASM-positive mice yielded rapid oocyte apoptosis, while ASM deficiency protected oocytes against apoptotis [47]. Oocytes were also rescued from radiation or chemotherapy-induced apoptosis by treatment with the ceramide metabolite sphingosine 1-phosphate, which serves as a functional antagonist of ceramide action [40].

12.5.2 Tumor Response

Spontaneous apoptosis can be observed in virtually all normal tissues and tumors. Following exposure to radiation, the number of apoptotic cells is further enhanced in a dose-dependent fashion, although the propensity of a cell population to die by apoptosis after radiation differs greatly from one tumor type to another [48,49]. Thymocytes, lymphocytes, and cells from the hematopoietic and germinal lineage are usually apoptosis sensitive, whereas cells from solid tumors are relatively resistant.

12.5.2.1 Experimental Studies with Rodent Tumors

In several rodent tumor models, including a variety of solid tumors, a correlation has been established between the level of apoptosis before irradiation and the tumor response following irradiation [48,49]. These findings raised the possibility that the pretreatment levels of apoptosis (the apoptotic index or AI) might predict tissue response following radiation. Tumors with a high level of spontaneous apoptosis might be expected to shrink more rapidly after initial radiation treatment but this does not necessarily predict for a long-term favorable response.

In most studies of radiation-induced apoptosis, radiation has been given as a large single dose and few studies have addressed the role of apoptosis in the response to fractionated radiation. Meyn and coworkers [50] evaluated several fractionation protocols in a murine ovarian tumor model. They found that five daily fractions of 2.5 Gy caused more apoptosis than either 2 doses of 12.5 Gy or a single dose of 25 Gy. From this it was concluded that following each fraction of radiation an apoptosis-sensitive sub-population reemerges and that the percentage of apoptotic cells induced by the second fraction increases with the time interval between fractions.

Experimental studies have demonstrated the role of the host in the apoptotic response of implanted tumors. It was found that implanted tumors grew 2–4 times as fast in ASM-deficient mice than in normal mice and that tumors grown in normal mice were radiosensitive, while those in ASM-deficient mice were radioresistant. The differential radiosensitivity reflects differential levels of apoptosis in the endothelial cells of the tumor vasculature which is derived from the host.

12.5.2.2 Human Tumor Systems

The suggestion from some of the rodent studies that the AI of solid tumors might be predictive of tumor response to therapy is not supported by preclinical and clinical studies with human tumor systems.

Tumors arising from normal tissues which are sensitive to the induction of apoptosis by DNA damaging agents, including T-cell lymphomas and some hematological tumors are often sensitive to the induction of apoptosis and experience marked overall response following treatment. Evidence of the influence of apoptosis on clonogenic survival in these tumors is provided by studies showing that over-expression of apoptosis-promoting or apoptosis-suppressing genes modifies radiation-induced cell survival and radiosensitivity. For instance, cells which are made resistant to radiation-induced apoptosis, either by over-expression of Bcl-2 or inactivation of p53 can show an increase in clonogenic cell survival [51–53].

For solid tumors, most of which are of epithelial origin, apoptosis is not the primary reason for cell death and the primary antitumor effect of radiation is the result of mitotic catastrophe or irreversible cell-cycle arrest. The apoptosis which can be observed in these tumors following radiation is probably secondary to mitotic catastrophe [29]. The relatively modest role of apoptosis in the tumor response to radiation can be explained by the fact that functional p53 and the ability to respond to stress by apoptosis is lost in a high proportion of tumor cells. The role of p53 as a guardian of the genome by inducing apoptosis to prevent continuing proliferation of cells with defects in DNA repair and involvement in damage checkpoints, or telomere function, is frequently abrogated when the cell becomes malignant. As a consequence, the cell's susceptibility to apoptosis is severely compromised and other forms of cell death become important for cell killing and tumor response to DNA-damaging agents [54].

A number of clinical studies have been done to determine whether the level of apoptosis itself or apoptosis-regulating proteins are predictive of tumor sensitivity to treatment. It was concluded by Brown and Attardi [54] that there was no clear evidence that AI or levels of p53, Bcl-2, or other Bcl-2 family members are predictive of the response of solid tumors to therapy. An exception is the high levels of the IAP survivin which are associated with poor prognosis. Since almost all studies have been done following surgery, this does not give a clear picture of the relationship of surviving to treatment sensitivity since survivin also has a role in promoting cell proliferation [55].

12.6 Summary

Apoptosis is a programmed sequence of reactions triggered by specific stimuli.

A hallmark of apoptotic death is the activation of the caspases, cysteine proteases which act as the executioners of the cell and are regulated at the translational level.

Protein degradation by caspases is responsible for the characteristic morphological features of apoptosis.

There are several routes by which radiation can induce apoptosis. The intrinsic apoptotic pathway is initiated by DNA damage and mediated by p53 via the mitochondrial route through a series of biochemical events culminating in the permeabilization of the outer mitochondrial membrane by the members of the Bcl-2 family of proteins. The Bcl-2 proteins can be pro-apoptotic or anti-apoptic. The extrinsic apoptotic pathway is activated by the engagement of DRs on the cell surface. Binding of ligands (FasL and TNF-α) to their receptors induces formation of the DISC which in turn recruits caspase-8 and promotes the caspase cascade that follows. Ionizing radiation can initiate Fas-related apoptosis through activation of p53 which up-regulates the expression of Fas and FasL proteins on the cell surface. Fas-mediated apoptosis can be suppressed by a number of mechanisms including those involving cytokines and that of PI3K mediated by PKB/Akt.

A third route by which ionizing radiation can cause apoptosis is by interaction of the plasma membrane with ROS to initiate extrinsic signaling mediated by the sphingomyelin, ceramide. Ceramide is generated by enzymatic hydrolysis of sphingomyelin via activation of SMase. Signals from the irradiated cell membrane through ceramide lead to apoptosis via the

SAPK/JNK cascade and terminate in the activation of transcription factors for apoptotic genes. Pro-apoptotic ceramide signals may be counterbalanced by anti-apoptotic signaling, the most important pathways being that mediated via DAG and PKC. Increasing the rate of metabolic removal of ceramide through action of the enzyme ceramidase also produces an anti-apoptotic phenotype.

Overall, the single most important determinant of the mode and the likelihood of AD is cell type. Other determinants which can influence the choice between p53-mediated cell-cycle arrest and apoptosis include again the cell type, the oncogenic composition of the cell, the intensity of the stress conditions, the level of p53 expression and its interaction with specific proteins, and the ability of the cells to repair defective DNA. In normal tissue, the apoptotic response to radiation correlates with the amount of p53 the cell can accumulate as a result of stress and those radiosensitive tissues have a high basal level of p53 mRNA. Most of the known p53-responsive genes are activated in a tissue-specific manner, the most striking responses being found in hemopoietic and lymphoid tissues, spermatogonia, small intestine, hair follicles, and early embryos which are highly radiosensitive sites.

The suggestion from some of the rodent studies that the AI of solid tumors might be predictive of tumor response to therapy is not supported by pre-clinical and clinical studies with human tumor systems. Tumors arising from normal tissues which are sensitive to the induction of apoptosis by DNA-damaging agents including T-cell lymphomas and some hematological tumors are often sensitive to the induction of apoptosis and show a marked overall response following radiation. For solid tumors, however, most of which are of epithelial origin, apoptosis is not the primary reason for cell death.

References

1. Kerr JF, Wyllie AH, and Currie AR. Apoptosis: A basic biological phenomenon with wide-ranging implications in tissue kinetics. *Br. J. Cancer* 26: 239–257, 1972.
2. Danial NN and Korsmeyer SJ. Cell death: Critical control points. *Cell* 116: 205–219, 2004.
3. Zimmermann KC, Bonzon C, and Green DR. The machinery of programmed cell death. *Pharmacol. Ther.* 92: 9257–9270, 2001.
4. Green DR and Reed JC. Mitochondria and apoptosis. *Science* 281: 1309–1312, 1998.
5. Yu J, Zhang L, Hwang PM, Kinzler KW, and Vogelstein B. PUMA induces the rapid apoptosis of colorectal cancer cells. *Mol. Cell.* 7: 673–682, 2001.
6. Nakano K and Vousden KH. PUMA, a novel proapoptotic gene, is induced by p53. *Mol. Cell* 7: 683–694, 2001.
7. Almasan A and Ashkenazi A. Apo2L/TRAIL: Apoptosis signaling, biology, and potential for cancer therapy. *Cytokine Growth Factor Rev.* 14: 337–348, 2003.

8. Sheard MA. Ionizing radiation as a response-enhancing agent for CD95-mediated apoptosis. *Int. J. Cancer* 96: 213–220, 2001.

9. Munsch D, Watanabe-Fukunaga R, Bourdon JC, Nagata S, May E, Yonish-Rouach E, and Reisdorf P. Human and mouse Fas (APO-1/CD95) death receptor genes each contain a p53-responsive element that is activated by p53 mutants unable to induce apoptosis. *J. Biol. Chem.* 275: 3867–3872, 2000.

10. Muller M, Wilder S, Bannasch D, Israeli D, Lehlbach K, Li-Weber M, Friedman SL, Galle PR, Stremmel W, Oren M, and Krammer PH. p53 activates the CD95 (APO-1/Fas) gene in response to DNA damage by anticancer drugs. *J. Exp. Med.* 188: 2033–2045, 1998.

11. Sheard MA, Vojtesek B, Janakova L, Kovarik J, and Zaloudik J. Up-regulation of Fas (CD95) in human p53 wild-type cancer cells treated with ionizing radiation. *Int. J. Cancer* 73: 757–762, 1997.

12. Chinnaiyan AM, Prasad U, Shankar S, Hamstra DA, Shanaiah M, Chenevert TL, Ross BD, and Rehemtulla A. Combined effect of tumor necrosis factor-related apoptosis-inducing ligand and ionizing radiation in breast cancer therapy. *Proc. Natl. Acad. Sci. U.S.A.* 97: 1754–1759, 2000.

13. Igney FH and Krammer PH. Death and anti-death: Tumour resistance to apoptosis. *Nat. Rev. Cancer* 2: 277–288, 2002.

14. Jarvis WD, Kolesnick RN, Fornari FA, Traylor RS, Gewirtz DA, and Grant S. Induction of apoptotic DNA damage and cell death by activation of the sphingomyelin pathway. *Proc. Natl. Acad. Sci. U.S.A.* 91: 73–77, 1994.

15. Obeid LM, Linardic CM, Karolak LA, and Hannun YA. Programmed cell death induced by ceramide. *Science* 259: 1769–1771, 1993.

16. Haimovitz-Friedman A. Radiation-induced signal transduction and stress response. *Radiat. Res.* 150: S102–S108, 1998.

17. Haimovitz-Friedman A, Kan CC, Ehleiter D, Persaud RS, McLoughlin M, Fuks Z, and Kolesnick RN. Ionizing radiation acts on cellular membranes to generate ceramide and initiate apoptosis. *J. Exp. Med.* 180: 525–535, 1994.

18. Kolesnick R and Fuks Z. Radiation and ceramide-induced apoptosis. *Oncogene* 22: 5897–5906, 2003.

19. Verheij M, Ruiter GA, and Zerp SF. The role of the stress-activated protein kinase (SAP/JNK) signaling pathway in radiation-induce apoptosis. *Radiother. Oncol.* 47: 225–232, 1998.

20. Zundel W and Giaccia A. Inhibition of the anti-apoptotic PI(3)K/Akt/Bad pathway by stress. *Genes Dev.* 12: 1941–1946, 1998.

21. Gulbins E and Kolesnick R. Raft ceramide in molecular medicine. *Oncogene* 22: 7070–7077, 2003.

22. Singer SJ and Nicolson GL. The fluid mosaic model of the structure of cell membranes. *Science* 175: 720–731, 1972.

23. Chmura SJ, Nodzensk IE, Weichselbaum RR, and Quintans J. Protein kinase C inhibition induces apoptosis and ceramide production through activation of a neutral sphingomyelinase. *Cancer Res.* 56: 2711–2714, 1996.

24. Santana P, Pena LA, Haimovitz-Friedman A, Martin S, Green DR, McLoughlin M, Cordon-Cardo C, Schuchman EH, Fuks Z, and Kolesnick R. Acid sphingomyelinase-deficient human lymphoblasts and mice are defective in radiation-induced apoptosis. *Cell* 86: 189–199, 1996.

25. Hallahan DE, Virudachalam S, Schwartz JL, Panje N, Mustafi R, and Weichselbaum RR. Inhibition of protein kinases sensitizes human tumor cells to ionizing radiation. *Radiat. Res.* 129: 345–350, 1992.

26. Uckun FM, Schieven GL, Tuel-Ahlgren LM, Dibirdik I, Myers DE, Ledbetter JA, and Song CW. Tyrosine phosphorylation is a mandatory proximal step in radiation-induced activation of the protein kinase C signaling pathway in human B-lymphocyte precursors. *Proc. Natl. Acad. Sci. U.S.A.* 90: 252–256, 1993.

27. Fuks Z, Alfieri A, Haimovitz-Friedman A, Seddon A, and Cordon-Cardo C. Intravenous basic fibroblast growth factor protects the lung but not mediastinal organs against radiation-induced apoptosis in vivo. *Cancer J. Sci. Am.* 1: 62–72, 1995.

28. Nikolova-Karakashian M, Morgan ET, Alexander C, Liotta DC, and Merrill AH. Bimodal regulation of ceramidase by interleukin-1beta. Implications for the regulation of cytochrome p450 2C11. *J. Biol. Chem.* 272: 18718–18724, 1997.

29. Gudkov AV and Komarova EA. The role of p53 in determining sensitivity to radiotherapy. *Nat. Rev. Cancer* 3: 117–129, 2003.

30. Offer H, Erez N, Zurer I, Tang X, Milyavsky M, Goldfinger N, and Rotter V. The onset of p53-dependent DNA repair or apoptosis is determined by the level of accumulated damaged DNA. *Carcinogenesis* 23: 1025–1032, 2002.

31. Lin Y and Benchimol S. Cytokines inhibit p53-mediated apoptosis but not p53-mediated G1 arrest. *Mol. Cell Biol.* 15: 6045–6054, 1995.

32. Fuks Z, Persaud RS, Alfieri A, McLoughlin M, Ehleiter D, Schwartz JL, Seddon AP, Cordon-Cardo C, and Haimovitz-Friedman A. Basic fibroblast growth factor protects endothelial cells against radiation-induced programmed cell death in vitro and in vivo. *Cancer Res.* 54: 2582–2590, 1994.

33. Haimovitz-Friedman A, Balaban N, McLoughlin M, Ehleiter D, Michaeli J, and Vlodavsky IZF. Protein kinase C mediates basic fibroblast growth factor protection of endothelial cells against radiation-induced apoptosis. *Cancer Res.* 54: 2591–2597, 1994.

34. Dragovich T, Rudin CM, and Thompson CB. Signal transduction pathways that regulate cell survival and cell death. *Oncogene* 17: 3207–3213, 1998.

35. Lowe SW, Schmitt EM, Smith SW, Osborne BA, and Jacks T. p53 is required for radiation-induced apoptosis in mouse thymocytes. *Nature* 362: 847–849, 1993.

36. Potten CS and Grant HK. The relationship between ionizing radiation-induced apoptosis and stem cells in the small and large intestine. *Br. J. Cancer* 78: 993–1003, 1998.

37. Song S and Lambert PF. Different responses of epidermal and hair follicular cells to radiation correlate with distinct patterns of p53 and p21 induction. *Am. J. Pathol.* 155: 1121–1127, 1999.

38. Haks MC, Krimpenfort P, van den Brakel JH, and Kruisbeek AM. Pre-TCR signaling and inactivation of p53 induces crucial cell survival pathways in pre-T cells. *Immunity* 11: 91–101, 1999.

39. Guidos CJ, Williams CJ, Grandal I, Knowles G, Huang MT, and Danska JS. V(D)J recombination activates a p53-dependent DNA damage checkpoint in scid lymphocyte precursors. *Genes Dev.* 10: 2038–2054, 1996.

40. Kolesnick R. The therapeutic potential of modulating the ceramide/sphingomyelin pathway. *J. Clin. Invest.* 110: 3–8, 2002.

41. Hendry JH, Cai WB, Roberts SA, and Potten CS. p53 deficiency sensitizes clonogenic cells to irradiation in the large but not the small intestine. *Radiat. Res.* 148: 254–259, 1997.

42. Hasegawa M, Zhang Y, Niibe H, Terry NH, and Meistrich ML. Resistance of differentiating spermatogonia to radiation-induced apoptosis and loss in p53-deficient mice. *Radiat. Res.* 149: 263–270, 1998.

43. Tron VA, Trotter MJ, Ishikawa T, Ho VC, and Li G. p53-dependent regulation of nucleotide excision repair in murine epidermis in vivo. *J. Cutan. Med. Surg.* 3: 16–20, 1998.

44. Hendry JH, Adeeko A, Potten CS, and Morris ID. P53 deficiency produces fewer regenerating spermatogenic tubules after irradiation. *Int. J. Radiat. Biol.* 70: 677–682, 1996.

45. Paris F, Fuks Z, Kang A, Capodieci P, Juan G, Ehleiter D, Haimovitz-Friedman A, Cordon-Cardo C, and Kolesnick R. Endothelial apoptosis as the primary lesion initiating intestinal radiation damage in mice. *Science* 293: 293–297, 2001.

46. Pena LA, Fuks Z, and Kolesnick RN. Radiation-induced apoptosis of endothelial cells in the murine central nervous system: Protection by fibroblast growth factor and sphingomyelinase deficiency. *Cancer Res.* 60: 321–327, 2000.

47. Morita Y, Perez GI, Paris F, Miranda SR, Ehleiter D, Haimovitz-Friedman A, Fuks Z, Xie Z, Reed JC, Schuchman EH, Kolesnick RN, and Tilly JL. Oocyte apoptosis is suppressed by disruption of the acid sphingomyelinase gene or by sphingosine-1-phosphate therapy. *Nat. Med.* 6: 1109–1114, 2000.

48. Meyn RE, Stephens LC, Ang KK, Hunter NR, Brock WA, Milas L, and Peters LJ. Heterogeneity in the development of apoptosis in irradiated murine tumours of different histologies. *Int. J. Radiat. Biol.* 64: 583–591, 1993.

49. Stephens LC, Ang KK, Schultheiss TE, Milas L, and Meyn RE. Apoptosis in irradiated murine tumors. *Radiat Res.* 127: 308–316, 1991.

50. Meyn RE, Stephens LC, Hunter NR, Ang KK, and Milas L. Reemergence of apoptotic cells between fractionated doses in irradiated murine tumors. *Int. J. Radiat. Oncol. Biol. Phys.* 30: 619–624, 1994.

51. Sentman CL, Shutter JR, Hockenbery D, Kanagawa O, and Korsmeyer SJ. Bcl-2 inhibits multiple forms of apoptosis but not negative selection in thymocytes. *Cell* 67: 879–888, 1991.

52. Lowe SW, Ruley HE, Jacks T, and Housman DE. p53 dependent apoptosis modulates the cytotoxicity of anticancer agents. *Cell* 74: 957–967, 1993.

53. Strasser A, Harris AW, Jacks T, and Cory S. DNA damage can induce apoptosis in proliferating lymphoid cells via p53-independent mechanisms inhibitable by Bcl-2. *Cell* 79: 329–339, 1994.

54. Brown JM and Attardi LD. The role of apoptosis in cancer development and treatment response. *Nat. Rev. Cancer* 5: 231–237, 2005.

55. Brown JM and Wilson G. Apoptosis genes and resistance to cancer therapy: What does the experimental and clinical data tell us? *Cancer Biol. Ther.* 2: 477–490, 2003.

13

Early and Late Responding Genes Induced by Ionizing Radiation

13.1 Gene Expression Is Induced by Ionizing Radiation

The observation that radiation treatment caused the induction of stress response genes in bacteria and yeast provided motivation for the early studies of changes in gene expression in irradiated mammalian cells. The demonstration that radiation induces transcription of the tumor necrosis factor-α (TNF-α) gene was the first evidence for activation of mammalian genes in the response to radiation, and it was subsequently found that radiation treatment is associated with induction of a number of early and late responding genes. These findings demonstrated that ionizing radiation activates nuclear signaling cascades and supported the notion that radiation-inducible genes contribute to the regulation of events such as DNA repair, mutagenesis, apoptosis, tissue repopulation, and the late sequelae of radiation damage, and that the effects of radiation on tumor and normal tissues are mediated in part by autocrine and paracrine responses to a number of secreted factors including TNF-α.

As has been described in several of the preceding chapters, ionizing radiation initiates complex intracellular signaling pathways that activate cell-cycle checkpoints, DNA repair, and apoptosis. These pathways include the activation of the plasma membrane receptors, ceramide synthesis, modulation of protein kinase cascades, and regulation of transcriptional pathways. This chapter focuses on transcriptional factors induced by ionizing radiation, on the late responding genes that are activated by these factors, and on the gene products that are of extreme importance in shaping the radiation response of the organism. The term "activation" will refer to the conversion of an inert protein to an active form that can stimulate transcription of the gene. Expression of a new transcript or protein as a result of the activation of these transactivation factors is then the product of an induced gene.

Much of the information about radiation-induced genes and the impact these genes produce on the radiation response of the organism was gathered before the advent of all-inclusive genetic screening methods such as cDNA microarray technology. The application of these methods has demonstrated that many hundreds of genes are regulated following radiation exposure with the responses being time, dose, and cell-type specific and have also provided support for the conclusions drawn from earlier studies.

13.1.1 Transcription Factors

The earliest nuclear targets, which are induced in the absence of de novo protein synthesis, are referred to as the immediate-early genes. A subset of these genes encodes transcription factors that constitute the first step in a cascade of protein–DNA interactions. The human genome contains more than 2000 hypothetical genes encoding putative transcriptional regulators serving as nuclear couplers of cytoplasmic events to the longer-term cellular response.

Transcription of genes is catalyzed by the enzyme RNA polymerase II and other supporting molecules that are collectively termed transcription factors. (The process of transcription and the involvement of general transcription factors are described in Chapter 3.) These proteins influence the regulation of genes by binding to specific DNA recognition sequences, which are found in the promoter regions, usually at the start of genes, but occasionally in introns (for example in the PCNA gene). A schematic of gene structure is shown in Figure 13.1. The formation of RNA transcripts is influenced by the interaction of these gene-specific factors with elements of a common group of molecules regulating the activity of RNA polymerase II. The activity of transcription factors can be modified, usually by phosphorylation, through the activity of a multitude of signaling pathways, some of which have already been described.

Based on the structure of their DNA-binding domains, transcription factors can be placed into helix-turn-helix (HTH), helix-loop-helix (HLH), zinc-finger, or leucine-zipper groupings (Figure 13.2).

- HTH transcription factors are made up structurally of three α-helices. Two helices mediate interaction with signaling molecules

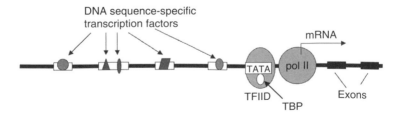

FIGURE 13.1
Transcription of the gene into RNA begins at the binding site of RNA polymerase II (pol II). The basal promotor contains the TATA box bound by a complex of approximately 50 proteins including transcription factor IID (TFIID) and the TATA-binding protein (TBP). Other transcription factors bind upstream of the basal promotor site.

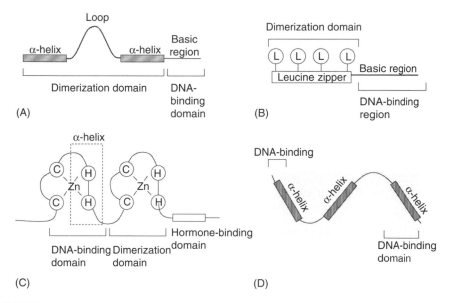

FIGURE 13.2
Schematic of the structure of each of the main classes of transcription factor. (A) HLH; (B) Leucine Zipper; (C) Zinc finger; (D) HTH.

or other transcription elements, while the third interacts with specific sequences in target DNA.

- HLH transcription factors, such as the product of the c-*myc* proto-oncogene, contain two α-helices separated by a short DNA-binding peptide loop and become activated after heterodimerization.

- Zinc-finger transcription factors contain a sequence of 20–30 amino acids having two paired cysteine or histidine residues that are coordinated by a zinc ion. The sequence between the paired cysteines protrudes as a finger, giving these transcription factors their name. DNA-binding specificity is provided by the sequence at the base of the loop.

- Leucine-zipper transcription factors contain helical regions with leucine residues occurring at every seventh amino acid, all protruding on the same side of the α-helix. These leucines interact hydrophobically with leucines of similar proteins in an antiparallel fashion. These factors, which exist as homo- or heterodimers, include the fos/jun pair (AP-1 transcription factor), which becomes activated by cellular stress.

A relatively short list of transcription factors are reported to be activated by radiation [1]. These include p53, NF-κB, the retinoblastoma control proteins (RCPs), the early growth response 1 transcription factor (Egr-1), the c-Fos/c-Jun AP-1 complex, and the octamer-binding protein (Oct-1) (Table 13.1).

TABLE 13.1

Radiation-Activated Transcription Factors

Transcription Factor	Cell Type	Dose Range (Gy)	Target Genes
p53	Epithelial and lymphocytic cancer cells	0.2–20	>100 known genes, growth arrest and apoptosis
NF-κB	Epithelial and lymphocytic cancer cells	0.5–20	Inflammatory cytokines ± regulation of radiosensitivity
AP-1	Epithelial and lymphocytic cancer cells, keratinocytes	2–20	Cytokines, in collaboration with other TFs
Egr-1	Normal and cancer lymphocytic, some epithelial cancer cells	4–20	Growth factors (bFGF, PDGF, Cytokines [TNF-α, IL-1])
SP1(RCPs)	Melanoma, head and neck SCC, cerebral cortex (rat)	0.1–10	

13.1.2   Important Transcription Factors Activated by Radiation

This part of the chapter is partially based on reviews by Criswell et al. [1], Chastel et al. [2], and the articles referenced therein.

13.1.2.1   p53

The signal upstream of p53 stabilization and activation after exposure to ionizing radiation is believed to originate from DNA DSBs (Chapter 8). In response to this DNA damage, p53 is rapidly stabilized through posttranslational modifications that include phosphorylation, acetylation, glycosylation, and ribosylation. Most important in terms of the response to ionizing radiation is the activation (stabilization) of p53 mediated through phosphorylation by the ATM kinase.

p53 plays a critical role in maintaining genomic integrity after cellular stress by acting as a transcription factor for a number of downstream genes that may mediate in growth arrest, DNA repair, apoptosis, and the inhibition of angiogenesis. The response to p53 stabilization is cell-type dependent, as well as DNA-damage dependent. The p53 protein consists of four major domains, a transactivation domain at the N-terminus, a proline-rich domain, a DNA-binding domain, and a carboxy-terminal tetramerization domain. The transactivation domain contains the Mdm2 binding site as well as binding sites for members of the transcription machinery. The proline-rich domain is thought to be important for effective growth arrest after DNA damage. p53 sequence-specific DNA binding occurs through the core DNA-binding domain. After stress-induced stabilization, p53 forms homotetramers that bind to two copies of a 10 base pair nucleotide sequence divided by a 13 base pair spacer. The importance of p53 sequence-specific DNA binding is emphasized by the fact that most mutations found in human tumors within the p53 gene occur at "hot spots" within this DNA-binding domain. These mutations result in a dominant-negative form of the protein that can heterodimerize with wild-type p53 protein and disrupt its DNA binding, thus inhibiting downstream gene activation. The C-terminal tetra-merization domain, as its name implies, is responsible for tetramer formation, as well as for regulation of DNA binding by the core domain.

Signaling between radiation-induced DNA damage and the downstream effectors of cell-cycle arrest, DNA repair, and apoptosis in which p53 plays a pivotal role has been described in detail in Chapter 8.

13.1.2.2 Nuclear Factor Kappa B

The nuclear factor kappa B (NF-κB) pathway is generally thought to be the primary oxidative stress-response pathway in which reactive oxygen species (ROS) act as second messengers. It acts as a key regulator of numerous genes involved in inflammatory processes, stress response, and programmed cell death. The NF-κB transcription factor consists of two subunits, NF-κB$_1$ (p50) and RelA (p65). NF-κB is sequestered in the cytoplasm in association with its inhibitor IκBα in the form of an inactive complex. Exposure to a variety of stimuli results in the activation of IκB kinases (IKKα, β, and γ) and phosphory-lation of IκBα on Ser32 and Ser36. The phosphorylated form of IκBα is ubiqui-tinated and degraded through the ubiquitin-proteasome system unmasking the nuclear localization sequence on NF-κB, which then translocates into the nucleus, where it binds to the consensus sequence 5'-GGGACTTTCC-3' in the promoters of its target genes (Figure 13.3).

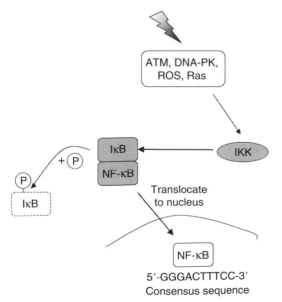

FIGURE 13.3

Molecular events in the activation of NF-κB by ionizing radiation.

Radiation activates NF-κB within 2–4 h of exposure, and activity returns to basal levels by 8 h [3,4]. Ionizing radiation has been shown to activate IKK through a mechanism involving degradation of radiation-induced (phosphorylated) IκBα. Inhibition of IκBα phosphorylation on Ser32 and Ser36 was shown to sensitize human glioma cells to DNA damaging agents. This is consistent with activation of NF-κB by oxidative stress and radiation proceeding through a common cascade involving ROS generation, but the exact mechanism that activates NF-κB is not completely understood. It has been reported that the regulation of NF-κB is impaired in cells with the human genetic disease ataxia telangiectasia (AT) and that the ataxia-telangiectasia mutated (ATM) kinase phosphorylates IκBα in vitro, suggesting a role for ATM in the induction of NF-κB by radiation [5]. The role of ROS in the activation process is supported by fact that antioxidants (e.g., *N*-acetylcysteine [NAC]) have been consistently shown to prevent NF-κB activation by radiation for a number of cell lines and a wide range of doses.

An important role of NF-κB in biological response to radiation is the modulation of radiosensitivity. Activation of NF-κB by ionizing radiation has been observed in a number of different cell systems at clinically relevant doses. Indication that NF-κB activation causes radioresistance comes from experiments where the inhibition of IκBα phosphorylation sensitized human glioma cells to DNA damaging agents [6]. Similarly, blocking radiation-induced NF-κB activation in irradiated colorectal cancer cells increased apoptotic response and decreased growth and clonogenic survival [7]. Other investigators have reported alteration of NF-κB activation to have no effect on radiation response. Although its role in radiosensitization appears to be dependent on the cancer cell line, there is a consensus regarding the involvement of NF-κB in mediating inflammatory responses to radiation exposures, and for this reason NF-κB is considered to be a promising target for modulation of response to radiotherapy.

Results of experiments in the human glioblastoma U251 and SH539 cell lines [4] indicated that the radiation-sensitive pool of IκBα/NF-κB is located at the plasma membrane. It was proposed that activated Ras protein recruits IκBα/NF-κB complexes to the cell membrane, leading to the selective degradation of IκBα and the translocation of NF-κB into the nucleus. In this way radiation-induced IκBα degradation is spatially segregated from that induced by cytokines.

13.1.2.3 Activating Protein-1 Transcription Factor

The activating protein-1 (AP-1) transcription factor consists of a complex of proteins from the c-Jun and c-Fos families, which form homo- or hetero-dimers binding to AP-1 regulatory elements in the promoter and enhancer regions of numerous genes. The components of the functional AP-1 transcription activator complex can be c-Jun, Jun-B, Jun-D, c-Fos, Fos-B, and other proteins of the Fos family (Figure 13.4). Jun proteins can form homo-dimers or heterodimers with Fos proteins, whereas Fos does not form

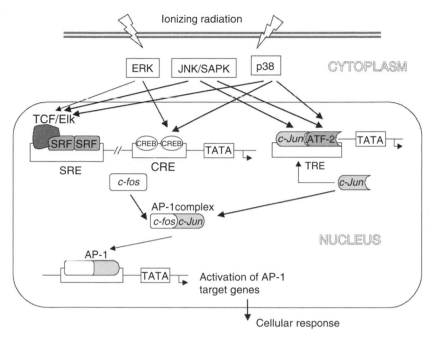

FIGURE 13.4

Regulation of immediate early genes *c-Fos* and *c-Jun* by ionizing radiation via MAP kinase pathway activation. Radiation induces phosphorylation of specific MAP kinases, the activities of which converge on transcription factors that activate the expression of immediate early genes *c-Fos* and *c-Jun*. Fos and Jun proteins dimerize to form a complex named AP-1, which activates the expression of target genes involved in radiation response. Other proteins, such as JunB, JunD, or members of the ATF subfamily, can also heterodimerize to form AP-1 complexes. ATF-2, activating transcription factor-2; SRF, serum response factor; SRE, serum response element; CREB, cAMP responsive element binding protein; CRE, cAMP responsive element; TRE, TPA responsive element; TATA, TATA box; AP-1, activating protein-1.

homodimers and requires heterodimerization to bind to DNA. A common feature of proteins that form the AP-1 transcription factor is the presence of a leucine zipper that consists of heptad repeats of leucine residues aligned along one face of an α-helix (Figure 13.2). Basic regions adjacent to the leucine zippers serve as the DNA-binding domain of the AP-1 factor and the subunit composition determines the affinity of the AP-1 factor for the respective DNA sequence. Differential expression of AP-1 proteins in response to extracellular stimuli is suggested to be the major mechanism that modulates AP-1 activity.

c-Jun and a number of Jun family members were isolated from mammalian cells based on their abilities to homo- or heterodimerize with themselves, with Fos family members and with related proteins. *c-Jun* is considered to be a proto-oncogene, because its ectopic kinases (described in Chapter 11) act to enhance the transactivation function of c-Jun and ATF2 over-expression can lead to cellular transformation and fibrosarcoma development. The members of the *Jun* gene family are *c-Jun*, *JunB*, and *JunD*. Intracellular signaling through stress-activated protein kinase cascades results in phosphorylation of c-Jun and related ATF (activating transcription factor-like) proteins on serines 63 and 73 of c-Jun and threonines 69 and 71 of ATF2. Oxidative damage stimulates the JNK/SAPK family. In response to the same stimulus, ATF2 can be a substrate of either the p38 MAPK cascade or the JNK/MAPK cascade depending on the cell line.

Some studies have suggested that induction of c-Jun was dependent on PKC activation. This conclusion is based on evidence that depletion of PKC by pretreatment with TPA inhibited post-IR induction of the transcription of AP-1 components c-Jun and c-Fos [8].

c-Fos is a nuclear phosphoprotein of about 55 kDa molecular weight. Fos family members are expressed in most cells and *c-Fos* is considered to be a proto-oncogene because over-expression of the gene can result in cellular transformation and leads to development of osteosarcomas. c-Fos forms exclusively heterodimers with its binding partners, members of the Jun family, and is unable to form homodimers. As in the case of c-Jun, c-Fos transcriptional activity is regulated by phosphorylation.

c-Fos is inducible at the level of transcription by a large number of agents including growth factors, lipopolysaccharides (LPS), cytokines, and UV and ionizing radiation. Mutational analysis identified several regions in the human *c-Fos* promoter that responded to extracellular signals in transient transfection systems or in in vitro studies. These key sequences are located within a 400 base pair (bp) segment directly upstream from the transcription initiation site. In this region, a combination of four regulatory elements, the sis-inducible element (SIE), the serum response element (SRE), the AP-1 site (FAP), and the cAMP response element (CRE), have been shown to be necessary for the correct transcriptional response of the promoter to any of the inducers investigated. The SRE, also known as the cArG box, is located ∼300 bp upstream from the transcription start site and acts as a binding site for the transcription factor Elk-1. A prerequisite for Elk-1 binding to the SRE of *c-Fos* is interaction with a serum response factor (SRF) dimer. Differentially regulated, but partly overlapping MAPK cascades converge to phosphorylate Elk-1 on phosphorylation sites located in the transcriptional activation domain of the protein leading to enhanced DNA binding of Elk-1 and to more effective transactivation of the target genes.

The CRE consensus sequence is located downstream from the cArG box at position-63 from the transcription start site of the *c-Fos* gene. The CRE controls the basal level of transcription as well as the response of the promoter to cAMP and calcium. Interaction of this site with the CRE binding proteins (CREBs) also enhances the transactivation function of the complex.

The AP-1 complex is activated by many effectors, including growth factors, cytokines, hormones and mechanical and osmotic stress, hypoxia, UV, and oxidative stress. Radiation-induced ROS activate AP-1 through the JNK pathway, modifying the c-JUN and ATF-2 constituents of the complex and promoting the induction of the transcription factor ATF-3. Because a number of AP-1 binding sites are located in the promoter region of the c-Jun gene itself, the phosphorylation and the subsequent activation of c-Jun result in its auto-activation. This regulation of c-JUN activity leads to long-term activation of AP-1 and to a steady increase in the expression levels of its target genes under oxidative stress conditions.

Among the genes that have been clearly shown to be activated by AP-1 after IR are various cytokines, including transforming growth factor β1 (TGF-β1), vascular endothelial growth factor (VEGF), and Interleukin 6 (IL-6). However, AP-1 activation by itself is not sufficient for the induction of TGF-β1 and in fact none of the cytokine genes have been shown to be induced by the action of AP-1 alone. In the case of IL-6, it was shown that AP-1 acts synergistically with NF-κB to induce IL-6 transcription.

13.1.2.4   Early Growth Response Factor Egr-1

Early studies of radiation-induced gene expression showed an increase in *Egr-1* mRNA after exposure of cells and it was subsequently confirmed that the increase in mRNA was mediated largely by a transcriptional mechanism. *Egr-1* encodes a nuclear phosphoprotein that interacts with DNA to regulate transcriptional activity of other genes and a number of investigators have studied the control of *Egr-1* as a model for gene regulation by ionizing radiation.

The *Egr-1* gene product is unique among transcription factors in that both the transcription factor and the gene transcript itself are activated and induced respectively by supra-lethal doses of radiation. *Egr-1* encodes a transcription factor containing a DNA-binding domain formed by three zinc-finger motifs. Egr-1 preferentially binds to the GC-rich consensus sequence GCGGGGGCG. This sequence is found in the regulatory regions of many genes including *TNF-α*, *IL-1*, platelet-derived growth factor- (*PDGF-*), and basic fibroblast growth factor (*bFGF*).

Transcription of the *Egr-1* gene is rapidly and transiently induced by radiation in a variety of cell types. The response occurs within 15 min and lasts for at most 3 h after exposure [8]. The promoter sequence of the human *Egr-1* contains at least five copies of the serum response element (SRE) also known as the cARG box. This is a 22 bp segment that contains the inner core sequence CCATATTAGG. Datta et al. [9] found that radiation activates CArG boxes in the *Egr-1* promoter and that the three distal SRE elements are responsible for its radio-inducibility and thereby for *Egr-1* gene transcription. As is the case for the induction of the *c-fos* gene by extracellular stimuli, SRF and the ternary complex factors Elk-1, Sap1, or Sap2 are required for SRE-mediated activation of the *Egr-1* promotor. Elk1 is phosphorylated by MAP kinases, which leads to its enhanced DNA-binding activity and ternary complex forming ability and thus to enhanced SRE-mediated transcription. In response to various stressors, including ionizing radiation, the stress kinase pathways p38 and JNK/SAPK become activated by phosphorylation and induce *Egr-1* promotor activity through modification of Elk-1 [10]. It is likely that radiation modulates the binding of the phosphorylated activated complex SRF/Elk-1 to the SRE element in the Egr-1 promotor through activation of the p38 and JNK/SAPK pathways and that this ultimately leads to the transactivation of the *Egr-1* gene [11].

Another intracellular pathway, this time activated by PKC, appears to contribute to the regulation of the *Egr-1* gene by radiation. Prolonged stimulation with TPA, known to deplete PKC, was shown to block x-ray inducibility of *Egr-1* in a squamous cell carcinoma cell line, and in the same cell line pretreatment with a nonspecific inhibitor of protein kinases, markedly attenuated x-ray inducibility of *Egr-1* [8]. It is thus likely that activation of PKC contributes to the phosphorylation of Elk-1 and therefore to the induction of *Egr-1* expression in response to radiation.

13.1.2.5   SP Family of Transcription Factors

SP1 was the first identified member of a family of transcription factors composed of six different proteins that bind to similar sequences in the promoter regions of human genes. SP1 is ubiquitously expressed in mammalian cells and binds with high affinity to GC-rich sequences called GC-boxes, and with a lower affinity to CACC-elements called GT-boxes. SP1 is thought to play a role in various cellular processes, including cell-cycle regulation and chromatin remodeling. The importance of SP1-mediated transcription is underscored by the fact that SP1-null mice are embryonic lethal by day 10.

SP1 and SP3 are retinoblastoma (Rb) control proteins (RCPs) that bind to promoter regions, known as retinoblastoma control elements (RCEs). RCPs are regulated by the Rb protein and control induction of several immediate early growth-response genes (IEGs) after cell stress (i.e., c-Fos, c-Jun, c-Myc, and TGF-1). There have been several reports of the inducibility of SP1 DNA binding after high doses of IR (>4 Gy) and a preliminary report of changes in SP1 posttranslational modification after as little as 2 cGy IR [1].

13.1.3   Radiation Gene Therapy

The realization that gene expression could be controlled by ionizing radiation prompted an approach to anticancer therapy based on the hypothesis that external beam radiation could be used to localize the transcription of a deleterious or lethal gene in the target cell. A therapeutic strategy by which gene therapy could be targeted to the tumor by radiotherapy involved inserting a cDNA encoding a toxic or an immune-modulator protein downstream from a radiation-inducible promoter [12]. These studies have focused on the use of the radiation-inducible SRE (cArG) elements present in the promoter of the human *Egr-1* gene. A construct was generated,

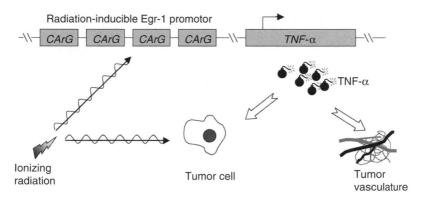

FIGURE 13.5

The Egr-1 promoter is upstream of cDNA encoding TNF-α. Ionizing radiation activates the CArG elements in the EGR1 promoter, leading to increased expression of the *TNF-α* gene in the tumor bed. The combination of greater TNF-α expression and ionizing radiation results in increased tumor-cell apoptosis and vascular destruction and improved antitumor activity.

consisting of a 425 bp region containing six SRE elements upstream of the transcription start site of the EGR-l gene, which was ligated to a cDNA encoding TNF-α. This construct was cloned into a replication-deficient adenoviral vector known as Ad.Egr.TNF, to increase the expression of TNF-α in irradiated tumors (Figure 13.5). (Adenoviruses have been widely used in human gene therapy; they show a high efficiency of infection and transfection of target cells, and little toxicity except at very high viral concentrations.)

Several factors recommend the use of TNF-α as an anticancer agent. TNF-α is an endogenous protein and hence unlikely to induce an immune response; the protein is secreted and diffuses from the transduced cells to neighboring cells, so transduction of all the tumor cells is not necessary; the cytokine is cytotoxic and radiosensitizing to tumor cells and at high concentrations, selectively destroys the tumor vasculature. In one experiment, mice xenografted with the radioresistant human epidermoid carcinoma SQ-20B cells were treated with Ad.Egr.TNF-α and radiation. Combined treatment resulted in greater tumor regression than occurred when Ad.Egr.TNF-α or radiation was used alone. Quantification of TNF-α levels within tumor xenografts demonstrated a seven- to eightfold induction of TNF-α within the irradiated field compared with the nonirradiated samples. No TNF-α was detectable in the circulation of experimental animals and no systemic cytotoxicity was observed. Similar experiments were performed on human prostate and glioma xenografts, which are partially resistant to the cytotoxic effect of TNF-α [13], and enhanced antitumor effects were obtained in both tumor types by the use of combined treatments.

Other experiments have been conducted using radiation-inducible "suicide-gene" therapy. One example of a radio-inducible suicide gene can be constructed by insertion of the *Egr-1* gene upstream of the thymidine kinase (*HSV-tk*) gene. The HSV-tk protein converts the nontoxic pro-drug ganciclovir (GCV) to a highly toxic phosphorylated GCV that acts as chain terminator of DNA synthesis and an inhibitor of DNA polymerase. Advantages of the HSV-tk system include both lower toxicity for normal cells and the action of a bystander effect by which *HSV-tk* negative cells may be killed by toxic phosphorylated GCV produced by their neighboring *HSV-tk* positive cells exposed to GVC [14]. The chimeric construct pEgr-TK, in combination with GCV, was found to be effective with radiation in the treatment of brain and liver tumors in mice [15].

It was later shown that multiple SRE elements gave greater inducibility by serum and specific growth factors than did a single SRE unit [16], and synthetic promoters composed of multiple SRE elements were developed on this basis [17]. A synthetic promoter containing four SRE elements was shown to be responsive to radiation doses as low as 1 Gy and was more radiation responsive than the wild-type *Egr-1* promoter at an optimal 3 Gy dose. Results of animal model studies showed that synthetic SRE promoters also function in vivo and significantly amplify tumor growth control when combined with radiation [18].

Promotors other than Egr-1 have been investigated for radio-inducibility. The p21/WAF1 promotor was inserted upstream to sequences encoding inducible nitric oxide synthase (iNOS, see Chapter 5) [19]. iNOS expression was induced fourfold after transfection of the construct into vascular cells and treatment with 4.0 Gy. In this study, iNOS was being induced as a mechanism to regulate vascular tone, an example of how genetic radiotherapy could be used in settings other than the treatment of cancer.

In human solid tumors, the presence of hypoxic regions is one of the most significant factors predictive of a lack of response to radiotherapy. Several groups of investigators have developed a gene-therapy approach based on the use of promoter elements responsive to tissue hypoxia. (Hypoxia-inducible transcription factors such as HIF-1α are discussed in detail in Chapter 16.) A combination approach exploited the use of chimeric promoters, containing radiation-responsive SRE elements in combination with hypoxia-responsive elements (HREs) [20]. This strategy restricts the activation of the therapeutic gene to hypoxic or irradiated tissues and ensures the selective expression of only the tumor-located vectors. These dual enhancer constructs functioned in response to radiation or hypoxia alone or to both stimuli. Results of in vivo studies [21] demonstrated that regional tumor hypoxia could be exploited to improve local tumor control. In this case, cDNA encoding the erythropoietin hypoxia-responsive element (Epo) was placed upstream from the Ad.Egr.TNF-α construct. Treatment of xenografted colon adenocarcinoma with combined regimen of Epo-Ad.Egr.TNF-α plasmid and radiation resulted in significant tumor growth delay. Tumor TNF-α content was increased by 30% in the combined treatment group compared with either treatment alone (Figure 13.6).

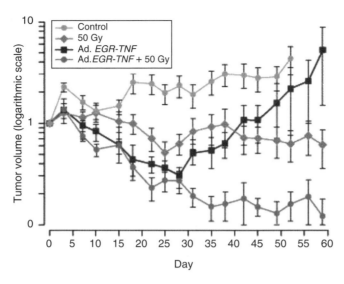

FIGURE 13.6

Tumour regression after treatment with Ad. EGR-TNF and ionizing radiation. SQ-20B xeno-grafts growing in nude mice were injected with 2×10^8 plaque-forming units of Ad.EGR-TNF (twice weekly for 2 weeks) and irradiated with 50 Gy (5 Gy daily, 4 days per week, for 10 fractions). Mean tumor volume was reduced to 16% of original volume at day 38 ($p < 0.05$) with no tumor regrowth. Tumors treated with Ad.EGR-TNF plus ionizing radiation shrank to approximately 10% of original volume and did not regrow for up to 60 days after the start of treatment. (From Weichselbaum, R.R., Kufe, D.W., Hellman, S., Rasmussen, H.S., King, C.R., Fischer, P.H., and Mauceri, H.J., *Lancet Oncol.*, 3, 665, 2002. With permission.)

13.1.3.1 Clinical Trials of Radiation-Targeted Gene Therapy

Clinical trials of radiation-targeted gene therapy have been reviewed by Kufe and Weichselbaum [22] and by Mezhir et al. [23]. The strong scientific rationale and encouraging experimental data motivated clinical testing of radio-genetic therapy protocols. Clinical trials were initiated with Ad. EGR-TNF in combination with ionizing radiation. Ad. EGR-TNF (TNFerade) is a second-generation adenoviral vector incorporating the TNF-α gene under the control of the radiation-responsive promoter Egr-1. In a phase I/II trial, TNFerade was administered intratumorally in combination with single, daily fractionated radiation therapy to patients with soft tissue sarcoma of the extremities. TNFerade was well tolerated with no dose-limiting toxicities noted. Eleven patients (85%) showed objective or pathological tumor responses and partial responses were achieved despite some of the tumors being very large. It was concluded that TNFerade plus radiation therapy was well tolerated in the treatment of patients with soft-tissue sarcoma of the extremity [22].

Another human study of TNFerade plus radiation was a phase I trial during which TNFerade was administered intratumorally at weekly intervals for 6 weeks with concomitant radiation (30–70 Gy). Thirty-six patients were assessable for toxicity and 30 for tumor response. No dose limiting toxicities were observed. Seventy percent of patients showed objective tumor response with five complete responses. In cases where direct comparison could be made, a differential response between tumors treated with TNFerade plus radiation and radiation alone was seen. On the basis of these findings other controlled prospective trials were planned [24].

13.2   Early and Late Response Genes

Genes can be categorized according to the level of expression that results from a particular stimulus. Constitutive genes are constantly expressed and have little inducibility. An example of a constitutively expressed gene is the Ku80 gene that is essential for DNA double-strand break repair. Genes that are inducible include those which have little or no constitutive expression, but their transcription is initiated in response to a stimulus. In some cases, products of inducible genes can be deleterious to the survival of the organism if they are continuously expressed, and it is a survival advantage that these genes are induced only when required. Examples of inducible genes are those that regulate apoptosis, cell proliferation, and genes whose products are involved in the inflammatory response. In some cases, gene expression may be reduced in response to a stimulus, an example being the Cyclin B gene, the expression of which is decreased during G_2 arrest.

Inducible genes can be divided further into genes that are immediately inducible, referred to as immediate early genes. The transcription of immediate early genes requires no de novo protein synthesis. Examples that have been just described are the radiation-mediated immediate early genes *c-jun* and *Egr-1*. Gene expression of immediate early genes is regulated either by the acceleration of transcription or after transcription has occurred by posttranscriptional regulation, which occurs when the half-life of the mRNA or the encoded protein is prolonged.

The products of radiation-induced immediate early response genes participate in subsequent events by binding to specific promoter elements of other genes referred, known as late responding genes since their expression is delayed and requires the prior activation of immediate early genes before they are induced (Figure 13.7).

13.2.1   Induction of Late Response Genes by Ionizing Radiation

The principle mediators of the effects of late responding genes and of radiation effects on normal tissue are cytokines which play a role in the inflammation process and growth factors that are involved in the response of tissue to radiation-induced damage. Cytokines are proteins that control cell proliferation differentiation and the function of cells of the immune system. Unlike hormones, cytokines are not stored in glands as preformed

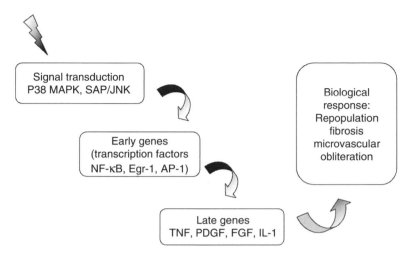

FIGURE 13.7
Sequence of events between radiation exposure and expression of late phenotypic events.

molecules, but are rapidly synthesized and secreted by different cells mostly after stimulation. Cytokines are pleiotropic in their biological activities and play pivotal roles in a variety of responses, including the immune response, hematopoiesis, neurogenesis, embryogenesis, and oncogenesis. Their role in these processes can be diverse; in some instances cytokines are pathogenic and in others protective.

Stress and genotoxic agents including radiation induce production of a number of cytokines. Cytokines in turn induce a cascade of additional cytokines, which may modulate the initial signal and up- or down-regulate cellular receptors and adhesion molecules on cell membranes. In this way, cytokines orchestrate cellular and tissue response to damage, promote damage repair, and participate in the elimination of damaged cells.

Cytokines that have been best documented as being produced in response to radiation are proinflammatory cytokines such as TNF-α; the interleukins, IL-1, IL-6, and IL-8; and the antiinflammatory IL-10, which have been demonstrated in in vitro and in vivo studies in animal models and in patients. For example, one of the targets of early response gene products is IL-6. IL-6 is a cytokine that regulates immune cell function and it can generate an anti-apoptotic signal, which protects cells from the killing effects of radiation. Radiation-induced expression of IL-6 depends upon prior activation of two of the transcription factors described above, NF-κB and AP-1. These cytokines can signal the radiation injury to the neighboring cells and in certain cases amplify the radiation damage (Figure 13.7).

Sources of cytokine secretion in response to radiation include epithelial cells, fibroblasts, and macrophages. Macrophages, derived from hemopoietic progenitors, are long-lived cells found in all organs of the body. They act as scavengers, engulfing large particles by a process called phagocytosis.

Macrophages also secrete cytokines and chemokines that orchestrate the immune response and play a central role in acute and chronic inflammation by secreting and responding to a wide range of inflammatory mediators. Macrophages mainly produce TGF-β, but can also produce PDGF, bFGF, and IL-1 cytokines, which are the main factors responsible for late irreversible radiation effects such as fibrosis. Cytokines produced after radiation exposure by tissue-specific macrophages in the lung, for instance contribute to the fibrosis in this tissue. However, the late response of the lung and other organs is not solely attributable to circulating macrophages; other cells including fibroblasts, epithelial cells, and pneumocytes are involved [25]. Table 13.2 lists some cytokines that are modulated after radiation in different cell types from in vivo and in vitro studies. Some of the characteristics of the major players are described below.

TABLE 13.2

Changes in Three Cytokines after Exposure to Ionizing Radiation in Different Cellular Models

Protein	Species	Cell Type	Dose (Gy)	Time after Radiation (h)	Conditions	Response
TNF-α	Human	PB monocytes	20	1.0	in vitro	Up [60]
		PBL	4	0.5		Up [61]
		PBL	5, 10, 15, 30	48		Down [62]
		HL-60 myeloid	2–50	1–3		Up [60]
		U-937 monocytic	20	1–3		Up [60]
		Sarcoma	5	3–6		Up [63]
		Ewing tumor	5–10	4–72		Up [64]
	Rat	Ileal muscularis	10	3–6	in vivo	Up [41]
		Hypothalamus	15	6		Up [65]
IL1β	Human	PBL	2	24	in vitro	Up [57]
		PBL	5, 10, 15, 30	48		Down [62]
		Ewing tumor	5–10	4–72		Up [64]
		Peritoneal Macrophages	0.04	4	in vivo in vitro	Up [66]
	Rat	Ileal muscularis	10	3–6	in vivo	Up [41]
		Hypothalamus	15	6		Up [65]
		Sertoli cells	6–21	6–24	in vitro	Up [67]
		Astrocytes	6	2–8	in vitro	Up [67]
			21	2–48		
TGFβ1	Mouse	Mammary	5	1	in vivo	Up [68]
		Endothelial	0.3–0.7	24	in vitro	Up [68]
	Rat (heart)	Fibroblasts	2	4–48		Up [69,70]
		Myocytes	8.5	4–48		Down [69]
		Myocytes	15	4–48		Up [69]
		Endothelial	2	4–48	in vitro	Up [69]

Note: PBL, peripheral blood lymphocytes.

13.2.1.1   Transforming Growth Factor-β1

Transforming growth factor-β1 is a ubiquitously expressed homodimeric protein. A wide variety of cells possess TGF-β1 receptors and respond to this ligand. Following irradiation, regulation of the *TGF-β1* gene has been shown to be AP-1 dependent [26] and TGF-β1 protein levels were found to be increased in fibroblasts, keratinocytes, and endothelial cells, as well as in many other cell types, after radiation. In nontransformed cells, TGF-β1 causes growth arrest and differentiation, whereas in tumor cells, it has been shown to bring about either apoptosis or cytoprotection, the latter being mediated through MAPK signaling or the PI3 kinase survival pathways [27]. The biological activity of TGF-β1 is constrained by its secretion as a latent complex consisting of TGF-β covalently associated with its processed N-terminal prosegment, the latency associated peptide (LAP). Release of LAP is a prerequisite for TGF-β to bind to its cell surface receptors. Irradiation activates latent TGF-β in a dose responsive manner releasing TGF-β to initiate tissue response to damage via several physiological processes, particularly inflammation, and fibrosis [28].

13.2.1.2   Platelet-Derived Growth Factor

Platelet-derived growth factor (PDGF) is composed of two polypeptide chains, A and B, which form homo- or heterodimers. Only the dimeric forms of PDGF react with the receptor. Because the *PDGF*-promoter contains Egr-1- and AP1-like binding sequences, these two transcription factors could be involved in the induction of *PDGF* after irradiation. PDGF induced by radiation has been shown to be secreted from the intima of blood vessels and may serve as a paracrine factor regulating the proliferation of smooth muscle cells that has been observed in irradiated small arterioles in vivo [29].

13.2.1.3   Basic Fibroblast Growth Factor

Basic fibroblast growth factor (b(FGF)) was shown to protect bovine aortic endothelial cells from lethal effects of irradiation through a PKC-dependent mechanism. Experiments in human breast carcinoma cells suggest that *bFGF* expression and regulation are dependent on AP-1 activation. Radiation stimulates endothelial cells to synthesize and secrete bFGF, resulting in an autocrine stimulation of a repair process, which improves cell survival [30].

13.2.1.4   Tumor Necrosis Factor α

Tumor necrosis factor α has been reviewed by Weichselbaum et al. [31]. TNF-α is one of a member of a family of trimeric cytokines, which includes Fas ligand and the TNF-α-related apoptosis-inducing ligand or APO2 ligand. TNF-α is produced by various cell types in response to infection, oxidative stress, and endotoxin, and has an important role in antibacterial and viral immune responses. NF-κB was found to be partly responsible for transactivation of *TNF-α* in rat astrocytes upon radiation exposure [7].

In addition to the NF-κB binding site, a putative AP-1 binding sequence and an Egr-1-like binding sequence are also present in the human *TNF-α* promoter, and it is possible that they cooperate in transactivation of the gene. TNF-α mediates its activity through binding of two receptors TNF-R1 and TNF-R2. TNF-R1 mediates cell death by association with the intracellular adaptors TRADD and FADD and the subsequent activation of caspases-8 and -3 (see Chapter 12). Alternatively, TNF-R1 binding can mediate an anti-apoptotic response by activating downstream genes after association with the intracellular adaptor modulators RIP1 and TRAF2. TNF-R2 is thought to mediate an anti-inflammatory or anti-apoptotic response by association with the intracellular adaptor TRAF1, although other studies have indicated that under some circumstances TNF-R2 can also mediate apoptosis and inflammation.

TNF-α can stimulate or inhibit cell proliferation in vitro, depending on the cell type. It is cytostatic or cytolytic to several human and murine tumor cell lines, but does not inhibit growth of normal cells. TNF-α has various effects on endothelial cells including stimulation of procoagulant activity and cell-surface antigen expression and inhibition of cell growth indicating that TNF-α might inhibit angiogenesis in vivo. The cellular response to radiation-induced *TNF-α* is thus the result of the integration of a number of mutually opposed signals.

13.2.1.5   Radiation-Inducible Interleukins

Interleukins are a subset of cytokines whose name is derived from the fact that they are not only secreted by leukocytes but are also able to affect cellular responses of other leukocytes. Some, but by no means all, of the interleukins are radiation-inducible; an example being IL-1, which has been the subject of a number of studies. IL-1 is involved in the acute phase radiation response and it can bind to receptors on the vascular epithelium to mediate the adhesion of lymphocytes and their subsequent emigration from the circulation.

IL-6, another proinflammatory cytokine, is a target of early response gene products that are up-regulated by radiation. In HeLa cells radiation-induced expression of IL-6 depends on prior activation of two transcription factors described earlier, NF-κB and AP-1 [32]. IL-6 is a cytokine that regulates immune cell function and it can also generate an anti-apoptotic signal, which protects cells from the killing effects of radiation. This protective effect has been proposed to be mediated by the survival pathway involving the phosphatidyl inositol PI3-kinases.

13.2.2   Cytokine-Mediated Responses in Irradiated Tissues

Acute and subacute sequelae of radiation therapy have multiple causes. Cytokine induction by radiation and radiation response have been correlated for a number of tissues of which only the most important can be described here.

13.2.2.1 Brain

The pathogenesis of cerebral edema includes increased vascular permeability and inflammatory cell infiltration into the irradiated tissue. Inflammatory mediators that are induced by ionizing radiation include the cytokines TNF-α and IL-1α, which bind to their respective receptors on endothelial cells and leukocytes to mediate inflammation. Cell adhesion molecules, including E-selectin and the intracellular adhesion molecule-1 (ICAM-1) are induced by stimulation of vascular endothelium with TNF-α, or IL-1. Cell adhesion molecules (CAMs) are required for extravasation of leukocytes from the circulation. E-selectin is a proteoglycan that is expressed on the luminal surface of the endothelium to initiate inflammatory cell adhesion while ICAM-1 regulates extravasation of leukocytes and their migration into inflamed tissue [33].

Levels of both cytokine (TNF-α, IL-1β) adhesion molecules and, to a lesser extent, IL-1α messenger RNA were found to be increased in the brain after irradiation, regardless of whether the dose was delivered to the whole body or limited to the midbrain of mice [34]. Responses were radiation dose-dependent, but were not found below 7 Gy with the exception of ICAM-1, which was increased by doses as low as 2 Gy. The acute phase of response in the mouse brain was reported to involve elevation of TNF-α. Messenger RNA levels of this cytokine peak after 2–8 h and return to baseline after 24 h [35]. Also during the acute phase, radiation caused apoptosis of cells in the white matter areas and the putative stem cell areas of the brain [36]. After the acute phase, the TNF-α response in the mouse brain followed a cyclic pattern during the subacute and late phases with peaks occurring around 2–3 and 5–6 months after irradiation [35]. These peaks in TNF-α expression correlated with markers of brain damage and with neurological side effects [37].

13.2.2.2 Lung

The predominant early histological changes in irradiated lung are edema and leukocyte infiltration. Lung irradiation results in the expression of a number of genes that may participate in radiation-mediated pneumonitis. These include the genes for TNF-α, TGF-β, and the interleukins. Human lung fibroblasts respond to x-ray treatment with the release of IL-6. The acute response to radiation occurs predominantly within the field of irradiation and has little systemic effect (Figure 13.8).

The molecular events of radiation-induced lung injury occur earlier than histopathological or clinical changes. Radiation triggers a rapid cascade of genetic and molecular events that proceed during the clinical latent period [38]. This is an active process involving a variety of cytokines (IL-1, TNF-α, PDGF, and TGF-β), cell types, (macrophages, epithelial cells, pneumocytes, and fibroblasts), and transcription factors (NF-κB, Egr-1, c-jun, and c-fos). Recently, several radiation-specific gene loci on chromosomes 1, 17, and 18 have been implicated in radiation-induced fibrosis [39]. Although the majority of information has come from animal studies, TGF-β, IL-1α, and IL-6

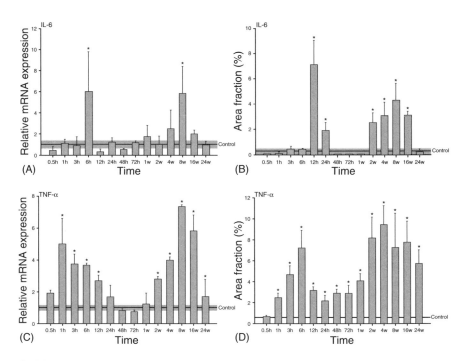

FIGURE 13.8
(A) Time course of relative IL-6 mRNA expression in the lung tissue of mice that underwent thoracic irradiation with 12 Gy, compared with nonirradiated lung tissue of control animals (control). (B) Time-course of IL-6 protein expression (expressed as positive area fraction in %) in the lung tissue of mice that underwent thoracic irradiation with 12 Gy, compared with non-irradiated lung tissue of control animals (control). (C) Time course of relative TNF-α mRNA expression in the lung tissue of mice that underwent thoracic irradiation with 12 Gy, compared with nonirradiated lung tissue of control animals (control). (D) Time course of TNF-α protein expression (expressed as positive area fraction in %) in the lung tissue of mice that underwent thoracic irradiation with 12 Gy, compared with nonirradiated lung tissue of control animals (control). In all cases data are mean ± SD of triplicate determinations from three different mice. *Statistically significant differences from control. (From Rube, C.E., Uthe, D., Schmid, K.W., Richter, K.D., Wessel, J., Schuck, A., Willich, N., and Rube, C., *Int. J. Radiat. Oncol. Biol. Phys.* 47, 1033, 2000. With permission.)

have been implicated in human effects as well [40]. Some recent findings suggest that tissue hypoxia present after RT plays a central role in generating a progressive nonhealing wound response that perpetuates lung injury through continuous generation of ROS and expression and activation of cytokines [25].

13.2.2.3 Intestine

In the irradiated intestine the inflammatory mediators that cause acute inflammation-induced changes in motor function and later intestinal fibrosis are located in the intestinal muscle layer including mesenchymal tissue, fibroblasts, and muscle cells. The cytokines that are thought to have a key role in gastrointestinal diseases are functionally grouped into proinflamma-tory (mainly IL-1β, IL-6, IL-8, and TNF-α) and anti-inflammatory (IL-10) cytokines [41]. They are activated mainly by the induction and translocation of the nuclear transcription factor NF-κB.

In a study in rats [41], it was found that IL-1β, TNF-α, and IL-6 mRNA increased at 3 and 6 h after radiation whereas expression of IL-6 and IL-8 was elevated at 3 days. In contrast, levels of antiinflammatory cytokine IL-10 were markedly lower on day 3. Caffeic acid phenethyl ester (CAPE), a specific inhibitor of NF-κB, given intra-peritoneally, did not prevent increase in primary proinflammatory cytokines but did reduce expression of mRNA for IL-8, IL-6, and IL-6 receptors.

13.2.3 Late Effects: Radiation-Mediated Fibrosis

Fibrosis is a late sequelae of radiation therapy that occurs in many normal tissues within the radiation field, the most severe response being seen in the lung, liver, and skin. A well-characterized mechanism of fibrosis is initiated by the induction of the gene encoding TGF-β, a cytokine associated with radiation-mediated injury to the lung and liver [42,43]. TGF-β is a prolifera-tive cytokine that is elevated at 2 weeks after lung irradiation, persists until 8 weeks, and then returns to baseline values [38]. Radiation-mediated fibro-sis involves collagen deposition within irradiated tissues with the collagen, subsequently interfering with the normal functioning of the involved organ.

TGF-β is probably the major cytokine responsible for the fibrotic reaction in normal tissue following radiation exposure [44] and because it is auto-inductive and chemotactic to monocytes and macrophages and may medi-ate further increases in the growth factor level at the site of injury. TGF-β is also a potent chemoattractant for fibroblasts and stimulates production of collagen [45,46]. TGF-β also increases extracellular matrix accumulation by inhibiting matrix degradation and induces premature terminal differenti-ation of progenitor fibroblasts into leading to accumulation of postmitotic fibrocytes capable of enhanced collagen synthesis.

In an experiment that compared radiation-sensitive mice (C57BL/6) with a radiation-resistant strain (C3H/HeJ), mice were irradiated with a single dose of 5 or 12.5 Gy to the thorax and the expression of collagens I, III, and IV; fibronectin; and TGF-β1 and -β3 genes were measured. Alterations in mRNA abundance were observed in the sensitive mice at all times, whereas levels in the resistant mice were unaffected until 26 weeks after irradiation. There was a biphasic response to radiation involving cytokine IL-1α and TGF-β paralleled by changes in expression of the collagen genes. The early and persistent elevation of cytokine production suggests a continuum of response underlying pulmonary radiation reactions and supports the concept of a perpetual cascade of cytokines initiated immediately after radiation [38].

TGF-β1 gene expression was also increased 19-fold in the irradiated skin during the early erythematous phase, which started 3 weeks after irradi-ation of mice. During the later phases of fibrosis, from 6 to 12 months after irradiation, the TGF-β1 gene is highly expressed in the repaired skin and the underlying muscular fibrotic tissue, with 10- and 8-fold maximal increases, respectively. Immunostaining for TGF-β1 revealed the presence of the protein in endothelial cells of capillaries, myofibroblasts, and the collagenous matrix of fibrotic tissue. Results suggest that TGF-β is a key cytokine in the cascade of events that leads to radiation-induced fibrosis of the skin at both early and late stages [47].

13.2.4   Gene Expression Associated with Radiation-Mediated Vascular Damage

Aberrant growth of cells within blood vessel walls leads the late sequelae of radiation including tissue necrosis and telangiectasias. Genes whose expression is increased after radiation and which are associated with aberrant vascular growth include PDGF, TNFα, bFGF, and E9 [48]. The E9 gene product is a surface protein expressed in response to radiation, which has approximately 70% homology with type-III cell surface receptor for TGF-β. Expression of E9 may be secondary to the production and release of mitogenic factors such as bFGF. TNF-α is produced by monocytes and macrophages after exposure to radiation and binds to receptors on endothelial cells to initiate angiogenesis. Growth factors also influence the response of blood vessels to radiation; PDGF and bFGF are induced in vascular endothelium and bind to receptors to protect vessels from apoptosis and stimulate growth [49].

13.3   Cytokines as Therapeutic Agents: Radioprotection and Radiosensitization

Cytokines as therapeutic agents for radioprotection and radiosensitization has been reviewed by Neta and Durum [50]. A possible clinical application of cytokines would be to selectively protect normal tissue from harmful effects of genotoxic agents. Alternatively, the cytokines might be targeted to enhance radiation tumor cytotoxicity. Radioprotection and radiosensitization of mice by systemically administered cytokines has been extensively investigated mostly in terms of lethality resulting from damage to the gastrointestinal or hematopoietic system. Both radioprotection and sensitization by cytokines have been observed, and in some cases whether the cytokine is protective or sensitizing may depend on the endpoint chosen. An observation common to both sensitization and protection is that the effects are dose, schedule, and organ specific.

13.3.1   Radiosensitization

Radio-modifying effects of cytokines on normal tissue are summarized in Table 13.3. Radiosensitization by cytokines has been studied in terms of the

TABLE 13.3

Radioprotection and Radiosensitization of Mice by Cytokines

	Bone Marrow			Intestine	
	Protection	Restoration	Sensitization	Protection	Sensitization
TNF-α	✓				✓
IL-1	✓	✓			✓
IL-3, IL-4		✓			
IL-6		✓	✓		
IL-11		✓			
IL-12	✓				✓
SCF	✓			✓	
bFGF	✓	✓			
TGF-β			✓		
IFN-α or -β			✓		
IFN-γ					✓
LIF, G-CSF, GM-CSF		✓			

Note: Mice were exposed to acute effects of whole body radiation on the hematopoietic and gastrointestinal systems. The endpoints were $LD_{100/6}$ (GI death) and $LD_{100/30}$ (hematopoietic death). Restoration describes the effect of cytokines that increase survival when injected after a mid-lethal dose.

SCF, stem cell factor; G-CSF, granulocyte colony stimulating factor; GM-CSF, granulocyte macrophage colony stimulating factor.

response of mice to whole-body radiation with a decrease in the radiation dose to kill 100% of mice in 6 days ($LD_{100/6}$) being indicative of sensitization to gastrointestinal (GI) death and a decrease in the radiation dose to kill 100% of mice in 30 days ($LD_{100/30}$) being indicative of sensitization to hematopoietic death. Radiosensitization is reported to occur in response to TGF-β and IL-6 for $LD_{100/30}$ and in response to IFNγ and TNF-α for $LD_{100/6}$.

Toxicity and radiosensitization of tumors by cytokines has been almost entirely concerned with the effects of TNF-α, which derives its name from potent antitumor activities observed in experimental systems (reviewed by Weichselbaum et al. [31]). After treatment with TNF-α, experimental tumors show hemorrhagic necrosis and regression with necrosis occurring in the center of the tumor, leaving a rim of viable tumor cells around the periphery. TNF-α also induces endothelial cell permeability that leads to vascular leak syndrome and, in fact, the antitumor effects of TNF-α are thought to be largely mediated through cytotoxic effects on the tumor endothelium. TNF-α-induced vascular thrombosis and subsequent tumor necrosis have been linked to a T-cell-mediated response.

When given systemically as an antitumor treatment, TNF-α is not well tolerated. High concentrations in the serum are associated with cachexia, shock, and death, consistent with vascular leak syndrome. Side effects of intravenously administered TNF-α include profound fatigue, nausea, systemic inflammatory-response syndrome, and vascular collapse. Clinical application of localized TNF-α treatment for liver metastases was investigated in a phase I trial. Results indicated that intratumoral injection of TNF-α could influence local tumor control [51]. Application of TNF-α in combination with melphalan in isolated limb perfusion was shown to be a successful treatment [52]. The most effective means of localizing TNF-α to the tumor is by the use of radiation-targeted gene therapy as described in Section 13.1.3.1.

A phase I trial of intravenous administration of TNF-α in combination with ionizing radiation has been reported. Some tumor response was seen but the toxicity profile was unacceptable, and further trials were not undertaken [53].

13.3.2 Radioprotection

The first report of a cytokine being used as a protector was for IL-1, which, administered a few hours before radiation, protects lethally irradiated mice from death [54]. Subsequently, other cytokines were tested in in vivo experiments but only IL-1 has been evaluated in phase I and II clinical trials [55].

To date, IL-11, TNF-α, SCF, IL-12, and bFGF have shown a protective effect on mice, when administered before irradiation, while IL-6, TNF-α, and IFN have a radiosensitizing effect [56]. The use of cytokines in clinic has been often a controversial topic because they might be beneficial in some cases but could lead to toxic effects in other treated patients. This is the case for IL-12, which has potent antitumor and anti-metastatic activity and is also a stimulator of hematopoietic progenitor cells. However, though IL-12 protects from the lethal hematopoietic syndrome, it radiosensitizes the gut leading to premature gastrointestinal syndrome and this toxicity limits clinical application. Cytokines have been tested also for restoration of bone marrow stem cell function following radiation. This group of factors includes granulocyte colony stimulating factor (G-CSF), granulocyte-macrophage colony stimulating factor (GM-CSF), IL-1, IL-3, and stem cell factor (SCF).

13.4 Radiation-Induced Genes and Gene-Products as Biomarkers of Radiation Exposure

Molecular biological markers of radiation response could potentially be of use for monitoring the progress of radiation therapy and even for predicting outcome early in a treatment regimen. Biomarkers of radiation exposure could also be an important tool for triage of individuals in potentially exposed populations after a radiologic accident or "dirty bomb" incident. Most currently available radiation exposure biomarkers, such as those based on cytogenetic assays, do not provide the rapid results that are required for such situations therefore creating interest in the development of rapid noninvasive tests for radiation exposure.

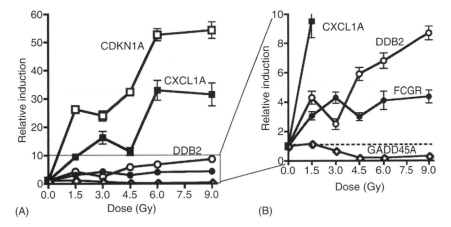

FIGURE 13.9

(A) Measurements by QRT-PCR response to a course of total body irradiation (TBI) treatment of a non-Hodgkin's lymphoma patient. Data points are the average of two (*DDB2*) or three (all other genes) independent PCR runs carried out in triplicate for a total of six or nine determinations. *Error bars,* SEMs. (B) The same data graphed as in (A), with the Y-axis expanded so that relative inductions of 10-fold can be more clearly seen. The dashed line is the level of basal expression before the start of TBI. (From Amundson, S.A., Grace, M.B., McLeland, C.B., Epperly, M.W., Yeager, A., Zhan, Q., Greenberger, J.S., and Fornace, A.J., *Cancer Res.* 64, 6368, 2004. With permission.)

One approach is the use of microarray hybridization analysis, which provides the technology for the discovery of potentially informative radiation-exposure gene expression profiles. This approach was used in a study of *ex vivo* irradiated human peripheral white blood cells. Dose–response relationships with little variability between donors were found for a number of genes. In another study, patients undergoing total body irradiation (TBI) in preparation for hematopoietic stem cell transplantation were the test population. In vivo induction of expression of the *CDKN1A*, *GADD45A*, and *DDB2* genes in these patients, along with the identification of additional potential in vivo exposure marker genes was reported [57,58] (Figure 13.9).

Another approach is the use of proteomic profiling of serum from irradiated individuals. In a recent study, mass spectrometry (MS) was used to generate proteomic profiles of unfractionated serum samples, which were analyzed for unique biomarker signatures. Computer-based analyses of the MS spectra distinguished irradiated from irradiated donors with a high degree of sensitivity. Twenty-three protein fragments were uniquely detected in the irradiated group including an IL-6 precursor protein [59].

13.5 Summary

Cellular exposure to ionizing radiation results in the activation of transcription factors, proteins that bind to specific DNA sequences and activate transcription of cytokines, growth factor, and cell-cycle-related genes. Transcription factors reported to be activated by radiation include p53, NF-κB, the retinoblastoma control proteins (RCPs), the early growth response 1 transcription factor (Egr-1), the c-Fos/c-Jun AP-1 complex, and the octamer-binding protein (Oct-1). Activation of NF-κB by ionizing radiation occurs at clinically relevant doses. Radiosensitization by NF-κB is cell line-dependent, but NF-κB-dependent mediation of the inflammatory response to radiation exposure occurs in all tissues. Transcription of the *Egr-1* gene is rapidly and transiently induced by radiation in a variety of cell types. The promoter sequence of the human *Egr-1* contains several copies of the serum response element (SRE) also known as the cARG box, which are responsible for its radio-inducibility.

Gene therapy can be targeted to the tumor by radiation if a radiation-inducible promoter is upstream of an inserted cDNA encoding a toxic or an immune-modulator protein. One such construct has the radiation-inducible SRE (cArG) elements present in the promoter of the human *Egr-1* gene ligated to a cDNA encoding tumor necrosis factor-α. Clinical trials with Ad. EGR-TNF (TNFerade) in combination with ionizing radiation have shown promising results. Other approaches to radiogenetic therapy have used radiation-inducible "suicide-gene" therapy or an approach based on the use of promoter elements responsive to tissue hypoxia.

The products of radiation-induced immediate early response genes participate in subsequent events by binding to specific promoter elements of other genes referred to as late responding genes. The principal mediators of the effects of late-responding genes and of radiation effects on normal tissue are cytokines that play a role in the inflammation process, and growth factors that are involved in the response of tissue to radiation-induced damage. Cytokines that have been best documented as being produced in response to radiation are proinflammatory cytokines such as TNF-α; the interleukins, IL-1, IL-6, and IL-8; and the antiinflammatory IL-10. Fibrosis is a late sequelae of radiation therapy that occurs in many normal tissues within the radiation field, the most severe response being seen in the lung, liver, and skin. A number of cytokines and growth factors have been shown to participate in radiation-mediated fibrosis including IL-1 and TNF-α and growth factors TGF β and bFGF. Growth factors that are radiation-inducible include TGF-β1 and bFGF. There is strong evidence that TGF-β1 plays the most important role in activating collagen deposition.

Radioprotection and radiosensitization of mice by systemically administered cytokines has been investigated mostly in terms of lethality resulting from damage to the gastrointestinal or hematopoietic system. An observation common to both sensitization and protection is that the effects are dose, schedule, and organ specific. Radiation-induced genes and gene products are also being investigated as markers of radiation exposure.

References

1. Criswell C, Leskov KS, Miyamoto S, Luo G, and Boothman DA. Transcription factors activated in mammalian cells after clinically relevant doses of ionizing radiation. *Oncogene* 22: 5813–5827, 2003.
2. Chastel C, Jiricny J, and Jaussi R. Activation of stress-responsive promoters by ionizing radiation for deployment in targeted gene therapy. *DNA Repair* 3: 201–215, 2004.
3. Prasad AV, Mohan N, Chandrasekar B, and Meltz ML. Activation of nuclear factor kappa B in human lymphoblastoid cells by low-dose ionizing radiation. *Radiat. Res.* 138: 367–372, 1994.
4. Russell JS and Tofilon PJ. Radiation-induced activation of nuclear factor-kappaB involves selective degradation of plasma membrane-associated I(kappa) B(alpha). *Mol. Biol. Cell* 13: 3431–3440, 2002.
5. Lee SJ, Dimtchev A, Lavin MF, Dritschilo A, and Jung M. A novel ionizing radiation-induced signaling pathway that activates the transcription factor NF-kB. *Oncogene* 17: 1821–1826, 1998.
6. Li N and Karin M. Ionizing radiation and short wavelength UV activate NF-kappaB through two distinct mechanisms. *Proc. Natl. Acad. Sci. USA* 95: 13012–13017, 1998.
7. Wang T, Zhang X, and Li JJ. The role of NF-kappaB in the regulation of cell stress responses. *Int. Immunopharmacol.* 2: 1509–1520, 2002.
8. Hallahan DE, Sukhatme VP, Sherman ML, Virudachalam S, Kufe D, and Weichselbaum RR. Protein kinase C mediates X-ray inducibility of nuclear signal transducers EGR1 and JUN. *Proc. Natl. Acad. Sci. USA* 88: 2156–2160, 1991.
9. Datta R, Rubin E, Sukhatme V, Qureshi S, Hallahan D, Weichselbaum R, and Kufe DW. Ionizing radiation activates transcription of the EGR1 gene via CArG elements. *Proc. Natl. Acad. Sci. USA* 89: 10149–10153, 1992.
10. Lim CP, Jain N, and Cao X. Stress-induced immediated-early gene, egr-1, involves activation of p38/JNK1. *Oncogene* 16: 2915–2926, 1998.
11. Meyer RG, Kupper JH, Kandolf R, and Rodemann HP. Early growth response-1 gene (Egr-1) promoter induction by ionizing radiation in U87 malignant glioma cells in vitro. *Eur. J. Biochem.* 269: 337–346, 2002.
12. Weichselbaum RR, Hallahan DE, Sukhatme VP, and Kufe DW. Gene therapy targeted by ionizing radiation. *Int. J. Radiat. Oncol. Biol. Phys.* 24: 565–567, 1992.
13. Chung TD, Mauceri HJ, Hallahan DE, Yu JJ, Chung S, Grdina WL, Yajnik S, Kufe DW, and Weichselbaum RR. Tumor necrosis factor-alpha-based gene therapy enhances radiation cytotoxicity in human prostate cancer. *Cancer Gene Ther.* 5: 344–349, 1998.
14. Kawashita Y, Ohtsuru A, Kaneda Y, Nagayama Y, Kawazoe Y, Eguchi S, Kuroda H, Fujioka H, Ito M, Kanematsu T, and Yamashita S. Regression of hepatocellular carcinoma in vitro and in vivo by radiosensitizing suicide gene therapy under the inducible and spatial control of radiation. *Hum. Gene Ther.* 10: 1509–1519, 1999.
15. Joki T, Nakamura Y, and Ohno T. Activation of the radiosensitive EGR-1 promoter induces expression of the herpes simplex virus thymidine kinase gene and sensitivity of human glioma cells to ganciclovir. *Hum. Gene Ther.* 6: 1507–1513, 1995.

16. Christy BA, Lau LF, and Nathans D. A gene activated in mouse 3T3 cells by serum growth factors encodes a protein with "zinc finger" sequences. _Proc. Natl. Acad. Sci. USA_ 85: 7857–7861, 1988.

17. Marples B, Scott SD, Hendry JH, Embleton MJ, Lashford LS, and Margison GP. Development of synthetic promoters for radiation-mediated gene therapy. _Gene Ther._ 7: 511–517, 2000.

18. Scott SD, Joiner MC, and Marples B. Optimizing radiation-responsive gene promoters for radiogenetic cancer therapy. _Gene Ther._ 9: 1396–1402, 2002.

19. Worthington J, Robson T, Murray M, O'Rourke M, Keilty G, and Hirst DG. Modification of vascular tone using iNOS under the control of a radiation-inducible promoter. _Gene Ther._ 7: 1126–1131, 2000.

20. Greco O, Marples B, Dachs GU, Williams KJ, Patterson AV, and Scott SD. Novel chimeric gene promoters responsive to hypoxia and ionizing radiation. _Gene Ther._ 9: 1403–1411, 2002.

21. Salloum RM, Mauceri HJ, Hanna NN, Gorski DH, Posner MC, and Weichselbaum RR. Dual induction of the Epo-Egr-TNF-alpha-plasmid in hypoxic human colon adenocarcinoma produces tumor growth delay. _Am. Surg._ 69: 24–27, 2003.

22. Kufe D and Weichselbaum R. Radiation therapy: Activation for gene transcription and the development of genetic radiotherapy-therapeutic strategies in oncology. _Cancer Biol. Ther._ 2: 326–329, 2003.

23. Mezhir JJ, Smith KD, Posner MC, Senzer N, Yamini B, Kufe DW, and Weichselbaum RR. Ionizing radiation: A genetic switch for cancer therapy 2006. _Cancer Gene Ther._ 13: 1–6, 2006.

24. Senzer N, Mani S, Rosemurgy A, Nemunaitis J, Cunningham C, Guha C, Bayol N, Gillen M, Chu K, Rasmussen C, Rasmussen H, Kufe D, Weichselbaum R, and Hanna N. TNFerade biologic, an adenovector with a radiation-inducible promoter, carrying the human tumor necrosis factor alpha gene: A phase I study in patients with solid tumors. _J. Clin. Oncol._ 22: 592–601, 2004.

25. Marks LB, Yu X, Vujaskovic Z, Small W, Folz R, and Anscher MS. Radiation-induced lung injury. _Semin. Radiat. Oncol._ 13: 333–345, 2003.

26. Martin M, Vozenin M-C, Gault N, Crechet F, Pfarr CM, and Lefaix J-L. Coactivation of AP-1 activity and TGF-b1 gene expression in the stress response of normal skin cells to ionizing radiation. _Oncogene_ 15: 981–989, 1997.

27. Chen RH, Su YH, Chuang RL, and Chang TY. Suppression of transforming growth factor-beta-induced apoptosis through a phosphatidylinositol 3-kinase/Akt-dependent pathway. _Oncogene_ 17: 1959–1968, 1998.

28. Barcellos-Hoff MH. How do tissues respond to damage at the cellular level? The role of cytokines in irradiated tissues. _Radiat. Res._ 150: S109–S121, 1998.

29. Witte L, Fuks Z, Haimovitz-Friedman A, Vlodavsky I, Goodman DS, and Eldor A. Effects of irradiation on the release of growth factors from cultured bovine, porcine, and human endothelial cells. _Cancer Res._ 49: 5066–5072, 1989.

30. Lee JY, Galoforo SS, Berns C, Erdos G, Gupta AK, Ways DK, and Corry PM. Effect of ionizing radiation on AP-1 binding activity and basic fibroblast growth factor gene expression in drug-sensitive human breast carcinoma MCF-7 and multidrug-resistant MCF-7/ADR cells. _J. Biol. Chem._ 270: 28790–28796, 1995.

31. Weichselbaum RR, Kufe DW, Hellman S, Rasmussen HS, King CR, Fischer PH, and Mauceri HJ. Radiation-induced tumour necrosis factor-alpha expression: Clinical application of transcriptional and physical targeting of gene therapy. _Lancet Oncol._ 3: 665–671, 2002.

32. Beetz A, Peter RU, Oppel T, Kaffenberger W, Rupec RA, Meyer M, van Beuningen D, Kind P, and Messer G. NF-kappaB and AP-1 are responsible for inducibility of the IL-6 promoter by ionizing radiation in HeLa cells. *Int. J. Radiat. Biol.* 76: 1443–1453, 2000.
33. Hallahan DE. Radiation-mediated gene expression in the pathogenesis of the clinical radiation response. *Semin. Radiat. Oncol.* 6: 250–267, 1999.
34. Hong JH, Chiang CS, Campbell IL, Sun JR, Withers HR, and McBride WH. Induction of acute phase gene expression by brain irradiation. *Int. J. Radiat. Oncol. Biol. Phys.* 33: 619–626, 1995.
35. Chiang CS, Hong JH, Stalder A, Sun JR, Withers HR, and McBride WH. Delayed molecular responses to brain irradiation. *Int. J. Radiat. Biol.* 72: 45–53, 1997.
36. Bellinzona M, Gobbel GT, Shinohara C, and Fike JR. Apoptosis is induced in the subependymoma of young adult rats by ionizing irradiation. *Neurosci. Lett.* 208: 163–166, 1996.
37. Chiang CS, McBride WH, and Withers HR. Radiation-induced astrocytic and microglial responses in mouse brain. *Radiother. Oncol.* 29: 60–68, 1993.
38. Rubin P, Johnston CJ, Williams JP, McDonald S, and Finkelstein JN. A perpetual cascade of cytokines postirradiation leads to pulmonary fibrosis. *Int. J. Radiat. Oncol. Biol. Phys.* 33: 99–109, 1995.
39. Haston CK, Zhou X, Gumbiner-Russo L, Irani R, Dejournett R, Gu X, Weil M, Amos CI, and Travis EL. Universal and radiation-specific loci influence murine susceptibility to radiation-induced pulmonary fibrosis. *Cancer Res.* 62: 3782–3788, 2002.
40. Chen Y, Williams J, Ding I, Hernady E, Liu W, Smudzin T, Finkelstein JN, Rubin P, and Okunieff P. Radiation pneumonitis and early circulatory cytokine markers. *Semin. Radiat. Oncol.* 12: 26–33, 2002.
41. Linard C, Marquette C, Mathieu J, Pennequin A, Clarencon D, and Mathe D. Acute induction of inflammatory cytokine expression after gamma-irradiation in the rat: Effect of an NF-kappaB inhibitor. *Int. J. Radiat. Oncol. Biol. Phys.* 58: 427–434, 2004.
42. Anscher MS, Murase T, Prescott DM, Marks LB, Reisenbichler H, Bentel GC, Spencer D, Sherouse G, and Jirtle RL. Changes in plasma TGF beta levels during pulmonary radiotherapy as a predictor of the risk of developing radiation pneumonitis. *Int. J. Radiat. Oncol. Biol. Phys.* 30: 671–676, 1994.
43. Geraci JP and Mariano MS. Radiation hepatology of the rat: Parenchymal and nonparenchymal cell injury. *Radiat. Res.* 136: 205–213, 1993.
44. Pelton RW and Moses HL. The beta-type transforming growth factor. Mediators of cell regulation in the lung. *Am. Rev. Respir. Dis.* 142: S31–S35, 1990.
45. Ritzenthaler JD, Goldstein RH, Fine A, and Smith BD. Regulation of the alpha 1 (I) collagen promoter via a transforming growth factor-beta activation element. *J. Biol. Chem.* 268: 13625–13631, 1993.
46. Roberts AB, Sporn MB, Assoian RK, Smith JM, Roche NS, Wakefield LM, Heine UI, Liotta LA, Falanga V, and Kehrl JH. Transforming growth factor type beta: Rapid induction of fibrosis and angiogenesis in vivo and stimulation of collagen formation in vitro. *Proc. Natl. Acad. Sci. USA* 83: 4167–4171, 1986.
47. Martin M, Lefaix JL, Pinton P, Crechet F, and Daburon F. Temporal modulation of TGF-beta 1 and beta-actin gene expression in pig skin and muscular fibrosis after ionizing radiation. *Radiat. Res.* 134: 63–70, 1993.

48. Hallahan D, Clark ET, Kuchibhotla J, Gewertz BL, and Collins T. E-selectin gene induction by ionizing radiation is independent of cytokine induction. *Biochem. Biophys. Res. Commun.* 217: 784–795, 1995.
49. Haimovitz-Friedman A, Kolesnick RN, and Fuks Z. Modulation of the Apoptotic Response: Potential for Improving the Outcome in Clinical Radiotherapy. *Semin. Radiat. Oncol.* 6: 273–283, 1996.
50. Neta R and Durum SK. Whole organism responses to DNA damage. Modulation by cytokines of damage induced by ionizing radiation. In Nickoloff JA and Hoekstra MF (Eds.), *DNA Repair in Higher Eukaryotes*, Vol. 2. Humana Press Inc, Totowa, New Jersey, 1998, pp. 587–601.
51. van der Schelling GP, IJzermans JN, Kok TC, Scheringa M, Marquet RL, Splinter TA, and Jeekel J. A phase I study of local treatment of liver metastases with recombinant tumour necrosis factor. *Eur. J. Cancer* 28A: 1073–1078, 1992.
52. Renard N, Lienard D, Lespagnard L, Eggermont A, Heimann R, and Lejeune F. Early endothelium activation and polymorphonuclear cell invasion precede specific necrosis of human melanoma and sarcoma treated by intravascular high-dose tumour necrosis factor alpha (rTNF alpha). *Int. J. Cancer* 57: 656–663, 1994.
53. Hallahan DE, Mauceri HJ, Seung LP, Dunphy EJ, Wayne JD, Hanna NN, Toledano A, Hellman S, Kufe DW, and Weichselbaum RR. Spatial and temporal control of gene therapy using ionizing radiation. *Nat. Med.* 1: 786–791, 1995.
54. Neta R, Douches S, and Oppenheim JJ. Interleukin 1 is a radioprotector. *J. Immunol.* 136: 2483–2485, 1986.
55. Smith JW, Urba WJ, Curti BD, Elwood LJ, Steis RG, Janik JE, Sharfman WH, Miller LL, Fenton RG, and Conlon KC. The toxic and hematologic effects of interleukin-1 alpha administered in a phase I trial to patients with advanced malignancies. *J. Clin. Oncol.* 10: 1141–1152, 1992.
56. Mori M and Desaintes C. Gene expression in response to ionizing radiation: An overview of molecular features in hematopoietic cells. *J. Biol. Regul. Homeost. Agents* 18: 363–371, 2004.
57. Amundson SA, Do KT, Shahab S, Bittner M, Meltzer P, Trent J, and Fornace AJ. Identification of potential mRNA biomarkers in peripheral blood lymphocytes for human exposure to ionizing radiation. *Radiat. Res.* 154: 342–346, 2000.
58. Amundson SA, Grace MB, McLeland CB, Epperly MW, Yeager A, Zhan Q, Greenberger JS, and Fornace AJ. Human in vivo radiation-induced biomarkers: Gene expression changes in radiotherapy patients. *Cancer Res.* 64: 6368–6371, 2004.
59. Menard C, Johann D, Lowenthal M, Muanza T, Sproull M, Ross S, Gulley J, Petricoin E, Coleman CN, Whiteley G, Liotta LA, and Camphausen K. Discovering clinical biomarkers of ionizing radiation exposure with serum proteomic analysis. *Cancer Res.* 66: 1844–1850, 2006.
60. Sherman ML, Datta R, Hallahan DE, Weichselbaum RR, and Kufe DW. Regulation of tumor necrosis factor gene expression by ionizing radiation in human myeloid leukemia cells and peripheral blood monocytes. *J. Clin. Invest.* 87: 1794–1797, 1991.
61. Weill D, Gay F, Tovey MG, and Chouaib S. Induction of tumor necrosis factor alpha expression in human T lymphocytes following ionizing gamma irradiation. *J. Interferon. Cytokine. Res.* 16: 395–402, 1996.

62. Weinmann M, Belka C, Scheiderbauer J, and Bamberg M. Soluble levels of CD-95, CD 95-L and various cytokines after exposing human leukocytes to ionizing radiation. *Anticancer Res.* 20: 1813–1818, 2000.
63. Hallahan DE, Spriggs DR, Beckett MA, Kufe DW, and Weichselbaum RR. Increased tumor necrosis factor alpha mRNA after cellular exposure to ionizing radiation. *Proc. Natl. Acad. Sci. USA* 86: 10104–10107, 1989.
64. Konemann S, Bolling T, Malath J, Kolkmeyer A, Janke K, Riesenbeck D, Hesselmann S, Diallo R, Vormoor J, Willich N, and Schuck A. Time- and dose-dependent changes of intracellular cytokine and cytokine receptor profile of Ewing tumour subpopulations under the influence of ionizing radiation. *Int. J. Radiat. Biol.* 79: 897–909, 2003.
65. Marquette C, Linard C, Galonnier M, Van Uye A, Mathieu J, Gourmelon P, and Clarencon D. IL-1beta, TNFalpha and IL-6 induction in the rat brain after partial-body irradiation: Role of vagal afferents. *Int. J. Radiat. Biol.* 79: 777–785, 2003.
66. Ibuki Y and Goto R. Contribution of inflammatory cytokine release to activation of resident peritoneal macrophages after in vivo low-dose gamma-irradiation. *J. Radiat. Res. (Tokyo)* 40: 253–262, 1999.
67. Brouazin-Jousseaume V, Guitton N, Legue F, and Chenal C. GSH level and IL-6 production increased in Sertoli cells and astrocytes after gamma irradiation. *Anticancer Res.* 22: 257–262, 2002.
68. Ewan KB, Henshall-Powell RL, Ravani SA, Pajares MJ, Arteaga CL, Warters R, Akhurst RJ, and Barcellos-Hoff MH. Transforming growth factor-beta1 mediates cellular response to DNA damage in situ. *Cancer Res.* 62: 5627–5631, 2002.
69. Boerma M, Bart CI, and Wondergem J. Effects of ionizing radiation on gene expression in cultured rat heart cells. *Int. J. Radiat. Biol.* 78: 19–25, 2002.
70. Rube CE, Uthe D, Schmid KW, Richter KD, Wessel J, Schuck A, Willich N, and Rube C. Dose-dependent induction of transforming growth factor beta (TGF-beta) in the lung tissue of fibrosis-prone mice after thoracic irradiation. *Int. J. Radiat. Oncol. Biol. Phys.* 47: 1033–1042, 2000.

14

Cell Death, Cell Survival, and Adaptation

The first part of this chapter presents some traditional radiation biology, describes the various mechanisms of radiation-induced cell death, and discusses the analysis of cell-survival curves. The second part is concerned with events that may occur after very low doses of radiation. Some of these latter findings have led to a shift in the perception of radiobiological phenomena from the traditional deterministic "hit-effect" model to a dynamic model of interactive cellular relationships.

14.1 Cell Death

Mammalian cells can die in a number of ways, from a programmed sequence of events culminating in cell suicide to the apparently passive cessation of metabolic activity associated with cell age or pathology.

14.1.1 Modes of Cell Death in Nonirradiated Cells

Modes of cell death in nonirradiated cells have been reviewed by Gudkov and Komarova [1].

14.1.1.1 Apoptosis

Apoptosis is a form of programmed cell death (PCD), which is a physiological "cell-suicide" program essential for embryonic development, immune-system function, and the maintenance of tissue homeostasis in multicellular organisms. Dysregulation of apoptosis has been implicated in numerous pathological conditions, including neurodegenerative diseases, autoimmunity, and cancer. Apoptosis in mammalian cells is mediated by a family of cysteine proteases known as the caspases. To keep the apoptotic program under control, caspases are initially expressed in cells as inactive procaspase precursors that are activated by oligomerization, and cleave the precursor forms of effector caspases. Activated effector caspases in turn cleave a specific set of cellular substrates, resulting in the well-documented constellation of biochemical and morphological changes associated with the apoptotic phenotype. Caspase activation can be triggered by extrinsic and intrinsic apoptotic pathways both of which may be initiated by radiation as described in detail in Chapter 12.

14.1.1.2 Senescence

Primary cells in culture initially undergo a period of rapid proliferation, during which the telomeres of their chromosomes become significantly shorter. Eventually, cell growth decelerates and the cells enter a form of permanent cell-cycle arrest that is known as replicative senescence. A senescent cell typically shows morphological changes, including flattened cytoplasm, increased granularity, biochemical changes in metabolism, induction of senescence-associated-galactosidase activity and, at the genetic level, alterations to chromatin structure and gene-expression patterns.

14.1.1.3 Necrosis

Necrosis is an apparently disorganized, unregulated process of traumatic cell destruction, which culminates in the release of intracellular components. A distinctive set of morphological features resulting from profound cellular damage is seen, including membrane distortion, organelle degradation, and cellular swelling. Necrosis is usually a consequence of a pathophysiological condition, including infection, inflammation, or ischaemia. The resulting trauma causes extensive failure of normal physiological pathways that are essential for maintaining cellular homeostasis, such as regulation of ion transport, energy production, and pH balance.

14.1.1.4 Autophagy

A form of nonapoptotic, nonnecrotic cell death is associated with a process that is known as autophagy. Autophagy is a mechanism by which long-lived proteins and organelle components are directed to and degraded within lysosomes. Cells that undergo excessive autophagy are induced to die in a nonapoptotic manner, and the morphology of autophagic cells is distinct from those that undergo either necrosis or apoptosis. Autophagy is activated in response to growth-factor withdrawal, differentiation, and developmental signals.

14.1.1.5 Mitotic Catastrophe

Mitotic catastrophe is used to explain the type of mammalian cell death that is caused by aberrant mitosis. It is associated with the formation of multinucleate, giant cells that contain uncondensed chromosomes and is morphologically distinct from apoptosis, necrosis, and autophagy.

The G_2 checkpoint of the cell cycle is responsible for blocking mitosis when a cell has sustained an insult to DNA. However, if the G_2 checkpoint is defective, the cell can enter mitosis prematurely, before DNA replication is complete or with DNA damage unrepaired. This aberrant mitosis causes the cell to undergo death by mitotic catastrophe. Mitotic catastrophe is also caused by agents that damage microtubules and disrupt the mitotic spindle. The drug paclitaxel, for instance, induces an abnormal metaphase in which the sister chromatids fail to segregate properly and the cells die by mitotic catastrophe.

14.1.2   Radiation-Induced Cell Death

The definition of cell death is important in the context of radiation response. As has been emphasized frequently in this book, much of the experimental demonstration of radiation killing in radiation biology is based on the clonogenic assay, the so-called gold standard of survival assays. This is quite reasonable since much of radiation biology research is focused on (and funded by) its application to cancer research. For tumor control, the bottom line is failure of the cancer cell to continue to proliferate and though the cell might not be metabolically dead it is, importantly, clonogenically dead.

At the molecular level, radiation-induced damage results in activation of DNA repair, coupled with arrest at cell-cycle checkpoints, which allows the cell to repair the damage before proceeding through mitosis. This mechanism is conserved in all eukaryotes. Multicellular organisms have acquired additional response mechanisms to genotoxic stress, which involve activation of the transcription factor p53 (Chapter 8). p53 can induce growth arrest or apoptosis, responses that maintain genomic stability. Failure of this system can result in cancer development and genomic instability. The ways in which cells respond to radiation are tissue specific and dependent on p53 status.

14.1.2.1   *Mechanisms of Cell Death in Normal Tissue*

Death of the organism following a lethal dose of radiation is caused by the failure of a few sensitive tissues, particularly the hematopoietic system and epithelium of the small intestine, to repair damage. Differential sensitivity of most tissues to radiation is directly linked to proliferation rates. Most tissues considered to be radioresistant tissues including brain, muscle, liver, lung, and kidney consist of terminally differentiated and nonproliferating cells. Nevertheless, some tissues with low proliferation rates, lung and kidney for instance, are also considered to be significantly dose limiting in terms of radiation treatment. This is because radiation-induced damage to such tissues can result in organ failure revealed only when cells with irreparable damage attempt mitosis and undergo mitotic catastrophe impairing the ability of the tissue to regenerate as stem-cell progenitors are depleted. In addition, these organs are severely affected by long-term effects of radiation such as fibrosis attributable to radiation-induced proliferation of fibroblasts and other long-lasting changes initiated by radiation-induced cytokines.

Tissues with high turnover rates, such as epithelia, spermatogonia, and hair follicles are consistently radiosensitive supporting the idea that radiosensitivity caused by radiation-induced damage is linked to cell-proliferation rates. However, there are exceptions. Among the most radiation-sensitive tissues are spleen and thymus, which in adults consist largely of nondividing cells and bone-marrow hematopoietic stem cells that are also predominantly quiescent.

The occurrence of apoptosis rather than growth arrest at very high doses of radiation eliminates severely damaged cells that might attempt repair if they had survived. Apoptosis prevails in tissues with a rapid turnover or a potential for rapid turnover, such as hematopoietic cells, embryonic tissues, hair follicles, and specific subpopulations of intestinal epithelium and dermis. Loss of damaged cells presents less risk to these populations than would the multiplication of genetically altered progeny, and growth arrest is not an option since their function is to actively proliferate.

The role of p53 is exemplified by the results of experiments comparing p53-deficient mice with wild-type p53 mice, showing that p53 is important for the rapid apoptosis that occurs in sensitive tissues after treatment with radiation. The most striking differences in radiosensitivity between p53-deficient and wild-type mice were found in hematopoietic and lymphoid tissues, spermatogonia, small intestine, hair follicles [2,3], and early embryos. The sizes of the spleen and thymus decrease greatly within 24 h after treatment with 9 Gy of γ-radiation in wild-type, but not in p53-deficient mice due to massive apoptosis [3]. Consistently, p53-deficient mice can survive doses of radiation that cause the development of lethal bone-marrow depletion—the hematopoietic syndrome—in wild-type animals [4], indicating that p53 can be a determinant of radiosensitivity. There is also a very high incidence of T-cell lymphomas in p53-deficient mice, which is likely to be a direct consequence of the lack of apoptosis in thymocytes.

Some cell types are p53 competent but do not respond to radiation by apoptosis, an example being fibroblasts that go into irreversible p53-dependent cell-cycle arrest independent of their ability to repair damage. In contrast, p53-deficient fibroblasts continue to grow after doses of radiation that cause arrest in p53 wild-type cells. Growth arrest for fibroblasts avoids the elimination of cells that are important to the structural integrity of connective tissue, which might disintegrate if too many fibroblasts were eliminated by apoptosis. At the same time, irreversible growth arrest obviates the risk of perpetuating carcinogenic change.

14.1.2.2 Cell Death in Tumors

Compared with normal tissue radiation, killing of tumor cells is less likely to occur by apoptosis and much more likely to result from mitotic catastrophe or in senescence-like irreversible growth arrest than is the case for normal tissue.

14.1.2.2.1 Apoptosis

As has been described, the mode of cell death in normal tissue cells is tissue specific: cells with a high proliferative capacity tend to apoptose; fibroblasts, the structural component of tissues tend to growth arrest. This suggests that tumor susceptibility to apoptosis might vary in accordance with tissue of origin. Comparison of the apoptotic response of primary mouse tumors that had developed from different tissues indicated that anticancer treatment by chemotherapeutic drugs and radiation could induce apoptosis in the tumor only if the tumors originated from tissues that were susceptible to p53-dependent apoptosis [5]. In fact, p53-dependent apoptosis is a determinant of treatment susceptibility only for tumors that originate from hematopoietic and reproductive tissues, which are naturally prone to apoptosis, and which retain wild-type p53. For most other cancers, the level of treatment-induced apoptosis did not correlate with a decrease in clonogenic survival or with a favorable prognosis [6].

p53 is inactivated by mutations in more than half of human malignancies and is not functional in many others due to alterations in other components of the p53 pathway [7], which reduces the role of p53-dependent apoptosis to an even smaller fraction of malignancies [6,8]. Those tumors that retain a functional p53 pathway frequently lose the basic apoptotic mechanisms as they progress toward unconstrained proliferation and independence from intrinsic and extrinsic negative growth-control mechanisms. Resistance to apoptosis occurs due to over-expression of anti-apoptotic Bcl-2 family members, such as Bcl-2 and Bcl-XL [9] or due to loss of important components of the apoptotic machinery, such as transcriptional silencing of Apaf-1 in melanomas [7] or inhibition of caspase-encoding genes [10,11]. Tumors also frequently lose sensitivity to extrinsic apoptotic stimuli (such as FAS, TNF, and TRAIL) and as a result escape from the host immune response [7].

Thus, apoptosis might be an important factor in tumor susceptibility at early stages of progression, but many tumors have lost their pro-apoptotic mechanisms during progression. p53 inactivation, resulting in resistance to apoptosis, is frequently observed to occur spontaneously in experimental mouse tumors that were originally sensitive to p53-dependent apoptosis as part of the selection of more malignant variants and is possibly mediated by hypoxia-induced factors (Chapter 16).

14.1.2.2.2 Mitotic Catastrophe in Radiation-Treated Tumors

Mitotic catastrophe is a significant cause of cell death after treatment with radiation or chemotherapy, particularly for cells of tumors of epithelial origin. Mitotic death was detected in more than 50% of cells of the HT1080 fibrosarcoma that had been treated with various anticancer drugs or ionizing radiation. In another study, only 2 out of 14 solid-tumor cell lines showed a predominantly apoptotic response, whereas the other 12 lines underwent mitotic death with or without apoptosis [12]. Mitotic catastrophe and apoptosis are not mutually exclusive, however, since mitotic catastrophe is frequently followed by apoptosis in apoptosis-competent cells [13]. In fact, strictly speaking, mitotic catastrophe is not a mode of cell death but a means to cell death. Cells that undergo mitotic catastrophe may remain stable for many hours being finally destroyed probably by apoptotic processes.

14.1.2.2.3   Radiation-Induced Irreversible Growth Arrest

Irreversible growth arrest has recently been recognized as one type of tumor response to anticancer treatment. It has many features of replicative senescence, including similar cell morphology, irreversibility, characteristic changes in gene-expression profiles, and acquisition of acidic β-galactosidase, a conventional senescence marker. As in the case of normal cells, this outcome depends primarily on the ability of p53 to transactivate the tumor growth suppressor WAF1/CIP1. In the absence of apoptosis, it is an alternative to mitotic catastrophe. Irreversible growth arrest is not limited to the tumors of connective tissue origin, and it can be a significant type of treatment outcome in lymphoid tumors if their apoptotic pathway is blocked by over-expression of Bcl-2 [14].

14.1.3   Role of p53

Inactivation of p53 has theoretically several possible outcomes in terms of tumor response to radiation treatment, some of which have already been discussed in Chapter 8. Wild-type p53 would be expected to make tumors more sensitive to treatment through the induction of apoptosis and conversely, p53 inactivation would lead to treatment resistance. This would only apply to those tumor cells that are capable of p53-dependent apoptosis, a property that is frequently lost in tumors. Another role of p53 is the prolonged arrest after radiation treatment, during which DNA repair can take place in the absence of an apoptotic response. In this case, p53 would be a pro-survival factor and tumors that had inactivated p53 during progression would be less capable of DNA repair and more sensitive to DNA-damage-induced mitotic catastrophe. This has been shown to be true in several experimental tumor models; knockout of functional p53 in the human colon carcinoma cell line HCT116, for example, resulted in sensitization to DNA-damaging treatments [15]. A marked increase in the number of cells undergoing mitotic catastrophe was also observed in human fibrosarcoma HT1080 cells after wild-type p53 was inactivated by a dominant-negative mutant [12]. For other tumor models, inactivation of p53 function has no effect on radiosensitivity.

Thus, while reduction of p53-dependent apoptosis would increase tumor cell survival, shortening of cell-cycle arrest would limit DNA damage repair and decrease cell survival and in some circumstances cell survival could be unaffected. In reality, loss of p53 is most frequently associated with an unfavorable prognosis [16,17]. The reason for this is that regardless of the influence of p53 on individual regulators of cell survival, the overriding effect of p53 loss is to make cells genetically unstable promoting increased proliferation, invasion, and metastatic progression characteristic of tumor progression.

14.2 Quantitating Cell Kill: Analysis of Cell-Survival Curves

Experimental radiation survival curves are based on the clonogenic assay, which does not distinguish between different modes of cells killed in an irradiated cell population. As has already been noted from the viewpoint of radiotherapy, loss of clonogenicity is the endpoint that represents the most significant consequence of exposure to ionizing radiation.

14.2.1 Target Theory

The target theory model of cell survival is based on the concept that a number of critical targets have to be inactivated for cells to be killed. The earliest models of radiation cell kill were based on the equation and parameters derived from this concept and are still sometimes used to describe the shape of cell-survival curves.

The number of targets (dN) inactivated by a small dose of radiation dD is proportional to the initial number of targets N and to dose dD so that

$$dN \propto NdD \text{ or } dN = -\frac{NdD}{D_0} \tag{14.1}$$

where $1/D_0$ is a constant of proportionality and the expression is negative because the number of targets decreases as the dose increases. The equation can be integrated to give

$$N = N_0 e^{-D/D_0} \tag{14.2}$$

where N is the number of targets at zero dose. If the cells contain only one target, which must be inactivated for them to be killed, then the survival (S) of a population of cells is represented by

$$S = \frac{N}{N_0} = e^{-D/D_0} \tag{14.3}$$

where
N_0 is the initial number of cells
N is the final number of cells surviving radiation dose D
D_0 is a constant

This expression represents a single-hit, single-target survival curve that is a straight line on a semilogarithmic plot (Figure 14.1A). Survival curves of this shape have been found for the inactivation of viruses and bacteria. They may also be applied to the response of certain very radiosensitive mammalian cells (normal and malignant), to the response to very low-dose rates, and to the response to high LET radiation.

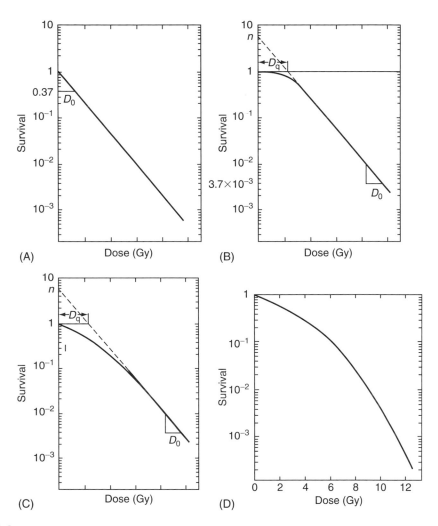

FIGURE 14.1
Survival curves for mammalian cells. (A) Single-hit, single-target survival curve (Equation 14.3); (B) Multitarget survival curve (Equation 14.4); (C) Two-component survival curve (Equation 14.5); (D) Linear quadratic model of cell killing (Equation 14.7).

The term D_0 represents the dose required to reduce the surviving fraction to 0.37 and is the dose required to give an average of one hit/target. The number of hits on a population of cells from a given dose of radiation and the number of hits/target is best described by a Poisson distribution. From Equation 14.3

$$P(\text{survival}) = p(0 \text{ hits}) = e^{-D/D_0}$$

where D_0 is defined as the dose that gives an average of 1 hit/target.

According to the Poisson formula

$$f(n) = e^{-a} \cdot a^{-n}/n!$$

where

n is the number of hits on a target
a is the average number of hits per target
$f(n)$ is the fraction of targets receiving n hits

When $a = 1$, the fraction of targets receiving 0 hits, i.e., the surviving fraction is $e^{-1} = 0.37$. The dose to reduce survival to 0.37 (the D_{37} or D_0 dose) is a very important parameter in radiation biology because it is the dose that will give an average of one hit per target or one inactivating event per cell.

If the cell contains more than one identical target, each of which must be inactivated by a single hit in order to inactivate the cell, survival is represented by the multitarget, single-hit survival equation:

$$S = \frac{N}{N_0} = 1 - \left(1 - e^{-D/D_0}\right)^n \tag{14.4}$$

Plotting this equation gives a survival curve with a shoulder at low doses and straight-line response at higher doses (Figure 14.1B). The straight-line portion extrapolates to n at zero dose and has a slope defined by D_0. The quasi-threshold dose, D_q, is the dose at which the extrapolated straight-line section of the dose curve crosses the dose axis and quantitatively defines the size of the shoulder. It can be calculated from $D_q = D_0 \ln n$.

The multitarget single-hit survival curve has an initial slope of zero. This is not representative of the situation for most mammalian cells since at low doses the survival curve usually has a negative initial slope. This can be incorporated into the equation as an additional single-target component D_1 to give a two-component model (Figure 14.1C).

$$S = \frac{N}{N_0} = e^{-D/D_1}\left(1 - \left[1 - e^{-D/D_0}\right]^n\right) \tag{14.5}$$

D_0 is the dose required to reduce the surviving fraction from 0.1 to 0.037 or from 0.01 to 0.0037 on the final, straight-line portion of the curve. In effect, over this dose range it is identical with the D_{37} dose.

Survival data is usually plotted on semi-$\log_{10}$ coordinates. The dose resulting in one decade of cell kill (the D_{10} dose) is related to D_0 by the expression

$$D_{10} = 2.3 \times D_0 \tag{14.6}$$

Target theory and the derivation of simple cell-survival relationships in terms of targets and hits dominated radiobiological thinking over a long period. A problem with this concept is that specific radiation targets have not been identified in mammalian cells. What is now understood is the importance of DNA strand breaks and of strand-break repair, the sites of damage and repair being dispersed throughout the cell nucleus. Another problem is that though the two-component model predicts cell killing in the low-dose region, the change in cell survival over the dose range 0 to D_q occurs almost linearly implying that there is no sparing as fraction size is reduced below 2 Gy, which has been found not to be the case either experimentally or clinically.

14.2.2 Linear Quadratic Model

A continually bending form of the survival curve can be fitted by a second-order polynomial, a formulation that is termed the linear-quadratic model (Figure 14.1D). The linear quadratic (LQ) model is based on the idea that multiple lesions induced by radiation interact in the cell to cause cell killing. The lesions that interact could be caused by a single ionizing track giving a direct dependence on dose or by two or more independent tracks giving a dependence on a higher power of dose. The resulting equation gives an equation that can fit most experimental survival curves at least over the first few decades of cell kill.

$$S = \frac{N}{N_0} = e^{-\left(\alpha D + \beta D^2\right)} \tag{14.7}$$

The parameters α and β are assumed to describe the probability of the interacting lesions being caused by a single particle track or by two independent tracks. This interpretation is supported by studies of the dose rate effect, which show that as dose rate is reduced survival curves become straight and extrapolate to the initial slope of the high-dose rate curve; in effect, the quadratic component of cell killing disappears leaving only the linear component. This would be expected for single-track events occurring at low-dose rate, which occur far apart with low probability of interaction. The nature of the interaction between separate tracks is unclear. Chadwick and Leenhouts [18] postulated that separate tracks might hit opposite strands of the DNA double helix. It now seems that this is unlikely because of the low probability of tracks interacting within the dimensions of the DNA molecule (diameter ~2.5 nm) with doses of a few grays. Interaction between more widely spaced regions of DNA structure or between DNA in different chromosomes seems more likely.

The linear quadratic model has taken precedence as the model of choice to describe survival curves. Work with chromosomes supports the idea that cell killing results from the interaction of two lesions to form lethal exchange type lesions. The cell-survival curve represented by the linear quadratic equation is continually bending with no straight-line portion. This does not coincide with some dose–response relationships determined experimentally, where survival curves determined through several decades of cell kill closely approximate a straight line. However, in the first 1–2 decades of cell kill and certainly for doses used for fractionated radiotherapy, the LQ model adequately represents the data. The model can be manipulated to predict response to fractionated radiation and it has the advantage of depending on only two unknown factors, α and β.

14.2.3 Lethal, Potentially Lethal Damage Model

The lethal, potentially lethal (LPL) model is a unified repair model of cell killing [19]. Ionizing radiation is considered to produce two different types of lesion, reparable, i.e., potentially lethal lesions and nonreparable, lethal lesions. The nonreparable lesions produce single-hit lethal effects and give rise to the linear component of cell killing. The effect of the reparable lesions depends on competing processes of repair and binary mis-repair, and this process gives rise to the quadratic component of cell killing. This model produces almost identical survival curves to the LQ equation down to a survival level of approximately 10^{-2} and could be taken to provide one mechanistic interpretation of the LQ equation. It predicts that as dose rate is reduced, the probability of binary interaction of potentially lethal lesions will fall and parameters can be used which accurately simulate the cell-survival data of human and animal cells at different dose rates.

14.2.4 Repair Saturation Models

The repair saturation models [20] propose that the shape of the survival curve depends only on a dose-dependent rate of repair. Only one type of lesion and single-hit killing are postulated and in the absence of repair the lesions produce a steep survival curve. The final survival curve results from repair of some of these lesions, but if the repair enzymes become saturated there is not enough enzyme to bind all damage sites simultaneously, so the rate of the repair reaction no longer increases with increasing damage. At higher doses, there is proportionally less repair during the time available before damage becomes fixed leading to greater residual damage and greater cell kill. An alternate version of the saturation hypothesis which leads to the same conclusion is that there is a pool of enzymes that is used up during repair so that at higher doses the repair system is depleted. Both repair saturation and lesion interaction models predict LQ survival curves in the clinically relevant dose region and provide explanations for split-dose recovery, the changing effectiveness of LET, and the dose rate effect. Whether lesion interaction or repair saturation really exist and if they do, the mechanisms of fixation of nonrepaired damage are not known, although it might be conjectured that fixation could occur when cells carrying unrepaired damage enter DNA synthesis or mitosis.

14.3   Cell Survival at Low Radiation Doses

The models of cell survival adequately describe, within a certain dose range, the relationship between radiation dose and cell kill. The survival data alone is not of sufficient quality to allow preference for one model over another, and selection of a model (in recent years the LQ model) is made on an operational basis as mentioned above.

The models have in common the following: they are based on the assumption that only those cells in the radiation path whose molecules sustained collisions with high energy particles and rays are damaged; the damage is proportional to the energy absorbed by each cell and to the number of cells absorbing energy; all cells had identical sensitivities to radiation. However, evidence has accumulated that cells respond to low-dose radiation in ways that appear to contradict one or more of these assumptions. Some of these phenomena currently studied are low-dose hypersensitivity (HRS), increased radiation radioresistance (IRR), the adaptive response (AR), the bystander effect (BE), and the death-inducing factor (DIE). The first three of these will be discussed here in the context of cell death and cell survival while BE and DIE will be dealt with in Chapter 15.

14.3.1   Low-Dose Hypersensitivity

Low-dose hypersensitivity has been reviewed by Marples et al. [21], Raaphorst and Boyden [22], Leskov et al. [23], and Joiner et al. [24]. Systematic studies of cell survival after low doses of radiation became possible with the development of automated, accurate cloning assays (described in Chapter 4), which overcame the imprecision associated with the classical clonogenic assay in the low-dose range where only a few cells are actually killed. Using the microscope relocation technique [25] or flow cytometry-based methods [26], it was possible to know the exact number of cells that were exposed to radiation. Using the former technique, V79 Chinese hamster cells were initially discovered to be hypersensitive to radiation doses below 25 cGy, a phenomenon that has since become known as low-dose hyper-radiosensitivity (HRS) [25]. It was shown that as the radiation dose was increased to 1 Gy, the cell population became increasingly resistant per unit dose (Figure 14.2), a phenomenon that was named increased radioresistance (IRR). Although these observations were novel for mammalian cells, atypical survival responses similar to HRS/IRR had been reported in lower cell systems (reviewed in Joiner et al. [27]). As is apparent from Figure 14.2 the interplay of HRS and IRR can lead to striking results, where, within a certain narrow dose range a lower dose leads to greater cell killing than does a higher dose.

Hyper-radiosensitivity at low doses is characterized by a slope α_s that is steeper than the back-extrapolate of the high-dose survival curve (α_r)

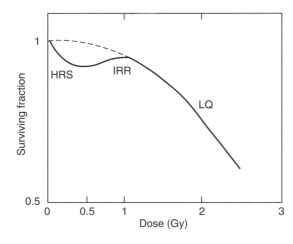

FIGURE 14.2
Response to radiation at low doses. Low dose hyper-radiosensitivity (HRR) is followed by an inducible radioprotective response (induced radioresistance, IRR). At higher doses, the data can be fitted by the linear quadratic model (LQ).

(Figure 14.3). The LQ model can be modified to take account of HRS/IRR to give the induced repair (IndRep) model [25]:

$$S = \frac{N}{N_0} = e^{\left\{-\alpha_r D\left[1+(\alpha_s/\alpha_r-1)e^{-D/D_c}\right]-\beta D^2\right\}} \qquad (14.8)$$

D_c is the dose at which the transition from HRS starts to occur (about 0.2 Gy). This equation tends to the LQ model with active parameters α_r and β. The IndRep model in fact comprises two LQ models with different α sensitivities merged into a single equation.

HRS has been reported for more than 40 human cell lines for different radiation qualities and for a variety of biological end points [25,28]. A minority of cell lines does not exhibit HRS (for example, Chinese hamster

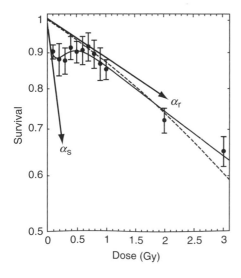

FIGURE 14.3
Low-dose clonogenic cell survival of V79 hamster fibroblasts irradiated with 240 kVp x-rays. The solid line shows the fit to the data set using the induced repair model. The dashed line represents the fit using the linear-quadratic model over the dose range of 0.5–10 Gy. The α_s represents the low-dose value (derived from the response at very low doses) while α_r is the value extrapolated from the conventional high-dose response. The ratio of α_s:α_r is used as a measure of the HRS response. (From Marples, B., Wouters, B.G., Collis, S.J., Chalmers, A.J., and Joiner, M.C., *Radiat. Res.*, 161, 247, 2004. With permission.)

ovary and human U373 glioma cells) and these have a survival response usually described by a pure linear-quadratic equation [29]. The majority of cell lines that do express the HRS response show considerable variation between different cell lines, the response being more prominent in malignant cells than normal tissues [24]. The α_s/α_r ratio (Figure 14.3) that compares the initial slopes of the fits for low- and high-dose radiation to the cell-survival curve, which has been adopted as quantitative measure of HRS, typically varies from 7 to 10. Hyper-radiosensitivity is observed after fractionated X irradiation in tumor model systems [24] and to a lesser extent in normal epithelial tissue during radiotherapy [30] indicating recoverability of HRS between fractions.

14.3.1.1 HRS Requires a Threshold Level of DNA Damage

Investigations of the adaptive response (described below) had established that for the full induction of repair processes it was necessary to exceed a threshold level of radiation injury. It was found, for instance, that if cells were pre-exposed to small priming doses of x-rays, the HRS response was eliminated in subsequently x-irradiated cell populations [31]. Similar findings had been reported for the adaptive response in lower organisms and in insect cells [27].

The threshold effect can be overcome by pre-radiation treatment with certain agents other than radiation. Low concentrations of hydrogen peroxide do not eliminate HRS, whereas a 100-fold higher concentration is effective [31]. These findings for H_2O_2 suggest that the formation of DNA double-strand breaks might be an important factor for activation of IRR. Cells that expressed HRS after small single exposures at acute dose rates were also reported to be sensitive to much larger doses of continuous radiation exposures given at very low-dose rates [32]. Lowering the dose rate from 1 Gy/h to 2–5 cGy/h enhanced net radiosensitivity by about a factor of 4, supporting the hypothesis that DNA damage must exceed some threshold to induce a protective pathway. This finding, however, is not in agreement with observations on the adaptive response (AR, Section 14.3.2). The AR can occur after the lowest possible dose, the passage through the cell of a single low LET track. At this dose level there is no cell killing, which is the endpoint measured by HRS and IRR.

14.3.1.2 HRS and DNA Repair

The requirement for nuclear-located radiation injury and by implication DNA damage in the HRS/IRR response was shown in charged-particle microbeam studies [33]. DNA double-strand breaks are the most relevant lesion in radiation-induced cell killing. Radiation-induced DNA DSBs are repaired efficiently by the process of nonhomologous end joining (NHEJ), with homologous recombination contributing to repair of breaks during the S and G_2 phase of the cell cycle. The importance of NHEJ in the phenomenon of HRS/IRR was indicated by the failure to detect a defined HRS response in the low-dose survival curves of Ku80-deficient XR-V15B cells [34]. Subsequently, an association was established between functionality of the NHEJ repair pathway and HRS/IRR in a panel of eight cell lines [35]. Further investigations indicated that initial response (HRS) is independent of DNA-PK activity, whereas the activation of IRR showed a dependence on the presence of PRKDC protein and functional DNA-PK activity.

As was described in Chapter 8, recognition of DNA damage after radiation exposure triggers downstream pathways which in turn activate cell-cycle arrest, DNA repair, and apoptosis. Activation of these biological responses has been linked with the production of strand breaks in DNA [31] and the role of initiating these responses has been attributed in mammalian cells to a small group of damage-sensing molecules that have been described. Ataxia telangiectasia mutated (ATM) protein kinase has emerged as a central protein in the recognition of DNA DSBs and the activation of several downstream biological pathways that control cell-cycle progression and DNA repair (Chapter 8). The kinase activity of ATM is rapidly induced in response of radiation-induced damage. A model has recently been described for the activation of ATM after DNA damage, involving the rapid dual autophosphorylation of the inactive 1981 serine of ATM dimers [36]. This process was shown to be exquisitely responsive to modifications in DNA structure or strand damage over the dose range of 0–1 Gy, with maximal phosphorylation occurring at 40 cGy. Half of the protein is activated within 15 min of radiation exposure, suggesting an early involvement of ATM in the damage signal transduction pathway. The rapid response and low damage threshold required to activate this system suggest that the ATM protein would be a credible candidate for involvement in the HRS/IRR response.

Another damage-sensing molecule is poly(ADP-ribose) polymerase (PARP), which is known to be efficiently activated by DNA strand breaks (Chapter 7), and loss of functional PARP activity negatively affects the ability of cells to repair radiation injury. The importance of PARP-mediated DNA damage-sensing pathways for inactivating HRS was initially suggested by the results of experiments in which the PARP inhibitor 3-aminobenzamide activity prevented the development of HRS in V79 Chinese hamster cells [37]. Later experiments using a more potent and specific PARP inhibitor, PJ34 (findings summarized by Marples et al. [21]) substantiated the importance of PARP as a damage sensor and activator of IRR.

14.3.1.3 *HRR/IRS in Relation to the Cell Cycle*

It has been shown that HRS is strongly dependent on cell-cycle phase [38,39] and, in fact, that hypersensitivity at low doses and the change in radiation response between 0.5 and 1 Gy appear to be limited to cells in the G_2 phase of the cell cycle.

Using hamster V79 and human cultures, Joiner and coworkers have shown that only the cell subpopulation in G_2 at the time of irradiation exhibits HRS/IRR, whereas clonal survival of cell populations in S and G_1 follow LQ functions. A population of V79 cells enriched in G_1 (92%) or S (60%)-phase cells did not show HRS, whereas G_2-phase-enriched showed a pronounced HRS response compared with asynchronous populations [21,39]. It has been shown that HRS is response specific to G_2 phase cells [39] and a direct link has been established between HRS and failure to activate a transient ATM-dependent early G_2 checkpoint [21]. Two distinct checkpoints in the G_2/M phase have been described. The first, which has been known for many years, is due to the accumulation of cells in G_2 that were irradiated in the G_1 or S phase of the cell cycle. The second, recently described checkpoint is believed to protect radiation-damaged G_2-phase cells from prematurely entering mitosis [40]. This checkpoint is independent of dose over the range 1–10 Gy with a distinct threshold for activation at 40 cGy and is transient over a short time-frame. The dependence of this checkpoint on ATM is reflected by the fact that the dose response for activation of ATM parallels the dose response for activation of the checkpoint. It has been hypothesized by Joiner and coworkers that failure to activate this checkpoint at doses less than 40 cGy would be manifest as a radiosensitive phenotype since radiation-damaged G_2-phase cells would proceed into mitosis carrying unrepaired breaks, a situation that would cause a survival response analogous to HRS. Moreover, the G_2-phase specificity of this highly threshold-dependent checkpoint would imply an enhanced HRS response for G_2-phase-enriched cell populations, as has been demonstrated [39].

In summary, activation of the G_2/M checkpoint occurs in the same dose range as induction of IRR and the dose required to change from the HRS to the IRR response in the survival experiments corresponds to the activation point of the G_2/M checkpoint; a relationship has been demonstrated between DNA repair by NHEJ and the occurrence of HRS/IRR. HRS/IRR is known to be predominant in G_2-phase cells, and may, in fact be exclusive to G_2-phase cells; an association has been established between the early G_2 cell-cycle checkpoint and the occurrence of low-dose hypersensitivity. Based on this evidence it has been hypothesized that the activation of ATM and the transition from an HRS to the IRR survival response due to increased NHEJ activity are linked through the G_2-phase-specfic cell-cycle checkpoint mechanism. It should be noted, however, that though the evidence implicating the G_2/M checkpoint in HRS/IRR is extensive, it is to date largely circumstantial.

14.3.2 Adaptive Response

The adaptive response has been reviewed by Broome et al. [41], Wolff [42], and Kadhim et al. [43]. The phenomenon of radiation-induced adaptive response was first described for chromosomal aberrations in human lymphocytes after radiation, where it was demonstrated that the level of chromosomal aberrations observed after radiation with x-rays was less than expected when cells were irradiated with low doses of β particles as a result of having been labeled with ³H-thymidine [44].

It is now known that enhanced resistance to the deleterious effects of ionizing radiation and other DNA-damaging agents is a commonly observed response after exposure to low doses of radiation. This adaptive response is evolutionarily conserved and has been observed in human and other mammalian cells, and in humans and animals, as well as in lower eukaryotes. In terms of radiation response, adaption increases the rate of DNA repair [45,46], reduces the frequency of radiation-induced and spontaneous neo-plastic transformation in rodent [47] and human cells [48], and has been shown to increase tumor latency in mice [49]. Adaption has been reported in response to both low-LET (x-rays, γ-rays, β particles) and to high-LET (neutrons, α particles) radiation [50].

14.3.2.1 Time and Dose Relationships for the Adaptive Response

At very low doses, the phenomenon known as hyper-radiosensitivity pre-faces the adaptive response. Adaptive responses have typically been detected by exposing cells to a low radiation dose and then challenging the cells with a higher dose of radiation and comparing the outcome to that seen with the challenge dose only (Figure 14.4). For an adaptive response to be seen, the challenge dose is usually delivered within 24 h of the inducing dose defining the time frame for delivery of a priming dose, i.e., the dose required to induce an effective adaptive response. The dose range is 0.01–0.2 Gy with low linear energy transfer radiation for radiation delivered at conventional dose rates. When the priming dose is over 0.2 Gy, adaptive responses

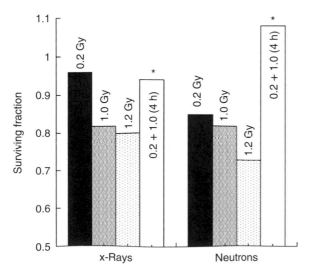

FIGURE 14.4
Survival values for V79 cells. Cells were treated with a priming dose of 0.2 Gy x-rays or neutrons followed by a challenge dose of 1.0 Gy x-rays. Treatment groups received 0.2 Gy x-rays or neutrons only. 1.0 Gy x-rays or neutrons only. 0.2 Gy x-rays or neutrons plus 1.0 Gy x-rays, interval 0 h. 0.2 Gy x-rays or neutrons plus 1.0 Gy x-rays, interval 4 h. (From Marples, B. and Skov, K.A., *Radiat. Res.*, 146: 382, 1996. With permission.)

are only barely induced and when it is over 0.5 Gy, adaptive responses are almost never induced [51]. In contrast, low-dose rate exposure (1–3 m Gy/min) to any dose between 1.0 and 500 m Gy of ^{60}Co γ-rays or ^{3}H β particles produced an adaptive response, which reduced the number of micronuclei seen in cells subsequently exposed to 4 Gy [41].

With regard to whole-body response, experiments with mice revealed that adaptive responses were not seen when mice were irradiated with less than a 0.025 cGy priming dose, the acquisition of radioresistance via the adaptive response occurred during a 2 week period at 2–2.5 months after irradiation when 0.05–0.1 Gy was used as a priming dose, adaptive responses were not found again when mice were subsequently irradiated with a 0.15–0.2 Gy priming dose, and the adaptive responses were observed again 2 weeks after a priming irradiation with 0.3–0.5 Gy [52].

14.3.2.2 Mechanisms of the Adaptive Response

A number of mechanisms have been proposed for the adaptive response, which are not necessarily mutually exclusive.

1. Increased DNA repair resulting from radiation-induced synthesis of DNA repair enzymes. The adaptive response is inhibited by preventing protein synthesis and particularly of the proteins that are involved in DNA damage responses including DNA repair. An important example is PARP, inhibition of which results in the abrogation of an adaptive response [53]. In mammalian cells, direct evidence for induction of DNA repair as part of the adaptive response comes from demonstration of repair of DNA base damage at a dose of 0.25 Gy. Transcription and translation of genes that participate in DNA repair and in cell-cycle regulation were required for the adaptive response in human lymphocytes [53]. In lower eukaryotes, the adaptive response is attributable to induced homologous repair [54].

2. Up-regulation of cytokine signaling pathways. The STAT1 proteins, components of the cytokine IFN signaling pathway were significantly up-regulated during acquired tumor radioresistance and were able to confer radioresistance when expressed in sensitive cells [55].

3. Homologous DNA repair: Sister-chromatid exchanges decreased in adapted cells, suggesting that either homologous, recombination repair between sister chromatids is being replaced by nonhomologous end joining or error-prone homologous recombination repair [43]. This might explain why an increase in the complexity-induced mutations was observed.

4. Activation of PKC in adapted cells exposed to low radiation doses was reported by a number of investigators (reviewed by Matsumoto

et al. [56]). Activation of PKC was required for radiation adaptive responses in murine m5S cells, and the intracellular signal transduction pathway induced by protein phosphorylation with PKC is a key step in the signal transduction pathways induced by low-dose irradiation. It was further shown that in cultured murine cells, adaptive responses were mediated by a rapid and robust feedback in the signal transduction pathway involving the activation of PKCα and p38MAPK with possible feedback via p38MAPK-associated PLCδ1.

5. Accumulation of p53 that occurs after irradiation with high-dose rate radiation (1.0 Gy/min, 5 Gy) was strongly suppressed by a priming dose of low-dose rate irradiation (0.001 Gy/min, 1.5 Gy). This finding led to the proposal that repression of p53-dependent responses is one of the mechanisms involved in the radiation adaptive response. p53-dependent apoptosis after exposure to high-dose rate radiation was found to be suppressed by a priming low-dose irradiation in cultured cells in vitro and in the spleens of mice in vivo [57]. In contrast, there was no suppression of p53-dependent apoptosis after a priming dose exposure in spleens of SCID mice, suggesting that DNA-PK activity might play a role in the radiation-induced adaptive response following priming with a low-dose exposure.

6. Exposure to low nontoxic doses of radiation (0.02–0.5 Gy) induced secretion of clusterin, in human cultured cells in vitro and in mice in vivo, leading to cytoprotective responses and suggesting a possible role for secreted clusterin in the adaptive response [58]. (Clusterin is a radiation-inducible protein that binds Ku-70 and triggers apoptosis when over-expressed in MCF-7 cells).

Overall it appears that adaptive responses (reviewed by Kadhim et al. [43]) lead to an increase in cloning efficiency and to an increased complexity of mutations, there was a decrease in transformation efficiency, mutation frequency, sister-chromatid exchanges, and micronucleus frequency. A decrease in micronucleus formation suggests that there are fewer chromosomal breaks in cells after a challenging dose, which would be expected if the radiation damage is repaired sooner or more efficiently in these cells.

14.4 Interactions of Adaptive Response and Bystander Effects

The bystander effect, which is defined as the observation of a biological response in cells that have not been directly traversed by ionizing radiation but which results from signals initiating in cells in which energy has been deposited, is described at greater length in Chapter 15. The bystander effect and the adaptive response are both measured by immediate or short-term effects, such as gene-expression alterations, apoptosis, sister-chromatid exchange, and micronucleus induction. The two processes would appear likely to have opposing effects on response. A contribution from bystander effects would tend to inflate the risk at low doses, whereas contribution of the adaptive response would reduce the risk. Possibly because the protocols used in assessment of bystander effect and adaptive response are fundamentally different, very few studies have attempted to directly relate these low-dose phenomena. One study that has measured bystander and adaptive responses in the same cell lines was that of Mothersill et al. [59], where it was demonstrated, using 13 different human cell lines, that a weak inverse relationship existed between the adaptive response and bystander effects. The cell lines showing the smallest bystander response and the largest adaptive response were least efficient at cell-to-cell communication as well as being the most malignant and the most rapidly dividing of all the cell lines. This suggests that the bystander effect was lost in more malignant cells due to a reduced ability to convey bystander signals.

In fact, the adaptive response itself constitutes a bystander effect. Using an assay of mutagenesis it has been shown that nonirradiated cells acquire mutagenesis through direct contact with cells whose nuclei had previously been traversed by either 1 or 20 α particles each (described in Chapter 15). Figure 14.5 shows results of a later experiment which showed that pretreatment with x-rays 4 h before α-particle irradiation significantly reduced this bystander mutagenic response [60].

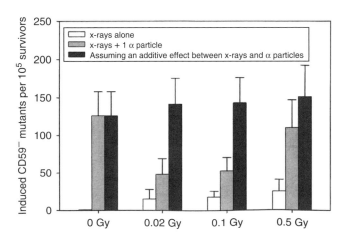

FIGURE 14.5
Effect of treatment with x-rays on bystander mutagenesis in A_L cells. Cells were pretreated with graded doses of x-rays 4 h before targeted nuclear radiation of 10% of randomly selected cells with a single α particle. Bars represent ± S.D. (From Zhou, H., Randers-Pehrson, G., Geard, C.R., Brenner, D.J., Hall, E.J., and Hei, T., *Radiat. Res.* 160, 512, 2003. With permission.)

14.5 Implications of Low-Dose Effects for Risk Assessment

14.5.1 Exposure to Background Radiation

An average dose to human cells from environmental sources is of the order of 0.01 m Sv/day (approximately 0.01 m Gy/day for γ-rays), generally in the form of high-energy γ-rays [61]. At this dose level about 1% of cells will be hit per day. The most likely outcome is single-strand breaks or base damage, but 3% of affected cells will sustain a DNA double-strand break. Since each human being is made up of 10^{14} cells, approximately 10^{10} of an individual's cells will sustain a DSB each day from background ionizing radiation. This compares with most studies of nontargeted radiation effects that have used exposures in the range of 0.01–1 Gy, corresponding to 0.3–30 DNA DSBs per cell. This implies that the dose attributed to radiation from environmental and occupational sources is presumably subject to the same modulation due to nontargeted effects as are the low doses used experimentally.

14.5.2 Adaptive Response and Neoplastic Transformation

Several investigations have examined the adaptive response using the end point of neoplastic transformation (reviewed by Redpath et al. [62]). Exposure to low doses of radiation has been reported to reduce the effectiveness of a subsequent challenge dose in inducing neoplastic transformation. In C3H 10T1/2 cells, exposure to doses in the 0.1–10 cGy range resulted in the suppression of the transformation frequency to a level significantly below that seen for spontaneous transformation of unirradiated cells. These studies were performed with a clone of C3H 10T1/2 cells that had an unusually high spontaneous transformation frequency indicative of genetic instability, but the result was subsequently confirmed for a dose of 1 cGy using the genetically stable HeLa x skin fibroblast human hybrid cell system. Studies with the hybrid cell system indicated an adaptive response against spontaneous transformation for doses up to 10 cGy γ-radiation.

This suggests that the adaptive response induced by a low preliminary dose of radiation might be effective in protecting against induction of neoplastic transformation. The relevance of in vitro studies on this topic is debatable since all quantitative studies of radiation-induced transformation in vitro use cell lines that are immortalized and thus already partly along the pathway to neoplastic transformation. It would be reasonable to assume that at high doses these cells may be more susceptible to neoplastic transformation than their primary counterparts and than normal tissues in vivo, but it is not known if this is the case at low doses. A more powerful adaptive response was seen for an unstable clone of C3H 10T1/2 cells [47] than for a more stable cell line with a much lower spontaneous transformation frequency [62]. On this limited basis it might be inferred that adaptive responses are more effective in a genetically unstable background i.e., in precancerous cells than in primary cells and tissues.

Risk estimations for radiation protection purposes apply a dose and dose-rate effectiveness factor to high-dose data, implying a linear-quadratic dose–response curve. It has been questioned whether an adaptive response component should be factored in for human in vivo risk estimation and there is, in fact, evidence for an adaptive response at the lowest doses analyzed in at least some epidemiological data including breast cancer [63] and leukemia [64].

14.6   Clinical Implications of Low-Dose Effects

Both the adaptive response and hyper-radiosensitivity are potentially important considerations where radiotherapy is concerned. Hyper-radiosensitivity, for example, might be exploited by using several very small doses to selectively kill sensitive populations, however since both genomic instability and bystander effects are effectively induced by very low doses of ionizing radiation (described in Chapter 15), this should be approached with care. As is the case for other low-dose effects, extensive variability in the adaptive response has been reported in human cell lines [27].

It has been proposed that ultra-fractionation, i.e., the application of multiple fractions of 0.5 Gy/day to the same total dose and within the same treatment time as used for conventional fractionation might improve the outcome of radiotherapy for radioresistant tumors that exhibit HRS (reviewed by Krause et al. [65]). Ultra-fractionation has been sporadically applied in patients with glioblastoma multiforme but no systematic clinical data is available. An in vivo study using ultra-fractionated treatment of tumors derived from a human tumor cell line that demonstrated HRS in vitro failed to demonstrate any evidence that HRS influenced sensitivity in vivo. In another in vivo experimental study, a rat rhabdomyosarcoma was irradiated with 126 fractions in 6 weeks. The results were equivocal since the effects of ultra-fractionation were only slightly greater than those of conventional fractionation, compared with historical controls. It is possible that HRS might only occur under the standardized growth conditions in vitro and not in tumors in vivo where interactions between tumor cells and stroma and microenvironmental factors such as hypoxia will affect the outcome. Other factors that might influence response are the cell-cycle distribution, which is likely to be different in monolayer cultures in vitro from that of tumors in vivo, and the possibility that HRS is only apparent after one or a small number of fractions. The failure to demonstrate a convincing effect in vivo with ultra-fractionation does not disprove HRS, nor is negation a possible role for ultra-fractionation. It does indicate, however, that simple extrapolation from experiments with single doses or few fractions in vitro would not be sufficient to predict the outcome of ultra-fractionation in vivo.

14.7 Summary

Mammalian cells die by a number of processes including apoptosis, senescence, necrosis, autophagy, and mitotic catastrophe. The principal modes of radiation-induced cell death or loss of clonogenicity are apoptosis, mitotic catastrophe, and prolonged cell-cycle delay. Following a lethal dose of radiation, death of the organism is caused by the failure of a few sensitive tissues, particularly the hematopoietic system and epithelium of the small intestine, to repair damage. The apoptosis that prevails in tissues with a rapid turnover or a potential for rapid turnover is p53-dependent. Some p53-competent cells do not respond to radiation by apoptosis, an example being fibroblasts that go into irreversible p53-dependent cell-cycle arrest.

Compared with normal tissue radiation, killing of tumor cells is less likely to occur by apoptosis and much more likely to result from mitotic catastrophe or in senescence-like irreversible growth arrest. Inactivation of p53, which occurs frequently in tumors, has several possible outcomes in terms of tumor response to radiation treatment. In reality, loss of p53 is most frequently associated with an unfavorable prognosis because the overriding effect of p53 loss is to make cells genetically unstable promoting increased proliferation, invasion, and progression.

The clonogenic cell-survival curve does not distinguish between different modes of cells killing the irradiated population. The target theory model of cell survival is based on the concept that a number of critical targets have to be inactivated for the cell to be killed.

The simplest survival model based on target theory is the single-hit, single-target survival curve that gives a straight line on a semilogarithmic plot. The dose to reduce survival to 0.37 (the D_{37} or D_0 dose) is the dose that will give an average of one hit per target or one inactivating event per cell. Other variants of target theory-based survival curves are the multitarget, single-hit survival and the two-component model. The linear quadratic model is based on the idea that multiple lesions induced by radiation interact in the cell to cause cell killing. The lesions that interact could be caused by a single ionizing track giving a direct dependence on dose or by two or more independent tracks giving a dependence on a higher power of dose. Survival curves are adequately fitted by the LQ model at least over the first few decades of cell kill. This model has taken precedence as the model of choice to describe survival curves both experimentally and clinically.

Evidence has accumulated that cells respond to low-dose radiation in ways contradicting the assumptions of classical radiation biology. One example is the fact that many cell lines are hypersensitive to radiation doses below 25 cGy, a phenomenon that become known as low-dose hyper-radiosensitivity (HRS). As the radiation dose increases to 1 Gy, the cell population becomes increasingly resistant per unit dose, a phenomenon called increased radioresistance (IRR). Manifestation of HRS/IRR requires a threshold level of DNA damage and functionality of the NHEJ repair pathway. The rapid response and low damage threshold required to activate the system suggest that the ATM protein might be involved as a damage sensor and the importance of PARP as a damage sensor and activator of IRR has also been demonstrated. HRS appears to be limited to cells in the G_2 phase of the cell cycle and it has been hypothesized that failure to activate this checkpoint at doses below 40 cGy would be manifest as a radiosensitive phenotype, since radiation-damaged G_2-phase cells would proceed into mitosis carrying unrepaired breaks, a situation that would give a survival response similar to HRS. It has been suggested that ultra-fractionation, i.e., the use of multiple low-dose fractions (0.5 Gy) would improve the outcome of radiotherapy, but it has not been possible to demonstrate an effect using in vivo tumor models.

The adaptive response is seen as enhancement of resistance to the effects of ionizing radiation when the challenge dose is preceded by a low dose of radiation; for low-LET radiation, protection is seen when the challenge dose is delivered within 24 h of the inducing dose (0.01–0.2 Gy). A number of mechanisms have been proposed for the adaptive response including increased DNA repair resulting from radiation-induced synthesis of DNA repair enzymes, up-regulation of cytokine signaling pathways, homologous DNA repair, and activation of PKC.

A comparison of the expression of adaptive response and bystander effects showed a weak inverse proportionality between the two effects. The two processes would be likely to have opposing effects on response. Exposure to low doses of radiation have been reported to reduce effectiveness of a subsequent challenge dose of radiation in inducing neoplastic transformation, and suppression of transformation frequency below that of spontaneous transformation has been reported. The importance of the adaptive response in terms of risk estimation is unclear although there is evidence of an adaptive response at very low doses from some epidemiological data.

References

1. Gudkov AV and Komarova EA. The role of p53 in determining sensitivity to radiotherapy. *Nat. Rev. Cancer* 3: 117–129, 2003.
2. MacCallum DE, Hupp TR, Midgley CA, Stuart D, Campbell SJ, Harper A, Walsh FS, Wright EG, Balmain A, Lane DP, and Hall PA. The p53 response to ionising radiation in adult and developing murine tissues. *Oncogene* 13: 2575–2587, 1996.
3. Komarova EA, Chernov MV, Franks R, Wang K, Armin G, Zelnick CR, Chin DM, Bacus SS, Stark G, and Gudkov AV. Transgenic mice with p53-responsive lacZ: p53 activity varies dramatically during normal development and determines radiation and drug sensitivity in vivo. *EMBO J.* 16: 1391–1400, 1997.
4. Komarov PG, Komarova EA, Kondratov RV, Christov-Tselkov K, Coon JS, Chernov MV, and Gudkov AV. A chemical inhibitor of p53 that protects mice from the side effects of cancer therapy. *Science* 285: 1733–1737, 1999.

5. Kemp CJ, Sun S, and Gurley KE. p53 induction and apoptosis in response to radio- and chemotherapy in vivo is tumor-type-dependent. *Cancer Res.* 61: 327–332, 2001.

6. Nieder C, Petersen S, Petersen C, and Thames HD. The challenge of p53 as prognostic and predictive factor in gliomas. *Cancer Treat. Rev.* 26: 67–73, 2000.

7. Johnstone RW, Ruefli AA, and Lowe SW. Apoptosis: A link between cancer genetics and chemotherapy. *Cell* 108: 153–164, 2002.

8. Brown JM and Wouters BG. Apoptosis: Mediator or mode of cell killing by anticancer agents? *Drug Resist. Update* 4: 135–136, 2001.

9. Sierra A, Castellsague X, Escobedo A, Lloveras B, Garcia-Ramirez M, Moreno A, and Fabra A. Bcl-2 with loss of apoptosis allows accumulation of genetic alterations: A pathway to metastatic progression in human breast cancer. *Int. J. Cancer* 89: 142–147, 2000.

10. Deveraux QL, Roy N, Stennicke HR, Van Arsdale T, Zhou Q, Srinivasula SM, Alnemri ES, Salvesen GS, and Reed JC. IAPs block apoptotic events induced by caspase-8 and cytochrome *c* by direct inhibition of distinct caspases. *EMBO J.* 17: 2215–2223, 1998.

11. Beere HM, Wolf BB, Cain K, Mosser DD, Mahboubi A, Kuwana T, Tailor P, Morimoto RI, Cohen GM, and Green DR. Heat-shock protein 70 inhibits apoptosis by preventing recruitment of procaspase-9 to the Apaf-1 apoptosome. *Nat. Cell Biol.* 2: 469–475, 2000.

12. Roninson IB, Broude EV, and Chang BD. If not apoptosis, then what? Treatment-induced senescence and mitotic catastrophe in tumor cells. *Drug Resist. Update* 4: 303–313, 2001.

13. Ianzini F and Mackey MA. Delayed DNA damage associated with mitotic catastrophe following X-irradiation of HeLa S3 cells. *Mutagenesis* 13: 337–344, 1998.

14. Schmitt CA, Fridman JS, Yang M, Lee S, Baranov E, Hoffman RM, and Lowe SW. A senescence program controlled by p53 and p16INK4a contributes to the outcome of cancer therapy. *Cell* 109: 335–346, 2002.

15. Bunz F, Hwang PM, Torrance C, Waldman T, Zhang Y, Dillehay LE, Williams JA, Lengauer C, Kinzler KW, and Vogelstein B. Disruption of p53 in human cancer cells alters the responses to therapeutic agents. *J. Clin. Invest.* 104: 263–269, 1999.

16. Cordon-Cardo C, Dalbagni G, Sarkis AS, and Reuter VE. Genetic alterations associated with bladder cancer. *Important Adv. Oncol.* 71–83, 1994.

17. Falette N, Paperin MP, Treilleux I, Gratadour AC, Peloux N, Mignotte H, Tooke N, Lofman E, Inganas M, Bremond A, Ozturk M, and Puisieux A. Prognostic value of p53 gene mutations in a large series of node-negative breast cancer patients. *Cancer Res.* 58: 1451–1455, 1998.

18. Chadwick KH and Leenhouts HP. A molecular theory of cell survival. *Phys. Med. Biol.* 13: 78–87, 1973.

19. Curtis SB. Lethal and potentially lethal lesions induced by radiation: A unified repair model. *Radiat. Res.* 106: 252–270, 1986.

20. Goodhead DT. Saturable repair models of radiation action in mammalian cells. *Radiat. Res.* 104: S58–S67, 1985.

21. Marples B, Wouters BG, Collis SJ, Chalmers AJ, and Joiner MC. Low-dose hyper-radiosensitivity: A consequence of ineffective cell cycle arrest of radiation-damaged G2-phase cells. *Radiat. Res.* 161: 247–255, 2004.

22. Raaphorst GP and Boyden S. Adaptive response and its variation in human normal and tumour cells. *Int. J. Radiat. Biol.* 75: 865–873, 1999.

23. Leskov KS, Criswell T, Antonio S, Li J, Yang CR, Kinsella TJ, and Boothman DA. When X-ray-inducible proteins meet DNA double strand break repair. *Semin. Radiat. Oncol.* 11: 352–372, 2001.

24. Joiner MC, Marples B, Lambin P, Short SC, and Turesson I. Low-dose hypersensitivity: Current status and possible mechanisms. *Int. J. Radiat. Oncol. Biol. Phys.* 49: 379–389, 2001.

25. Marples B and Joiner MC. The response of Chinese hamster V79 cells to low radiation doses: Evidence of enhanced sensitivity of the whole cell population. *Radiat. Res.* 133: 41–51, 1993.

26. Wouters BG and Skarsgard LD. Low-dose radiation sensitivity and induced radioresistance to cell killing in HT-29 cells is distinct from the "adaptive response" and cannot be explained by a subpopulation of sensitive cells. *Radiat. Res.* 148: 435–442, 1997.

27. Joiner MC, Lambin P, Malaise EP, Robson T, Arrand JE, Skov KA, and Marples B. Hypersensitivity to very-low single radiation doses: Its relationship to the adaptive response and induced radioresistance. *Mutat. Res.* 358: 171–183, 1996.

28. Joiner MC and Johns H. Renal damage in the mouse: The response to very small doses per fraction. *Radiat. Res.* 114: 385–398, 1988.

29. Bartkowiak D, Hogner S, Nothdurft W, and Rottinger EM. Cell cycle and growth response of CHO cells to X-irradiation: Threshold-free repair at low doses. *Int. J. Radiat. Oncol. Biol. Phys.* 50: 221–227, 2001.

30. Turesson I and Joiner MC. Clinical evidence of hypersensitivity to low doses in radiotherapy. *Radiother. Oncol.* 40: 1–3, 1996.

31. Marples B and Joiner MC. The elimination of low-dose hypersensitivity in Chinese hamster V79–379A cells by pretreatment with X rays or hydrogen peroxide. *Radiat. Res.* 141: 160–169, 1995.

32. Mitchell CR and Joiner MC. Effect of subsequent acute-dose irradiation on cell survival in vitro following low dose-rate exposures. *Int. J. Radiat. Biol.* 78: 981–990, 2002.

33. Schettino G, Folkard M, Prise KM, Vojnovic B, Bowey AG, and Michael BD. Low-dose hypersensitivity in Chinese hamster V79 cells targeted with counted protons using a charged-particle microbeam. *Radiat. Res.* 156: 526–534, 2001.

34. Skov KA, Marples B, Matthews JB, Joiner MC, and Zhou H. A preliminary investigation into the extent of increased radioresistance or hyper-radiosensitivity in cells of hamster cell lines known to be deficient in DNA repair. *Radiat. Res.* 138: S126–S129, 1994.

35. Vaganay-Juery S, Muller C, Marangoni E, Abdulkarim B, Deutsch E, Lambin P, Calsou P, Eschwege F, Salles B, and Bourhis J. Decreased DNA-PK activity in human cancer cells exhibiting hypersensitivity to low-dose irradiation. *Br. J. Cancer* 83: 514–518, 2000.

36. Bakkenist CJ and Kastan MB. DNA activates ATM through intermolecular autophosphorylation and dimer dissociation. *Nature* 421: 499–506, 2003.

37. Marples B and Joiner MC. Modification of survival by DNA repair modifiers: A probable explanation for the phenomenon of increased radioresistance. *Int. J. Radiat. Biol.* 76: 305–312, 2000.

38. Short SC, Woodcock M, Marples B, and Joiner MC. The effects of cell cycle phase on low dose hyper-radiosensitivity. *Int. J. Radiat. Biol.* 79: 99–105, 2003.

39. Marples B, Wouters BG, and Joiner MC. An association between the radiation-induced arrest of G2-phase cells and low-dose hyper-radiosensitivity: A plausible underlying mechanism? *Radiat. Res.* 160: 38–45, 2003.

40. Xu B, Kim ST, Lim DS, and Kastan MB. Two molecularly distinct G(2)/M checkpoints are induced by ionizing irradiation. *Mol. Cell Biol.* 22: 1049–1059, 2002.
41. Broome EJ, Brown DL, and Mitchel RE. Dose responses for adaption to low doses of (60)Co gamma rays and (3)H beta particles in normal human fibroblasts. *Radiat. Res.* 158: 181–186, 2002.
42. Wolff S. The adaptive response in radiobiology: Evolving insights and implications. *Environ. Health Perspect.* 106: 277–283, 1998.
43. Kadhim MA, Moore SR, and Goodwin EH. Interrelationship amongst radiation-induced genomic instability, bystander effect, and the adaptive response. *Mutat. Res.* 568: 21–32, 2004.
44. Olivieri G, Bodycote J, and Wolff S. Adaptive response of human lymphocytes to low concentrations of radioactive thymidine. *Science* 223: 594–597, 1984.
45. Azzam EI, de Toledo SM, Raaphorst GP, and Mitchel REJ. Réponse adaptative au rayonnement ionisant des fibroblasts de peau humaine. Augmentation de la vitesse de réparation de l'ADN et variation de l'expression des genes. *J. Chim. Phys.* 91: 931–936, 1994.
46. Broome EJ, Brown DL, and Mitchel REJ. Adaption of human fibroblasts to radiation alters biases in DNA repair at the chromosome level. *Int. J. Radiat. Biol.* 75: 681–690, 1999.
47. Azzam EI, de Toledo SM, Raaphorst GP, and Mitchel REJ. Low-dose ionizing radiation decreases the frequency of neoplastic transformation to a level below the spontaneous rate in C3H 10T1/2 cells. *Radiat. Res.* 146: 369–373, 1996.
48. Redpath JL and Antoniono RJ. Induction of an adaptive response against low-dose gamma radiation. *Radiat. Res.* 149: 517–520, 1998.
49. Mitchel REJ, Jackson JS, McCann RA, and Boreham DR. Adaptive response modification of latency for radiation-induced myeloid leukemia in CBA/H mice. *Radiat. Res.* 152: 273–279, 1999.
50. Marples B and Skov KA. Small doses of high-linear energy transfer radiation increase the radioresistance of Chinese hamster V79 cells to subsequent X irradiation. *Radiat. Res.* 146: 382–387, 1996.
51. Feinendegen LE. The role of adaptive responses following exposure to ionizing radiation. *Hum. Exp. Toxicol.* 18: 426–432, 1999.
52. Yonezawa M, Misonoh J, and Hosokawa Y. Two-types of X-ray-induced radioresistance in mice: Presence of 4 dose ranges with distinct biological effects. *Mutat. Res.* 358: 237–243, 1996.
53. Wolff S. Is radiation all bad? The search for adaptation. *Radiat. Res.* 131: 117–132, 1992.
54. Dolling JA, Boreham DR, Bahen ME, and Mitchel RE. Role of RAD9-dependent cell-cycle checkpoints in the adaptive response to ionizing radiation in yeast, *Saccharomyces cerevisiae*. *Int. J. Radiat. Biol.* 76: 1273–1279, 2000.
55. Khodarev N, Beckett M, Labay E, Darga T, Roizman B, and Weichselbaum R. STAT1 is overexpressed in tumors selected for radioresistance and confers protection from radiation in transduced sensitive cells. *Proc. Natl. Acad. Sci. USA* 101: 1714–1719, 2004.
56. Matsumoto H, Takahashi A, and Ohnishi T. Radiation-induced adaptive responses and bystander effects. *Biol. Sci. Space* 18: 247–254, 2004.
57. Takahashi A. Different inducibility of radiation- or heat-induced p53-dependent apoptosis after acute or chronic irradiation in human cultured squamous cell carcinoma cells. *Int. J. Radiat. Biol.* 77: 215–224, 2001.

58. Leskov KS, Klokov DY, Li J, Kinsella TJ, and Boothman DA. Synthesis and functional analyses of nuclear clusterin, a cell death protein. *J. Biol. Chem.* 278: 11590–11600, 2003.

59. Mothersill C, Seymour CB, and Joiner MC. Relationship between radiation-induced low-dose hypersensitivity and the bystander effect. *Radiat. Res.* 157: 526–532, 2002.

60. Zhou H, Randers-Pehrson G, Geard CR, Brenner DJ, Hall EJ, and Hei T. Interaction between radiation-induced adaptive response and bystander mutagenesis in mammalian cells. *Radiat. Res.* 160: 512–516, 2003.

61. Bonner WM. Phenomena leading to cell survival values which deviate from linear-quadratic models. *Mutat. Res.* 568: 33–39, 2004.

62. Redpath JL, Liang D, Taylor TH, Christie C, and Elmore E. The shape of the dose–response curve for radiation-induced neoplastic transformation in vitro: Evidence for an adaptive response against neoplastic transformation at low doses of low-LET radiation. *Radiat. Res.* 156: 700–707, 2001.

63. Miller AB, Howe GR, Sherman GJ, Lindsay JP, Yaffe MJ, Dinner PJ, Risch HA, and Preston DL. Mortality from breast cancer after irradiation during fluoroscopic examinations in patients being treated for tuberculosis. *N. Eng. J. Med.* 321: 1285–1289, 1989.

64. Little MP. A comparison of the degree of curvature in the cancer incidence dose-response in Japanese atomic bomb survivors with that in chromosome aberrations measured in vitro. *Int. J. Radiat. Biol. Relat. Stud. Phys. Chem. Med.* 76: 1365–1375, 2000.

65. Krause M, Hessel F, Wohlfarth J, Zips D, Hoinkis C, Foest H, Petersen C, Short SC, Joiner MC, and Baumann M. Ultrafractionation in A7 human malignant glioma in nude mice. *Int. J. Radiat. Biol.* 79: 377–383, 2003.

15

Bystander Effects and Genomic Instability

15.1 Dogma of Radiation Biology

If radiation biology has a dogma, it is that the deleterious effects of ionizing radiation result from the deposition of energy in the cell nucleus causing damage to a critical target (the nuclear DNA) with a major role assigned to the induction of DNA double-strand breaks. Radiation-induced cell death is attributable to failure to repair DNA damage resulting in cells which manifest residual DNA damage and die by mitotic failure or apoptosis. Surviving cells with misrepaired or unrepaired damage transmit changes to all descendant cells, i.e., the change is clonal. As malignant transformation is believed to be initiated by a gene mutation or a chromosomal aberration, the initiating lesion for malignant transformation has been similarly attributed to DNA damage in the directly irradiated cell. A litany of the results of early experiments which provided evidence for the nucleus as the sensitive target in the cell is part of the groundwork of any traditional radiation biology course or text as follows:

- Microbeam irradiation demonstrates the cell nucleus to be much more sensitive than the cytoplasm.
- Radioisotopes with short-range emissions (e.g., ^{3}H, ^{125}I) incorporated into the DNA cause cell killing at much lower absorbed doses than those incorporated into the cellular cytoplasm.
- Incorporation of thymidine analogues (e.g., iododeoxyuridine, bromodeoxyuridine) into DNA modifies cellular radiosensitivity.
- The level of chromatid and chromosomal aberrations following ionizing radiation correlates well with cell lethality.
- For different types of radiation, cell lethality correlates best with the level of radiation-induced DNA double-strand breaks rather than with other types of damage.
- The extreme radiosensitivity of some mutant cells is due to defects in DNA repair.

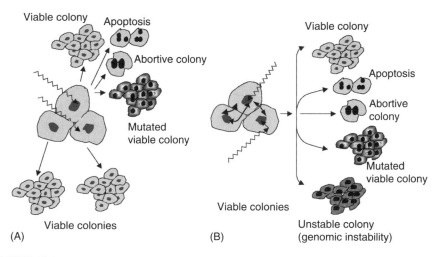

FIGURE 15.1

Two views of how radiation effects are perpetuated. (A) In the traditional clonal view, all the damage is introduced into the cell by the actual traversal of the nucleus by a track of ionizing radiation. If this causes a mutation, it is passed on to all the progeny. (B) In the new model, many different effects such as chromosome damage and mutations are introduced in cells that are not hit directly by the radiation beam. These effects may not be apparent until several generations after the actual exposure.

Like all dogmas, the radiation biology version is flawed in that it does not apply under all circumstances. The last 10 years has seen the publication of research describing biological responses which are observed in nonirradiated cells as a result of receiving signals from cells which have been irradiated. These responses have been called bystander effects, a term borrowed from the gene therapy literature where it referred to the killing of several types of tumor cells by targeting only one type of cell in a heterogenous population. In the radiation field, the bystander effect has come to be loosely defined as the induction of biological effects in cells that are not directly traversed by a charged particle but are in proximity to cells which have been traversed (Figure 15.1).

15.2   Bystander Effects

Reports going back to 1954 record that cells exposed to doses of low LET radiation can show an indirect effect, producing a plasma-borne factor which leads to chromosome breakage and cytogenetic abnormalities in human bone marrow or lymphocytes and causes tumors in rats (reviewed by Mothersill and Seymour [1]). These reports fall into two main categories: those concerning patients irradiated during the course of radiotherapy and those coming from survivors of atomic bomb blasts or individuals who had been exposed to radio-active fallout. Plasma from x-irradiated patients (30–40 Gy therapeutic x-rays) was found to cause chromosome damage in lymphocytes held in short-term culture and a transferable substance was produced by whole body irradiation which was capable of causing chromosome breaks in nonirradiated lymphocytes. Plasma from high-dose radiotherapy patients induced a variety of aberrations, including dicentrics and chromatid and chromosome breaks, in normal nonirradiated lymphocytes in short-term culture. The radiation-induced clastogenic factors in the plasma of irradiated patients were reported to be of low molecular weight and production of the factors involved lipid peroxidation and oxidative stress pathways. The factor was either very long lived or constantly regenerated since the presence of a clastogenic factor in plasma from atomic bomb survivors 31 years after exposure has been reported.

From the early 1990s, the development of stratagems for irradiation of single cells either with low fluences of α particles or with microbeams of charged particles has created both an extensive body of experimental data and an intense interest in the mechanisms and implication of bystander effects. Some of the evidence obtained using these techniques are reviewed in this chapter.

15.2.1 Bystander Effects In Vitro

One approach to the demonstration of bystander effects at the single cell level has been to observe the effect in nonirradiated cells of a single particle passing through an adjacent cell. Several techniques have been developed to achieve this.

15.2.1.1 Evidence for a Radiation-Induced Bystander Effect from Experiments Using Low Fluences of α Particles

Parts of this section are based on reviews by Little [2], Morgan [3,4], Mothersill and Seymour [1], and the references cited therein. One experimental system for the demonstration of bystander effects has used a broad beam of α particles at such a low fluence that, for a given exposure only a small proportion of cells are traversed by an α particle. Among the first reports of a bystander effect resulting from high LET irradiation was of sister chromatid exchanges in 30% of a cultured population Chinese hamster ovary (CHO) cells when less than 1% of cell nuclei were actually traversed by an α particle (a dose corresponding to 0.31 mGy) [5]. Similar results were later reported for normal human lung fibroblasts. An enhanced frequency of *hprt* mutations was also demonstrated in bystander CHO cells in cultures exposed to very low fluences of α particles. If cells were irradiated in the presence of an agent that disrupts lipid rafts and inhibits membrane signaling, the induction of both sister chromatid exchanges and mutations was suppressed. Resulting from this membrane-dependent bystander mechanism, there was a significantly higher frequency of mutations at low fluences than would be predicted from a linear extrapolation from the data for higher doses. The mutations arising in bystander cells were primarily point mutations, whereas those occurring in directly irradiated cells were largely partial and total gene deletions [6].

Changes in gene expression, notably up-regulation of the p53 damage-response pathway were shown to occur in bystander cells in monolayer cultures exposed to very low fluences of α particles [7]. This effect was mediated by gap-junction intracellular communication (GJIC). Signaling between cells in confluent monolayers was dependent on a connexin-mediated gap-junction transfer of the signal and no bystander effect was detectable in the presence of connexin inhibitors or when gap-junction null cells were used. The observation that p53 in bystander cells was phosphorylated on serine 15 suggested that the up-regulation of p53 in bystander cells is a consequence of DNA damage. The activation of the p53 damage-response pathways in bystander cells was confirmed by in situ immuno-fluorescence studies which showed that up-regulation of p21^{Waf1} occurred in clusters of cells in the monolayer population, of which only 12% of the cell nuclei had been traversed by an α particle (Figure 15.2A) [8].

The role of oxidative stress in modulating signaling between bystander and target cells was examined in confluent monolayer populations of human diploid cells exposed to low fluences of particles. Superoxide and hydrogen peroxide produced by flavin-containing oxidase enzymes mediated the activation of several stress-inducible p53 damage-response and the MAP kinase family signaling pathways as well as micronucleus formation in bystander cells. It was also reported that nitric oxide can initiate intercellular signal transduction pathways influencing the bystander response to radiation [8].

Cells defective in the NHEJ pathway (see Chapter 7), including mouse knockout cell lines for Ku80, Ku70, and DNA-PK$_{cs}$ were reported to be extremely sensitive to bystander-mediated induction of mutations and chromosomal aberrations [9,10]. The mutations in these repair-deficient bystander cells were primarily the result of partial and total gene deletions in contrast to those occurring in wild-type bystander cells which are mainly point mutations. The marked sensitization of repair-deficient bystander cells to the induction of mutations and chromosomal aberrations was attributed to un-rejoined DNA DSB resulting from clusters of unrepaired oxidative lesions and single-strand breaks, whereas mutations in wild-type cells in which DSB are repaired arose primarily from oxidative base damage.

In summary, results of the low fluence α particle experiments indicate that bystander effects are associated with up-regulation of oxidative metabolism and are mediated by ROS and possibly by NOS. Transmission of

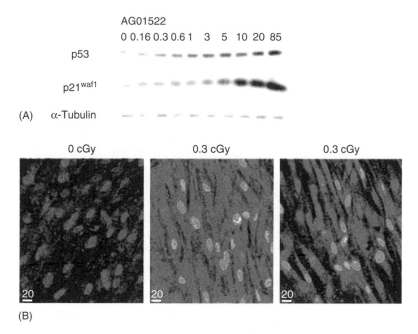

FIGURE 15.2
(A) Western blot analysis of p53 and p21^{Waf1} expression levels in low fluence-particle exposed normal human fibroblasts. Confluent density-inhibited AG1522 skin fibroblasts were exposed to α particles doses and held at 37°C for 3 h. Cell lysates from irradiated and control non-exposed cultures were prepared and examined. (B) In situ immunofluorescence detection of p21^{Waf1} in control and α particle exposed (0.3 cGy) AG1522 density-inhibited cultures where induction is seen to occur in aggregates of cells. About 1 cell in 50 would be traversed by an α particle at this mean dose. (From Azzam, E.I., de Toledo, S.M., and Little, J.B., *Proc. Natl. Acad. Sci. USA*, 98, 473, 2001. With permission.)

signals from irradiated to bystander cells is mediated by direct intercellular contact through gap junctions, and agents which disrupt membrane signaling also disrupt bystander signaling. The bystander effect is p53 dependent. In repair-competent cells, DNA damage resulting from bystander effects is largely point mutations whereas a direct damage from α particles is mainly total and partial gene deletions, except in the case of DNA repair-deficient cell lines which are sensitive to the induction of bystander-mediated DNA damage and where total and partial gene deletions are seen in bystander cells.

15.2.1.2 Evidence for Bystander Effects after Radiation with Charged-Particle Microbeams

Convincing demonstrations of the bystander effect have come from studies using charged-particle microbeams, which are capable of putting an exact number of particles through a specific subcellular compartment of a defined number of cells in a particular radiation environment.

15.2.1.3   Irradiation of the Cytoplasm

Experiments done using the microbeam at the Radiological Research Accelerator Facility of Columbia University targeted and irradiated the
cytoplasm of human–hamster hybrid (AL) cells [11]. Minimal cell kill was
observed, confirming the results of classical experiments in which radiation
was confined to the cytoplasm. However, there was a significant increase in
mutations at the CD59 (S1) nuclear gene locus. Cytoplasmic irradiation with
a single α particle doubled the spontaneous mutation frequency, while a
twofold to threefold increase was observed after four cytoplasmic traversals. The mutational spectrum was similar to the spontaneous mutation
spectrum in nonirradiated cells, but it was different from that observed
after targeted nuclear irradiation. Results of experiments in which irradiation was done in the presence of either the free-radical scavenger dimethyl
sulfoxide (DMSO) or the intracellular glutathione inhibitor buthionine-
S-R-sulfoximine (BSO) showed that the mutagenicity of cytoplasmic irradiation depends upon intercellular generation of reactive oxygen species
(ROS). Furthermore, while nuclear irradiation mutants were predominantly
large deletions, mutants induced by cytoplasmic irradiation consisted of
localized changes such as point mutations. From these experiments, it was
concluded that the target for genetic effects of radiation is larger than the
nucleus. Particle traversals of the cytoplasm may contribute a significant
proportion of overall mutant yield in the very low-dose region implying
that cytoplasmic traversal by α particles may be more dangerous than
nuclear traversal(s) because of the increased mutagenicity occurrence
where there is negligible killing of target cells [11].

15.2.1.4   Microbeams Targeted to the Nucleus

In the experiments using low fluences of α particles, the number of cells hit
by α particles was calculated from the fluence of α particles and the cross-
sectional area of the cell nucleus. Greater precision was achieved by the
development of a single-particle microbeam, which was able to put a predefined number of α particles through a target cell nucleus (Figure 15.3). In
a typical experiment, the cells in a monolayer are attached to a thin poly-
propylene base of a cell culture dish. Each cell is identified and located by
using an image analysis system, and its coordinates are stored in a computer. The cell dish is moved under computer control in such a way that the
target of irradiation; nucleus or cytoplasm is positioned over a highly
collimated shuttered beam of α particles generated by a Van de Graaff
accelerator. Each cell is exposed to a predetermined exact number of α
particles. This system has been used to demonstrate the bystander effect
for α particle irradiation for a variety of biological end points, including cell
lethality, gross damage to DNA, mutagenesis, and oncogenic transformation (reviewed by Hall and Hei [12]). Using an experimental protocol
which made it possible to irradiate a population of cells and then obtain
separate survival curves for hit and nonhit cells, it was shown that there is a

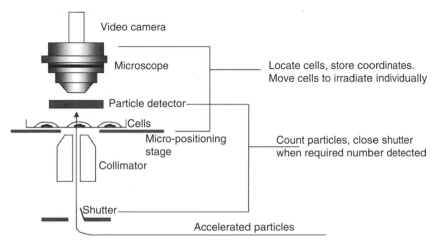

FIGURE 15.3
Schematic drawing showing the overall layout of a microbeam with integrated control system.

considerable degree of cell killing in the nonhit cells, implying a substantial bystander effect (Figure 15.4) [12].

Another end point investigated with the Columbia microbeam was mutagenesis. Twenty percent of randomly selected human–hamster hybrid (AL) cells were exposed to a near-lethal dose of 20 α particles directed to the nucleus. The surviving fraction was less than 1% but when assayed for mutations in human chromosome 11, the mutation yield was four times that of background. Since the irradiated cells were exposed to lethal doses of radiation, these mutations must have arisen in nonexposed bystander cells. The mutation spectrum observed in bystander cells was significantly different from the spontaneous spectrum and the spectrum seen after cytoplasmic irradiation, suggesting that different mutagenic mechanisms are involved [13]. Other experiments identified the importance of cell-to-cell communication via gap junctions as a mechanism of the bystander effect. When AL cells were transfected with a dominant-negative connexin 43 vector (DN6), which eliminates gap-junction communication, the bystander effect essentially disappeared. This effect was significantly reduced in cells pretreated with octanol, which inhibits gap-junction-mediated intercellular communication [14].

In studies of bystander-mediated oncogenesis mouse fibroblast (C3H 10T12), cells were plated in a monolayer and either every cell or every tenth cell selected at random was irradiated with one to eight α particles directed at the cell nucleus. The cells were later replated at low density and transformed foci, identified on the basis of morphology were counted 6 weeks later. The results indicated that more cells were inactivated than were actually traversed by α particles and that when 10% of the cells on a dish were exposed to two or more α particles; the resulting frequency of

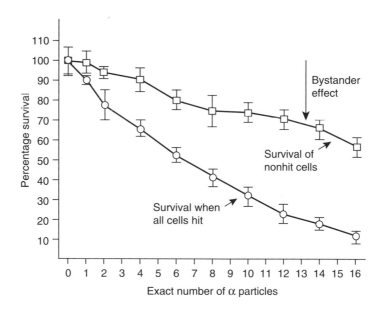

FIGURE 15.4
The bystander effect for cell survival in V79 cells. Each data point (mean ± SE) on the line with circles refers to the survival of cells when all cell nuclei on each dish were exposed to the exact number of α particle traversals using the microbeam system. The squares show survival for various numbers of α particles, from 1 to 16, traversing 10% of the cell population. The extent to which this falls below the 100% survival for nonhit cells is an indication of the magnitude of the bystander effect. (From Sawant, S.G., Zheng, W., Hopkins, K.M., Randers-Pehrson, G., Lieberman, H.B., and Hall, E.J., *Radiat. Res.*, 157, 361, 2002. With permission.)

induced oncogenic transformation was indistinguishable from that seen when all the cells on the dish were exposed to the same number of α particles (Figure 15.5) [12].

Results have been published from other microbeam facilities. At the Gray Cancer Institute in the United Kingdom, a proton beam with a LET of 100 keV/mm used for targeted cellular irradiation was adapted so that the effects of ultrasoft x-rays could be investigated. Using this microbeam, there was evidence for a bystander effect in primary human fibroblasts. Measurement of micronucleus formation and cell killing revealed significantly more damage than would be expected on the basis of direct effects of radiation [15]. In contrast to the Columbia microbeam experiments which were done at high cell density with 50%–60% of the cells in contact, cells irradiated in these experiments were hundreds of microns apart. In this case, it is unlikely that communication through gap junctions contributes to the effect which may more probably be mediated by radiation-induced soluble factors, which are released into the culture medium and affect nonirradiated cells. This suggested that there are at least two different

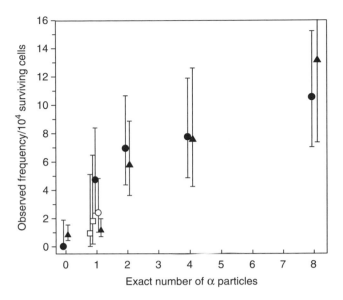

FIGURE 15.5

Yield of oncogenically transformed cells per 10^4 surviving C3H 10T1/2 cells produced by nuclear traversals by 5.3 MeV α particles. Triangles represent exposure of all cell nuclei on each dish to exact number of α particles, using the microbeam system. Solid circles represent exposure of 1 in 10 cell nuclei on each dish to exact number of α particles. Open squares represent subsequent repeats of the experiment in which 1 in 10 cell nuclei were exposed to exactly one α particle. Open circles represent combined data for all the experiments in which 1 in 10 cell nuclei were exposed to one α particle including these repeat experiments. (From Sawant, S.G., Randers-Pehrson, G., Geard, C.R., Brenner, D.J., and Hall, E.J., *Radiat. Res.*, 155, 397, 2001. With permission.)

types of bystander effect, those mediated by cell-to-cell gap-junction communication and those induced by secreted soluble factors.

15.2.2 Bystander Effects Seen after Transfer of Medium from Irradiated Cells

As described earlier in this chapter, there is a long historical record suggesting that irradiated individuals release into their plasma clastogenic factors that will induce chromosomal damage when transferred to cultured cells from nonirradiated donors. Such factors have since been reported to occur in populations of exposed individuals and to be associated with hereditary chromosomal breakage syndromes and pathological conditions, where they are thought to be biomarkers of oxidative stress.

Although evidence for these factors has been accumulating over the past half century, their exact nature has remained unknown, as have the mechanisms by which they could be continually produced for many months or years after radiation exposure. Recently, considerable evidence for the toxic effects of bystander signals has come from studies where culture media

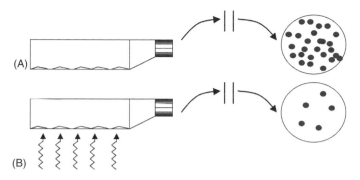

FIGURE 15.6
Procedure for demonstration of bystander effect by medium transfer. Medium from monolayer culture is filtered and transfered to nonirradiated cells and clonogenic cell survival is measured. (A) Transfer of medium from nonirradiated cells. (B) Transfer of medium from irradiated cells.

from irradiated established cell lines have been transferred to cultures of nonirradiated cell lines (reviewed by Mothersill and Seymour [1], Morgan [3], and Little [2]).

The Mothersill and Seymour group demonstrated that irradiated epithelial cells, but not fibroblasts, secreted a toxic substance, a so-called bystander factor, into the culture medium that could kill nonirradiated cells resulting in a significant reduction in plating efficiency in nonirradiated cells that received culture medium from irradiated cultures (shown schematically in Figure 15.6) [16]. Results indicated that medium irradiated in the absence of cells had no effect on survival when transferred to nonirradiated cells. Not all cells are capable of producing the toxic bystander factor, nor do all cells respond to the secreted signal. The effect was dependent on the cell number at the time of irradiation, could be observed as soon as 30 min postirradiation, and was still effective when medium transfer occurred 60 h after irradiation. The effect was independent of dose between 0.5 and 5 Gy (Figure 15.7).

The factor involved appears to be a protein since it is heat-labile but stable if frozen and does not require cell-to-cell contact to induce its effect in the recipient cells. The first detectable effect of medium containing the factor on recipient cells was a rapid (1–2 min) calcium pulse, which was followed 30 min–2 h later by changes in mitochondrial membrane permeability and the induction of ROS. The critical role of mitochondrial metabolism was suggested by the lack of signal production by cells that did not have a functional glucose-6-phosphate dehydrogenase (G6PD) enzyme [17].

Evidence that conventional, broad field 250 kVp x-irradiation can induce medium-mediated bystander responses in normal human fibroblast cells was recently described by Held and coworkers [18] who used a transwell insert culture dish system to show that x-irradiation induces medium-mediated bystander effects in AGO1522 normal human fibroblasts. The frequency of micronuclei formation in nonirradiated bystander cells was found to be approximately doubled from a background value of about 6.5% at all doses from 0.1 to 10 Gy to the target cells. Induction of p21^{Waf1} protein

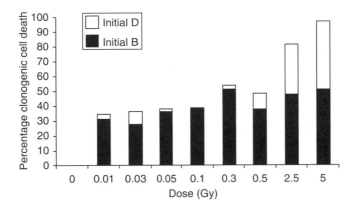

FIGURE 15.7
Clonogenic cell death measured in human keratinocytes. The total bar represents the total death detected after exposure of cells to the radiation dose. The death measured after exposure to ICM (irradiated conditioned medium) (B) is represented by the black portion of the bar, and the remaining death determined by subtraction is represented by the white portion of the bar, giving a value (D) for death not attributable to bystander effects of radiation. (From Seymour, C.B. and Mothersill, C., *Radiat. Res.*, 153, 508, 2000. With permission.)

and foci of γH2AX in bystander cells is also independent of dose to the irradiated cells above 0.1 Gy. Levels of ROS were increased persistently in directly irradiated cells up to 60 h after irradiation and in bystander cells for 30 h (Figure 15.8). Adding Cu Zn superoxide dismutase (SOD) and catalase

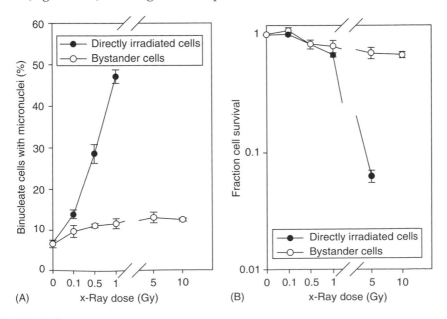

FIGURE 15.8
(A) Dose response for frequency of micronuclei formation in directly irradiated and bystander AGO1522 cells. (B) Clonogenic survival of directly irradiated and bystander AGO1522 cells. (From Yang, H., Asaad, N., and Held, K.D., *Oncogene*, 24, 2096, 2005. With permission.)

to the medium decreased the formation of micronuclei and induction of p21^{Waf1} and γH2AX foci in bystander cells, supporting a role for oxidative metabolism in the signaling pathways. The results of clonogenic assay of bystander cells showed that survival of bystander cells to be decreased following doses up to 0.5 Gy, but independent of the dose above 0.5 Gy. In contrast to the response with p21^{Waf1} expression, γH2AX foci and micronuclei, SOD, and catalase had no effect on the survival of bystander cells. This data suggests that irradiated cells must release toxic factors in addition to ROS into the medium.

15.2.2.1 Connection to Radiation Hypersensitivity

In the medium transfer experiments described ^{60}Co γ-ray doses of 0.01–0.5 Gy reduced clonogenic survival of bystander cells after transfer of medium from irradiated cells. Cell killing due to the bystander effect was relatively constant and appeared to saturate at doses in the range of 0.03–0.05 Gy [16]. At doses greater than 0.5 Gy, cell killing was predominantly the result of the direct action of radiation on the target cell. The conventional radiation cell survival curve indicates little, if any, cell killing by very low radiation doses to directly irradiated cells. However, as described in Chapter 14, low dose hypersensitivity to low LET radiation can be demonstrated in some cell lines by the use of special techniques [19]. Low-dose hypersensitivity occurs in the same dose range as bystander effects. A recent analysis of the relationship between radiation-induced low-dose hypersensitivity and the bystander effect indicated a weak inverse correlation between these two low-dose phenomena. Specifically, those cells exhibiting a large bystander effect did not show hyper-radiosensitivity [16].

15.2.2.2 Bystander-Mediated Adaptive Response

A number of studies have shown that very low doses of ionizing radiation can induce mechanisms whereby a cell becomes better able to deal with the adverse effects of subsequent exposures to higher doses, the so-called adaptive response (described in Chapter 14). The adaptive response is generally attributed to the induction in the irradiated cell of an efficient chromosome break repair mechanism, which, if active at the time of subsequent challenge with high doses leads to a reduction in residual damage.

In contrast to the bystander-mediated detrimental effects described by others, Iyer and Lehnert [20] have observed that radio-adaptation can be transferred to nonirradiated cells along with medium from irradiated cells. In experiments in which supernatants from normal human lung fibroblasts (HFL-1) irradiated with 1.0 cGy dose of γ-rays were transferred to nonirradiated HFL-1 cells, the nonirradiated bystander cells had increased clonogenic survival compared with directly irradiated cells following after γ-irradiation with 2 and 4 Gy (Figure 15.9).

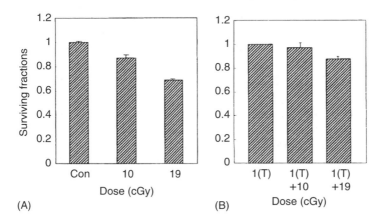

FIGURE 15.9
(A) Effect of direct irradiation of cells with 10 or 19 cGy of α particles. Ability to form colonies is decreased in a dose-dependent manner relative to the colony-forming ability of nonirradiated control cells (con). (B) Nonirradiated cells were incubated with supernatants harvested from α-irradiated (1 cGy) cells before exposing them to 10 or 19 cGy dose of α particles, 1(T) + 10 and 1(T) + 19 cGy, respectively. Radiosensitivity was decreased. (From Iyer, R. and Lehnert, B.E., *Radiat. Res.*, 157, 3, 2002. With permission.)

This radio-adaptive bystander effect was found to be preceded by early decreases in cellular levels of p53 protein, increase in intracellular ROS, and increase in the DNA repair protein AP-endonuclease. In other experiments, similar results were obtained when HFL-1 cells were irradiated with 1 cGy of α particles and their supernatants transferred to nonirradiated HFL-1 cells that were later exposed to 10 and 19 cGy of α particles [21]. TGF-β-1 was implicated in this effect since the cytokine, at concentrations detected in the supernatants, was able to induce increases in intracellular ROS in non-irradiated cells. Furthermore, TGF-β-1 induced decreased cellular levels of p53 and CDKN1A/p21, effects that were also induced by the addition of supernatants.

Overall, a clear picture has yet to emerge from the experience with medium transfer experiments. There is convincing evidence that factors are released into the medium by irradiated cells that can lead to changes in the viability of nonirradiated cells. The results however, are not consistent, with some workers reporting a reduction in cloning efficiency of the recipient cells while others find that it is enhanced or dependent on cell type. The effect appears likely to be mediated by cytokines or ROS, but the exact nature of the factor or factors responsible for the biological effects that arise in the nonirradiated, bystander cells is, as yet, unclear.

15.2.2.3 Death-Inducing Effect

Clonal cell populations derived from single cells which survived radiation exposure frequently show persistent genomic instability. Conditioned medium harvested from these unstable cells has been found to be highly cytotoxic to normal cells. The effect was lost when the medium was heated or frozen but was retained when the conditioned medium was diluted with fresh medium [22]. One explanation for this is that radiation induces conditions and factors that stimulate the production of ROS and that these reactive intermediates contribute to a chronic pro-oxidant environment that persists over multiple generations, promoting chromosomal recombination, and other phenotypes associated with genomic instability.

In a typical experiment, filtered medium from a chromosomally unstable clone of GM10115 hamster–human hybrid cells was used to culture nonirradiated cells. None of the nonirradiated GM10115 cells were able to survive and form colonies in conditioned medium from the unstable clone. This observation was interpreted to indicate that unstable clones of cells secrete factors that are not toxic to the unstable clone itself but are toxic to stable clones from a nonirradiated lineage. This effect is called the death-inducing effect or DIE (Figure 15.10) [23]. The DIE is different from the bystander effect which is observed after transferring medium from irradiated cells to nonirradiated cells. Reduction in plating efficiency was not seen when medium was transferred from directly irradiated cells to nonirradiated GM10115 cells, indicating that these cells either do not secrete a cytotoxic bystander factor or are not susceptible to it. It has been reported that within 30 min of transfer of unstable medium, there is a significant increase in γH2AX foci in the recipient cells. γH2AX foci formation is associated with the induction of DNA double-strand breaks (Chapter 8). The induction of γH2AX decreases with time after addition of unstable

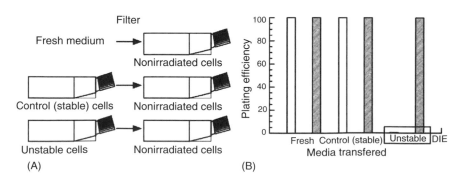

FIGURE 15.10
Strategy for identification of the death-inducing effect (DIE). (A) Fresh media, media from nonirradiated cells or irradiated but chromosomally stable cells, or media from unstable cells is filtered and transferred to nonirradiated parental cells. (B) The results of cell survival in this transferred medium as measured by colony forming ability. All cells survived in fresh medium or media from nonirradiated cells or irradiated but chromosomally stable cells. However, because of DIE (boxed) no surviving colonies were seen in the flasks receiving media from two of three chromosomally unstable clones. (From Sowa Resat, M.B. and Morgan, W.F., *J. Cell Biochem.*, 1, 92, 1013, 2004. With permission.)

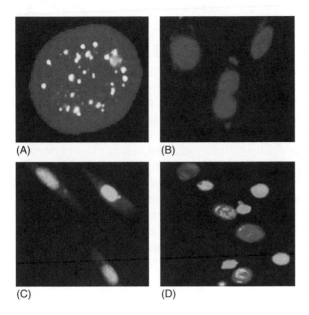

(A) (B)

(C) (D)

FIGURE 15.11 (See color insert following page 288.) The proposed sequence by which the DIE kills cells. (A) Within 30 min of transferring unstable media, a significant number of γH2AX foci are seen in recipient cells indicative of induced DNA double-strand breaks. (B) After 24 h of growth in unstable media, many cells show induced micronuclei. (C) Evidence of apoptosis by Annexin V staining, note the micronuclei. (D) Evidence of apoptosis by TUNEL assay. (From Sowa Resat, M.B. and Morgan, W.F., *J. Cell Biochem.*, 1, 92, 1013, 2004. With permission.)

medium and the formation of micronuclei followed by cell fragmentation is observed. Ultimately, the cell dies by a mitotic-linked or apoptotic cell death (Figure 15.11) [23].

15.2.3 Bystander Effects In Vivo

Bystander effects in vivo have been reviewed by Morgan [4]. The existence of radiobiological bystander effects in vitro are well established and supported by incontrovertible experimental evidence, but the possibilities for experimental demonstration of bystander effects in vivo are limited.

One of the clearest demonstrations came from a series of experiments described by Brooks and coworkers [24]. Chinese hamsters were injected with different-sized particles of the internally deposited α particle emitter plutonium. The radioactive particles concentrated in the liver and caused chronic low-dose radiation exposure with the dose and dose rate being highest for cells located closest to the largest particles. However, analysis of induced chromosome damage in these livers revealed increased cytogenetic damage but no change in aberration frequency as a function of local dose distribution. These observations were interpreted to indicate that all the cells in the liver were at the same risk of induced chromosome damage despite only a small fraction of the total liver being exposed to the radiation. The cumulative incidence of liver cancer as a function of time after plutonium injection and total dose was also determined. Neither the time of tumor onset nor the tumor incidence varied with particle size, indicating that the number of cells hit by α particles was not a factor in tumor induction in irradiated livers [25].

These findings suggest that radiation-induced genetic damage and ultimately tumor induction is related to the total dose to the organ, (i.e., the whole liver, rather than the dose to individual cells or the number of cells traversed by the α particle) and that the target for induced bystander effects may be limited to the specific organ irradiated.

In ingenious experiments modeling the action of bystander effects in tissue, Belyakov and coworkers [26] used reconstructed, normal human three-dimensional skin tissue systems. One of the systems used was a full-thickness skin model corresponding to the epidermis and dermis of normal human skin. Endpoints were induction of micronucleated and apoptotic cells. A charged-particle microbeam was used, which allowed irradiation of cells in defined locations in the tissue while guaranteeing that no cells located more than a few micrometers away receive any radiation exposure. Nonirradiated cells up to 1 mm distant from irradiated cells showed a significant enhancement in effect over background, with an average increase in effect of 1.7-fold for micronuclei and 2.8-fold for apoptosis [26] (Figure 15.12).

15.2.3.1 Abscopal Effects of Radiation

An abscopal effect has been defined as a significant response to radiation in tissues which are clearly separated from the radiation-exposed area with a measurable response occurring at a distance from the portals of irradiation large enough to rule out possible effects of scattered radiation.

An example of a radiation-induced abscopal effect is the study of early DNA damage induced in rat lung cells after single-dose, partial-volume

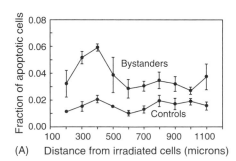 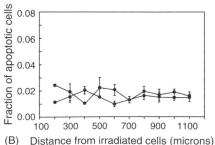

FIGURE 15.12
Fraction of apoptotic cells in nonirradiated keratinocyte layers at different distances from a plane of irradiated cells in a three-dimensional, full-thickness human skin model (EFT-300). Controls refer to sham irradiations, with conditions otherwise identical. Each data point (and SEM) is derived from studies with three independent tissues. (A) Microbeam irradiation of a plane of cells only in the epidermal layer, showing a significant bystander response. (B) Microbeam irradiation of a plane of cells only in the dermal layer, showing no evidence of a bystander response in the nonirradiated epidermal layer. (From Belyakov, O.V., Mitchell, S.A., Parikh, D., Randers-Pehrson, G., Marino, S.A., Amundson, S.A., Geard, C.R., and Brenner, D.J., *Proc. Natl. Acad. Sci. USA*, 102, 14203, 2005. With permission.)

irradiation (lung base and lung apex) described by Khan and coworkers [27]. When the lungs were removed 16–18 h after whole-lung irradiation, 0.85 micronuclei per binucleate cell were observed in the lungs of irradiated animals, compared to 0.02 micronuclei per binucleate cell in the lungs from control animals. When only the lung base was irradiated, the frequency of micronuclei in cells from the irradiated field was again 0.85; however, nonirradiated cells from the out-of-field lung apex also showed a significant increase in the frequency of micronuclei over the control level to 0.43 micronuclei per binucleate cell. Cells from the lungs of rats injected with SOD within 1 h before irradiation of the lung base and processed 16–18 h after irradiation showed a reduction in the number of micronuclei in the shielded lung apex, suggesting the involvement of oxygen radicals.

Abscopal reactions have been described in patients with chronic leukemias, where irradiation of an enlarged spleen or liver will induce a generalized remission with return of the bone marrow, WBC count, and peripheral smear to normal ranges and in fact, there are a number of well-recognized effects described by radiation therapists that occur beyond the radiation field. Goldberg and Lehnert [28] have recently reviewed these bystander-type effects and concluded that the clinical literature does not provide strong evidence for or against the existence of radiation abscopal effects in vivo. While many studies can be interpreted to suggest nontargeted effects in vivo, prospective clinical trials with detailed field and dose information, and well-documented patient risk factors would be needed for a firm conclusion to be reached.

15.2.3.2 Clastogenic Factors Induced by Ionizing Radiation

This section is based on review articles by Morgan [4], Mothersill and Seymour [1], and references cited therein. Clastogenic factors (or clastogenic plasma factors) were first described as early as 1954 when bone marrow damage was observed in the sternum of children with chronic granulocytic leukemia, whose spleens had been irradiated. Later it was shown that rats injected with plasma from irradiated animals developed mammary tumors at a significantly higher rate than rats exposed to plasma from nonirradiated animals. It was also reported that culturing normal human peripheral blood lymphocytes in medium containing plasma obtained from irradiated individuals resulted in significantly more chromosomal aberrations than were seen when lymphocytes were cultured in medium with plasma from nonirradiated individuals. These observations led to the suggestion that after in vivo irradiation, exposed individuals can possess clastogenic factors in their blood plasma which, when transferred to cell cultures from unexposed individuals, can induce chromosome damage [29].

Clastogenic factors have since been described in plasma from atomic bomb survivors, in salvage personnel involved in the Chernobyl accident and in children exposed as a consequence of the Chernobyl accident. In addition, clastogenic factors have been reported in human and rat blood plasma after in vitro irradiation. They can be induced within 15 min of irradiation, and they are either very persistent or are continuously regenerated, since they have been found at times of 10 weeks postirradiation for rats, 7–10.5 years for irradiated humans, and more than 30 years for the atomic bomb survivors. Clastogenic factors neither reflect radiation-induced depletion of protective factors nor radiation-induced changes in normal plasma components, but represent products secreted or excreted by cells as a result of irradiation.

The precise nature of clastogenic factors is unknown, but endogenous viruses, interference with DNA repair and the increased production of free radicals have all been implicated. On the basis of inhibitor studies, the bulk of evidence suggests that the mechanism of action of clastogenic factors is probably mediated by free radicals since free-radical scavengers such as SOD, penicillamine, cysteine, and various antioxidant plant extracts all reduce or abolish clastogenic factor activity. Both bystander factors and clastogenic factors are induced by ionizing radiation, and produce genetic damage in nonirradiated cells leading to the speculation that the two phenomena are related.

Clastogenic factors are not unique to radiation exposure. Transferable clastogenic effects have been described in blood plasma after whole body stresses such as asbestos exposure and ischemia–reperfusion injury and they occur spontaneously in patients with hepatitis C, Crohn's disease, and scleroderma. Clastogenic activity has also been observed in the plasma of patients with certain inherited chromosome breakage syndromes, including Bloom's syndrome and Fanconi's anemia. Individuals with these syndromes show an increased incidence of cancer, suggesting a role for clastogenic factors in creating a cellular environment predisposing to increased genomic instability and to neoplastic transformation.

15.2.4 Mechanisms Underlying Radiation-Induced Bystander Effects

There are two classes of experiments in which bystander responses have been demonstrated. One has exploited the ability of bystander signals to be transferred from irradiated cells to nonirradiated cells by medium transfer and the other has investigated response of cells to low fluences of α particles, where the majority of cells have not been irradiated. These two approaches also seem to broadly reflect the two types of mechanism; one dependent on the release of diffusible signaling molecules, the other dependent on GJIC. In both mechanisms, oxidative metabolism has been implicated in the signaling process and it seems probable that the observed differences could be a function of cell type, cell density, and other characteristics of the experimental systems used.

Medium transfer experiments demonstrated that irradiation can lead to secretion of a factor or factors by irradiated cells that can reduce cloning efficiency, increase neoplastic transformation, or induce genomic instability in nonirradiated cells. The first detectable effect on recipient cells after transfer of medium containing the bystander factors from irradiated cells was a rapid calcium pulse (1–2 min) followed 30–120 min later by changes in mitochondrial membrane permeability and the induction of ROS. Cell-to-cell contact during irradiation was not required to induce killing of bystander cells, but medium from cell cultures irradiated at high densities resulted in the greatest amount of bystander-induced cell death. Furthermore, the use of apoptosis inhibitors or medium from lactate dehydrogenase or G6PD mutant cells reduced or prevented the bystander effects. Treatment with the antioxidants such as L-lactate and L-deprenyl prevented bystander factor-associated cell killing, suggesting that energy/redox metabolism may be involved in the medium-mediated bystander response. Characteristics of the bystander effect and the possible nature of the bystander signaling molecules can be summarized as follows [30]:

- Multiple pathways likely.
- *p53* gene function need not be involved in the process.
- In confluent cultures, gap-junction-mediated cell-to-cell communication is essential. Molecules of size <1000 Da that can pass through gap junctions.
- In sparse cultures, secreted signaling molecules, particularly, ROS and RNS are involved.
- Other possible signaling molecules are long-lived organic molecules, protein hydroperoxides, and the cytokine TGFβ.
- Not clear if ROS are only the initiating signaling event which triggers other downstream, more stable secondary signaling pathways.

15.2.5 Implications in Risk Assessment

Results of the experiments described above indicate that bystander-type effects can occur in vivo in a nonirradiated tissue. This damage is in addition to damage directly induced by the deposition of energy in the irradiated cell. A detrimental bystander effect such as generation of chromosome aberrations will amplify the biological effectiveness of a given radiation dose by effectively increasing the number of cells that experience adverse effects over the number directly exposed to the radiation. Bystander effects appear to be limited to the organ irradiated and radiation risk estimates are also organ specific so that any bystander effect induced in vivo should be accounted for in models of organ risk evaluation.

Radiation protection guidelines are based on prediction of the biological effects of low doses of radiation by extrapolating from known epidemiological data sets. These data sets mainly relate to high-dose effects, the main source of information being the Japanese atomic bomb survivors. Standards and guidelines for acceptable doses to the general public and to radiation workers are based on an assumption of a linear no-threshold (LNT) hypothesis, which relates dose to biological effect. The LNT hypothesis states that the dose–effect relationship is linear even at very low doses, meaning that in theory the lowest dose imaginable has a finite probability, however small, of causing a biological effect. The occurrence of bystander effects means that there is no direct correlation between the number of cells that are exposed to radiation and the number of cells that are at risk of showing effects such as mutation, chromosomal damage, or apoptosis. Instead, biological effects depend on complex, interactions between the irradiated cell and the bystander cells. It is no longer a single cell that is at risk from radiation damage, instead, the risk is spread among bystander cells and a simple dose–effect relationship cannot be assumed.

If results obtained with laboratory model systems prove to be applicable in vivo, they could have significant consequences in terms of extrapolation of radiation risks from high to low doses, since they imply that the relevant target for radiation oncogenesis is larger than an individual cell. Figure 15.13 [12] combines the data from experiments in which only a proportion of cells are irradiated with a single particle (allowing the bystander effect to be manifest) with data from experiments where all cells were exposed to various numbers of particles from one to four. From the curve it appears that a linear extrapolation of risks from high to low doses (an average less than one particle per cell) would underestimate the risks at low doses. These experiments were done using α particle irradiation in vitro and it is not clear if the same results would be obtained with radiation of other qualities.

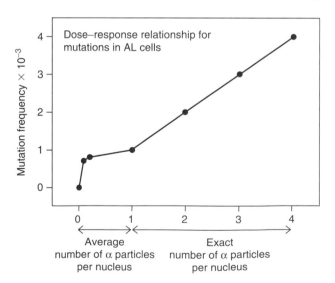

FIGURE 15.13
Mutation frequency as a function of the number of α particles per nucleus (data for the average number of particle traversals were calculated from cell population in which defined proportion of cells were exposed to a single α particle). Due to the bystander effect, which is evident when only a proportion of the population is exposed, the risk at low doses is higher than predicted by a linear extrapolation from high doses. (From Hall, E.J. and Hei, T.K., *Oncogene*, 22, 7034, 2003. With permission.)

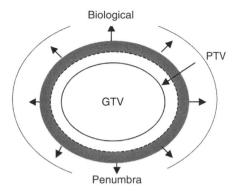

FIGURE 15.14
The biological penumbra. Due to the bystander effect, the biological effects of radiation spread beyond the physical penumbra of the field that gets directly hit by ionizing radiation tracks. GTV, Gross Target Volume; PTV, Planning Target Volume.

Radiation-induced bystander effects also have implications for response to radiotherapy because the biological effects of the radiation beam will cover a wider area than the physical beam suggesting the concept of a biological in addition to a physical penumbra (Figure 15.14) [31]. Multi-field radiotherapy treatments such as intensity-modulated radiotherapy (IMRT) increase the number of radiation fields, thus increasing the volume of tissue that is exposed to low levels of radiation while decreasing the size of the dose to some normal tissues. The effect of the increase in irradiated-tissue volume is not predictable by the current LNT hypothesis, which only relates dose to effect and not irradiated volume to effect. There will also be implications for the effect of fractionated-radiotherapy treatments since, for each fraction of radiotherapy given, there will be a combination of direct and bystander effects, each controlled by different mechanisms.

15.3   Genomic Instability

Genomic instability is an all-embracing term to describe the increased rate of acquisition of alterations in the genome. Radiation-induced instability is observed in cells at delayed times after irradiation and is manifest in the progeny of exposed cells multiple generations after the initial insult (Figure 15.15) [32]. Instability is measured as chromosomal alterations, changes in ploidy, micronucleus formation, gene mutations and amplifications, micro-satellite instabilities, and decreased plating efficiency (delayed cell death). This part of Chapter 15 is partly based on reviews by Morgan [3,4], Little [2], Coates et al. [32], and the references cited therein.

15.3.1   Genomic Instability In Vitro: Delayed Responses to Radiation Exposure

Among the earliest demonstrations of delayed responses to radiation exposure were experimental results showing that the progeny of irradiated cells

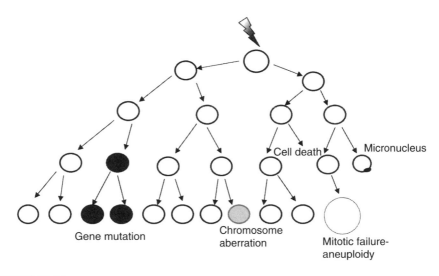

FIGURE 15.15
Genomic instability. Schematic representation of types of delay responses characteristic of the nonclonal effects in the clonal descendents of irradiated cells with characteristic mutations and chromosomal aberrations.

may exhibit an enhanced death rate and loss of reproductive potential that persists for many generations and possibly indefinitely in established cell lines.

A number of laboratory studies have demonstrated nonclonal chromosome aberrations arising many generations after irradiation in the progeny of irradiated cells, consistent with the induction of a state of genomic instability (reviewed by Little [2]). A variety of cell types were affected after exposures to both high and low LET radiations. A dose–response relationship was not seen as the responses saturate at low doses. The instability-induced cytogenetic abnormalities reflect those that arise spontaneously as a consequence of endogenous damaging processes. These unstable aberrations may result in apoptosis and thus account for a component of the delayed death phenotype in some cell systems. Most significant are gross chromosomal rearrangements, particularly chromosomal duplications, and partial trisomies which appear to involve amplification and recombination of large chromosomal regions by a currently unknown mechanism.

An increased mutation rate persisting over many generations has been observed in a number of studies. In one set of experiments, approximately 10% of clonal populations derived from single cells surviving radiation exposure showed a significant elevation in the frequency of spontaneously arising mutations as compared with clonal populations derived from nonirradiated cells. This increased mutation rate persisted for approximately 30 generations postirradiation, then gradually subsided. These late-arising mutants resemble spontaneous mutations in that the majority of them are point mutations indicating that they arise by a different mechanism from the direct x-ray-induced mutations that involve primarily deletions [33,34].

Radiation-induced delayed mutations in the *p53* and *hprt* genes have been demonstrated in primary murine mammary epithelium and hemopoietic tissues, respectively. In a study of hprt mutations arising as a consequence of induced instability, less than one-third were associated with large deletions and the majority were small-scale changes [33]. This mutation spectrum again more closely resembles that of spontaneously arising mutations, than it does conventional radiation-induced mutations, where three-quarters of hprt mutations induced directly by x-rays involve partial or total gene deletions and only 25% small-scale or point mutations [35].

15.3.2 Demonstration of Genomic Instability In Vivo

To demonstrate genomic instability in vivo, a number of hybrid experimental systems have been devised in which radiation is either given in vivo with the analysis done in vitro or conversely irradiation of cells is done in vitro, and genomic instability is determined in vivo.

Cytogenetic analysis of mammary glands from irradiated mice which were removed and put into culture at intervals postirradiation indicated that instability developed and persisted in situ in a mature, fully differentiated tissue after in vivo irradiation. There was a dose-dependent increase in the frequency of delayed aberrations at low doses (10–1 Gy) that reached a plateau at higher doses. In addition there was evidence of different strain susceptibilities to in vivo irradiation. For instance, mammary cells from BALB/c mice were more susceptible to radiation-induced genomic instability than were those from C57BL/6 or F1 hybrid crosses of C57BL/6 and BALB/c mice. DNA repair studies in the radiosensitive BALB/c mice revealed inefficient end joining of γ-ray-induced DNA double-strand breaks. Increased instability is apparently due to reduced expression of the DNA-PK$_{cs}$ protein and lowered DNA-PK activity in these mice [36].

In other experiments, genomic instability induced in vitro was found to be transmitted in vivo after transplantation of irradiated cells into recipient animals. In one experiment, C3H 10T1/2 cells were irradiated and one half was put in cell culture while the other half was transplanted into syngeneic C3H mice. Higher frequencies of minisatellite instability were observed in the irradiated cells which had been injected into mice than in those which had been maintained in culture. In a similar experiment, induction and long-term persistence of chromosomal instability were observed in murine bone marrow cells which had been irradiated in vitro, then transplanted into radiation-ablated recipient CBA/H mice [37].

The first definitive demonstration of in vivo radiation-induced chromosomal instability following whole body irradiation used direct karyotyping of bone marrow cells obtained from mice for up to 24 months after whole body irradiation with either x-rays or neutrons. The data clearly demonstrated that chromosomal instability was being initiated and expressed in vivo. When compared to results of in vitro studies, the in vivo data showed less damage per cell and fewer cells demonstrating chromosomal instability. This difference could be due to the cellular defense mechanisms that have evolved in the organism to recognize and remove aberrant cells. Significant interindividual variation was observed within the same inbred strain with a few animals exhibiting little or no induced instability and results of whole body irradiation were consistent with in vitro studies with respect to intra-strain differences since induced instability was not observed in C57BL/6 mice.

While there are many apparently contradictory reports in the literature regarding radiation-induced instability in mouse model systems, overall, the results indicate that genetic factors can play a major role in the instability phenotype and that analysis of radiation-induced genomic instability in vivo is significantly more complicated than in vitro.

15.3.3   Genomic Instability and Cancer

Genomic instability is considered a hallmark of the process of carcinogenesis and most human tumors express instability as multiple, unbalanced chromosomal aberrations. It has been hypothesized that as many as six to eight separate genetic events may be required to transform a normal cell into one with a fully malignant phenotype with specific genetic events associated with increasing levels of malignant change [38]. If each mutation is an independent event occurring with a probability of around 10^{-5}, the question arises as to how so many genetic events can develop in a single cell lineage within the lifetime of an individual. One answer would be if the development of genomic instability acts as an accelerant in the initiation and progression of cancer with the instability rendering the potential cancer cells more susceptible to the accumulation of multiple genetic abnormalities [39].

The timing of the development of genomic instability in the progress of the cell toward the malignant state is controversial, particularly as to whether it arises later in the process of neoplastic transformation and carcinogenesis or is an early or initiating event. If a clonal population of potential cancer cells develops a selective growth advantage, it could accumulate mutations more rapidly owing to enhanced cell turnover without invoking instability as a necessary event. This would be consistent with the observations that loss of p53 function which is believed to lead to genomic instability occurs late in the development of many tumors. On the other hand, experimental studies of radiation-induced cancer and malignant transformation in vitro suggest that instability may be an early event that facilitates the occurrence of p53 mutations.

Characterization of certain rare genetic disorders characterized by chromosomal instability and a heritable predisposition to the development of cancer has provided insight as to the importance of instability in carcinogenesis. The genes from these disorders have now been cloned and characterized, and their functions determined (see also Chapter 8). They include *ATM*, an important sensor of DNA damage; *NBS*, *BRCA1*, and *BRCA2*, which encode proteins involved in recombinational DNA repair, the seven cloned Fanconi's anemia (*FA*) genes, which have been linked with *BRCA1*, *ATM*, and *NBS1* in cell-cycle checkpoint and DNA repair pathways and the Bloom's syndrome (*BS*) gene, which codes for a helicase involved in DNA replication and repair. The genes responsible for these genetic disorders are involved in pathways that ensure the fidelity of DNA replication and repair in a number of cases, involving the nonhomologous end joining (NHEJ) DNA repair pathway. Defects in this pathway in transgenic mice have been associated specifically with the induction of chromosomal instability. Study of the cancer-prone disorders has provided a link between defects in DNA replication and repair, genomic instability, and the development of cancer [2,3].

Radiation-induced leukemia in mice has been used as a model system for the study of the relationship between radiation-induced genomic instability and cancer (reviewed by Plumb and coworkers [40]). The genetic lesions detected in radiation-induced mouse leukemias include nonclonal chromosomal aberrations, loss of heterozygosity, and minisatellite/microsatellite mutations. These are similar to those detected in de novo leukemias and cancers and in other in vivo models of radiation-induced instability. However, susceptibility to radiation-induced leukemia is genetically separable from sensitivity to radiation-induced genomic instability and a genetic relationship between susceptibility to radiation-induced leukemia and sensitivity of hemopoietic stem cells to induced instability could not be established [41].

Cytogenetic analysis of leukemia patients among the Japanese atomic bomb survivors revealed that patients exposed to more than 2 Gy exhibited a higher incidence of chromosomal aberrations and more complex chromosomal rearrangements than did patients exposed to lower radiation doses and unexposed patients. The cytogenetic observations are supported by studies demonstrating high frequencies of microsatellite instability in those atomic bomb survivors with acute myelocytic leukemia and a history of heavy exposure. The conclusion was that this persistent instability might strongly influence the development of leukemia in humans exposed to ionizing radiation [42].

15.3.4 Mechanisms Underlying Radiation-Induced Genomic Instability

While the multiple phenotypes associated with radiation-induced genomic instability are relatively well characterized, the molecular, biochemical, and cellular events that initiate and perpetuate instability remain unknown. It is believed that directly induced DNA damages, e.g., induced DNA double-strand breaks, are probably not responsible, although the range of environmental mutagens and cytotoxic drugs which induce a delayed death phenotype clearly link substances that induce DNA strand breaks with the induction of instability. Deficiencies in cellular responses to DNA damage, changes in gene expression, or perturbations in cellular homeostasis are more likely to be involved and to provide a rational explanation as to why the unstable phenotype can persist. Importantly, all the substances that were effective in inducing instability (ionizing radiation, nonionizing ultraviolet radiation, nickel, cadmium, hydrogen peroxide, and bleomycin) are known to induce oxidative stress. While the nucleus may be the ultimate target, there is evidence for a persistent increase in ROS in cultures of cells showing radiation-induced genomic instability suggesting a role for enhanced oxidative stress in perpetuating the unstable phenotype. Involvement of ROS in nontargeted radiation-induced bystander effects has also been described, suggesting that processes involved in these delayed effects of exposure to ionizing radiation have features in common. Overall, the results of many different investigations are consistent with a free-radical/oxidative stress model of radiation-induced instability [43].

15.3.5 Relationship between Radiation-Induced Bystander Effects and Genomic Instability

Bystander effects and genomic instability have many common features, including induced chromosomal rearrangements, micronuclei, increased mutation, increased transformation, and increase in cell kill, providing circumstantial evidence for a connection between the two phenomena. The up-regulation of oxidative stress in bystander cells is similar to the enhanced oxidative stress associated with radiation-induced genomic instability suggesting that the preponderance of point mutations which characterize both conditions are the result of oxidative damage. Persistently increased intracellular ROS have been reported in chromosomally unstable cells, and they provide an attractive mechanism for perpetuating instability over a long period.

Wright and coworkers [37] have provided convincing evidence that the induction of genomic instability after in vitro irradiation and in vivo expression can result from a nontargeted bystander-like effect. Nonirradiated cells were mixed with neutron-irradiated cells and transplanted into recipient mice; chromosomal markers enabled the investigators to distinguish between the irradiated and nonirradiated transplanted cells and the cells derived from the host mouse. There was unambiguous evidence for chromosomal instability in the progeny of nonirradiated cells implicating an in vivo bystander mechanism in the induction of chromosomal instability. Later it was proposed that a potential mechanism for in vivo radiation-induced bystander effects could be the persistent macrophage activation combined with neutrophil infiltration seen after whole body irradiation of mice. The inflammatory nature of the observed responses could provide a mechanism for the long-term production of genetic damage by a bystander effect ultimately contributing to radiation-induced instability and potentially leukemogenesis.

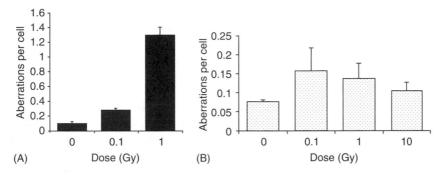

FIGURE 15.16
(A) Frequencies of chromosomal aberrations observed in human fibroblasts at the first cell division following α-irradiation (mean ± S.D). (B) Frequencies of chromatid-type aberrations in fibroblasts that were bystanders to α-irradiated fibroblasts at the first cell division postirradiation (mean ± S.D). (From Ponnaiya, B., Jenkins-Baker, G., Bigelow, A., Marino, S., and Geard, C.R., *Mutat. Res.*, 568, 41, 2004. With permission.)

That cells exhibiting chromosome instability could be derived from nonirradiated cells, was later confirmed by a direct experiment in which a shielding grid was interposed between the radiation source and the cells so that shielded clonogenic stem cells were protected whereas the majority of exposed cells were killed. [44]. The number of stem cell-derived colonies expressing chromosomal instability were similar with or without the grid, indicating that most (if not all) cells exhibiting chromosome instability were derived from nonirradiated stem cells and the induction of the instability must be attributed to an indirect intercellular communication mechanism, i.e., a bystander mechanism (Figure 15.16).

Increasing evidence suggests that induced genomic instability and bystander effects are linked and, given the common features of the end points observed, it is likely that both phenomena could be manifestations of the same nontargeted processes. A significant contribution from bystander-type factors could help explain the reportedly high frequency of radiation-induced instability.

15.4 Summary

The dogma of radiation biology that the lethal and genetic effects of radiation result solely from the deposition of energy in the nucleus of the targeted cell is challenged by observations that radiation induces conditions and factors that result in nonirradiated cells exhibiting a wide range of responses usually associated with direct DNA damage. These nontargeted effects are collectively referred to as radiation-induced bystander effects and radiation-induced genomic instability. Observed bystander responses include damage-inducible stress responses, sister chromatid exchanges, micronucleus formation, apoptosis, gene mutation, chromosomal instability, and transformation of rodent cells in vitro.

Results of experiments with low fluences of α particles indicate that bystander effects are associated with up-regulation of oxidative metabolism and are mediated by ROS and possibly by NOS. Transmission of signals from irradiated to bystander cells is mediated by direct intercellular contact through gap junctions and agents which disrupt membrane signaling also disrupt this mode of bystander signaling. Bystander effects have been demonstrated for many biological end points using charged-particle microbeams. These studies showed that irradiated cells can induce a mutagenic response in neighboring cells not directly traversed by α particles. Cell-to-cell communication processes play a critical role in the bystander phenomenon and there is evidence of bystander-mediated oncogenic transformation.

Another type of bystander effect which does not involve cell-to-cell contact has been demonstrated after transfer of medium from irradiated cells. Media transfer experiments showed that irradiation can lead to secretion of a factor or factors by irradiated cells which can reduce cloning efficiency, increase neoplastic transformation or induce genomic instability in bystander cells.

Bystander effects have been demonstrated in vivo and results suggest that genetic damage and tumor induction may be related to the dose to a whole organ rather than to individual cells. Abscopal effects, i.e., significant tissue response to radiation in tissues definitely separate from the irradiated area have been described in a number of species and in the clinical literature. However, firm clinical evidence is lacking, which would unequivocally support the occurrence of abscopal effects and by extension bystander effects. Plasma from irradiated animals and humans contains clastogenic factors which can cause chromosome aberrations in lymphocytes from non-irradiated individuals. Clastogenic factors can be induced within 15 min of irradiation and can persist or be regenerated over a long period.

Genomic instability is an all-embracing term to describe increased rate of acquisition of alterations in the genome. Radiation-induced instability is observed in cells at delayed times after radiation and appears in the progeny of irradiated cells many generations after the initial insult. It is measured as chromosome aberrations, changes in ploidy, micronucleus formation, gene mutations, and decreased plating efficiency. All substances that were effective in inducing instability are also known to induce oxidative stress and there is evidence for a persistent increase in ROS in cultures of cells showing radiation-induced genomic instability suggesting a role for enhanced oxidative stress in perpetuating the unstable phenotype.

Genomic instability has been shown to result from irradiation in vivo with genetic factors playing a major role in the instability phenotype. Genomic instability is considered a hallmark of the process of carcinogenesis and most human tumors display signs of instability such as multiple unbalanced chromosome aberrations. It has been suggested that development of genomic instability acts as an accelerant in the initiation and progression of cancer with the instability rendering the potential cancer cells more susceptible to the accumulation of multiple genetic abnormalities. Correlation between radiation-induced genetic instability has been observed in animal models, in leukemia patients and among the Japanese atomic bomb survivors.

There is evidence suggesting that induced instability and bystander effects are linked, and it is likely, given the commonality of the end points observed, that both phenomena could be manifestations of the same non-targeted processes. The increase in oxidative stress observed in bystander cells parallels the persistently increased intracellular ROS reported in chromosomally unstable cells.

The occurrences of bystander effects and genomic instability have implications for radiation protection and for radiotherapy. The occurrence of bystander effects means that there is no direct correlation between the number of cells that are exposed to radiation and the number of cells that are at risk of damage. This may call for new approaches to risk assessment which are not based on the assumption of targeted effects. Radiation-induced bystander effects also have implications for response to radiotherapy because the biological effects of the radiation beam will cover a wider area than the physical beam.

References

1. Mothersill C and Seymour C. Radiation-induced bystander effects: Past history and future directions. *Radiat. Res.* 155: 759–767, 2001.
2. Little J. Genomic instability and bystander effects: A historical perspective. *Oncogene* 22: 6978–6987, 2003.
3. Morgan WF. Non-targeted and delayed effects of exposure to ionizing radiation: I. Radiation-induced genomic instability and bystander effects *in vitro. Radiat. Res.* 159: 567–580, 2003.
4. Morgan WF. Non-targeted and delayed effects of exposure to ionizing radiation: II. Radiation-induced genomic instability and bystander effects in vivo, clastogenic factors, and transgenerational effects. *Radiat. Res.* 159: 581–596, 2003.
5. Nagasawa H, Cremesti A, Kolesnick R, Fuks Z, and Little JB. Involvement of membrane signaling in the bystander effect in irradiated cells. *Cancer Res.* 62: 2531–2534, 2002.
6. Huo L, Nagasawa H, and Little JB. HPRT mutants induced in bystander cells by very low fluences of alpha particles result primarily from point mutations. *Radiat. Res.* 156: 521–525, 2001.
7. Azzam EI, de Toledo SM, and Little JB. Direct evidence for the participation of gap junction-mediated intercellular communication in the transmission of damage signals from a particle irradiated to non-irradiated cells. *Proc. Natl. Acad. Sci. USA* 98: 473–478, 2001.
8. Azzam EI, de Toledo SM, Spitz DR, and Little JB. Oxidative metabolism modulates signal transduction and micronucleus formation in bystander cells from α-particle irradiated normal human fibroblasts. *Cancer Res.* 62: 5436–5442, 2002.

9. Little JB, Nagasawa H, Li GC, and Chen DJ. Involvement of the nonhomologous end joining DNA repair pathway in the bystander effect for chromosomal aberrations. *Radiat. Res.* 159: 262–267, 2003.

10. Nagasawa H, Huo L, and Little JB. Increased bystander mutagenic effect in DNA double-strand break repair-deficient mammalian cells. *Int. J. Radiat. Biol.* 79: 35–41, 2003.

11. Wu L-J, Randers-Pehrson G, Xu A, Waldren CA, Geard CR, Yu Z, and Hei TK. Targeted cytoplasmic irradiation with alpha particles induces mutation in mammalian cells. *Proc. Natl. Acad. Sci. USA* 96: 4959–4964, 1999.

12. Hall EJ and Hei TK. Genomic instability and bystander effects induced by high LET radiation. *Oncogene* 22: 7034–7042, 2003.

13. Zhou H, Randers-Pehrson G, Waldren CA, Vannais D, Hall EJ, and Hei TK. Induction of a bystander mutagenic effect of alpha particles in mammalian cells. *Proc. Natl. Acad. Sci. USA* 97: 2099–2104, 2000.

14. Zhou H, Suzuki M, Randers-Pehrson G, Vannais D, Chen G, Trosko JE, Waldren CA, and Hei TK. Radiation risk to low fluences of alpha particles may be greater than we thought. *Proc. Natl. Acad. Sci. USA* 98: 14410–14415, 2001.

15. Folkard M, Vojnovic B, Prise KM, Bowey AG, Locke RJ, Schettino G, and Michael BD. A charged-particle microbeam: I. Development of an experimental system for targeting cells individually with counted particles. *Int. J. Radiat. Biol.* 72: 375–385, 1997.

16. Seymour CB and Mothersill C. Relative contribution of bystander and targeted cell killing to the low-dose region of the radiation dose–response curve. *Radiat. Res.* 153: 508–511, 2000.

17. Mothersill C, Stamato TD, Perez ML, Cummins R, Mooney R, and Seymour CB. Involvement of energy metabolism in the production of 'bystander effects' by radiation. *Br. J. Cancer* 82: 1740–1746, 2000.

18. Yang H, Asaad N, and Held KD. Medium-mediated intercellular communication is involved in bystander responses of X-ray-irradiated normal human fibroblasts. *Oncogene* 24: 2096–2103, 2005.

19. Joiner MC, Marples B, Lambin P, Short SC, and Turesson I. Low-dose hypersensitivity: Current status and possible mechanisms. *Int. J. Radiat. Oncol. Biol. Phys.* 49: 379–389, 2001.

20. Iyer R and Lehnert BE. Factors underlying the cell growth-related bystander response to α-particles. *Cancer Res.* 60: 1290–1298, 2000.

21. Iyer R and Lehnert BE. Alpha-particle-induced increases in the radioresistance of normal human bystander cells. *Radiat. Res.* 157: 3–7, 2002.

22. Nagar S, Smith LE, and Morgan WF. Characterization of a novel epigenetic effect of ionizing radiation: The death inducing effect. *Cancer Res.* 63: 324–328, 2003.

23. Sowa Resat MB and Morgan WF. Radiation-induced genomic instability: A role for secreted soluble factors in communicating the radiation response to non-irradiated cells. *J. Cell. Biochem.* 92: 1013–1019, 2004.

24. Brooks AL, Retherford JC, and McClellan RO. Effect of 239PuO2 particle number and size on the frequency and distribution of chromosome aberrations in the liver of the Chinese hamster. *Radiat. Res.* 59: 693–709, 1974.

25. Brooks AL, Benjamin SA, Hahn FF, Brownstein DG, Griffith WC, and McClellan RO. The induction of liver tumors by 239Pu citrate or 239PuO2 particles in the Chinese hamster. *Radiat. Res.* 96: 135–151, 1983.

26. Belyakov OV, Mitchell SA, Parikh D, Randers-Pehrson G, Marino SA, Amundson SA, Geard CR, and Brenner DJ. Biological effects in unirradiated human tissue induced by radiation damage up to 1 mm away. *Proc. Natl. Acad. Sci. USA* 102: 14203–14208, 2005.
27. Khan MA, Hill RP, and Van Dyk J. Partial volume rat lung irradiation: An evaluation of early DNA damage. *Int. J. Radiat. Oncol. Biol. Phys.* 40: 467–476, 1998.
28. Goldberg Z and Lehnert BE. Radiation-induced effects in unirradiated cells: A review and implications in cancer. *Int. J. Oncol.* 21: 337–349, 2002.
29. Morgan WF. Is there a common mechanism underlying genomic instability, bystander effects and other nontargeted effects of exposure to ionizing radiation? *Oncogene* 22: 7094–7099, 2003.
30. Hei TK, Persaud R, Zhou H, and Suzuki M. Genotoxicity in the eyes of bystander cells. *Mut. Res.* 568: 111–120, 2004.
31. Seymour CB and Mothersill C. Radiation-induced bystander effects—implications for cancer. *Nat. Rev. Cancer* 4: 158–164, 2004.
32. Coates PJ, Lorimore SA, and Wright EG. Damaging and protective cell signalling in the untargeted effects of ionizing radiation. *Mut. Res.* 568: 5–20, 2004.
33. Grosovsky AJ, Parks KK, Giver CR, and Nelson SL. Clonal analysis of delayed karyotypic abnormalities and gene mutations in radiation-induced genetic instability. *Mol. Cell. Biol.* 16: 6252–6262, 1996.
34. Little JB, Nagasawa H, Pfenning T, and Vetrovs H. Radiation-induced genomic instability: Delayed mutagenic and cytogenetic effects of X rays and alpha particles. *Radiat. Res.* 148: 299–307, 1997.
35. Ponnaiya B, Cornforth MN, and Ullrich RL. Radiation-induced chromosomal instability in BALB/c and C57BL/6 mice: The difference is as clear as black and white. *Radiat. Res.* 147: 121–125, 1997.
36. Okayasu RK, Suetomi Y, Yu Y, Silver A, Bedford JS, Cox R, and Ullrich RL. A deficiency in DNA repair and DNA-PKcs expression in the radiosensitive BALB/c mouse. *Cancer Res.* 60: 4342–4345, 2000.
37. Watson GE, Lorimore SA, Macdonald DA, and Wright EG. Chromosomal instability in unirradiated cells induced in vivo by a bystander effect of ionizing radiation. *Cancer Res.* 60: 5608–5611, 2000.
38. Kinzler KW and Vogelstein B. Cancer-susceptibility genes. Gatekeepers and caretakers. *Nature* 386: 761–763, 1997.
39. Cheng KC and Loeb LA. Genomic instability and tumor progression: Mechanistic considerations. *Adv. Cancer Res.* 60: 121–156, 1993.
40. Plumb M, Cleary H, and Wright E. Genetic instability in radiation-induced leukaemias: Mouse models. *Int. J. Radiat. Biol.* 4: 711–720, 1998.
41. Boulton E, Cleary H, Papworth D, and Plumb M. Susceptibility to radiation-induced leukaemia/lymphoma is genetically separable from sensitivity to radiation-induced genomic instability. *Int. J. Radiat. Biol.* 77: 21–29, 2001.
42. Nakanishi M, Tanaka K, Takahashi T, Kyo T, Dohy H, Fujiwara M, and Kamada N. Microsatellite instability in acute myelocytic leukaemia developed from A-bomb survivors. *Int. J. Radiat. Biol.* 77: 687–694, 2001.
43. Morgan WF, Hartmann A, Limoli CL, Nagar S, and Ponnaiya B. Bystander effects in radiation-induced genomic instability. *Mutat. Res.* 504: 91–100, 2002.
44. Lorimore SA, Kadhim MA, Pocock DA, Papworth D, Stevens DL, Goodhead DT, and Wright EG. Chromosomal instability in the descendants of unirradiated

surviving cells after alpha-particle irradiation. *Proc. Natl. Acad. Sci. USA* 95: 5730–5733, 1998.

45. Sawant SG, Zheng W, Hopkins KM, Randers-Pehrson G, Lieberman HB, and Hall EJ. The radiation-induced bystander effect for clonogenic survival. *Radiat. Res.* 157: 361–364, 2002.

46. Sawant SG, Randers-Pehrson G, Geard CR, Brenner DJ, and Hall EJ. The bystander effect in radiation oncogenesis: I. Transformation in C3H10T1/2 cells in vitro can be initiated in the unirradiated neighbors of irradiated cells. *Radiat. Res.* 155: 397–401, 2001.

16

Tumor Radiobiology

16.1 Tumor Radiobiology

In the context of radiobiology, the term "tumor environment" is usually equated with tumor hypoxia. In fact, the presence of foci of low oxygen tension in solid tumors has multiple consequences for tumor progression and treatment outcome, some of which are only recently being explored. The realization that human tumors contained viable cells that were radiobiologically hypoxic, and consequently refractory to radiation treatment, had a massive impact on the practice and development of radiation oncology and focused the attention of clinicians and researchers almost exclusively on the problem of hypoxia for many years. In fact, the solid tumor is an environment in which a number of physiological and metabolic factors combine to create a unique environment, one aspect of which, albeit a very important one, is hypoxia.

16.2 Unique Tumor Microenvironment

The microenvironment of a solid tumor has a number of characteristics distinguishing it from the corresponding normal tissue (reviewed by Cairns et al. [1] and Vaupel [2]). These characteristics result from the interaction of the poorly formed tumor vasculature and the physiology of the tumor cells. Three recognized microenvironmental hallmarks of solid tumors are high interstitial fluid pressure (IFP), low oxygen tension or hypoxia, and low extracellular pH (pHe). The mechanisms responsible for the establishment of these conditions are shown schematically in Figure 16.1.

In normal tissues, the vascular system is regulated by a balance of pro-angiogenic and anti-angiogenic molecules, which ensures that an efficient and orderly network of blood vessels is maintained to meet the metabolic demands of the tissue. In addition, a network of lymphatics is available to drain fluid and cellular byproducts from the interstitium. When this balance of growth factors in the tumor is perturbed the result is the development of

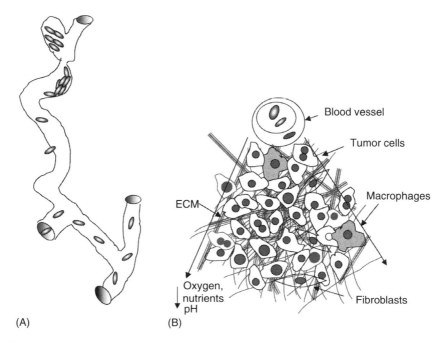

(A)   (B)

FIGURE 16.1

Microenvironmental conditions in a solid tumor. (A) Schematic representation of a tumor blood vessel. Chaotic tumor vasculature contains unstable endothelium, leaky vessels, dead-ends, and instability in RBC flux. (B) Schematic representing gradients in oxygen and nutrients around a tumor blood vessel. Poor oxygen delivery by the defective vasculature and oxygen consumption by the tumor cells result in hypoxic areas. Oxygen deprivation and activation of HIF-1 mediates adaptation of tumor cells to hypoxia by increasing glucose import and utilization in the cytoplasm (glycolysis and anaerobic conversion to lactate). Induction of carbonic anhydrase IX and XII contributes to the acidification of the Extracellular matrix (ECM). Increased numbers of activated fibroblasts and macrophages contribute to the formation of highly contractile ECM, rich in collagen fibers, raising the IFP.

a disorganized vasculature with multiple structural and functional abnormalities and a complete lack of lymphatics. Tumor vessels often have an incomplete endothelial lining and lack a layer of pericytes or an intact basement membrane, making them more permeable than those in normal tissues. The architecture of the vasculature is also highly disorganized. Tumor vessels are generally long and highly chaotic, and many vessels have an unusually large diameter. The mean vascular density of tumors is generally lower than in normal tissues, and diffusion distances are greater and the vascular network often contains blind-ends, arteriolar-venous shunts, and plasma channels that are devoid of RBC (Figure 16.2).

Functionally, the ability of the tumor vasculature to deliver nutrients and remove waste products is badly compromised. However, vessel organization is highly heterogeneous and hotspots may occur where the vascular density is higher than in normal tissues.

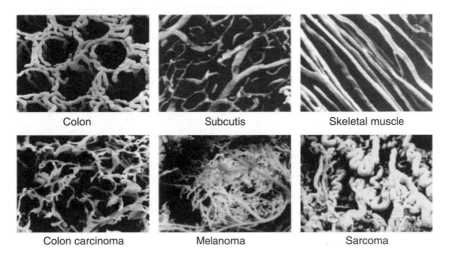

Colon   Subcutis   Skeletal muscle

Colon carcinoma   Melanoma   Sarcoma

FIGURE 16.2
Scanning electron micrographs of corrosion casts of tumor vasculature compared with normal tissue of origin. (From Vaupel, P., *Semin. Radiat. Oncol.*, 14, 198, 2004. With permission.)

16.2.1 Interstitial Fluid Pressure

A direct result of the porous tumor vasculature and lack of lymphatics is the inappropriate accumulation of vascular fluid contents in the tumor and the resulting elevation of IFP. The interstitium of a tissue is the space between the cells and the vascular compartment that represents a large fraction of the total tissue volume, and it is essential for molecular transport from the blood vessel to the cells and back again. Normal interstitial pressure is close to atmospheric, and normal transcapillary pressure is slightly higher (1–3 mmHg) within the vessels to aid in the convective flow of solutes from the vascular space to the interstitial space. Three factors combine to elevate the IFP within the tumor: the diminished function of the tumor blood vessels and lymphatics, the osmotic forces that draw solutes into the tissue, and the contractile characteristics of the tumor stroma [3]. Perturbation of all of these factors in the tumor combines to elevate tumor IFP to levels of up to 100 mmHg. Unlike oxygen tension (pO$_2$) and pHe, which are heterogeneous, tumor IFP is relatively uniform throughout the center of the tumor, dropping to normal, slightly negative values at the extreme periphery. Consequently, transcapillary fluid flow and convective transport of therapeutic molecules in tumors are greatly impaired [4].

16.2.2 Tumor Hypoxia

Impaired blood flow through the tumor also leads to reduced oxygen delivery, which is a major factor contributing to regions of tumor hypoxia. The delivery of oxygen to the tumor cannot meet the demand in the tumor cells, so large regions of the tumor exist in a chronic state of inequality between supply and demand. Normal tissue pO_2 ranges between 10 and 80 mmHg depending on the tissue type, whereas tumors contain regions where the pO_2 is <5 mmHg [5]. These chronically hypoxic regions arise because a large proportion of tumor cells lie beyond the distance that oxygen can diffuse from the nearest vessel (usually 100–200 μm [6]). The diffusion distance is determined primarily by the rate of oxygen consumption of the tumor cells, a variable that varies widely both within and between tumors. Variations in oxygen consumption may have a large effect on the hypoxic fraction of solid tumors and experimental data suggests that the physiologic regulation of oxygen consumption within the tumor cells is as important as the reduced oxygen delivery in determining tumor hypoxia. In addition, fluctuations in tumor RBC flux can produce temporal changes in pO_2 and intermittent episodes of hypoxia even in cells relatively close to vessels. Thus, within individual tumors, tissue oxygenation is highly heterogeneous both spatially and temporally.

16.2.3　Tumor Acidosis

Direct measurement of the pH within a solid tumor with electrode-based techniques has shown that the tumor compartment usually has a relatively acidic pH when compared with the corresponding normal tissue. The overall net accumulation of acid within the tumor was originally thought to be simply the result of lactate accumulation due to the high rate of aerobic and anaerobic glycolysis within the tumor cells. New data generated from genetically modified model tumors have shown that even tumors that are impaired in their ability to execute glycolysis and generate lactate are still able to generate an acidic environment [7]. It has been determined that the acidity within the tumor is a combination of the pHe and the internal, intracellular pH (pHi). Cells within the tumor are able to maintain a reasonably neutral cytosolic pH in the face of external acidosis, the pH gradient across the plasma membrane being maintained by the activity of a variety of ion pumps, including the monocarboxylate H^+ cotransporter, the vacuolar H^+ ATPase, the Na^+/H^+ exchanger, and the Na^+-dependent Cl^-/bicarbonate exchanger [8]. The spatiotemporal pH gradients within the tumor are due to a complex interaction between the metabolic state of the tumor cells and the ion-pumping characteristics of the cells and can vary significantly in spontaneous tumors; these patterns do not necessarily coincide with the regions of the tumor that are hypoxic. Whatever may be the source of H^+ ions, the impaired clearance of the products of metabolism by the tumor vasculature significantly contributes to the low pH levels observed in solid tumors.

16.2.4　Tumor Metabolism: Aerobic and Anaerobic Glycolysis

A common property of invasive cancers is altered glucose metabolism (reviewed by Gatenby and Gillies [9]). Glycolysis, literally the lysis of glucose, is the initial process of many pathways of carbohydrate catabolism.

The overall reaction of glycolysis is

$$\text{Glucose} + 2\text{NAD}^+ + 2\text{ADP} + 2\text{P}_i \rightarrow 2\text{NADH} + 2\text{Pyr} + 2\text{ATP} + 2\text{H}_2\text{O} + 2\text{H}^+$$

In aerobic organisms, pyruvate produced by glycolysis typically enters the mitochondria where it is fully oxidized to carbon dioxide and water by pyruvate decarboxylase and the set of enzymes of the citric acid cycle (also known as the TCA or Krebs cycle). The products of pyruvate are sequentially dehydrogenated as they pass through the cycle conserving the hydrogen equivalents via the reduction of NAD^+ to NADH. NADH is ultimately oxidized by an electron transport chain using oxygen as final electron acceptor to produce a large amount of ATP, a process known as oxidative phosphorylation. A small amount of ATP is also produced by substrate-level phosphorylation during the TCA cycle (Figure 16.3).

In mammalian cells, the flux through the glycolytic pathway is lower during aerobic conditions since the full oxidation of one molecule of pyruvate (equivalent to one-half molecule of glucose) can lead to 18 times more ATP. Malignant rapidly growing tumor cells, however, have glycolytic rates that are up to 200 times higher than those of their normal tissues of origin, despite the ample availability of oxygen. Conversion of glucose to lactic acid in the presence of oxygen is known as aerobic glycolysis or the Warburg effect and increased aerobic glycolysis is a unique property of cancer cells first reported by Warburg in the 1920s, leading him to the hypothesis that cancer results from impaired mitochondrial metabolism. The Warburg

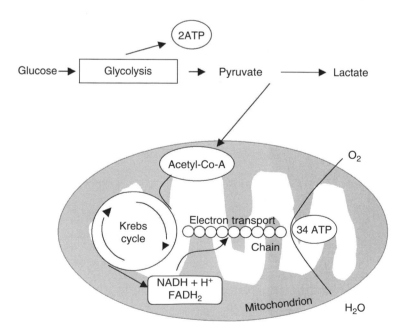

FIGURE 16.3
Glucose metabolism.

hypothesis was proven incorrect, but the experimental observations of increased glycolysis in tumors even in the presence of oxygen have been repeatedly verified. In addition to aerobic glycolysis, energy production in hypoxic foci can be maintained over a short period by anaerobic glycolysis.

Interest in tumor metabolism has revived recently, in part because of the widespread clinical application of the imaging technique positron-emission tomography (PET) using the glucose analogue tracer 18-fluorodeoxyglucose (FdG). FdG PET imaging of many thousands of cancer patients has unequivocally shown that most primary and metastatic human cancers show significantly increased glucose uptake (Figure 16.4). For many cancers, the specificity and sensitivity of FdG PET to identify primary and metastatic lesions is near 90% [10].

The increased glucose uptake imaged with FdG PET is largely dependent on the rate of glycolysis. FdG uptake and trapping occur because of up-regulation of glucose transporters (notably Glut1 and Glut3) and Hexokinases I and II. A key regulator of the glycolytic response is the transcription factor hypoxia-inducible factor-1 (HIF-1). This factor mediates a pleiotropic response to hypoxic stress by inducing survival genes, including glucose transporters, angiogenic growth factors (e.g., VEGF), and hexokinase II and hematopoietic factors (e.g., transferrin and erythropoietin). In some systems, constitutively increased HIF-1 levels are associated with constitutively high glucose consumption rates.

Up-regulation of glycolysis is a successful adaptation to hypoxia/anoxia but it also has significant negative consequences in that it partially

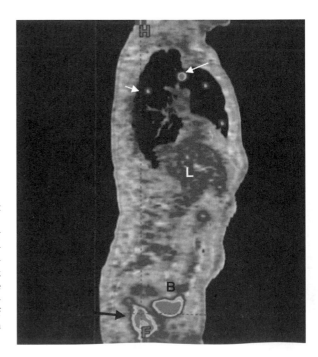

FIGURE 16.4 (See color insert following page 288.)
Positron-emission tomography imaging with 18-fluorodeoxyglucose (FdG) showing primary prostate tumor (black arrow) and metastases in the lung (white arrows). B—bladder, L—lung. (Courtesy of Dr Slobodan Devic. With permission.)

contributes to significant decreases in local extracellular pH; and prolonged exposure of normal cells to an acidic microenvironment typically results in necrosis or apoptosis through p53- and caspase-3-dependent mechanisms. Constitutive up-regulation of glycolysis requires additional adaptation to the negative effects of extracellular acidosis through resistance to apoptosis or up-regulation of membrane transporters to maintain normal intracellular pH.

16.3 Tumor Microenvironment Creates Barriers to Conventional Therapies

16.3.1 Chemotherapy

Several aspects of the tumor environment just described can be obstructive to cancer treatment:

1. Many studies in human and rodent tumors have observed a relationship between decreased drug uptake and elevated IFP, and several clinical studies have also shown that elevated IFP is a poor prognostic factor for patient outcome. Patients with lymphoma or melanoma have been shown to have better response to chemotherapy if they have IFP that drops during treatment, and cervical cancer patients treated with radiotherapy can be predicted to have a poor outcome if their tumors have elevated IFP. The high IFP in tumors can present a barrier to efficient drug-delivery by decreasing transcapillary fluid flow and convective transport of compounds from the bloodstream into the tumor interstitium. This is especially important for large molecules, such as antibodies and other proteins, as they are more heavily dependent on convection as opposed to simple diffusion for transport [4].

2. Although the mechanisms are not as well defined, tumor hypoxia also causes resistance to several types of chemotherapy. Oxygen concentration can have a direct effect on the effectiveness of drugs, such as mephalan, bleomycin, and etoposide, which require molecular oxygen for maximal efficiency [11]. Extreme hypoxia is also known to decrease the rate of cell division and ultimately cause cell-cycle arrest, which can decrease the effectiveness of drugs that are more active against proliferating cells. Finally, hypoxia has also been shown to select for apoptosis-resistant cells in model tumors [12].

3. Low pHe and the resulting pH gradient across the plasma membrane of tumor cells also present a barrier to drug delivery for many chemotherapeutic agents. Large lipophilic molecules pass through both plasma membranes and intracellular membranes

most efficiently in an uncharged state. Many commonly used chemotherapeutic agents are either weak bases (pKa 5.5–6.8) or weak acids (pKa 7.8–8.8) and in an acidic extracellular environment the weak bases, such as doxorubicin, mitoxantrone, or vinblastine, are more likely to exist in a charged state that inhibits their transport across the plasma membrane and significantly reduces their cellular uptake in vitro. Low pHe also impairs the effectiveness of paclitaxel and topotecan although their chemical structures do not predict pH-dependent ionizations [13].

16.3.2 Radiotherapy

16.3.2.1 Tumor Hypoxia and Radioresistance

Oxygen is a potent radiosensitizer, with well-oxygenated cells requiring one-third of the dose of anoxic cells to achieve a given level of cell killing. This relative resistance of hypoxic cells is thought to explain the clinical correlation between tumor oxygenation and patient response to radiotherapy. A large number of clinical trials have established the relationship between tumor hypoxia and poor clinical outcome for patients treated with radiotherapy for several different types of cancer including cervical carcinoma, soft tissue sarcoma, and head and neck cancer.

16.3.2.2 Radiosensitization by Oxygen

The response of cells to ionizing radiation is strongly dependent upon oxygen [14,15]. At high doses the enhancement of radiation damage by oxygen is dose-modifying, i.e., the radiation dose that gives a particular level of survival is reduced by the same factor at all levels of survival (Figure 16.5A). This allows us to calculate an oxygen enhancement ratio (OER).

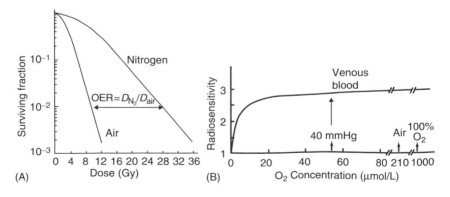

FIGURE 16.5
(A) Diagrammatic survival curve for cultured mammalian cells irradiated under oxic or hypoxic conditions illustrating the dose modifying effect of oxygen. (B) Change in the oxygen enhancement ratio (OER) with oxygen tension.

OER = Radiation dose in hypoxia/Radiation dose in air for the same level of biological effect. For most cells the OER for x-rays is 2.5–3.0. However, at clinically relevant radiation doses of <3.0 Gy the OER is reduced [16,17]. The OER is also dependent on the type of radiation, declining to a value of 1.0 for radiation with linear energy transfer values greater than about 200 keV.

The dependence of the extent of sensitization on oxygen tension is shown in Figure 16.5B. By definition, the OER under anoxic conditions is 1.0. As the oxygen level increases, there is a steep increase in radiosensitivity (and in OER) with the greatest change occurring between 0 and about 20 mmHg. Further increase in oxygen concentration, up to that of air (155 mmHg) or 100% oxygen (760 mmHg), produces a small but not trivial increase in radiosensitivity. From a radiobiological standpoint most normal tissues can be considered to be well oxygenated (venous blood has a pO_2 of approximately 45 mmHg); however, moderate hypoxia can be a feature of some normal tissues such as cartilage and skin.

It has been shown that the oxygen effect only occurs if oxygen is present during irradiation or within a few milliseconds (Michael et al. [17]), implying that the mechanism is mediated by short-lived free radicals. It appears that O_2 can interact with target radicals formed by radiation, resulting in damage to DNA that is more difficult for the cell to repair. The mechanism that has been believed to be responsible for the enhancement of radiation damage by oxygen is generally referred to as the oxygen-fixation hypothesis (OFH) and was developed in the late 1950s from the work of Alexander and Charlesby [18] (reviewed by Ewing [19]). The OFH is represented in the equations

$$R \rightarrow {}^\bullet R$$
$${}^\bullet R + O_2 \rightarrow {}^\bullet RO_2$$

where R represents a target radical formed as a direct result of exposure to ionizing radiation and ${}^\bullet RO_2$ is a peroxy radical formed by addition of O_2 to ${}^\bullet R$. The OFH assumes that the addition of O_2 to the target radical fixes damage so that it cannot be repaired or restored. A development from the OFH was to consider DNA to be the target, to link sensitization by oxygen to the reaction of OH radicals, and to make the end product less explicit [20]. This is summarized in the equations:

$$DNA + {}^\bullet OH \rightarrow {}^\bullet R$$
$$R + O_2 \rightarrow \text{inactive product}$$

This version, which states that DNA lesions formed with participation of O_2 are difficult or impossible to repair, is widely accepted as the best explanation of the function of oxygen. This hypothesis was however formulated before the extensive knowledge of DNA repair processes was accumulated and is subject to criticism on the grounds that if enzymic repair is successful, the issue of chemical restorability is not important. As it stands, OFH is not a totally successful explanation of why O_2 is a radiosensitizer, even though the basic chemistry is correct [19].

16.3.2.3   Hypoxia in Tumors

The first reports the hypoxia could exist in solid tumors were published by Thomlinson and Gray in 1955, based on their observations of histological sections of fresh specimens from human carcinoma of the bronchus [21]. They observed viable tumor regions surrounded by vascular stroma. As the tumor foci expanded, areas of necrosis appeared at the center. The thickness of the cylindrical zone of viable tumor tissue (100–180 µm) was calculated to be similar to the diffusion distance of oxygen in respiring tissue. This suggested that as oxygen diffused from the stroma it was consumed by cell respiration. Cells beyond the diffusion distance became necrotic but cells bordering on necrotic regions were viable although hypoxic.

Similar observations were made by Tannock (1968) in a mouse mammary tumor system. In this case there was extensive necrosis and each blood vessel was surrounded by a cord of viable tumor cells [22]. Cells at the edge of the cord were hypoxic and these cells are in the category that is sometimes described as chronically hypoxic. These cells have a short life-span being continually displaced as cells are pushed away from the blood vessel and, in turn, become hypoxic as the tumor proliferates.

16.3.2.4   Acute and Chronic Hypoxia

As has already been described, the microenvironment of solid tumors is characterized by regions of nutrient deprivation, low extra-cellular pH, high IFP, and hypoxia. The oxygen concentration (pO_2) in most normal tissues ranges between 10 and 80 mmHg, depending on the tissue type, whereas tumors often contain regions where the pO_2 is less than 5 mmHg. These conditions in solid tumors are due primarily to the abnormal vascularity that develops during tumor angiogenesis. A proportion of tumor cells may lie in hypoxic regions beyond the diffusion distance of oxygen where they are exposed to chronically low oxygen tensions. These regions of chronic hypoxia in which cells remain viable but exist under a pO_2 that will make them radiobiologically hypoxic and hence radioresistant, were first described in the classic paper of Thomlinson and Gray [21].

Tumor cells may also be exposed to shorter (often fluctuating) periods of (acute) hypoxia due to intermittent flow in individual blood vessels. Tumor hypoxia has been found to be heterogeneous both within and among tumors, even those of identical histopathological type, and it does not correlate simply with standard prognostic factors such as tumor size, stage, and grade [23]. Studies with both extrinsic and intrinsic markers of hypoxia [24] have shown that hypoxic cells can occur close to blood vessels, presumably due to fluctuation in blood flow in individual vessels resulting

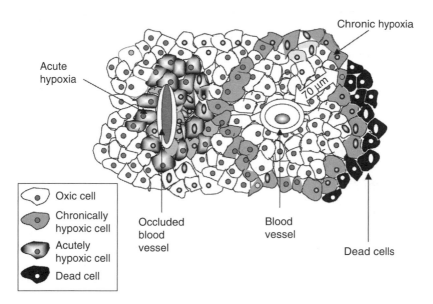

FIGURE 16.6
Acute and chronic hypoxia. Acute hypoxia develops around the temporarily occluded blood vessel on the left. Chronic hypoxia is found in cells 70 μm or more from the blood vessel on the right. Necrotic cells are found distal to the zone of chronic hypoxia.

in regions of hypoxia for short periods of time (minutes to hours). Studies involving the intravenous injection of diffusible fluorescent dyes as markers of functional blood vessels and measurements of microregional blood flow and tissue oxygenation have given direct evidence for this effect in experimental tumors [25]. Acute and chronic hypoxia can coexist in the same tumor, and hypoxic regions in tumors are often diffusely distributed throughout the tumor and rarely concentrated only around a central core of necrosis (Figure 16.6).

Oxic–hypoxic cycles in tumors have been observed to occur with periodicities of minutes. Using a magnetic-resonance imaging (MRI) technique sensitive to oxygenation status [26], fluctuations in signal intensity (oxygenation) were shown to occur with discrete periodicities of 1 and 20 cycles per hour in xenograft tumors while periodicities of about 1–2 cycles per minute [27] have been shown using microelectrodes. These studies served to show that oxygen delivery to tumors is inconsistent and that there is an acutely fluctuating hypoxic population in addition to the chronic hypoxia of tumor cells distant to the blood supply.

16.3.2.5 Re-Oxygenation

If tumors are irradiated with a large single dose of radiation, most of the sensitive aerated cells in the tumor are killed and the cells that survive are those which are resistant because of hypoxia. As a result, the hypoxic fraction immediately after irradiation is close to 100%. It has been shown in a number of studies that the hypoxic fraction does not remain at this high level indefinitely but subsequently falls and may return close to its starting value (shown schematically in Figure 16.7). This phenomenon, termed re-oxygenation, has been reported to occur in a variety of tumor systems, although the time frame can vary widely, re-oxygenation occurring within a few hours in some tumors and taking several days in others. The final level of hypoxia after re-oxygenation can also be higher or lower than its value before irradiation.

Re-oxygenation has important implications in clinical radiotherapy. If no re-oxygenation occurs, then each dose of radiation would be expected to kill only a small number of the hypoxic cells, and at the end of treatment the overall tumor response would be dominated by the hypoxic cell population. However, if re-oxygenation occurs between fractions, then the radiation killing of initially hypoxic cells will be greater and the hypoxic cells then have less impact on response. It has not been possible to detect re-oxygenation directly in human tumors, but its existence is supported by the fact that for many tumors the dose required for local control by fractionated radiation is consistent with the known SF_2 values for oxic tumor cells.

It has been shown in experimental animals that some tumors take several days to re-oxygenate whereas in others the process appears to be complete within 1 or 2 h. In a few tumors, both fast and slow components to re-oxygenation are evident. The differences of timescale probably reflect the time to reverse the different modes of hypoxia. The slow component

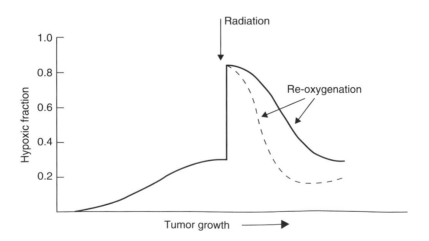

FIGURE 16.7
Re-oxygenation: Time course of changes in the hypoxic fraction during the growth of a tumor. Small tumors are generally well oxygenated but as the tumor grows, the hypoxic fraction increases. A large single dose of radiation kills nearly all the oxic cells and briefly raises the hypoxic fraction close to 100%. The hypoxic fraction then falls in a time frame and to an extent that is dependent on the type of tumor.

of re-oxygenation that takes place over a period of days involves restructuring or a revascularization of the tumor as the cells killed by the radiation are broken down and removed from the population. As this occurs, the tumor shrinks in size and surviving cells that were previously beyond the range of oxygen diffusion are closer to a blood supply and so re-oxygenate. The fast component of re-oxygenation is caused by the re-oxygenation of acutely hypoxic cells and is complete within hours.

16.3.3 Measurement of Tumor Hypoxia

16.3.3.1 Experimental Tumors

Since the first demonstration of a hypoxic fraction in a mouse tumor by Powers and Tolmach [28], the proportion of hypoxic cells has been measured in many other rodent and xenograft tumors. The most frequently used method has been to obtain paired survival curves. The presence of hypoxic cells in experimental tumors can be demonstrated by comparing cell-survival response of tumor cells irradiated under three sets of conditions: In situ in air-breathing mice, nitrogen-asphyxiated (i.e., hypoxic) mice, and single-cell suspension in vitro under fully oxic conditions. Cell-survival curves can be prepared using a single-cell suspension of disaggregated tumor cells from animals sacrificed immediately after irradiation. Survival is estimated by the excision or lung colony assays (Chapter 4).

The survival curves for tumor cells in air-breathing mice are biphasic. At low radiation doses, the response is dominated by the aerobic cells and the curves are similar to the oxic curve; whereas at large radiation doses, the presence of hypoxic cells influences the response, and the survival curve eventually becomes parallel to the hypoxic curve. The proportion of hypoxic cells (the hypoxic fraction) can be calculated from the vertical separation between the hypoxic and air-breathing survival curves in the region where they are parallel (shown schematically in Figure 16.8).

The hypoxic fraction can also be obtained from clamped tumor growth-delay assay, which involves comparing the time taken for clamped (i.e., hypoxic) and unclamped tumors to grow to a specific size after irradiation. A similar procedure with a different endpoint is the clamped tumor control assay, in which the radiation doses required to produce 50% local tumor control for clamped and unclamped tumors are compared.

For both of these techniques, it is necessary to produce full radiation dose–response curves under normal and clamped conditions and the hypoxic fractions can then be calculated from the displacement of these curves.

All of these techniques are based on several assumptions, the most significant being that cells made artificially hypoxic have the same radiosensitivity as those which reached this condition as a result of tumor metabolism and that the tumor is composed of two populations, one aerated and the other hypoxic with no cells having an intermediate status. Measurements of hypoxic fractions by these methods should thus be regarded as an indicator

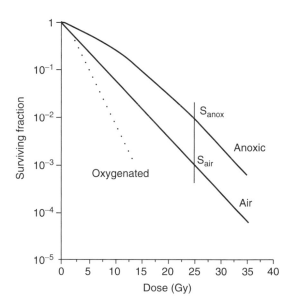

FIGURE 16.8

Paired survival curves: The curves shown are for a well-oxygenated population of cells, cells from tumors irradiated under air-breathing conditions, and cells from tumors irradiated under anoxic conditions. The hypoxic fraction can be estimated by taking the ratio of the survival obtained under air-breathing conditions (S_{air}) to that obtained under anoxic conditions (S_{anox}) at a dose level where the survival curves are parallel.

of hypoxic status rather than absolute values. Using the methods described above, it has been estimated that most experimental solid tumors contain radiation-resistant hypoxic cells with estimates of the hypoxic fractions ranging from below 1% to well over 50% of the total viable cell population (reviewed by Moulder and Rockwell [29]).

16.3.3.2 Measurement of Hypoxia in Human Tumors

A variety of techniques have been used to determine the oxygen status of human tumors:

1. Histomorphometric assays: Measurement of distance between blood vessels and zones of necrosis and of vascular density. This is a relatively simple procedure that can be applied to archived tissue. It can be combined with cryo-spectrophotometry to measure hemoglobin saturation within frozen tissue sections. The disadvantage is that it is an indirect measure of tumor oxygenation, which does not take into account perfusion in blood vessels.

2. Measurements using an oxygen electrode is the technique that has been most widely used and has received the most attention. The Eppendorf histograph succeeded the older oxygen electrodes employing more robust, reusable electrodes plus an automatic stepping motor making it possible to obtain a large number of oxygen measurements along several track within a short period of time. Using this method a direct relationship between electrode estimates of tumor oxygenation and the actual percentage of hypoxic clonogenic cells has been shown in some studies. The pO_2

electrode is stepped through tissue, and electrode signals are converted to pO_2 values providing a real-time measure of tissue oxygenation with multiple sampling. Measurements of pO_2 can be made over time in the same location. Disadvantages of this technique are that it is invasive and does not differentiate the pO_2 values in necrotic vs. viable regions or between tumor and normal tissue.

3. Luminescent probe: The probe is inserted into the tissue and interrogated with light pulses to measure lifetime of luminescence, which is proportional to O_2 concentration. Measurements can be made over time at the same location.

4. DNA strand-break assays: The comet assay is used to measure DNA strand breaks in irradiated tumor cells. Hypoxic cells are distinguished on the basis that they exhibit fewer breaks. This assay directly measures DNA damage in cells under hypoxic conditions in situ. The disadvantages are that the assay is invasive and indirect and requires rapid preparation of a cell suspension from a tissue biopsy post-irradiation. It is subject to sampling errors.

5. Extrinsic hypoxic cell markers: Preferential binding or uptake by hypoxic cells of radioactive or fluorescent-tagged 2-nitro-imidazoles (originally developed as hypoxic radiosensitizers), which bind irreversibly to hypoxic cells, requires the injection or ingestion of the marker drug. These compounds are administered systemically but are only metabolized to form adducts under hypoxic conditions. The two most commonly used are pimonidazole and EF5 (fluoromidonidazole). Nitro-imidazole binding can be detected by immunohistochemistry of tumor biopsies or external scanning.

   a. Tumor biopsy and immunohistochemistry: Microscopy analysis can visualize the hypoxic cells at the cellular level. This approach has a number of advantages over the use of oxygen electrodes in terms of the sensitivity of the technique. It is possible to detect hypoxia in a cell-by-cell basis, viable cells can be distinguished from necrotic tissue and distinction can be made between chronic and acute hypoxia. The disadvantages are that binding can be affected by metabolic factors and diffusion limitations, and microscopic analysis is subject to sampling error. Serial sampling is not practical (Figure 16.9).

   b. Noninvasive hypoxia imaging using external scanning procedures: Following injection of radio-active 2-nitro-imidazoles or other redox sensitive compounds imaging can be done by PET (positron emission spectroscopy) or SPECT. Noninvasive imaging has lower resolution than invasive procedures, however, changes in oxygenation of primary tumors and metastases can be visualized in the same individual although analysis

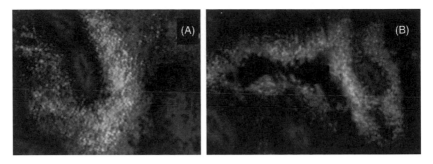

FIGURE 16.9 (See color insert following page 288.)
Antibody detection of chemical and endogenous hypoxia markers in frozen sections of tumors.
(A) Expression of CA9 (red, membranous), HIF-1α (green, nuclear), and Hoechst 33342 (blue,
nuclear) in a representative section from a SiHa cervical carcinoma xenograft. After Hoechst
33342 intravenous injection, frozen sections were stained sequentially with CA9 primary
antibody followed by Alexa-594 conjugated antimouse immunoglobulin G, and then HIF-1α
primary antibody followed by Alexa-488 conjugated antimouse immunoglobulin G. (B) Pimonidazole (green) and HIF-1α (red nuclei) staining are shown in a SiHa xenograft tumor. (From
Olive, P.L. and Aquino-Parsons, C., *Semin. Radiat. Oncol.*, 14, 241, 2004. With permission.)

cannot be done in "real time" because the data have to be
reconstructed. With PET-CT it is possible to average signals
over the entire tumor mass and, by simultaneous CT scan,
define the tumor mass. Frequently used imaging agents include
^{18}F-miso, ^{18}F-EF5, ^{60}CuATSM, and ^{123}I-IAZA.

6. Intrinsic hypoxic cell markers visualized by antibody staining of
   proteins up-regulated in cells by hypoxic exposure (e.g., HIF-l,
   CA-IX, GLUT-l, VEGF). Markers can be assessed in archived tissue. Markers may be up-regulated by factors other than hypoxia.
   The assay is subject to sampling error and uniform fixation of
   tissue is important for reliable quantitation. Endogenous hypoxia
   markers do not always correlate with hypoxia as they may be
   up-regulated by other stresses.

7. Imaging of blood oxygenation and flow: MRI-BOLD (blood oxygen level dependent) imaging or functional MRI or CT using
   contrast agents introduced into the blood. Near-infrared light
   spectroscopy measures HbO_2 saturation by absorption at different
   wavelengths. These procedures are noninvasive but do not measure tumor oxygenation directly and have relatively poor spatial
   resolution.

Results obtained with these techniques strongly support the view that
hypoxia is a common feature of human solid tumors, which can influence
both malignant progression and response to therapy. The most frequently
used technique has been the use of the oxygen probe, the so-called gold

standard for measuring tumor pO_2. This method has usually been applied to groups of tumors, on the basis of median pO_2 values ignoring tumor heterogeneity. Important information about tumor hypoxia on an individual patient basis is likely to come in the future from the noninvasive methods that are currently under development.

16.3.4 Radio-Sensitization by Modifying Tumor Oxygenation

A variety of strategies designed to improve tumor oxygenation that have been investigated are described in this section.

16.3.4.1 Modification of Physiology and Metabolism

1. Increasing oxygen-carrying capacity in the blood, through transfusion or erythropoietin administration, has had mixed results. The most promising approach still under investigation employs carbogen-breathing in combination with nicotinamide infusion during radiotherapy. This approach aims to reduce both chronic, diffusion-limited hypoxia through the use of carbogen gas breathing and acute, perfusion-limited hypoxia through the use of the vasoactive drug nicotinamide to render tumors more radiosensitive. In early clinical testing, increased tumor control rates have been observed in phase II trials in bladder cancer and head and neck cancer, and phase III trials are ongoing [30].

2. Blood vessel normalization occurs in response to anti-vascular endothelial growth factor (VEGF) therapies, which may cause significant changes to tumor physiology, including increased tumor oxygenation. Although initially counterintuitive, it has been proposed on the basis of studies in tumor models that during the early phases of anti-VEGF therapy, tumor blood vessels acquire a more normal structure and organization, which results in increased perfusion, drug delivery, and oxygenation [31].

3. Modeling of tumor oxygen levels suggests that decreasing tumor oxygen consumption could be more effective than increasing oxygen delivery as a means to reducing tumor hypoxia. These models predict that decreasing oxygen consumption by 30% would have an effect equivalent to increasing oxygen delivery by fourfold. The use of the general mitochondrial inhibitor *m*-iodobenzylguanidine to increase tumor oxygenation and response to radiation has been reported. Similar approaches based on the "crabtree effect" have shown increases in tumor oxygenation following bolus glucose administration and the resulting decrease in mitochondrial function.

4. The utility of cytotoxic hyperthermia as a primary therapy modality for solid tumors has not been shown convincingly, but mild

hyperthermia as a possible approach to improving tumor oxygenation before radiotherapy could be a useful procedure [32]. Cytotoxic hyperthermia (>42°C–43°C) impairs blood flow and increases the extent of hypoxia. Mild hyperthermia (41°C–42°C) on the other hand has been shown in a number of studies to improve oxygenation of experimental and human tumors.

5. A more targeted approach to modify tumor oxygenation has emerged through the study of farnesyl transferase inhibitors to block Ras signaling (described in Chapter 11). Several groups have reported that treatment of xenografted tumors with farnesyl transferase inhibitors can increase tissue oxygenation as measured by a decrease in hypoxic marker binding. This effect is greatest in tumors bearing oncogenic Ras mutations and it also results in increased sensitivity to radiotherapy [33].

16.3.4.2 Drugs Targeting Hypoxic Cells

16.3.4.2.1 Hypoxic Sensitizers

One approach to reducing the influence of tumor hypoxia on treatment outcome has involved the use of drugs that mimic the radiosensitizing properties of oxygen. Development of this class of radiosensitizer was based on the idea that the radiosensitizing properties of oxygen are due to its electron affinity and that other electron-affinic compounds might act as sensitizers [34,35]. A family of compounds, the nitromidazoles, was found to contain members that sensitize hypoxic cells both in vitro and in animal tumors (Table 16.1). The most extensively studied of these compounds is misonidazole, which can sensitize hypoxic cells in vitro in a dose-dependent fashion and does not sensitize oxygenated cells. The extent of the sensitization can be assessed in terms of a sensitizer enhancement ratio (SER) that is analogous to the OER.

For large acute radiation doses, good correspondence was found between the values obtained for tumors and the results from in vitro studies. However, when misonidazole was combined with fractionated radiation doses, the SER was reduced both because of re-oxygenation occurring between the fractions and because lower individual doses of the drug had to be given as fractionated treatments. A large number of sensitizers have been investigated of which nine reached clinical evaluation. Overall, results from the trials using misonidazole have been disappointing, possibly because the dose of misonidazole was limited by a dose-dependent peripheral neuropathy. Studies using drugs that are less toxic, such as etanidazole and nimorazole, revealed conflicting results. Although nimorazole has been associated with improved tumor control in head and neck cancer in the Danish Head and Neck Cancer Study (DAHANCA) trial [36], a benefit was not demonstrated in two multicenter trials for head and neck cancer using etanidazole [37]. Although most trials with nitroimidazoles have failed to

TABLE 16.1

Hypoxic Radiosensitizers and Bioreductive Drugs

Hypoxic Radiosensitizers

Misonidazole

A 2-nitro-imidazole. More active than metronidazole. Has shown some benefit in sub-groups. Toxicity (peripheral neuropathy).

Etanidazole

A 2-nitro-imidazole. Less toxic than misonidazole. No benefit shown in clinical trials when added to conventional radiotherapy.

Nimorazole

A 5-nitro-imidazole. Less effective radiosensitizer than misonidazole or etanidazole but much less toxic so much larger doses can be given. Shows significant improvement in loco-regional control and survival for head and neck cancer.

Hypoxic Cytotoxins

Tirapazamine

An organic nitroxide. A bioreductive drug reduced intracellularly to cytotoxic agent. Has shown a large hypoxic/oxic toxicity ratio in experimental studies. Clinically it is used as an adjunct to chemotherapy.

---

demonstrate a significant benefit, a meta-analysis of results from over 7000 patients included in 50 randomized trials indicated a small but significant improvement in local control and survival, with most of the benefit attributed to an improved response in patients with head and neck cancer [38]. This suggests that the apparent lack of clinical benefit in most of the individual trials may be due to the small number of patients involved rather than lack of the biologic importance of tumor hypoxia.

The overall conclusion is that though the nitro-imidazoles have not lived up to their original promise in terms of clinical efficacy, they may be of value in the treatment of certain conditions.

16.3.4.2.2   Hypoxic Cytotoxins

Another approach to reducing the influence of hypoxia on the radiation response of tumors is to use (bioreductive) drugs that are toxic under hypoxic conditions. Complementary effects of radiation (against aerobic cells) and of the drug (against hypoxic cells) might then increase the therapeutic ratio. The principal bioreductive drug of current clinical interest is tirapazamine [39], a benzotriazine di-N-oxide. Tirapazamine is cytotoxic to hypoxic cells at oxygen concentrations up to about 10 μM/L (equivalent to pO_2 of 6 mmHg) (Table 16.1). Under hypoxia, it is metabolized to an oxidizing radical that produces DNA damage including double-strand breaks, probably by interacting with topoisomerases. In the presence of oxygen, the radical is converted (by oxidation) back to the parent compound. The drug also interacts with the chemotherapeutic agent cisplatin to increase its toxicity. Tirapazamine is being evaluated in clinical trials and has shown efficacy in a phase III trial with cisplatin in non-small-cell carcinoma of the lung [40] and in early studies with advanced head and neck cancers treated with combination radiation and cisplatin therapy [41].

16.4   Effect of Hypoxia on Tumor Development and Progression

This section is partly based on reviews by Leo et al. [42], Le et al. [43], Semenza [44], and the references cited therein.

Hypoxia plays an important role in treatment outcome for many tumor types because of the drug and radiation-resistant phenotype of hypoxic cells. It has become apparent in recent years that the role of hypoxia is not simply as a passive instigator of treatment resistance but as an active provocateur of tumor progression, metastasis, and drug and radiation resistance. Over the last several years, clinical studies have shown that hypoxia is an independent prognostic indicator of poor patient survival in different tumor types, including cervical carcinoma, head and neck cancer, and soft-tissue sarcomas [45–48]. This observation also holds true for surgically treated patients, suggesting that there are fundamental biological differences between hypoxic and non-hypoxic tumor cells [45]. In addition hypoxia can affect the metastatic ability of some tumor cells (for review see Rofstad [49] and Subarsky and Hill [50]).

16.4.1   Regulation of Gene Expression by Hypoxia

A major mechanism by which hypoxia confers its effects is by differential regulation of gene expression and the most robust hypoxia-induced transcription factor is HIF-1. The changes in gene expression in the hypoxic tumors are thought to be the same changes that help normal cells to adapt to a hypoxic microenvironment under noncancerous conditions such as wound healing. Similar changes when they occur in tumor tissue may render tumor cells more aggressive or resistant toward different treatment modalities. In addition to epigenetic changes in gene expression, hypoxia will select resistant cells with genetic features of the hypoxic phenotype that contribute to tumor progression. The expression of as much as 1.5% of the genome may be modified by exposure to hypoxia. Many of these genes are involved in cellular functions such as anaerobic respiration and include glycolytic enzymes and cell membrane proteins such as glucose transporters (e.g., GLUT-I) and enzymes that control carbonate levels (carbonic anhydrase IX [CA-lX]). Genes that modify the oxygen carrying capacity of blood (e.g., erythropoietin) or increase vascularity, such as angiogenic growth factors like VEGF are also up-regulated, as are survival factors and invasive factors. A partial list of genes regulated by hypoxia is given in Table 16.2.

16.4.2   Hypoxia-Induced Factor I: Regulator of Oxygen Homeostasis

The transcription factor with the most sensitive and specific induction in hypoxic conditions is HIF-1, a heterodimeric basic helix-loop-helix transcription factor composed of two subunits, HIF-1α and HIF-1β (also called ARNT). HIF-1β/ARNT is stably expressed in cells but the HIF-1α protein is unstable in the presence of oxygen. The level of HIF-1 expression is determined by the rates of protein synthesis and protein degradation. HIF-1 synthesis is regulated by O_2-independent mechanisms (Figure 16.10), whereas HIF-1 degradation is regulated primarily by O_2-dependent mechanisms (Figure 16.11), described below.

16.4.2.1   Oxygen-Dependent Regulation of HIF-1

Cells transduce decreased O_2 concentration into increased HIF-1 activity via a novel O_2-dependent post-translational modification [51] (Figure 16.10). Three prolyl hydroxylases—known as prolyl hydroxylase-domain protein (PHD) 1–3 or, alternatively, as HIF-1 prolyl hydroxylase (HPH) 1–3—modify proline-402 and -564 of HIF-1. Hydroxylation of these prolines is required for interaction of HIF-1 with the von Hippel-Lindau (VHL) tumor-suppressor protein. VHL is the recognition component of an E3 ubiquitin-protein ligase that targets HIF-1 for proteasomal degradation. The prolyl hydroxylases use molecular oxygen and 2-oxoglutarate (-ketoglutarate) as substrates in a reaction that generates prolyl-hydroxylated HIF-1 and succinate. Under physiological conditions, O_2 is a limiting substrate, thereby providing a mechanism for O_2-dependent regulation of HIF-1 expression [51]. Acetylation of HIF-1 at lysine-532 by the ARD1 acetyltransferase enhances the interaction of VHL with HIF-1, promoting its ubiquitylation and degradation. Oxygen also regulates the interaction of HIF-1 with transcriptional coactivators. The O_2-dependent hydroxylation of asparagine residue 803 in HIF-1 by the enzyme FIH-1 (factor inhibiting HIF-1) blocks the binding of p300 and CBP

TABLE 16.2

Genes That Are Transcriptionally Activated by HIF-1

Functional Category	Gene
Cell proliferation	Cyclin G_2
	Insulin-like growth factor-2 (IGF2)
	IGFBP1, IGFBP2, IGFBP3
	Cyclin-dependent kinase inhibitor 1A (CDKN1A)
	TGF-α
	TGF-β3
Glucose metabolism	Hexokinase 1, 2
	Enolase 1
	Glucose Transporter 1, 3
	GAPDH
	Lactate dehydrogenase
	Pyruvate kinase
	Phosphofructokinase L
	Triosephosphate isomerase
Angiogenesis	VEGF A, B, C, D
	Endothelin 1, 2
	Nitric oxide synthase
	Cyclo-oxygenase 1, 2
Apoptosis resistance	Adrenomedullin (ADM)
	Erythropoietin (EPO)
	Endothelin-1 (ET1)
	Insulin-like growth factor-2 (IGF2)
	Nitric oxide synthase (NOS)
	TGF-α
Invasion, metastases	Autocrine motility factor (AMF)
	Cathepsin D (CATHD)
	c-met
	Fibronectin-1 (FN1)
	Keratin 14 (KRT14)
	KRT18
	KRT19
	Matrix metalloproteinase 2 (MMP2)
	Urokinase plasminogen activator receptor (uPAR)
	Vimentin (VIM)

to HIF-1, and as a result inhibits HIF-1-mediated gene transcription. Under hypoxic conditions, these reactions do not occur because the rate of asparagine and proline hydroxylation is decreased and, since VHL cannot bind to HIF-1 that is not prolyl-hydroxylated, the rate of HIF-1 degradation is depressed leading to HIF-1 accumulation.

16.4.2.2   Oxygen-Independent Regulation of HIF-1

Growth-factors, cytokines, and other signaling molecules stimulate HIF-1α through activation of the phosphatidylinositol 3-kinase (PI3K) and mitogen-activated protein kinase (MAPK) pathways. (Signal transduction pathways

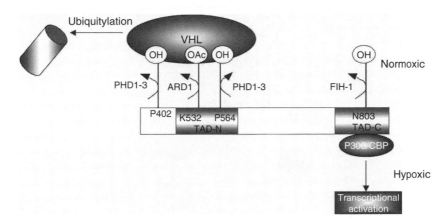

FIGURE 16.10
O_2-dependent regulation of HIF-1 activity. O_2 regulates the rate at which HIF-1α protein is degraded. In normoxic conditions, O_2-dependent hydroxylation of proline (P) residues 402 and 564 in HIF-1 by the enzymes PHD (prolyl hydroxylase-domain protein) 1–3 is required for the binding of the von Hippel–Lindau (VHL) tumor-suppressor protein, which is the recognition component of an E3 ubiquitin-protein ligase. VHL binding is also promoted by acetylation of lysine (K) residue 532 by the ARD1 acetyltransferase. Ubiquitylation of HIF-1α targets the protein for degradation by the 26S proteasome. O_2 also regulates the interaction of HIF-1 with transcriptional coactivators. O_2-dependent hydroxylation of asparagine (N) residue 803 in HIF-1 by the enzyme FIH-1 (factor inhibiting HIF-1α) blocks the binding of p300 and CBP to HIF-1 and therefore inhibits HIF-1α-mediated gene transcription. Under hypoxic conditions, the rate of asparagine and proline hydroxylation decreases. VHL cannot bind to HIF-1α that is not prolyl-hydroxylated, resulting in a decreased rate of HIF-1α degradation. In contrast, p300 and CBP can bind to HIF-1 that is not asparaginyl-hydroxylated, allowing transcriptional activation of HIF-1α target genes.

are described in Chapter 11.) PI3K activates the downstream serine/threonine kinases AKT (also known as protein kinase B [PKB]) and mTOR, the mammalian target of rapamycin. In the MAPK pathway, the extracellular-signal-regulated kinase (ERK) is activated by the upstream MAP/ERK kinase (MEK), and ERK then activates MNK. ERK and mTOR phosphorylate p70 S6 kinase (S6K)—which, in turn, phosphorylates the ribosomal S6 protein—and the eukaryotic translation initiation factor 4E (eIF-4E) binding protein (4E-BP1). This reaction is important because binding of 4E-BP1 to eIF-4E inactivates the latter, inhibiting cap-dependent mRNA translation, and phosphorylation of 4E-BP1 prevents its binding to eIF-4E. MNK phosphorylates eIF-4E and stimulates its activity directly (shown schematically in Figure 16.11). The downstream effect of growth factor signaling is an increase in the rate at which a subset of mRNAs within the cell, which includes HIF-1 mRNA, is translated into protein.

The regulation of HIF-1α is summarized as follows. The HIF-1α protein is labile under oxic conditions, being rapidly targeted for degradation by the VHL protein via a proteasomal pathway. Under hypoxic conditions HIF-1α is stabilized, leading to its rapid accumulation and translocation to the nucleus (Chapter 8).

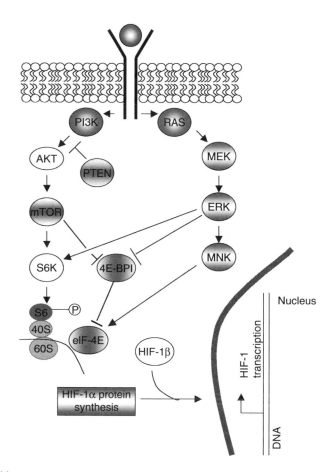

FIGURE 16.11

O₂-independent regulation of HIF-1 activity. Growth-factor binding to a cognate receptor tyrosine kinase activates the phosphatidylinositol 3-kinase (PI3K) and mitogen-activated protein kinase (MAPK) pathways. PI3K activates the downstream serine/threonine kinases AKT (also known as protein kinase B [PKB]) and mammalian target of rapamycin (mTOR). In the MAPK pathway, the extracellular-signal-regulated kinase (ERK) is activated by the upstream MAP/ERK kinase (MEK). ERK, in turn, activates MNK. ERK and mTOR phosphorylate p70 S6 kinase (S6K)—which, in turn, phosphorylates the ribosomal S6 protein and the eukaryotic translation initiation factor 4E (eIF-4E) binding protein (4E-BP1). Binding of 4E-BP1 to eIF-4E inactivates the latter, inhibiting cap-dependent mRNA translation. Phosphorylation of 4E-BP1 prevents its binding to eIF-4E. MNK phosphorylates eIF-4E and stimulates its activity directly. The effect of growth-factor signaling is an increase in the rate at which a subset of mRNAs within the cell (including HIF-1α mRNA) is translated into protein.

In the nucleus, HIF-1α heterodimerizes with the HIF-1β subunit, binds to specific DNA sequences within so-called hypoxia-response elements, and starts to transactivate specific target genes. The effect of signaling by growth factors is to increase the level of HIF-1 protein independent of oxygen level. Other transcription factors can be induced by hypoxia, such as NF-κB, AP-1, and early growth response factor-1 (Egr-1) (Chapter 13). However, these are more general stress-responsive transcription factors whose response to hypoxia is less specific than that of HIF-1. NF-κB has been particularly implicated as a mediator of the effects of hypoxia and re-oxygenation in tumor cells. It transactivates an extensive number of genes, among them cytokines and growth factors, acute-phase response proteins, leucocyte adhesion molecules, transcription/growth control factors, and immunoregulatory molecules.

16.4.3 Genes That Are Transcriptionally Activated by HIF-1

Based on an extrapolation from expression profiling data, it has been estimated that approximately 1.5% of the genome is transcriptionally responsive to hypoxia [52]. Hypoxia-regulated genes are involved in diverse biological processes, including regulation of metabolism, apoptosis, angiogenesis, and invasion. As a result, hypoxic cancer cells undergo adaptive changes that allow them not only to survive but even to proliferate or leave the adverse tumor micromilieu.

16.4.3.1 Metabolism

As already described, cancer cells are adapted to supply their energy needs under hypoxic conditions. To adapt to hypoxia, cells switch from the (aerobic) citric acid cycle to anaerobic glycolysis to generate energy. Because of the reduced energy yield (2 ATP molecules from 1 glucose molecule by glycolysis in comparison to 38 ATP molecules from the citric acid cycle), the overall glucose consumption must increase. HIF-1α activates many enzymes of the glycolytic pathway as well as transporters responsible for accumulating glucose within the cell. Another metabolic characteristic of cancer cells is their capability to survive and grow in low-pH environments. The anaerobic consumption of glucose leads to the accumulation of lactic acid, which results in intracellular and extracellular acidosis. The HIF-1-regulated induction of carbonic anhydrases IX and XII contributes to the regulation of the low extracellular pH environment of the tumor.

16.4.3.2 Apoptosis

Apoptosis has been reviewed by Hammond and Giaccia [53]. Prolonged exposure to hypoxia can lead to cell death by apoptosis. In p53-competent cells, the hypoxia-induced p53-dependent apoptosis pathway requires Apaf-1 and Caspase-9. Cells that have a mutated p53 gene have been found to acquire genetic resistance to hypoxia-mediated apoptosis [12]. This suggests that hypoxia could promote tumor progression by selecting for cells with p53 mutations. This acquired resistance to apoptosis can in turn contribute to the aggressive phenotype that is characteristic for many hypoxic tumors.

In contrast to DNA damaging agents, hypoxia-induced p53 has little or no transcriptional activation capability and instead functions largely as a trans-repressor in order to induce apoptosis [53]. p53 accumulates at oxygen concentrations low enough to induce replication arrest and complete cessation of DNA synthesis. Hypoxia-induced replication arrest leads to ATR phosphorylation of p53, which acts as a transrepressor of anti-apoptotic proteins. Some experimental data indicates that p53 can form a repression complex on the promoters of target genes such as the survivin promotor. Survivin is a member of the inhibitor of apoptosis (IAP) family and has been shown to inhibit apoptosis by binding to and inhibiting terminal effectors in the apoptotic cascade such as caspase 9. The "traditional" role of p53 in response to damage and stress is to accumulate and induce apoptosis through transcriptional activation of such pro-apoptotic proteins as Bax. The p53 response to hypoxia is different although the end point is still apoptosis mediated by mitochondrial-based processes.

16.4.3.3 Angiogenesis and Oxygen Delivery

In cancer, hypoxia results from the inexorable oxygen demand of the growing tumor, which sooner or later exceeds its blood supply. To respond to this condition, hypoxia coordinately induces several pro-angiogenic growth factors and represses anti-angiogenic factors. The key to the synthesis of new vessels is the coordinated expression of the numerous angiogenic factors, most notably the HIF-1 target gene VEGF. Other genes involved in the growth of new vessels in a hypoxic environment are listed in Table 16.2. The large number of pro-angiogenic genes suggests that a coordinated pattern of expression is necessary for a functional vessel. In addition, hypoxia inhibits anti-angiogenic factors such as thrombospondin I and II.

As already described, the tumor vasculature is characterized by a pathologic and chaotic architecture leading to an impaired function that maintains the hypoxic micromilieu and thus hypoxic gene activation. The sprouting of pathological vessels, thromboembolism, and fibrosis has been thought to lead to areas of transient re-oxygenation. The gene for COX-2 was shown to be induced by hypoxia via the transcription factor NF-κB. COX-2 is of interest because it supports angiogenesis by up-regulating VEGF and inhibits apoptosis by inducing Bcl-2. Over-expression of COX-2 has been shown in a variety of tumor entities and is generally associated with poor treatment outcome. Besides inducing angiogenesis, HIF-1 activates genes that are involved in increasing oxygen supply to peripheral tissue, by transactivation of erythropoietin, transferrin, the transferrin receptor, and heme oxygenase.

16.4.3.4 Invasion and Metastasis

Hypoxia can directly increase tumor cell invasiveness and metastasis. Studies have shown that hypoxia in combination with re-oxygenation can result in a dramatic but temporary increase in the metastatic potential of murine tumor cells. Several hypoxia-responsive genes involved in the process of invasion and metastasis have been identified.

It has been shown that hypoxia sensitizes cells to the invasive qualities of hepatocyte growth factor by increasing the levels of the growth factor receptor, the *c-met* proto-oncogene [54]. The hypoxic induction of met mRNA is activated by cooperation between HIF-1 and AP-I. Up-regulated met stimulates growth, promotes cell shape changes, causes cell division, increases cell mobility, and produces proteases that lead to matrix degradation, all characteristics of the invasive growth cascade that leads to metastatic spread of the hypoxic tumor.

Another potential contributor of the invasive process is the hypoxia-responsive plasminogen activator inhibitor-1 (PAI-1) gene. Studies in PAI-1 knockout mice revealed that PAI-1 deficiency prevented local invasion and that by restoring the PAI-1 genotype the invasive phenotype was regained. A third candidate for hypoxia-induced metastasis and invasion is the urokinase plasminogen activator receptor (uPAR), which catalyzes the conversion of plasminogen to plasmin, thereby leading to the degradation of the extracellular matrix. A group of other extracellular matrix/adhesion molecules, including cathepsin D, fibronectin, and matrix metalloproteinase 2, have also been reported to be hypoxia-inducible.

16.4.4 Endogenous Hypoxia Markers in Human Tumors

The association between intratumoral pO_2 and the expression or endogenous hypoxia genes in human cancer is still somewhat unclear although immunodetection of endogenous HIF-1α has been realized in a variety of human tumor types. In lymph node-positive and -negative breast cancer, HIF-1 over-expression is associated with an unfavorable prognosis. Similar data exists for early-stage invasive cervical cancer. In contrast, it was shown that HIF-1 over-expression in squamous cell cancer of the head and neck was associated with improved survival.

The only hypoxia-regulated marker genes whose expression has been shown to be dependent on intratumoral oxygenation status are Glut-1 and carbonic anhydrase CA IX. Glut-1 correlated with intratumoral hypoxia, as measured by pimonidazole binding and by Eppendorf needle electrodes, in cervical cancer. The absence of Glut-1 was associated with metastasis-free survival. CA IX expression correlates positively to the level of intratumoral hypoxia measured in cervical cancer and is associated with poor survival in cervical cancer, breast cancer, and lung cancer [55]. In contrast, VEGF, the prototypical example of a hypoxia-induced protein, did not show an association with the intratumoral oxygenation measured with the Eppendorf electrode in a clinical study [56]. These apparently conflicting results suggest that multiple factors including oncogenes, tumor suppressor genes, and growth factors interact with hypoxia to regulate the expression pattern, in vivo.

16.4.5 Therapeutics Based on Tumor-Specific Targeting

The difference between tumors and normal tissues is that the former have mutated genes that affect the signaling pathways that control cell proliferation, differentiation, or death. These mutations allow tumor cells to circumvent positional control mechanisms and survive in a state of relative positional independence. Signal transduction pathways become re-equilibrated, and the cells become addicted to specific pathways for survival. The therapeutic advantage is that tumor cells are very sensitive to blockade of these pathways and they may serve as an "Achilles heel" for that cancer. Identification and characterization of these pathways in individual cancers is therefore very important for selection of an appropriate specific therapy.

In an earlier chapter (Chapter 11), drugs that inhibit or blockade growth factors, growth factor receptors, and signal transduction intermediates, which are preferentially over-expressed or activated in cancer cells, were described. In this section another class of tumor specific therapeutic agents, the drugs that target tumor-stromal cell responses, are discussed.

16.4.5.1 Targeting Cells Which Express HIF-1

Targeting cells which express HIF-1 have been reviewed by Semenza [44]. A large number of drugs are in clinical trials at present as anticancer agents based on their ability to inhibit angiogenesis [44]. A concern about this strategy is that inhibition of angiogenesis might select for cancer cells that are adapted to hypoxic conditions, as these are the cells that are most likely to survive a reduction in oxygen perfusion. Recent studies have shown, however, that HIF-1 mediates resistance to chemotherapy and radiation, and inhibition of HIF-1 activity could therefore represent an important component of combination anti-angiogenesis therapies. In addition to drugs that have been developed specifically as anti-angiogenesis agents, it is clear that many novel therapeutic agents that target signal-transduction pathways have anti-angiogenic effects. Some of these are listed in the first part of Table 16.3. The second part lists small molecule inhibitors of HIF-1.

Tumors that might benefit from treatment by HIF-1 inhibitors are those characterised by HIF-1 over-expression and VHL mutations. Experimental data indicate that HIF-1 is involved in the pathophysiology of renal cell carcinoma (RCC) and the up-regulation of HIF-1 activity seems to occur in the earliest detected neoplastic lesions [57]. HIF-1 target genes that are activated in RCC include platelet-derived growth factor-β and TGFα, encoding proteins that activate the PDGF and EGF receptors—two of the main receptor tyrosine kinases that signal via the PI3K and MAPK pathways to stimulate cell proliferation and survival.

Another tumor that might benefit from HIF-1 targeted therapy is glioblastoma multiforme (GBM). Like RCC, GBM is an intractable disease, and patients typically survive for less than one year after diagnosis. Among gliomas, there is a strong correlation between HIF-1 expression, tumor grade, and tumor vascularization [58]. PTEN loss of function and EGFR gain of function are commonly observed in primary GBM and are known to increase HIF-1 levels [59].

TABLE 16.3

Inhibitors of HIF-1 Activity

Drug	Target	Mechanism, Comments
Camptothecin	Topoisomerase I	Clinically approved agents
Topotecan		Mechanism unknown [65]
YC-1		Mechanism unknown
(3-[5'-hydroxy-methyl-2'-furyl]-1-benzylindazole)		Reduces HIF-1 level and xenograft growth [66]
17-AAG (17-allyl-aminogel-danamycin)	HSP90	Induces HIF-1α degradation in a VHL-independent manner [67]
Pleurotin (1-methylpropyl 2-imidazolyl disulphide)	Thioredoxin 1	Thioredoxin inhibitor Blocks HIF-1α expression and xenograft growth. Mechanism unknown [68]
2ME2 (2-methoxyestradiol)	Microtubule polymerization	Decrease HIF-1α and VEGF expression. Reduced vacularization and xenograft growth in vivo. In clinical trials Mechanism unknown [69]

16.4.6 HIF-1 and Radioresistance

Clinical and preclinical studies have also implicated HIF-1 in radiation resistance. In patients with oropharyngeal cancer, over-expression of HIF-1α in tumor biopsy is associated with an increased risk of failure to achieve complete remission after radiation therapy [60]. Tumor xenografts of HIF-1α null transformed mouse embryo fibroblasts manifest increased radiation sensitivity [61].

A study by Moeller and Dewhirst (2004) established a connection between HIF-1, tumor vasculature, and radiation resistance [62]. The authors demonstrated that irradiation of tumor xenografts induces HIF-1 activity, leading to the expression of VEGF and basic fibroblast growth factor (bFGF), which acted to prevent radiation-induced endothelial cell (EC) death. The induction of HIF-1 did not start until 12 h and peaked at 48 h after radiation. It was shown that radiation-induced re-oxygenation of hypoxic tumor cells results in the production of reactive oxygen species (ROS) that induce HIF-1 activity.

16.5 Targeting the Ubiquitin/Proteasome System

Aberrant expression of signal transduction molecules in pathways controlling cell survival, proliferation, death, or differentiation is a common feature of all tumors. The identification of the molecules that are involved allows the development of novel tumor-specific strategies, and targeting these pathways often also results in radiosensitization. The efficacy of such directed therapies may, however, be limited by the heterogeneity and the multiple mutations that are associated with the cancerous state. A more robust alternative may be to target global mechanisms of cellular control. The ubiquitin (Ub)/proteasome degradation pathway is one candidate for such therapeutic intervention [63]. This pathway is the main post-transcriptional mechanism that controls levels of many short-lived proteins involved in regulation of cell-cycle progression, DNA transcription, DNA repair, and apoptosis. This is in addition to its more established roles in the removal of misfolded, damaged, and effete proteins. (The role of the Ub/proteasome system in the regulation of p53 is described in Chapter 8.)

The Ub/proteasome system is involved in processes underlying the classical effects of irradiation on cells, such as radiation-induced gene expression, DNA repair and chromosome instability, oxidative damage, cell-cycle arrest, and cell death. Other evidence suggests that the proteasome is a redox-sensitive target for ionizing radiation and other oxidative stress signals, so that in effect the Ub/proteasome system may not simply be a passive player in radiation-induced responses, but may modulate them. Cell types vary in the Ub/proteasome structures they possess and the level at which they function, and this changes as they go from the normal to the cancerous condition. Cancer-related functional changes within the Ub/proteasome system may therefore present unique targets for cancer therapy, especially when targeting agents are used in combination with radiotherapy or chemotherapy. The peptide boronic acid compound PS-341, which was designed to inhibit proteasome chymotryptic activity, is in clinical trials for the treatment of solid and hematogenous tumors [64]. It has shown some efficacy on its own and in combination with chemotherapy. Preclinical studies have shown that PS-341 will also potentiate the cytotoxic effects of radiation therapy. In addition, other drugs in common clinical use have been shown to affect proteasome function, and their activities might be reconsidered from this perspective.

16.6 Summary

The microenvironment of a solid tumor is distinguished from the corresponding normal tissue by high interstitial fluid pressure (IFP), low oxygen tension or hypoxia, and low extracellular pH. Another common property of invasive cancers is a shift in glucose metabolism from respiration to glycolysis, which allows for the maintenance of energy production through a range of oxygen concentrations. Elevated IFP, tumor hypoxia, and low pHe can all contribute to resistance to chemotherapy.

Oxygen is a potent radiosensitizer, with well-oxygenated cells requiring one-third of the dose of anoxic cells to achieve a given level of cell killing.

The oxygen effect only occurs if oxygen is present within a few milliseconds of irradiation, implying that the mechanism is mediated by short-lived free radicals. Radiosensitization by oxygen is explained on the basis of the oxygen fixation hypothesis (OFH), which proposes that oxygen makes permanent radiation-inflicted damage to DNA. Most solid tumors have regions beyond the diffusion distance of oxygen where there are foci of chronically low oxygen tension. Tumor cells may also be exposed to shorter periods of acute hypoxia due to intermittent flow in individual blood vessels. Immediately after irradiation, the hypoxic fraction of a tumor will be close to 100%, but this high level is not maintained indefinitely. Instead, the hypoxic fraction declines over a period of hours or days and may return close to its starting value, a phenomenon called re-oxygenation.

Measurement of the proportion of hypoxic cells in human tumors indicates that hypoxia is a common feature. The most frequently used technique has been the use of the oxygen probe, the so-called gold standard for measuring tumor pO_2. Noninvasive methods are currently under development. A variety of strategies designed to improve tumor oxygenation have been investigated including increase in the oxygen carrying capacity in the blood by anti–vascular endothelial growth factor (VEGF) therapies, decrease of tumor oxygen consumption, mild hyperthermia, and inhibitors that block Ras signaling. Great efforts have been invested in the development of drugs that would radiosensitize hypoxic cells, and some nitroimidazoles were identified which sensitized hypoxic cells both in vitro and in animal tumors. Results from a large number of clinical trials indicate that though the nitroimidazoles have not lived up to their original promise in terms of clinical efficacy, they may be of value in the treatment of a limited number of conditions. Bioreductive drugs are pro-drugs that become toxic under hypoxic conditions. The principal bioreductive drug of current clinical interest is tirapazamine, which is being evaluated in clinical trials.

Hypoxia plays an important role in treatment outcome for many tumor types both because of the drug and radiation-resistant phenotype of hypoxic cells and because hypoxia has been shown to play an important role in malignant progression, with clinical studies having shown that hypoxia is an independent prognostic indicator of poor patient survival. A major mechanism by which hypoxia confers its effects is by differential regulation of gene expression, and the most robust hypoxia-induced transcription factor is hypoxia-inducible factor-1 (HIF-1). The level of HIF-1 expression is determined by the rates of protein synthesis and protein degradation. Under hypoxic conditions, the normally labile HIF-1α protein becomes stabilized, leading to its rapid accumulation. On translocation to the nucleus, it heterodimerizes with the HIF-1β subunit, binds to specific DNA sequences within so-called hypoxia-response elements, and starts to transactivate specific target genes. Under normoxia, HIF-1α is rapidly targeted for degradation by the von Hippel-Lindau protein (VHL) via a proteasomal pathway. HIF-1α activates many enzymes of the glycolytic pathway and glucose transporters. Synthesis of new vessels as the tumor grows results from the coordinated expression of the numerous angiogenic factors, most notably the HIF-1 target gene VEGF. Hypoxia can directly increase tumor-cell invasiveness and metastasis by increasing the levels of the growth factor receptor, *c-met* proto-oncogene. Other hypoxia responsive contributors to the invasive process are plasminogen activator inhibitor-1 (PAI-1) and the urokinase plasminogen activator receptor (uPAR). Prolonged exposure to hypoxia can lead to cell death by apoptosis. In p53-competent cells, the hypoxia-induced p53-dependent apoptosis pathway requires Apaf-1 and Caspase-9. Cells with mutated p53 gene have been found to acquire genetic resistance to hypoxia-mediated apoptosis contributing to the aggressive phenotype that is characteristic of hypoxic tumors.

A robust approach to anticancer therapy would be to target global mechanisms of cellular control. One such system is the Ub/proteasome system, which is involved in processes basic to the effects of irradiation on cells and is a redox-sensitive target for ionizing radiation and other oxidative stress signals. A compound designed to inhibit proteasome chymotryptic activity, is in clinical trials for the treatment of solid and hematogenous tumors.

References

1. Cairns R, Papandreou I, and Denko N. Overcoming physiologic barriers to cancer treatment by molecularly targeting the tumor microenvironment. *Mol. Cancer Res.* 4: 61–70, 2006.
2. Vaupel P. Tumor microenvironmental physiology and its implications for radiation oncology. *Semin. Radiat. Oncol.* 14: 198–206, 2004.
3. Heldin CH, Rubin K, Pietras K, and Ostman A. High interstitial fluid pressure—An obstacle in cancer therapy. *Nat. Rev. Cancer* 4: 806–813, 2004.
4. Jain RK. Physiological barriers to delivery of monoclonal antibodies and other macromolecules in tumors. *Cancer Res.* 50: 814s–819s, 1990.
5. Vaupel P, Schlenger K, Knoop C, and Hockel M. Oxygenation of human tumors: Evaluation of tissue oxygen distribution in breast cancers by computerized O_2 tension measurements. *Cancer Res.* 5: 3316–3322, 1991.
6. Koch CJ. Measurement of absolute oxygen levels in cells and tissues using oxygen sensors and 2-nitroimidazole EF5. *Methods Enzymol.* 352: 3–31, 2002.
7. Helmlinger G, Sckell A, Dellian M, Forbes NS, and Jain RK. Acid production in glycolysis-impaired tumors provides new insights into tumor metabolism. *Clin. Cancer Res.* 8: 1284–1291, 2002.
8. Izumi H, Torigoe T, Ishiguchi H, Uramoto H, Yoshida Y, Tanabe M, Ise T, Murakami T, Yoshida T, Nomoto M, and Kohno K. Cellular pH regulators: Potentially promising molecular targets for cancer chemotherapy. *Cancer Treat. Rev.* 29: 541–549, 2003.
9. Gatenby RA and Gillies RJ. Why do cancers have high aerobic glycolysis? *Nat. Rev. Cancer* 4: 891–899, 2004.
10. Czernin J and Phelps ME. Positron emission tomography scanning: Current and future applications. *Annu. Rev. Med.* 53: 89–112, 2002.

11. Koch S, Mayer F, Honecker F, Schittenhelm M, and Bokemeyer C. Efficacy of cytotoxic agents used in the treatment of testicular germ cell tumours under normoxic and hypoxic conditions in vitro. *Br. J. Cancer* 89: 2133–2139, 2003.
12. Graeber TG, Osmanian C, Jacks T, Housman DE, Koch CJ, Lowe SW, and Giaccia AJ. Hypoxia-mediated selection of cells with diminished apoptotic potential in solid tumours. *Nature* 379: 88–91, 1996.
13. Vukovic V and Tannock IF. Influence of low pH on cytotoxicity of paclitaxel, mitoxantrone and topotecan. *Br. J. Cancer* 75: 1167–1172, 1997.
14. Gray LH, Conger AD, Ebert M, Hornsey S, and Scott OC. The concentration of oxygen dissolved in tissues at the time of radiation as a factor in radiotherapy. *Br. J. Radiol.* 26: 638–648, 1953.
15. Wright EA and Howard-Flanders P. The influence of oxygen on the radiosensitivity of mammalian tissues. *Acta Radiol.* 48: 36–42, 1957.
16. Palcic B and Skarsgaard LD. Reduced oxygen enhancement ratio at low doses of ionizing radiation. *Radiat. Res.* 100: 328–339, 1984.
17. Michael BD, Adams GE, Hewitt HB, Jones WB, and Watts ME. The post effect of oxygen in irradiated bacteria: A millisecond fast mixing study. *Radiat. Res.* 54: 239–251, 1973.
18. Alexander P and Charlesby A. Energy transfer in macromolecules exposed to ionizing radiations. *Nature* 173: 578–579, 1954.
19. Ewing D. The oxygen fixation hypothesis: A reevaluation. *Am. J. Clin. Oncol.* 21: 355–361, 1998.
20. Johansen I and Howard-Flanders P. Macromolecular repair and free radical scavenging in the protection of bacteria against X-rays. *Radiat. Res.* 24: 184–200, 1965.
21. Thomlinson RH and Gray LH. The histological structure of some human lung cancers and the possible implications for radiotherapy. *Br. J. Cancer* 9: 539–549, 1955.
22. Tannock IF. The relation between cell proliferation and the vascular system in a transplanted mouse mammary tumour. *Br. J. Cancer* 22: 258–273, 1968.
23. Vaupel P, Thews O, and Hoeckel M. Treatment resistance of solid tumors: Role of hypoxia and anemia. *Med. Oncol.* 18: 243–259, 2001.
24. Bussink J, Kaanders JH, and van der Kogel AJ. Tumor hypoxia at the microregional level: Clinical relevance and predictive value of exogenous and endogenous hypoxic cell markers. *Radiother. Oncol.* 67: 3–15, 2003.
25. Dewhirst MW. Concepts of oxygen transport at the microcirculatory level. *Semin. Radiat. Oncol.* 8: 143–150, 1998.
26. Baudelet C and Gallez B. Effect of anesthesia on the signal intensity in tumors using BOLD-MRI: Comparison with flow measurements by laser Doppler flowmetry and oxygen measurements by luminescence-based probes. *Magn. Reson. Imaging* 22: 905–912, 2004.
27. Braun RD, Lanzen JL, and Dewhirst MW. Fourier analysis of fluctuations of oxygen tension and blood flow in R3230Ac tumors and muscle in rats. *Am. J. Physiol.* 277: H551–H568, 1999.
28. Powers WE and Tolmach LJ. A multicomponent x-ray survival curve forlymphosarcoma cells irradiated in vivo. *Nature* 197: 710–711, 1963.
29. Moulder JE and Rockwell S. Hypoxic fractions of solid tumors: Experimental techniques, methods of analysis, and a survey of existing data. *Int. J. Radiat. Oncol. Biol. Phys.* 10: 695–712, 1984.

30. Kaanders JH, Bussink J, and van der Kogel AJ. Clinical studies of hypoxia modification in radiotherapy. *Semin. Radiat. Oncol.* 14: 233–240, 2004.
31. Jain RK. Normalization of tumor vasculature: An emerging concept in antiangiogenic therapy. *Science* 307: 58–62, 2005.
32. Dewhirst MW, Vujaskovic Z, Jones E, and Thrall D. Re-setting the biologic rationale for thermal therapy. *Int. J. Hyperthermia* 21: 779–790, 2005.
33. Brunner TB, Hahn SM, Gupta AK, Muschel RJ, McKenna WG, and Bernhard EJ. Farnesyltransferase inhibitors: An overview of the results of preclinical and clinical investigations. *Cancer Res.* 63: 5656–5668, 2003.
34. Adams GE, Clarke ED, Gray P, Jacobs RS, Stratford IJ, Wardman P, Watts ME, Parrick J, Wallace RG, and Smithen CE. Structure-activity relationships in the development of hypoxic cell radiosensitizers. II. Cytotoxicity and therapeutic ratio. *Int. J. Radiat. Biol. Relat. Stud. Phys. Chem. Med.* 35: 151–160, 1979.
35. Adams GE, Stratford IJ, Bremner JC, Edwards HS, and Fielden EM. Nitroheterocyclic compounds as radiation sensitizers and bioreductive drugs. *Radiother. Oncol.* 20 (Suppl 1): 85–91, 1991.
36. Overgaard J, Hansen HS, Overgaard M, Bastholt L, Berthelsen A, Specht L, Lindelov B, and Jorgensen K. A randomized double-blind phase III study of nimorazole as a hypoxic radiosensitizer of primary radiotherapy in supraglottic larynx and pharynx carcinoma. Results of the Danish Head and Neck Cancer Study (DAHANCA) Protocol 5–85. *Radiother. Oncol.* 46: 135–146, 1998.
37. Lee DJ, Moini M, Giuliano J, and Westra WH. Hypoxic sensitizer and cytotoxin for head and neck cancer. *Ann. Acad. Med. Singapore* 25: 397–404, 1996.
38. Overgaard J and Horsman MR. Modification of hypoxia-induced radioresistance in tumors by the use of oxygen and sensitizers. *Semin. Radiat. Oncol.* 6: 10–21, 1996.
39. Brown JM and Giaccia AJ. Tumour hypoxia: The picture has changed in the 1990s. *Int. J. Radiat. Biol.* 65: 95–102, 1994.
40. von Pawel J, von Roemeling R, Gatzemeier U, Boyer M, Elisson LO, Clark P, Talbot D, Rey A, Butler TW, Hirsh V, Olver I, Bergman B, Ayoub J, Richardson G, Dunlop D, Arcenas A, Vescio R, Viallet J, and Treat J. Tirapazamine plus cisplatin versus cisplatin in advanced non-small-cell lung cancer: A report of the international CATAPULT I study group. Cisplatin and tirapazamine in subjects with advanced previously untreated non-small-cell lung tumors. *J. Clin. Oncol.* 18: 1351–1359, 2000.
41. Rischin D, Peters L, Hicks R, Hughes P, Fisher R, Hart R, Sexton M, D'Costa I, and von Roemeling R. Phase I trial of concurrent tirapazamine, cisplatin, and radiotherapy in patients with advanced head and neck cancer. *J. Clin. Oncol.* 19: 535–542, 2001.
42. Leo C, Giaccia AJ, and Denko NC. The hypoxic tumor microenvironment and gene expression. *Semin. Radiat. Oncol.* 14: 207–214, 2004.
43. Le QT, Denko NC, and Giaccia AJ. Hypoxic gene expression and metastasis. *Cancer Metastasis Rev.* 23: 293–310, 2004.
44. Semenza GL. Targeting HIF-1 for cancer therapy. *Nat. Rev. Cancer* 3: 721–732, 2003.
45. Hockel M, Schlenger K, Aral B, Mitze M, Schaffer U, and Vaupel P. Association between tumor hypoxia and malignant progression in advanced cancer of the uterine cervix. *Cancer Res.* 56: 4509–4515, 1996.
46. Nordsmark M, Hoyer M, Keller J, Nielsen OS, Jensen OM, and Overgaard J. The relationship between tumor oxygenation and cell proliferation in human soft tissue sarcomas. *Int. J. Radiat. Oncol. Biol. Phys.* 35: 701–708, 1996.

47. Fyles A, Milosevic M, Hedley D, Pintilie M, Levin W, Manchul L, and Hill RP. Tumor hypoxia has independent predictor impact only in patients with node-negative cervix cancer. *J. Clin. Oncol.* 20: 680–687, 2002.
48. Brizel DM, Scully SP, Harrelson JM, Layfield LJ, Bean JM, Prosnitz LR, and Dewhirst MW. Tumor oxygenation predicts for the likelihood of distant metastases in human soft tissue sarcoma. *Cancer Res.* 56: 941–943, 1996.
49. Rofstad EK. Microenvironment-induced cancer metastasis. *Int. J. Radiat. Biol.* 76: 589–605, 2000.
50. Subarsky P and Hill RP. The hypoxic tumour microenvironment and metastatic progression. *Clin. Exp. Metastasis* 20: 237–250, 2003.
51. Jiang BH, Semenza GL, Bauer C, and Marti HH. Hypoxia-inducible factor 1 levels vary exponentially over a physiologically relevant range of O_2 tension. *Am. J. Physiol.* 271: C1172–C1180, 1996.
52. Denko NC, Fontana LA, Hudson KM, Sutphin PD, Raychaudhuri S, Altman R, and Giaccia AJ. Investigating hypoxic tumor physiology through gene expression patterns. *Oncogene* 22: 5907–5914, 2003.
53. Hammond EM and Giaccia AJ. The role of p53 in hypoxia-induced apoptosis. *Biochem. Biophys. Res. Commun.* 331: 718–725, 2005.
54. Pennacchietti S, Michieli P, Galluzzo M, Mazzone M, Giordano S, and Comoglio PM. Hypoxia promotes invasive growth by transcriptional activation of the met protooncogene. *Cancer Cells* 3: 347–361, 2003.
55. Swinson DE, Jones JL, Richardson D, Wykoff C, Turley H, Pastorek J, Taub N, Harris AL, and O'Byrne KJ. Carbonic anhydrase IX expression, a novel surrogate marker of tumor hypoxia, is associated with a poor prognosis in non-small-cell lung cancer. *J. Clin. Oncol.* 21: 473–482, 2003.
56. West CM, Cooper RA, Loncaster JA, Wilks DP, and Bromley M. Tumor vascularity: A histological measure of angiogenesis and hypoxia. *Cancer Res.* 61: 2907–2910, 2001.
57. Mandriota SJ, Turner KJ, Davies DR, Murray PG, Morgan NV, Sowter HM, Wykoff CC, Maher ER, Harris AL, Ratcliffe PJ, and Maxwell PH. HIF activation identifies early lesions in VHL kidneys: Evidence for site-specific tumor suppressor function in the nephron. *Cancer Cells* 1: 459–468, 2002.
58. Zagzag D, Zhong H, Scalzitti JM, Laughner E, Simons JW, and Semenza GL. Expression of hypoxia-inducible factor 1alpha in brain tumors: Association with angiogenesis, invasion, and progression. *Cancer* 88: 2606–2618, 2000.
59. Zundel W, Schindler C, and Haas-Kogan DA. Loss of PTEN facilitates HIF-1-mediated gene expression. *Genes Dev.* 14: 391–396, 2000.
60. Aebersold DM, Burri P, Beer KT, Laissue J, Djonov V, Greiner RH, and Semenza GL. Expression of hypoxia-inducible factor-1alpha: A novel predictive and prognostic parameter in the radiotherapy of oropharyngeal cancer. *Cancer Res.* 61: 2911–2916, 2001.
61. Unruh A, Ressel A, Mohamed HG, Johnson RS, Nadrowitz R, Richter E, Katschinski DM, and Wenger RH. The hypoxia-inducible factor-1 alpha is a negative factor for tumor therapy. *Oncogene* 22: 3213–3220, 2003.
62. Moeller BJ and Dewhirst MW. Raising the bar: How HIF-1 helps determine tumor radiosensitivity. *Cell Cycle* 3: 1107–1110, 2004.
63. Pervan M, Pajonk F, Sun JR, Withers HR, and McBride WH. Molecular pathways that modify tumor radiation response. *Am. J. Clin. Oncol.* 24: 481–485, 2001.

64. Wright J, Hillsamer VL, Gore-Langton RE, and Cheson BD. Clinical trials referral resource. Current clinical trials for the proteasome inhibitor PS-341. *Oncology* 14: 1593–1594, 2000.
65. Rapisarda A, Uranchimeg B, Scudiero DA, Selby M, Sausville EA, Shoemaker RH, and Melillo G. Identification of small molecule inhibitors of hypoxia-inducible factor 1 transcriptional activation pathway. *Cancer Res.* 62: 4316–4324, 2002.
66. Yeo EJ, Chun YS, Cho YS, Kim J, Lee JC, Kim MS, and Park JW. YC-1: A potential anticancer drug targeting hypoxia-inducible factor 1. *J. Natl. Cancer Inst.* 95: 516–525, 2003.
67. Isaacs JS, Jung YJ, Mimnaugh EG, Martinez A, Cuttitta F, and Neckers LM. Hsp90 regulates a von Hippel Lindau-independent hypoxia-inducible factor-1 alpha-degradative pathway. *J. Biol. Chem.* 277: 29936–29944, 2002.
68. Welsh SJ, Williams RR, Birmingham A, Newman DJ, Kirkpatrick DL, and Powis G. The thioredoxin redox inhibitors 1-methylpropyl 2-imidazolyl disulfide and pleurotin inhibit hypoxia-induced factor 1alpha and vascular endothelial growth factor formation. *Mol. Cancer Ther.* 2: 235–243, 2003.
69. Mabjeesh NJ, Escuin D, LaVallee TM, Pribluda VS, Swartz GM, Johnson MS, Willard MT, Zhong H, Simons JW, and Giannakakou P. 2ME2 inhibits tumor growth and angiogenesis by disrupting microtubules and dysregulating HIF. *Cancer Cells* 3: 363–375, 2003.

17

Radiation Biology of Nonmammalian Species:
Three Eukaryotes and a Bacterium

17.1   Introduction: Lower Eukaryotes in Radiation Research

Radiation biologists have a long history of working with unconventional models, using whatever was most accessible and most appropriate for their research. These have included virus, bacteria, slime molds, amoeba, the yeasts *Saccharomyces* and *Neurospora*, eggs of the parasitic wasp *Habrobracon*, the fruit fly *Drosophila melanogaster*, the unicellular alga *Acetabularia*, the ciliates *Tetrahymena* and *Paramecium*, fern spores, and frogs' eggs.

Molecular biologists and geneticists working with mammalian systems have turned to lower organisms, particularly lower eukaryotes, for specific reasons. Eukaryotic cells have typically 3–30 times as many genes as prokaryotes, and often thousands of times more noncoding DNA allowing for the complex replication of gene expression required for the construction of multicellular organisms. Some eukaryotes are unicellular, however, among them the yeast *Saccharomyces cerevisiae* one of the simplest model organisms for eukaryotic cell biology, revealing the molecular basis of conserved fundamental processes such as the cell division cycle. A small number of other organisms have been chosen as primary models for multicellular plants and animals and the sequencing of their entire genomes has opened the way to systematic and comprehensive analysis of gene functions, gene regulation, and genetic diversity. As a result of gene duplications during vertebrate evolution, vertebrate genomes contain multiple closely related homologues of most genes. This genetic redundancy has allowed diversification and specialization of genes for new purposes, but it also makes gene functions harder to decipher. There is less genetic redundancy in, for instance, the nematode *Caenorhabditis elegans* and the fruit fly *Drosophila melanogaster* which have played a key role in revealing the universal genetic mechanisms of animal development.

17.2   Yeast, a Single-Celled Eukaryote

Yeasts have greater genetic complexity than bacteria, containing 3.5 times more DNA than *Escherichia coli* cells, but they share many of the technical advantages that permitted rapid progress in the molecular genetics of prokaryotes and their viruses. Some of the properties that make yeast particularly suitable for biological studies include rapid growth, dispersed cells, the ease of replica plating and mutant isolation, a well-defined genetic system, and importantly, a highly versatile DNA transformation system.

Two yeast species are of particular interest in the context of molecular genetics, the budding yeast, *Saccharomyces cerevisiae*, and the fission yeast, *Schizosaccharomyces pombe*. Although widely separated in evolutionary terms, there are large areas of structure and function which are conserved between the two yeasts. The most obvious difference between the two is in the life cycle. *S. cerevisiae* divides by budding, whereas *S. pombe* divides by medial fission. These two methods of cell division are reflected in the nuclear division cycle. Most of this section of Chapter 17 will be concerned with *S. cerevisiae* but reference will be made to fission yeast in the description of mechanisms regulating checkpoint pathways, an area where *S. pombe* appears to have greater homology with higher eukaryotes than does *S. cerevisiae*.

Saccharomyces cerevisiae, the budding yeast, has been a friend to mankind for millennia through its participation in brewing, winemaking, and baking. More recently, it has become the special friend of researchers as the work-horse of genetics and molecular biology. Unlike most other microorganisms, strains of *S. cerevisiae* have both a stable haploid and diploid state. Thus, recessive mutations can be conveniently isolated and manifested in haploid strains and complementation tests can be carried out in diploid strains. The development of DNA transformation has made yeast particularly accessible to gene cloning and genetic engineering techniques. Structural genes corresponding to virtually any genetic trait can be identified by complementation from plasmid libraries. Plasmids can be introduced into yeast cells either as replicating molecules or by integration into the genome. In contrast to most other organisms, integrative recombination of transforming DNA in yeast proceeds exclusively via homologous recombination. Exogenous DNA with at least partial homologous segments can therefore be directed to specific locations in the genome. Also, homologous recombination, coupled with yeast's high levels of gene conversion, has led to the development of techniques for the direct replacement of genetically engineered DNA sequences into their normal chromosome locations. Thus, normal wild-type genes, even those having no previously known mutations, can be conveniently replaced with altered and disrupted alleles. The phenotypes arising after disruption of yeast genes have contributed significantly toward understanding of the function of certain proteins in vivo.

There are two forms in which yeast can survive and grow: haploid and diploid. The haploid cells have a simple life cycle of mitosis and growth, and under conditions of high stress will simply die. The diploid cells also have a life cycle of mitosis and growth, but under conditions of stress can undergo sporulation, entering meiosis and producing a variety of haploid spores, which can later mate (conjugate), reforming the diploid. Yeast has two mating types, a and α which show primitive aspects of sex different-iation. The mating of yeast only occurs between haploids which can be either a or α. Mating type is determined by a single locus, *MAT*, which governs sexual behavior of both haploid and diploid cells. Through a form of genetic recombination, haploid yeast can switch mating type as often as every cell cycle.

Saccharomyces cerevisiae contains a haploid set of 16 well-characterized chromosomes, ranging in size from 200 to 2200 kb. The total sequence of chromosomal DNA, constituting 12,052 kb, was released in April, 1996 [1]. A total of 6183 open reading frames (ORF) of over 100 amino acids long were reported, and approximately 5800 of them were predicated to corres-pond to actual protein-coding genes. A larger number of ORFs were pre-dicted by considering shorter proteins. In contrast to the genomes of multicellular organisms, the yeast genome is highly compact, with genes representing 72% of the total sequence. The average size of yeast genes is 1.45 kb, or 483 codons, with a range from 40 to 4910 codons. A total of 3.8% of the ORF contain introns. Approximately 30% of the genes have already been characterized experimentally.

17.2.1 Radiation Biology of Yeast

Radiation biology of yeast has been reviewed by Game [2,3] and Jablonovich [4]. *Saccharomyces* cells containing only one DNA copy per nucleus show a response to ionizing radiation that differs sharply from that of cells contain-ing two or more copies (Figure 17.1). Diploid cells generate a survival curve with a definite shoulder, whereas haploid cells irradiated in logarithmic growth show a biphasic curve. At low doses, haploid killing is exponential, but at higher doses (starting at about 11 krad) a highly resistant subpopula-tion or cells are revealed by the existence or a characteristic tail on the survival curve. Many studies have shown that in haploids, the resistant fraction or cells are those in the G_2 stage of the cell cycle and that the tail is absent from survival curves when stationary phase cells (which typi-cally arrest in G_1) are irradiated. Evidence of the relationship of the diploid shoulder and the haploid tail on survival curves also comes from radiation-sensitive mutants, which show neither G_2 resistance nor the diploid shoul-der. It is now understood that both phenomena result from DNA repair processes that require the presence or two copies of the DNA. These obser-vations, combined with the discovery that mutants defective in DNA double-strand break (DSB) repair fail to carry out homologous recombination, supported the concept that DNA DSBs are repaired by a recombinational mechanism in yeast.

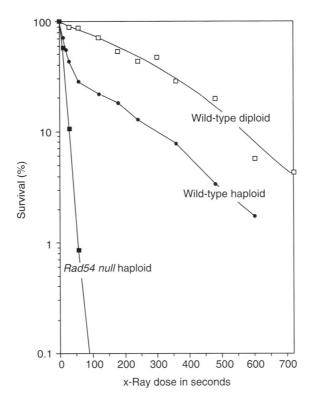

FIGURE 17.1
x-Ray survival of haploid and diploid wild-type yeast and a radiation-sensitive mutant, *Rad54 null*. Cells in logarithmic growth were plated on nutrient medium and immediately irradiated for the times indicated. The dose rate was approximately 188 rad/s. Note that after elimination of a sensitive G_1 fraction, sensitivity of G_2 haploid cells resembles that of the congenic diploid strain. (From Game, J.C., *Semin. Cancer Biol.*, 4, 73, 1993. With permission.)

An additional aspect of radiation resistance in yeast relates to the mating-type locus. Usually, diploid cells are heterozygous at this locus because they arise by conjugation of haploids that carry opposite mating-type alleles. However, when diploids are homozygous for either one of the two mating-type alleles are constructed, they are found to be significantly radiation sensitive compared to heterozygous diploids. Mutations in loci implicated in DSB repair largely abolish this mating-type effect. Since this is not true for other studies of radiation-sensitive mutants, it seems that mating-type heterozygosity potentiates DSB repair, probably by potentiating cellular recombination mechanisms.

17.2.2   Radiosensitive Mutants for the Study of DNA Repair

Much of the understanding or DNA repair in yeast comes from the analysis of radiosensitive mutants, a wide variety of which have been isolated, genetically mapped and phenotypically characterized. Radiation-sensitive mutants in yeast can be loosely classified into three main groups based on genetic analysis and on mutant phenotypes, although the groups are not rigidly defined and may themselves be somewhat heterogeneous. The groups (known as epistasis groups) were originally identified on the basis of double-mutant interactions. In these experiments, if a yeast strain carried mutations in two or more genes conferring radiation sensitivity but showed no further increase in sensitivity beyond that shown by the single mutants then the genes were placed in the same group; whereas, if increased sensitivity was seen in the double mutant then the loci involved were assigned to different groups. This is based on the rationale that increased sensitivity in the double mutant probably implies that the two mutations conferred sensitivity via different mechanisms, whereas no increase in sensitivity would be seen if both mutants were in the same pathway.

It was agreed at an international yeast genetics conference in 1970 that genes which mutate to confer significant sensitivity to radiation should be designated by the symbol *RAD*. Those primarily affecting UV repair or both UV and IR repair were to be numbered from *RAD1* upward, with numbers 1–49 set aside for them, while genes controlling IR sensitivity with only a minimal role in UV repair were to be numbered from *RAD50* upward. A standardized set of *RAD* locus numbers from *RAD1* to *RAD22*, was published in 1971 followed in 3 years later by a study of x-ray sensitive mutants [5] that designated eight loci numbered from *RAD50* to *RAD57*. Although almost all the allelic relationships reported in these original papers remain valid today, many other mutants have since been isolated and some of the original UV-sensitive mutants have been lost or can no longer be shown to segregate.

The three groups of mutants have been identified on the basis of epistasis analysis:

- Nucleotide excision repair (*RAD3*) group of genes, which is mainly involved in the repair of UV-irradiated DNA
- "Error-prone" or post-replicative repair (*RAD6*) group
- Recombinational repair (*RAD52*) group

Ionizing radiation repair studies are complicated by the differing radiation biology of haploid and diploid cells, and especially the fact that recombinational repair cannot operate in G_1 haploids. Sister chromatid repair in G_2 haploids may also differ from interchromosomal repair in G_1 diploids. However, a number of reports indicated that the *RAD6/RAD18* group and the *RAD50* upward genes do form separate ionizing radiation epistasis groups resembling those seen for UV sensitivity. Data in Figure 17.2A and B confirm this in two sets of inbred strains, and also indicate that the excision repair loci play a significant role in x-ray repair that is independent of the other two groups. This role is apparent from the response of triple

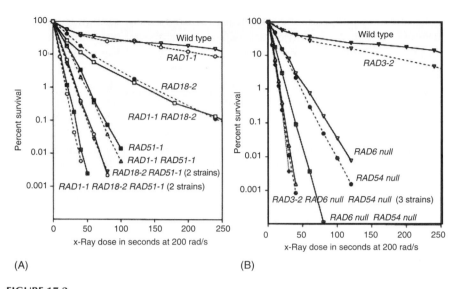

FIGURE 17.2

(A) x-Ray survival of inbred haploid yeast strains carrying single, double, or triple mutations in the *RAD1*, *RAD18*, and *RAD51* genes. (B) x-Ray survival of inbred haploid yeast strains carrying single, double, or triple mutations in the *RAD3*, *RAD6*, and *RAD54* genes. (From Game, J.C., *Mutat. Res.*, 451, 277, 2000. With permission.)

mutants where the other two pathways are also blocked. Figure 17.2A and B present haploid x-ray survival data for some single, double, and triple mutant strains that are mutant for *RAD1-1* (UV excision repair), *RAD18-2* (post-replicational UV repair), and *RAD51-1* (recombinational repair). It can be seen that the two triple mutant strains show good agreement with each other and are significantly more sensitive than the two *RAD18-2 RAD51-1* double mutants, which are in turn more sensitive than either single mutant. Hence, the three loci probably represent three epistasis groups for x-ray sensitivity.

This last group of genes is essential for the repair of the DSBs produced by ionizing radiation. This repair makes use of homologous information present in the genome, to restore the broken chromosome in a recombination process that is essentially error-free (described in Chapter 7). The recombinational repair group includes the *RAD50*, *RAD51*, *RAD52*, *RAD54*, *RAD55*, *RAD57*, *RAD59*, *XRS2*, and *MRE11* genes. A mutation in any of these genes causes sensitivity to ionizing radiation and other DSB-inducing agents. Most of these mutations also affect meiotic recombination with the result that the mutants either do not sporulate or produce inviable spores. The recombination machinery shows great conservation throughout evolution. The RAD51 protein, a member of the recombination repair group, shares extensive sequence homology with the prokaryotic RecA proteins, which play essential roles in homologous recombination and in the SOS response. Homologues of most of the genes in the recombinational repair group have been found in higher eukaryotes and some of them play essential roles in mammalian cells.

Two other repair genes of great interest that have been identified recently are the *Saccharomyces* homologues of the mammalian Ku70 and Ku80 loci, named in yeast *HDF1* [6] and *YKU80* [7], respectively. In mammals, these gene products form a complex with the catalytic subunit of DNA protein kinase (DNA PK$_{cs}$, Chapter 7). This complex mediates a major route for DSB repair via an end-joining mechanism that is different from the recombinational repair mechanism mediated by *RAD51* and related genes. The yeast *KU80* homologue was identified by screening the yeast genome for homology with the relevant mammalian and *Caenorhabditis* sequences.

In spite of the great progress which has been made in the study of DNA repair, especially with mammalian cells, yeast continues to be an excellent paradigm for repair in humans. Almost, all the known repair pathways and most of the individual loci are significantly homologous from *Saccharomyces* to mammals. Among the differences, the most notable is the apparently greater importance of recombinational repair compared to the Ku-mediated end-joining pathway in processing ionizing radiation-induced DSBs in yeast compared to mammals. Surprisingly, this apparent difference does not seem to affect the overall efficiency of radiation repair in the two systems, since there is a good correspondence between yeast and mammalian cells in survival per unit dose, per unit of DNA, i.e., when the much smaller DNA target in yeast is taken into account. A possible reason for the greater emphasis on recombinational repair in *Saccharomyces* may be that yeast has little noncoding DNA, hence almost all effective DSB repair must be accurate, whereas mammalian cells may be better able to tolerate short deletions or additions in noncoding DNA arising from end-joining mechanisms that are error-prone. Another factor may be the different lifestyles involved. Yeast cells must be capable of ongoing division; hence repair must be accurate, and sister chromatids as well as homologous chromosomes are available for recombinational repair in G$_2$. Less accurate repair may be tolerated in nondividing tissues of complex eukaryotes, and recombinational repair between large nonreplicating chromosome homologues may be harder to effect.

17.2.3 DNA-Damage Checkpoints

DNA-damage checkpoints have been reviewed by Carr [8], Lowndes and Murguia [9], and Humphrey [10]. Checkpoint proteins are well conserved from yeast to human cells, indicating that the basic organization of these pathways has been preserved throughout evolution (Table 17.1). In *S. cerevisae*, at least six pathways are required for radiation checkpoint control and the genes involved in sensing DNA damage have been proposed to function in two distinct groups that sense DNA damage additively, primarily in G$_1$ and G$_2$ (Figure 17.3). These are defined by the *RAD9* gene and by the *RAD24* subclass of genes. Included in the *RAD24* subclass are *RAD17*, *MEC3*, and *DDC1*. RAD24 and its homologues in other species have been shown to have homology with all five subunits of replication factor C (RFC) while RAD17

TABLE 17.1

Partial List of Checkpoint Proteins Which Are Structurally
and/or Functionally Conserved

S. cerevisiae	S. Pombe	Human	Function or Structural Motif
Mec 1	Rad 3	Atm, Atr	Protein kinase
Rad 17	Rad 1	Rad 1	RFC-like
Ddc 1	Rad 9	Rad 9	PCNA-related
Mec 3	Hus 1	Hus 1	PCNA-related
Rad 24	Rad 17	Rad 17	RFC-like
Dpb11	Rad4/Cut5		BRCT, zinc-finger
Rad 9	Crb2/Rhp9	Brca1	BRCT domain
Chk1	Chk1	Chk1	Protein kinase
Rad53	Cds1	Chk2	Protein kinase
Bmh1, Bmh2	Rad 24, Rad 25	14-3-3	14-3-3

and its homologues exhibit weak homology throughout their entire lengths with PCNA (proliferating cell nuclear antigen), another replication factor (Table 17.1). It is not known how Rad9 senses DNA damage. The primary sequence of Rad9, contains a tandem domain in its carboxyl terminus the BRCT which is also present in other proteins with roles in DNA metabolism, including Crb2sp/Rhp9sp, a possible fission yeast homologue of Rad9, and the tumor suppressor Brca1 (breast cancer associated 1). The BRCT domain of Rad9 has been reported to be required for its DNA damage-dependent oligomerization, which is necessary for checkpoint function. Brca1 and the yeast proteins are similarly regulated; both are subjected to Cdk-dependent phosphorylation in $S/G_2/M$. Furthermore, they are hyperphosphorylated after DNA damage and this regulation is dependent on the Rad3sp/Mec1/ataxia telangiectasia mutated (ATM) class of phosphoinositide kinases. These regulatory similarities are consistent with the possibility that Brca1, Rad9, and Crb2sp/Rhp9sp might be functional analogues.

As the Rad9 and Rad24 branches of the checkpoint pathway appear to function additively it is likely that these branches "sense" DNA damage via distinct mechanisms. Given that purified Rad9 does not interact with naked DNA one possibility is that DNA damage is sensed indirectly via perturbations to chromatin structure. Consistent with this, Rad9 can be co-purified with chromatin.

In budding yeast, three principal protein kinases are known to be involved in signal transduction (Figure 17.3). One of these, Mec1, primarily because of its similarities with another member of the phosphoinositide kinases DNA-PK, has also been implicated in DNA-damage sensing. Recent information on the role of Rad53 in G_2/M checkpoint regulation has indicated that it is required to keep Clb–Cdc28 kinase activity high, thereby preventing not only anaphase but also mitotic exit. By examining the kinetics of anaphase entry after DNA damage, Cdc5 (a member of the polo family

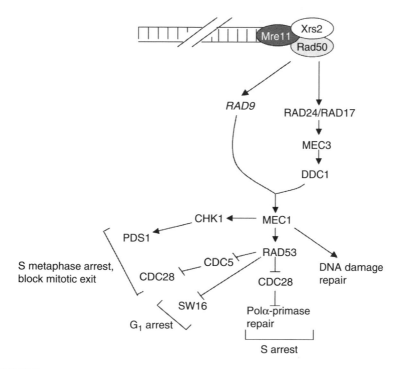

FIGURE 17.3

Organization of the DNA-damage-dependent checkpoint pathways of budding yeast. The Rad50/Mre11/Xrs2 complex is involved in DSB repair. *RAD9* and *RAD24* define upstream "sensor" branches of the pathway and seem to respond, primarily outside of S phase, to multiple types of DNA damage. No other members of the *RAD9* branch have been identified. The *RAD9* and *RAD24* branches converge on *MEC1*, a member of the PI3 kinase family. Within S phase, one or more independent pathways for sensing DNA damage exist. The S phase-specific sensing pathways are also thought to converge on *MEC1*. The *MEC1*, *RAD53*, and *CHK1* genes encode transducers of the checkpoint signal. Effectors for Rad53 include Swi6 which encodes a transcription factor targeting the G_1 checkpoint; polyα-primase, targeting the S-phase checkpoint; and Cdc5, a "polo-like" kinase which inhibits Cdc28 kinase. Chk1 seems to primarily target Pds1, a regulator of sister chromatid cohesion.

of protein kinases) was found to be required for the Rad53-dependent anaphase delay. As Cdc5 is known to be required for mitotic exit, the implication is that Rad53 may inhibit Cdc5 activity to prevent both anaphase and mitotic exit. Consistent with this idea, overproduction of Cdc5 can drive DNA-damaged cells through both anaphase and mitotic exit, indicating that Cdc5 is rate limiting for both of these events. These data are consistent with a role for Rad53 in inhibition of the Cdc5 pathway which, in turn, functions to inhibit Clb–Cdc28 kinase activity. Accordingly, Cdc5 is itself phosphorylated in a Rad53-dependent fashion. A third PK involved in signal transduction is Dun1 which is involved in G_2M arrest and DNA damage regulon induction (not shown in Figure 17.3).

In *S. cerevisiae*, the mechanism by which mitosis is inhibited does not require phosphorylation of p34 to inhibit mitosis in response to a block in DNA replication or DNA damage. In *S. pombe*, tyrosine phosphorylation is necessary for the inhibition of mitosis following a block to bulk DNA synthesis but it is not necessary following DNA damage. Budding yeast cells arrested in G_2/M by the DNA-damage checkpoint will eventually proliferate, even if the damage is not repaired. This latter option, termed "adaptation," makes sense for a unicellular organism as it is better to resume cell cycling, with the possibility of survival, than to die without making an attempt. For multicellular organisms, however, this would not be the best strategy, as cells that adapt to the presence of irreparable DNA damage will do so by fixing mutations in the cell's lineage.

17.2.4 Genome-Wide Screening for Radiation Response-Associated in Yeast

New approaches to determining genes involved in response to damage by ionizing radiation have been described recently using resources such as the yeast genome deletion project which contains sets of isogenic haploid and diploid yeast strains containing deletions of nonessential genes. Bennett and coworkers [11] sought to identify new genes that might be associated with toleration of ionizing radiation damage in homozygous diploid mutants by screening diploid yeast strains in this set. A genome-wide screen of diploid mutants, which are homozygous with respect to deletions in 3670 nonessential genes, revealed 107 new loci that influenced γ-ray sensitivity. Many loci affected replication, recombination, and checkpoint functions. Most loci represented new functional classes of genes having new categories of cross sensitivities to other DNA damaging agents. Approximately 30 of these genes were implicated in the repair of radiation-induced DSBs by recombination and nonhomologous end-joining repair (NHEJ) pathways.

In other experiments, Game and coworkers [12] used a set of all homozygous diploid deletion mutants in budding yeast, *S. cerevisiae*, to screen for new mutants conferring sensitivity to ionizing radiation. In each strain, a different ORF had been replaced with a cassette containing unique 20 mer sequences that allow the relative abundance of each strain in a pool to be determined by hybridization to a high-density oligonucleotide array. Putative radiation-sensitive mutants in the pool of 4627 individual deletion strains were identified as having a reduced abundance after irradiation. Of the top 33 strains most sensitive to radiation in this assay, 14 contained genes known to be involved in DNA repair. Eight of the remaining deletion mutants were studied. Only one of these (which was named *RAD61*), conferred reproducible radiation sensitivity in both the haploid and diploid deletions and had no problem with spore viability when the haploid was backcrossed to wild type. The remaining strains showed only marginal sensitivity as haploids, and had problems with spore viability when backcrossed, suggesting the presence of gross aneuploidy or polyploidy in strains initially presumed haploid. This screen did not confirm many of the genes which had been identified as involved in radioresponse by a screen of diploid yeast strains [11]. The authors emphasize that secondary mutations or deviations from euploidy can be a problem in screening this resource for sensitivity to ionizing radiation.

17.3 *Caenorhabditis elegans*

The nematode *Caenorhabditis elegans* is an animal model with distinct advantages for the study of normal tissue function and pathophysiology and is also ideal for genetic studies.

Caenorhabditis elegans is a self-fertilizing, effectively isogenic hermaphrodite which produces approximately 300 progeny from a single individual. A diagram of the adult *C. elegans* gonad is shown in Figure 17.4. It is easy to culture in the laboratory being normally maintained on agar culture plates and fed nonpathogenic bacteria. The hermaphroditic lifestyle facilitates maintaining animals over several generations, on for e.g., long space missions. The complete cell lineage is known, including those cells genetically programmed to die by apoptosis. Since the worms are transparent at all developmental stages, it is possible to examine cell division and development in real time. *C. elegans* is a metazoan with a number of tissue types, such as nervous system, muscle, intestine, and gonads and many of the developmental and biochemical pathways of *C. elegans* are conserved in mammals.

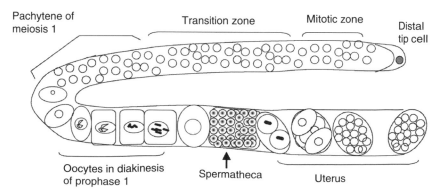

FIGURE 17.4

DNA damage responses in an adult hermaphrodite *C. elegans* germ line. Mitotic stem cells proliferate throughout adulthood in the distal end of each gonad arm, then pass through a characteristic set of morphological stages as they undergo meiosis and descend toward the uterus. Following DNA damage germ cells in the mitotic region undergo proliferation arrest, whereas meiotic germ cell nuclei undergo apoptosis.

Caenorhabditis elegans has five pairs of autosomes and one pair of sex chromosomes. Hermaphrodites have two X chromosomes, while males which are XO, are produced spontaneously as a result of X-chromosome loss or nondisjunction. The genome has been completely sequenced [13] and consists of approximately 20,000 predicted protein-encoding genes at least 500 of which have human homologues. A powerful approach has recently been developed for studying loss of gene function by introducing double-stranded RNA into the worm either by injection or by feeding. The most widely used approach involves feeding worms bacteria which are producing double-stranded RNA for the gene of interest. The ingestion of the dsRNA causes inhibition of gene expression (RNAi). Libraries containing bacterial strains for many of the predicted protein-encoding genes have been constructed, and the phenotype for each of the genes observed.

The radiobiology of the *C. elegans* model is at an early stage of development. However, since much of the machinery for the sensing of DNA DSBs, induction of cell cycle checkpoints and induction of apoptosis appears to be evolutionarily conserved; the worm holds great promise for modeling of radiation response in higher eukaryotes, including man, particularly since, as noted, this is a low-maintenance experimental model.

17.3.1 Apoptosis in *C. elegans*

The significant progress which has been made in the understanding of apoptotic events in higher eukaryotes owes a considerable debt to studies in *C. elegans*. In this species, two waves of apoptotic deaths have been found. The first wave occurs largely during embryogenesis and helps to sculpt the cell lineages that produce all the animal's somatic cells. The second wave of death occurs in the adult female germ line, where several hundred cells are eliminated during oogenesis.

17.3.1.1 Developmental Cell Death

During the somatic development of the animal, 131 of the 1090 cells generated undergo programmed cell death in a highly reproducible way: It is always the same cells that die, and each cell dies at a characteristic point in development. Extensive genetic analysis of these cell deaths led to the identification of an evolutionarily conserved core apoptotic pathway that regulates all programmed cell deaths in *C. elegans*. Two genes, *ced-3* and *ced-4* (ced = cell death abnormal), are required for the killing process. The product of the former is a member of the caspase family of cysteine proteases; the product of the latter is homologous to mammalian Apaf-1 and functions genetically as a positive regulator of ced-3. A third gene, *ced-9*, protects cells that normally survive from undergoing programmed cell death. The ced-9 protein is homologous to the oncoprotein Bcl-2, which similarly promotes cell survival in mammals. Finally, a protein called EGL-1 is required for all developmental cell deaths in the animal. It belongs to the subset of Bcl-2 family members that contain only one of four Bcl-2 homology domains, the BH3 domain, and are thus known as BH3 domain only proteins (described in Chapter 12). The BH3 domain allows EGL-1 to associate with and inhibit ced-9.

Biochemical studies have suggested that the key event required for apoptotic cell death in *C. elegans* is the processing of ced-3 from the inactive zymogen state into the active caspase. While the activation of ced-3 appears to be autocatalytic in nature, it does require association of the zymogen with oligomerized ced-4, probably forming a worm version of the apoptosome, which in mammals consists of at least three proteins caspase-9, Apaf-1, and cytochrome *c*. In cells programmed to survive, formation of the apoptosome is prevented, at least in part because ced-4 is bound to and sequestered by ced-9 in a stable complex on the surface of mitochondria. Cells fated to die appear to be marked for apoptosis through the expression of EGL-1. Binding of EGL-1 to ced-9 causes the release of ced-4, which is then free to trigger the lethal proteolytic action of ced-3.

17.3.1.2 C. elegans p53 Homologue CEP-1

A homologue of the mammalian p53 tumor suppressor protein has been identified in *C. elegans* that is expressed ubiquitously in embryos [14]. The gene encoding this protein, *cep-1*, promotes DNA damage-induced apoptosis and is required for normal meiotic chromosome segregation in the germ line. Moreover, although somatic apoptosis is unaffected, *cep-1* mutants show hypersensitivity to hypoxia-induced lethality and decreased longevity in response to starvation-induced stress. Overexpression of CEP-1 promotes widespread caspase-independent cell death. These findings show that *C. elegans* p53 mediates multiple stress responses in the soma, and mediates apoptosis and meiotic chromosome segregation in the germ line.

17.3.1.3 Radiation-Induced Apoptosis

Radiation-induced apoptosis has been reviewed by Gartner and coworkers [15] and Stergiou and Hengartner [16]. DNA damage activates germ cell apoptosis through a conserved checkpoint pathway that includes the *rad-5* and *mrt-2* genes and the gene altered by the *op241* mutation; however, none of these genes is required for physiological germ cell death. The role of p53 as coordinator of cellular responses to DNA damage is undertaken in *C. elegans* by CEP-1 which regulates apoptosis in the germ line cells in response to genotoxic stress. Cells with mutated *cep-1(w40)* are resistant to ionizing radiation-induced apoptosis and *cep-1(RNAi)* mimics this effect of *w40*. This block in activation of the germ line cell death program may be general to DNA damage because *cep-1(w40)* mutants, like *rad-5*, *mrt-2*, and *op241* mutants, fail to undergo germ cell death induced by the DNA modifying compound N-ethyl-N-nitrosourea.

The truncated CEP-1(w40) protein has been shown to interfere with the pro-apoptotic activity of wild-type CEP-1. Both the heterozygous *w40* mutation and overexpression of the *cep-1(w40)* gene from a heat shock promoter in a wild-type background confer resistance to radiation-induced germ cell apoptosis, confirming that *w40* dominantly attenuates wild-type *cep-1* function. Unlike *rad-5*, *mrt-2*, and *op241* mutants, which are defective in both germ cell apoptosis and cell cycle checkpoint arrest induced by DNA damage, *cep-1(w40)*, and *cep-1(RNAi)* germ cells undergo a transient cell cycle arrest in response to radiation that is indistinguishable from that of the wild type. Furthermore, ectopic expression of CEP-1 in early embryos fails to cause cell division arrest. This ability to activate apoptosis but not arrest the cell cycle is a property shared by *Drosophila* p53, but not vertebrate homologues.

DNA damage-induced apoptosis requires the core apoptotic machinery. There is no cell death induced in the germ line of animals homozygous for the *ced-3* or *ced-4* loss-of-function alleles. However, in contrast to physiological germ cell death, DNA damage-induced cell death is blocked by *ced-9* gain-of-function and is severely reduced in the *egl-1* loss-of-function mutation, indicating that the DNA damage response machinery is genetically distinct from the pathways that control somatic cell death and physiological germ cell death. Similarly, mutations in the checkpoint genes *rad-5*, *him-7*, and *mrt-2* only prevent DNA damage-induced apoptosis; physiological germ cell death continues unchecked in these mutants.

17.3.2   Cell Cycle Checkpoints in *C. elegans*

Genetic screens for mutants defective in different signaling pathways, such as *rad* mutants and mutants with a high rate of chromosome nondisjunction, have revealed three strains defective for radiation-induced apoptosis: *hus-s*, *mrt-2*, and *rad-5*. All three mutations abrogate cell cycle arrest and apoptosis induced by DNA damage without affecting developmental or physiological germ cell death. Consistent with the situation in yeast and mammalian checkpoint gene mutants, they also show increased genomic instability and reduced long-term survival following genotoxic insults, as demonstrated by the high rates of embryonic lethality.

Molecular characterization of *mrt-2* and *hus-1* showed that these genes encode the *C. elegans* homologues of *S. pombe* Rad1 and Hus1 (*S. cerevisiae* Rad17 and Mec3) checkpoint proteins indicating that at least part of the DNA damage response pathway characterized in yeasts is conserved through evolution and also functions in nematodes (Table 17.2). In fission yeast, Hus1, Rad1, and Rad9 form a heterotrimeric complex that resembles the PCNA complex in structure. Similarly, *C. elegans* HUS-1 is a nuclear protein that requires MRT-2 and the Rad9 homologue HPR-9 for proper localization to the nucleus.

A third checkpoint gene, *rad-5*, was originally identified in a screen for mutations that show reduced long-term survival following ionizing

TABLE 17.2

Homologies between Proteins of Mammals and Lower Eukaryotes

Function	Protein	*Saccharomyces cerevisiae*	*Schizosaccharomyces pombe*	*Caenorhabditis elegans*	Mammals
Sensors	RFC-like	Rad24	Rad17	HPR-17	RAD17
	PCNA-like	Ddc1	Rad9	HPR-9	RAD9
		Mec3	Hus1	HUS-1	HUS1
		Rad17	Rad1	MRT-2	RAD1
	BRCT-containing	Rad9	Rhp9/Crb2	HSR-9	BRCA1
	DSB recognition, repair	Mre11	Rad32	MRE11	MRE11
		Rad50		RAD-50	RAD50
Transducers	PI3-kinases	Tel1	Tel1	ATM-1	ATM
		Mec1	Rad3	ATL-3	ATR
	Rad3 regulatory subunit		Ddc2	Rad26	
	Effector kinases	Chk-1	Chk-1	Chk1	CHK2
Downstream effectors				CEP-1	p53

radiation. The mutation isolated in this screen, *rad-5(mn159)*, was later found to be allelic to *clk-2(qm37)* (clk-2 = clock gene), a mutation characterized by slow growth and a mild increase in animal life span. As with *mrt-2* and *hus-1*, DNA damage-induced cell cycle arrest and apoptosis are abrogated in *rad-5clk-2* mutants. RAD-5 is an essential protein and complete elimination of *rad-5* gene function results in developmental arrest and embryonic lethality.

17.3.2.1 Radiosensitivity of Checkpoint Mutants

Embryos derived from checkpoint mutants have a dramatically decreased survival rate as compared to those from wild-type worms. Embryos die necrotically with no signs of increased apoptotic cell death in a radiation dose-dependent manner at various stages of embryonic development, presumably as a delayed consequence of unrepaired DNA damage. This necrotic death of embryos, which occurs several cell generations after exposure to radiation, is reminiscent of the necrotic death of yeast cells and of the clonogenic death of mammalian cells following treatment with radiation or DNA damage-inducing chemotherapeutic agents.

Interestingly, the apoptotic mutant *ced-3* shows an intermediate sensitivity to radiation. Because *ced-3* animals show normal cell proliferation arrest following DNA damage, the increased sensitivity of *ced-3* mutants is likely due to the absence of DNA damage-induced apoptosis; consistent with this hypothesis, the apoptosis following genotoxic insults significantly contributes to the removal of cells with unrepaired DNA.

17.3.3   DNA Repair in *C. elegans*

Caenorhabditis elegans has a large number of repair genes which function in highly conserved pathways with protein sequences that have been conserved with both yeast and higher organisms, including human. Genes, for which mutant phenotypes have been described, include components of the pathways for nucleotide excision repair, mismatch repair, DNA-damage checkpoint, NHEJ, homologous recombination repair, and chromosomal structure surveillance. A broad perspective on genes involved in DNA repair has been gained using high throughput, genome-wide analysis of RNAi phenotypes and protein interactions.

17.3.3.1   Genes Involved in Meiotic DNA Recombination

Double-strand breaks occur not only following genotoxic stress, but also normally during meiotic prophase to initiate meiotic recombination events. Two coupled processes, the formation and the processing of DSBs are involved in meiotic recombination. The sporulation, meiosis-specific protein (*SPO11* gene product), is responsible for enzymatic DNA cleavage to create DSBs, and Mre11 later processes these through its intrinsic exonuclease activity. Rad51, a member of the recombinase A (RecA)-strand exchange protein family, catalyzes the invasion of the single-strand DNA overhangs generated by Mre11 into a recipient homologous double-strand DNA molecule, thereby initiating the formation of D loops and the later steps of meiotic recombination. Silencing of the *rad-51* gene via RNAi in *C. elegans* results in high levels of embryonic lethality and increased frequency of males, phenotypes encountered commonly in meiotic mutants. Moreover, a dramatic increase in germ cell apoptosis is observed when *rad-51* is inactivated. When meiotic recombination is blocked by the absence of sporulation, meiosis-specific protein (SPO-11), this increase no longer occurs, suggesting that the resulting deaths are triggered by accumulation of recombination intermediates, which are perceived as a form of DNA damage.

Mutations in *hus-1*, *mrt-2*, and *rad-5* checkpoint genes suppress the *rad-51*-dependent apoptotic death, suggesting that disruption of the DNA-damage checkpoint pathway compromises the ability of the cells to sense and respond to the incurred damage. Damage caused by radiation treatment in late embryonic stages, in animals depleted of RAD-51, resulted in several developmental defects in vulva and gonad formation. This effect is likely due to inability of repair of radiation-induced DSBs, and implies that *rad-51* also functions in DNA damage response in the soma.

17.3.4   DNA Damage Responses in *C. elegans*

In the germ line cells of *C. elegans*, DNA damage-induced signaling induces two clear responses—cell cycle arrest and apoptosis—that are spatially separated. Exposure of worms to ionizing radiation causes a transient halt in cell cycle progression in the proliferating zone, resulting in a decreased number of mitotic germ cells, while the volume of the arrested nuclei as well as that of the surrounding cytoplasm becomes enlarged, since cellular and nuclear growth continue to occur. In the meiotic compartment, after the exit from the pachytene region, increasing doses of ionizing radiation cause a dramatic increase in the number of apoptotic cell deaths, which appear as early as 2–3 h after the insult, and persist for 20–60 min before being engulfed and degraded by the surrounding somatic sheath cells. There are two waves of apoptosis, one early and one late, in response to irradiation. The second wave of deaths might be caused by the delayed removal of cells that had been damaged in the mitotic region and failed to be repaired properly (Figure 17.5).

17.3.4.1 Genes Protecting against Effects of Ionizing Radiation

Caenorhabditis elegans is an ideal model organism for identifying genes that protect cells against ionizing radiation; it is multicellular and experiences canonical cell cycle and apoptotic responses to DNA damage; the genome contains orthologs of virtually all known key DNA-damage signaling and repair genes and most importantly, systematic gene knockdown is possible in complete animals on a genome-wide scale. Worm genes can be inactivated by feeding animals on bacteria expressing double-stranded RNA homologous to a specific worm gene. Van Haaften and coworkers used an RNAi bacterial library that targets 86% of the 19,000 *C. elegans* genes [17]. After synchronization in the L1 larval stage, worms were irradiated with a dose of 140 Gy and after 4 days, these cultures were scored for the presence or absence of progeny. A nonirradiated set of cultures was analyzed in parallel to control for radiation-independent sterility resulting from knockdown of essential genes.

A total of 45 *C. elegans* genes were identified in a genome-wide RNA interference screen for increased sensitivity to ionizing radiation in germ cells. These genes include orthologs of well-known human cancer predisposition genes as well as novel genes, including human disease genes not previously linked to defective DNA damage responses. Knockdown of 11 genes also impaired radiation-induced cell cycle arrest, and seven genes were essential for apoptosis upon exposure to irradiation.

17.3.5 Radiation-Induced Mutation

The most common type of easily identified mutation is that affecting expression of an essential gene (lethals). An extensively used experimental approach has been the *eT1*-system which determines the frequency of recessive lethals in a sizable region of the *C. elegans* genome [18]. This system has been applied to measurement of the accumulated mutation rate following ionizing radiation (Figure 17.6).

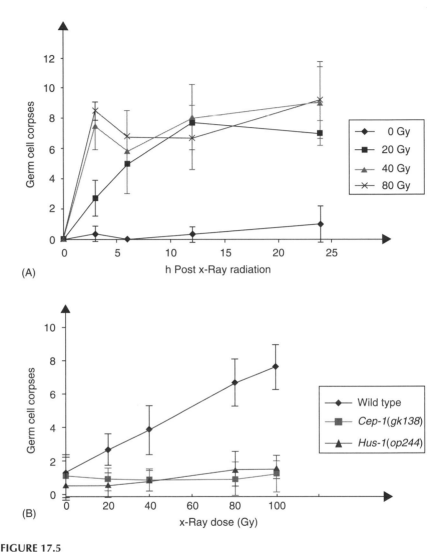

FIGURE 17.5
(A) Wild-type worms respond with an early and a late wave of apoptotic deaths following exposure to ionizing radiation, in a dose-dependent manner. (B) The checkpoint mutants *cep-1* and *hus-1* are defective in DNA damage-induced apoptotic cell death. (From Stergiou, L. and Hengartner, M.O., *Cell Death Differ.*, 11, 21, 2004. With permission.)

The *eTI* method can also be used to distinguish between different types of mutations such as point mutations or small or large rearrangements. Mutations induced by ionizing radiation are mainly rearrangements. These can be efficiently analyzed for the entire genome using the method of comparative genomic hybridization which involves the construction of a DNA microarray containing overlapping cosmid clones that cover the entire regions of interest. These microarrays are sensitive enough to detect single copy change and so can detect heterozygous deficiencies and duplications [19].

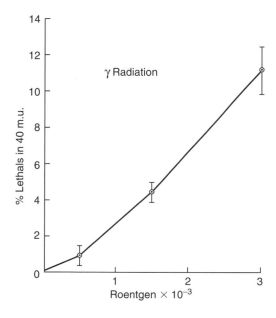

FIGURE 17.6
Dose–response curve for the induction of recessive lethals in the regions balanced by *eT1(III;V)*. (From Rosenbluth, R.E., Cuddeford, C., and Baillie, D.L., *Mutat. Res.*, 110, 39, 1983. With permission.)

17.3.6 Worms in Space

Caenorhabditis elegans also has many characteristics that make it an excellent model system for use in space. It reproduces as a self-fertilizing hermaphrodite; it is small (adults are approximately 1 mm long), and easily grown in a small space. The life cycle is short, approximately 1 week under conditions at the International Space Station (ISS) and the progeny numbers are high, a few hundred per hermaphrodite. The normal food source for *C. elegans* is bacteria; however, for space experimentation it can be fed a chemically defined medium. The worms will survive for several months in chemically defined medium and can be transferred to fresh medium and maintained apparently indefinitely. In addition, under conditions of overcrowding and limited food, larvae can enter a dormant "dauer larva" stage. These dauers can survive for several months at ambient temperature and will resume their normal life cycle when introduced to fresh medium.

Johnson and Nelson [20] first proposed using *C. elegans* as a model system for space biology studies. Since then *C. elegans* has flown on several missions to Earth orbit and was shown to develop and reproduce normally, confirming that it is an excellent model system for biological research in space. In a study of mutations induced by cosmic rays in *C. elegans* on spacelab in low Earth orbit for a short-term (8 days) exposure, dormant worms were suspended in buffer or on agar or immobilized next to CR-39 plastic nuclear track detectors to correlate fluence of HZE particles with genetic events. This configuration was used to isolate mutations in a set of 350 essential genes as well as in the *unc-22* structural gene. From flight samples, 13 mutants in the *unc-22* gene were isolated along with 53 lethal mutations from the autosomal regions balanced by the translocation *eT1(III;V)*.

Preliminary analysis suggested that mutations which correlated with specific cosmic ray tracks may have a higher proportion of rearrangements than those isolated from tube cultures on a randomly sampled basis. Flight sample mutation rate was approximately eightfold higher than ground controls which exhibited laboratory spontaneous frequencies [21].

Currently nothing is known about longer-term exposure to the different types of radiation in space, nor about the effects of exposure to the range of radiation in the space environment. A unique feature of the space radiation environment is the dominance of high-energy charged particles (HZE or high LET radiation) which present a significant hazard to space flight crews. Accelerator-based experiments are underway to quantify the health risks due to unavoidable radiation exposure. In a recent study using *C. elegans*, the effect of different types of radiation, γ-rays, accelerated protons, and Fe ions at the same physical dose was examined. Using RT-PCR, differential display and whole genome microarray hybridization experiments, unique transcriptional profiles for the different radiation treatments were defined [22].

The biology of *C. elegans* and the variety of research resources available for its study make it an outstanding candidate for future space missions. A high priority would be the development of an accumulating dosimeter which would function to determine the effects of long-term exposure to the space environment over multiple generations.

17.4 Zebrafish

The zebrafish, *Danio rerio*, has many of the advantages of simpler model organisms such as yeast and *C. elegans*, but unlike these organisms, it has a full complement of vertebrate organs, including a brain and spinal cord, chambered heart, digestive and excretory organs, and a hematopoietic system similar in many respects to that of mammals. Thus it is possible to investigate organ-specific effects of radiation that cannot be studied in lower eukaryotic models.

The human and zebrafish genomes share considerable homology, including conservation of most DNA repair-related genes. Embryonic development is rapid, with major organ systems such as the eyes, brain, heart, liver, muscles, bone, and gastrointestinal tract evident within 48 h post-fertilization (hpf). Because embryogenesis occurs outside the mother and in water, the effects of drug and radiation exposure are easy to assess and do not require parental sacrifice. Genetic manipulations are relatively simple in the zebrafish because genetic knockouts and transgenic animals can be created by the microinjection of morpholino-modified oligonucleotides complementary to the target gene or mRNA expressed from the target gene, respectively. These factors together contribute to the growing popularity of zebrafish for cancer-related studies and suggest that large-scale screening to identify new genes and pathways that influence organ-specific susceptibility to radiation injury would be a fruitful area of research.

17.4.1 Zebrafish for the Evaluation of Genotoxic Stress

Zebrafish embryos are ideal for evaluating genotoxic stress including that of ionizing radiation. The effects of radiation on zebrafish embryos were noted in initial reports that found greater radiosensitivity before the midblastula transition (MBT), but these studies did not assess the effects of different radiation doses or categorize the types and severity of mutations [23].

More detailed investigations [24] revealed that the lethal effects of ionizing radiation in zebrafish were directly proportional to radiation dose given and inversely proportional to embryonic age, with gastrula stage embryos (1–2 hpf) showing greater sensitivity to low (≤4 Gy) doses of radiation than blastula stage embryos (4 hpf). By the segmentation stage (8–24 hpf), embryos were essentially resistant even to higher radiation doses (8–10 Gy). Effects on survival were similar to dose-and time-dependent effects on development. This result was reflected in the 50% lethal dose response rates, measured at 144 hpf (the latest time point studied), which increased with advancing embryonic age at time of x-ray exposure (Figure 17.7).

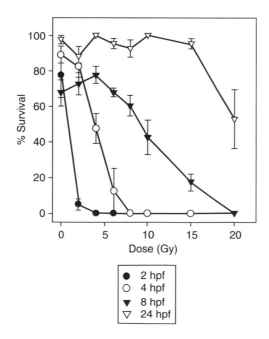

FIGURE 17.7
Effect of ionizing radiation on zebrafish development and survival. The effect of x-ray dose on survival of zebrafish embryos exposed at 2 (●), 4 (○), 8 (▼), and 24 (▽) hpf was evaluated at 144 hpf to determine 50% response values. Data represent the mean (± standard error of the mean) of at least three replicate experiments. (From McAleer, M.F., Davidson, C., Davidson, W.R., Yentzer, B., Farber, S.A., Rodeck, U., and Dicker, A.P., *Int. J. Radiat. Oncol. Biol. Phys.*, 61, 10, 2005. With permission.)

Irradiation of zebrafish embryos at or before MBT (approximately 4 hpf) produces multiple morphologic defects, including microcephaly, microphthalmia, micrognathia, distal notochord and segmental abnormalities, pericardial edema, and inhibition of yolk sac resorption, which may be considered as common, nonspecific manifestations of embryonic mutagenesis [25]. These findings are consistent with the concept that proteins required to repair radiation-induced damage are not present in the embryo before MBT when development is directed by preexisting maternal proteins present in the yolk. Transcription by the embryo proper as initiated after MBT appears to be necessary to express necessary repair genes to correct DNA damage incurred via radiation.

17.4.2 Effects of Ionizing Radiation on Brain and Eye Development

Radiation of zebrafish embryos resulted in a number of specific morphologic abnormalities, including curvature of the spine, shortening of the overall length of the body, pericardial edema, inhibition of yolk sac resorption, microphthalmia, and microcephaly (Figure 17.8). Of particular interest are the effects of radiation on the eye and brain. These organs are often dose limiting for clinical therapeutic radiation since radiation-related effects such as cataract formation, retinal degeneration/atrophy, blindness, and microcephaly are complications that have been reported after inadvertent or therapeutic exposure to radiation. In addition, the developing nervous system of the embryo is particularly sensitive to radiation injury and zebrafish may afford a model for the effect of radiation on neurogenesis in the mammalian brain.

It was found that radiation led to marked reduction in the volume of the tectum of the brain, such that a large portion of the skull case was devoid of brain matter. This was evident at 10 Gy and was even more marked after 20 Gy. The telencephalon and cerebellum were considerably less affected by radiation. It was difficult to judge the effect on the diencephalon because it is contiguous with and lacks obvious anatomic boundaries that separate it from the tectum.

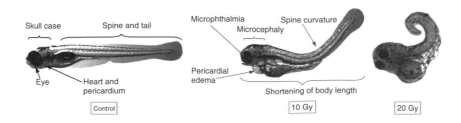

FIGURE 17.8
Morphological effects of ionizing radiation in the developing zebrafish. Embryos were irradiated with 0, 10, or 20 Gy, then incubated at 28°C and photographed at 144 hpf. (From Geiger, G.A., Parker, S.E., Beothy, A.P., Tucker, J.A., Mullins, M.C., and Kao, G.D., *Cancer Res.*, 66, 8172, 2006. With permission.)

The effect of radiation on the eye was also marked. The diameter of the eyes of control unirradiated embryos at 72 hpf was 185 ± 21 μm, compared with 141 ± 26 μm in embryos irradiated with 10 Gy and 170 ± 19 μm in embryos treated with 10 Gy in the presence of amifostine [25]. The cellular organization of unirradiated control eye shows concentric layers representing the zebrafish retina (rods and cones), which is immediately interior to the pigmented epithelium followed sequentially by concentric rings representing the outer plexiform layer, the bipolar cell layer, the inner plexiform layer, the ganglion cell layer, and finally the lens. In contrast to these distinct cellular layers detectable in the control eye, the organization of the eye after irradiation was considerably disrupted. Although thinner than in the control, the pigmented epithelium could still be discerned in the irradiated eye, but it was difficult to distinguish most of the remaining cellular layers. The rod and cone layer was not concentrically organized and the inner plexiform layer, which was conspicuous in the control eye, was nearly absent after radiation and the prominent round nuclei of the ganglion cell layer also could not be discerned. Finally, radiation resulted in lens opacification.

The profound effects of radiation on the development of the eye and brain may be due to increased cell death, decreased growth, or to both causes. The terminal transferase dUTP nick end labeling (TUNEL) assay which detects double-strand DNA breaks resulting from DNA-damaging agents or apoptosis has been used to demonstrate apoptosis in the zebrafish hindbrain and the peripheral nervous system after irradiation [26]. It was noted that while unirradiated, control brain showed no TUNEL; 10 Gy ionizing radiation led to substantial TUNEL throughout the tectum. In addition, radiation resulted in extensive TUNEL within the eye as well as in the tissues outside of the eye [25].

Another method to detect apoptosis is measurement of caspase expression. The small size of zebrafish embryos renders them suitable for microplate-based assays and the effects of radiation on caspase activation of irradiated embryos in 96 well plates have been investigated. Radiation resulted in a dose-dependent increase in caspase activation which was blocked by amifostine [25].

17.4.3 Modulation of Radiation Response

Modulation of radiation response has been reviewed by Geiger and coworkers [25]. Having established a reproducible phenotype of radiation injury in zebrafish embryos, modulation of this effect was examined by pretreating embryos with known radioprotective and radiosensitizing agents.

1. Radioprotectors: Amifostine is a free-radical scavenging thiol that is currently in clinical use as a radio and chemo-protector. Amifostine, at concentrations ≤4 mM, had no adverse effect on normal zebrafish development or gross morphology. However, after exposure to 4–6 Gy x-rays at 4 hpf, embryos pretreated with amifostine demonstrated both improved morphologic

development and survival relative to the irradiated embryos not administered the drug. This effect was statistically significant in embryos given 4 mM of amifostine before 6 Gy x-ray exposure compared with irradiated embryos not pretreated with amifostine at 144 hpf, the latest time point examined, implicating potential sustained benefit from the use of the radioprotector. Lesser protection by amifostine was observed after 4 Gy x-ray exposure.

Unlike the embryos treated with radiation alone, the skull case of embryos irradiated with amifostine was not devoid of brain tissue in the area of the tectum and the other regions of the brain also seemed to be less perturbed. For the eye, the presence of amifostine at the time of irradiation partially preserved the cellular organization. The layer of rods and cones could be identified immediately internal to the pigmented epithelium. Both the outer and inner plexiform layers could be readily identified albeit thinner than in the unirradiated control eye. The ganglion cell layer was retained after irradiation and amifostine; there seemed to be fewer cells than in the control eye and the lens remained opacified. Amifostine partially prevented these changes but neither protected against cellular depletion and thinning of specific structures, nor against lens opacification.

Amifostine resulted in markedly diminished TUNEL of tissues outside of the eye, but labeling of cells within the eye could still be detected although the intensity of the staining for most of the cells seemed to be less than in eyes irradiated in the absence of amifostine.

2. Radiosensitizers: The epidermal growth factor receptor tyrosine kinase inhibitor AG1478 alone disrupted normal development and survival of zebrafish at the highest concentration tested. Lower AG1478 concentrations (2.5–5 mM), however, were well tolerated by the embryos. When exposed to radiation ≤4 Gy at 4 hpf, the AG1478-treated zebrafish demonstrated enhanced teratogenicity (and lethality) as early as 72 hpf. At radiation doses ≥6 Gy, this radiosensitizing effect was masked by the embryotoxic effect of the radiation itself.

The results of these experiments indicate that zebrafish embryos should be considered as a valid preclinical model to examine the effects of modifiers of the radiation response in vertebrates. The results have implications for development of a rapid screen of novel radiation protectors and sensitizers with potential human therapeutic use.

17.4.4 Gene Function during Embryonic Development

The rapid and external embryonic development of zebrafish enables researchers to readily visualize dynamic developmental processes in the presence of extrinsic manipulation. In general, MO-based knockdown of targeted genes in zebrafish embryos is highly penetrant throughout early development during the first 2 days. This period is particularly important since this is when the critical processes of somitogenesis and organogenesis occur and it is prominently susceptible to genotoxic stress.

17.4.4.1   Expression of Zebrafish ATM

Zebrafish ATM (zATM) mRNA was observed to be ubiquitously expressed during gastrulation and early neurulation [27]. Differential tissue expression was evidenced by increased mRNA levels at later developmental stages in the eye, brain, and somites with relatively weak expression in the trunk and tail. The profile of zATM expression was consistent with previous observations in *Xenopus* and mice which also show increased expression of ATM in CNS, implicating a common importance in early development of the nervous system among vertebrates.

Functional studies using MO-mediated knockdown of zATM in zebrafish embryos further supported a significant role for zATM in general embryogenesis and in specific organogenesis. Dose-dependent effects of zATM MO on early developmental defects after ionizing radiation were observed suggesting that the amount of intrinsic zATM expression may act as a guardian or a teratological suppressor for genotoxic stress, consistent with the findings in ATM-knockout mice.

17.4.4.2   Ku80

The embryonic function of the *XRCC5* gene, which encodes Ku80, an essential component of the NHEJ pathway of DNA repair has been characterized. After the onset of zygotic transcription, Ku80 mRNA accumulates in a tissue-specific pattern, which includes proliferative zones of the retina and central nervous system. In the absence of genotoxic stress, zebrafish embryos with reduced Ku80 function develop normally. However, low-dose irradiation of these embryos during gastrulation leads to marked apoptosis throughout the developing central nervous system. Apoptosis is p53 dependent, indicating that it is a downstream consequence of unrepaired DNA damage. These results suggest that NHEJ components mediate DNA repair to promote survival of irradiated cells during embryogenesis [26]. The Ku80 protein is an essential component of the NHEJ pathway of DSB repair which requires at least four other genes: *LIG4* and *XRCC4*, which encode two subunits of a DNA ligase, *PRKDC*, which encodes the DNA-dependent protein kinases catalytic subunit (PK_{cs}), and *G22P1*, which encodes the Ku70 protein. Putative orthologs for each of these can be identified in the zebrafish EST database. The presence of these genes, together with the functional characterization of the *Ku80* gene, suggests that an intact functional NHEJ pathway operates in the zebrafish.

The finding that Ku80 function is necessary for radioprotection during embryogenesis is relevant to human experience where the consequences of radiation exposure to brain development are well documented. The dose range of 15–150 cGy, used in the majority of these experiments, is equivalent to a much smaller dose in humans who have a twofold larger genome. This corresponds with the range of prenatal exposures that resulted in sharply increased risk of microcephaly, severe mental retardation, lower IQ, and seizure disorders in the Japanese atomic bomb survivors exposed in utero.

17.4.5 Hematological Studies with Zebrafish

Studies with mammals had shown that the acute irradiation syndrome could be rescued by bone marrow transplant (BMT) and that the mean lethal dose of radiation (MLD) specifically causes death by hematopoietic failure. Despite the evolutionary divergence between zebrafish and mammals, the major consequences following graded doses of ionizing irradiation have been found to be quite similar.

17.4.5.1 Hematopoietic Syndrome in Zebrafish

In zebrafish, the MLD is approximately 40 Gy, and as in mammals the zebrafish hematopoietic system is the first to fail following increasing doses of irradiation and can be rescued by transplantation of whole kidney marrow (WKM) cells [28].

The kinetics of hematopoietic cell depletion over the first 2 days after irradiation are quite similar to those initially reported in mouse bone marrow and spleen following the MLD of 9 Gy. Daily sampling performed following 40 Gy doses in zebrafish showed that the vast majority of hematopoietic cells were depleted in the kidney and thymus. Compared with unirradiated controls, mean cell counts of each of the kidney myeloid, lymphoid, and precursor populations showed over 7-fold, 14-fold, and 16-fold decreases in absolute cell numbers, with the total kidney leukocyte number dropping from an average of 6.3×10^5 to 6.8×10^3 by day two. All cell lineages continued to decline over time until death. Continually decreasing cellularity was observed only in hematopoietic tissues, suggesting that a dose of 40 Gy leads specifically to hematopoietic failure.

One major difference between the zebrafish results and the results of studies done with mammals is the larger dose of irradiation required for the MLD. The lethal dose for mice is generally from 9 to 10 Gy, whereas that for zebrafish was found to be approximately 40 Gy. One contributing difference may be genome size. The zebrafish genome is approximately 1.7×10^9 bp, which is half the size of the mouse genome. Since a smaller genome presents a smaller target, the irradiation dose necessary to elicit the same biologic response would be larger in zebrafish than in mice. Another important difference in terms of the biologic response to irradiation between teleosts and mammals is temperature. Mammals maintain a constant temperature relative to their environments, whereas teleosts are poikilothermic and dependent on ambient conditions. In general, cold-blooded animals have been shown to be more radioresistant than warm-blooded animals. Reports on the effects of temperature on the MLD in many organisms, including teleosts, have shown a strong negative correlation between temperature and radioresistance.

17.4.5.2 Hematopoietic Cell Transplantation in Zebrafish

To enable the study of hematopoietic stem cell (HSC) biology, immune cell function, and leukemogenesis in zebrafish, a system for hematopoietic cell transplantation (HCT) into adult recipient animals conditioned by irradiation has been developed [28]. Dose–response experiments had shown that while the MLD of 40 Gy led to the specific ablation of hemato-lymphoid cells and death by 14 days after irradiation, sublethal irradiation doses of 20 Gy predominantly ablated lymphocytes and enabled the transplant of transplantation of donor hematopoietic cells.

Transplantation of hematopoietic cells carrying transgenes yielding red fluorescent erythrocytes and green fluorescent leukocytes showed that HCT is sufficient to rescue the MLD, that recipient hematolymphoid tissues were repopulated by donor-derived cells, and that donor blood cell lineages could be independently visualized in living recipients. These results established transplantation assays to test for HSC function in zebrafish.

The transplantation model was also of value for the study of oncogenesis in zebrafish. It was shown that transient ablation of lymphocytes by sublethal irradiation doses was both necessary and sufficient for the transfer of Myc-induced leukemias from *rag2-EGFP-mMyc* transgenic zebrafish to wild-type recipients. When transplanted into irradiated animals, doses as low as 5×10^3 leukemic cells could confer lethal disease, demonstrating frank cellular transformation, and this model of acute leukemia is particularly aggressive. Transplant doses as high as 5×10^6 cells could not transfer leukemia in unconditioned recipients suggesting that leukemic cells are recognized as foreign and rejected by the host immune system.

17.5 Deinococcus radiodurans

It is appropriate to end this chapter with a brief description of the radiation biologists' favorite microbe—*Deinococcus radiodurans*.

Deinococcus radiodurans was isolated from canned ground meat that had been supposedly sterilized by exposure to a dose of 4000 Gy. This new species was named at the time *Micrococcus radiodurans* because of its superficial morphological similarity to members of the genus *Micrococcus* but was later reclassified, along with its closest relatives, into a distinct phylum within the domain bacteria. The genus name—*Deinococcus*—was based on the Greek adjective "deinos," which means strange or unusual. *D. radiodurans* is also known to some of its admirers as "Conan the bacterium."

Deinococcus radiodurans is a nonpathogenic, nonsporulating, obligate aerobic bacterium that typically grows in undefined rich medium as clusters of

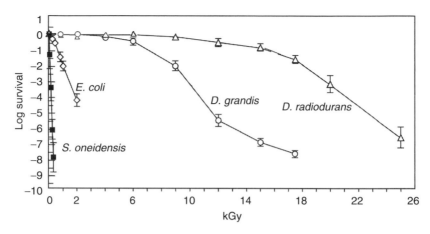

FIGURE 17.9

Survival of bacterial strains of varying degrees of radiosensitivity exposed to an acute dose of ionizing radiation. (From Daly, M.J., Gaidamakova, E.K., Matrosova, V.Y., Vasilenko, A., Zhai, M., Venkateswaran, A., Hess, M., Omelchenko, M.V., Kostandarithes, H.M., Makarova, K.S., Wackett, L.P., Fredrickson, J.K., and Ghosal, D., *Science*, 306, 1025, 2004. With permission.)

two cells (diplococci) in the early stages of growth and as four cells (tetracocci) in the late stages. *D. radiodurans* is extremely resistant to ionizing radiation and desiccation and maintains four to eight genomic copies per cell. The D_{37} dose for *D. radiodurans* R1 is approximately 6500 Gy, at least 200-fold higher than the D_{37} dose of *Escherichia coli* cultures irradiated under the same conditions (Figure 17.9). The energy deposited by 6500-Gy γ radiation would be expected to introduce thousands of DNA lesions, including hundreds of DSBs. In *D. radiodurans*, the presence of this amount of DNA damage does not result in a catastrophic loss of genetic information, instead, a number of passive and or interactive protective and repair mechanisms combine to neutralize the damage.

17.5.1 Origins of Extremophiles

There are no naturally occurring environments where ionizing radiation exposures are known to exceed 400 mGy per year, so it is unlikely that species have evolved mechanisms to protect against the effects of high-dose ionizing radiation by a process of selection. It seems more likely that the damage introduced by γ irradiation shares features with the damage that results from other stresses to which bacteria have adapted. One example is desiccation which introduces many DNA DSBs into the genomes of *D. radiodurans* and members of the extremophile cyanobacterial genus *Chroococcidiopsis*. Both organisms are tolerant to desiccation and are resistant to the potentially lethal effects of ionizing radiation, which might indicate that the radioresistance of these species is a fortuitous consequence of their ability to tolerate desiccation-induced strand breaks.

Radiation resistance is not restricted to the domain bacteria. Several hyperthermophilic archaea show an extreme ionizing radiation resistance in some cases (e.g., *Thermococcus gammatolerans*) comparable to that of *D. radiodurans*. These organisms are sometimes called extremophils because of their ability to withstand extreme environments. The most likely explanation of the scattered appearance of ionizing-radiation resistance among distinct prokaryotic lineages is that this phenotype has arisen independently in unrelated species exposed to extreme conditions of heat and desiccation at some time in their respective evolutionary histories.

17.5.2 Genetics of *D. radiodurans*

To date, *D. radiodurans* has received more attention than the other deinococci. The genome of *D. radiodurans* strain R1 has been sequenced [29]. The *D. radiodurans* chromosome is 3.28 Mb, with a GC content of 66.6%. There are nine types of short nucleotide repeats, ranging in size from 60 to 215 bp, found at 295 sites that are randomly scattered in the genome. The genome is segmented and consists of a 2.64 Mb chromosome (chromosome I), a 0.41 Mb chromosome (chromosome II), a 0.18 Mb megaplasmid, and a 0.045 Mb plasmid. *D. radiodurans* has between four and ten genome copies per cell, depending on the stage of the bacterial growth.

17.5.3 Characteristics of *D. radiodurans* Predisposing to Radiation Resistance

The γ irradiation survival curves of actively growing cultures of *D. radiodurans* R1 have a shoulder of resistance up to 5000 Gy, and until this dose is achieved there is no measurable loss of viability in the irradiated culture. Under these conditions, the D_{37} dose for *D. radiodurans* R1 is approximately 6500 Gy. Assuming that there are eight genome copies per cell in these cultures, a 5000 Gy dose will introduce approximately 1600 DSBs per cell. As this dose is sublethal, the implication is that unrestituted, potentially lethal deletions and genome rearrangements occur at low frequencies. Although there is no formal proof, it seems highly probable that the process of DSB repair in *D. radiodurans* is error-free although this has not been formally demonstrated.

Based on the amount of DNA in the cell and the rate of DNA DSBs accumulated when *E. coli* and *D. radiodurans* are irradiated under identical conditions [30,31]; it can be calculated that at the D_{37} dose, *D. radiodurans* has approximately 30 times the number of DSBs compared to *E. coli*. This eliminates the possibility that the formation of strand breaks in the genome is blocked by a protective mechanism. The only other possibility is that the phenotype and the constitutive and radiation-induced genotype of *D. radiodurans* have evolved in such a way as to effectively and accurately repair or bypass massive damage to the genome. Some of the possible mechanisms will be summarized in the remainder of this chapter.

17.5.3.1 Genome Copy Number

Cells with increased numbers of genome copies have enhanced resistance to ionizing radiation. The extra genetic material protects the cell in two ways. When multiple genomes are present, there are additional copies of crucial loci increasing the probability of the cell surviving irradiation. Thus, if 100 inactivating lesions are randomly introduced into a single copy of the *D. radiodurans* genome (2897 genes), the probability of inactivating any specific gene on that genome copy is 3.4%. The probability of inactivating all the copies of the same gene is reduced to 0.12% when two genomes are present, and to 0.004% when there are three genomes [32].

Repair of DNA DSBs by recombination requires the presence of more genomes than one, so it seems probable that genome redundancy is necessary for genome repair. However, while the number of genome copies in *D. radiodurans* is never less than 4, the cells have the same survival rate whether they contain 4 or 10 genome copies, so the extent of the contribution of genome redundancy to radioresistance in *D. radiodurans*, relative to other mechanisms, is unclear. Many bacteria contain more than one genome copy, particularly during exponential growth. For example, *E. coli* contains between 5 and 18 genome equivalents during exponential-phase growth and *Azotobacter vinelandii* can accumulate more than 100 genome copies. Neither species is radioresistant, so genetic redundancy alone cannot account for radioresistance.

17.5.3.2 Nucleoid Organization

The nucleoids of stationary-phase *D. radiodurans* cells are arranged as a tightly structured ring that remains unaltered by high-dose irradiation. It has been suggested [33] that this structure passively contributes to *D. radiodurans* radioresistance by preventing the fragments that are formed by DSBs from drifting apart during repair thus maintaining the linear continuity of the genome even when it is fragmented [33]. This hypothesis has been questioned because the presence of ring-like nucleoids under different growth conditions does not always correlate with resistance to ionizing radiation. The nucleoids of members of the radioresistant genera *Deinococcus* and *Rubrobacter* display a high degree of genome condensation relative to the more radiosensitive species *E. coli* and *Thermus aquaticus*, indicating that some species with a condensed genome are better protected from ionizing radiation, regardless of nucleoid shape.

17.5.3.3 Manganese Content

The manganese content has been reviewed by Daly and coworkers [34]. The existence of high intracellular Mn/Fe concentration ratios in phylogenetically distant, radiation-resistant bacteria, but not in sensitive cells, supports the idea that Mn(II) accumulation in conjunction with low Fe might be a widespread strategy that facilitates survival. A positive relationship between the high concentrations of Mn(II) that can accumulate in *D. radiodurans* and the capacity of these cells to survive irradiation is supported by the observation that when *D. radiodurans* cultures are starved of Mn(II), their resistance to ionizing radiation decreases. The numbers of DNA DSBs formed as a result of radiation exposure are the same in the presence or absence of Mn(II), so Mn(II) does not prevent DNA damage. Instead, cellular damage that results from exposure to high radiation doses is better tolerated if Mn(II) is present.

Intracellular Mn(II) might act as a protector by scavenging reactive oxygen species (ROS). In another bacterial species, *Lactobacillus plantarum* which lacks the protective enzyme superoxide dismutase (SOD), the potential for radiosensitization is countered by intracellular Mn(II) concentrations of 20–25 mM. Since levels of DSBs in *D. radiodurans* are unaffected by the presence of Mn(II), the scavenging of ROS must be protective of macromolecules other than DNA. It has been proposed that Mn(II) accumulation prevents superoxide and related ROS that are produced during irradiation from damaging proteins.

Another possibility is that the increased Mn(II) concentration could contribute to the condensation of the *D. radiodurans* genome. An aqueous solution of DNA can be condensed in vitro by addition of multivalent cations which act to neutralize the repulsion of phosphate groups in the DNA backbone. This hypothesis is attractive in that it creates a role for manganese in the nucleoid organization scenario of radioresistance.

17.5.4 Regulation of Cellular Responses to Extensive Radiation Damage

When *D. radiodurans* is exposed to ionizing radiation, a well-characterized sequence of physiological events takes place, including rapid cessation of DNA replication. The observation of inhibition of DNA replication is reminiscent of the damage checkpoints of eukaryotes—mechanisms that sense DNA damage and initiate a delay in the cell cycle until the damage is repaired (Chapter 8) but the existence of DNA-damage checkpoints operating in *D. radiodurans* has not yet been established.

Repair of the genome occurs in two distinct stages. Substantial chromosome repair through RecA-independent repair processes occurs during the first 1.5 h after *D. radiodurans* is exposed to high doses of ionizing radiation. Approximately one-third of the DSBs are repaired in this phase. RecA-dependent recombinational DNA repair becomes important several hours after irradiation, and predominates in the later stages of genome reconstitution.

17.5.4.1 *Radiation-Induced DNA Repair Genes*

A search for novel genes that are induced in response to ionizing radiation and desiccation, using genome-based microarrays, revealed evidence for both RecA-independent and RecA-dependent pathways of DSB repair [35]. In exponentially growing cells, 72 genes were found to be induced threefold or more after γ irradiation. In comparison, 73 loci were induced during recovery from extended desiccation. There were 33 genes induced in common following irradiation and desiccation. The five genes most highly induced in response to both stresses were identical and encoded proteins of unknown function. Inactivation of these loci—*ddrA, ddrB, ddrC, ddrD,* and *pprA*—produced phenotypes that suggested the involvement of the gene product in genome repair. Genetic analyses defined three epistasis groups that affected ionizing radiation resistance, and established that two of the loci (*ddrA* and *ddrB*) contribute to radioresistance through different RecA-independent processes. The *pprA* and *recA* loci form a third epistasis group, indicating that the *pprA* gene product interacts with RecA.

17.5.4.2 DNA End Protection

The *ddrA* locus was the first of the five new genes to be characterized. This locus is one of the most highly induced genes in *Deinococcus* following γ irradiation, with expression increasing 20- to 30-fold relative to an untreated control. An evolutionary relationship has been detected between *ddrA* gene product DdrA and the important eukaryotic recombination protein Rad52. Deletion of *ddrA* function results in a modest increase in radiation sensitivity in cells grown in rich media. However, when cells are irradiated and then starved, deletion of *ddrA* results in a 100-fold loss of viability over 5 days compared with wild-type cells. The loss in viability of the *ddrA* mutant is accompanied by a dramatic decrease in genomic DNA content as a result of nucleolytic degradation.

The protein, DdrA appears to function as a DNA end protection system by binding to the 3′ ends of single-stranded DNA in vitro and protecting them from nucleolytic degradation. As DSBs occur, DdrA (and possibly other proteins) binds to the exposed DNA ends and prevents nuclease digestion of the chromosomal DNA. This strategy is particularly valuable in the genome repair that occurs after desiccation. DNA repair uses a lot of metabolic energy so cells recovering from desiccation in an environment that lacks nutrients would not have the means to repair DNA damage. By protecting the broken DNA ends from being degraded by nucleases, cells would be able to preserve genomic DNA until conditions become suitable for cell growth and DNA repair.

17.5.4.3 RecA-Independent Double-Strand Break Repair

Both NHEJ and single-strand annealing (SSA) pathways have been hypothesized to function in *D. radiodurans*. It has been suggested that NHEJ repair of DSBs would be facilitated in the condensed chromosome where the ends might not be free to diffuse away from each other. NHEJ has been identified in *Bacillus subtilis* and may be present in other bacteria where presence of the PprA and PolX proteins is consistent with the existence of NHEJ. However, eukaryote NHEJ systems are generally error-prone and seem to be unsuited to the accurate genome repair that is observed in *Deinococcus* species. In addition, patterns of recombination between plasmids and the re-circularization of integrated plasmids in irradiated *Deinococcus* cells are not consistent with NHEJ.

Plasmid repair and re-circularization of genome-integrated plasmids during the RecA-independent phase of DNA DSB repair in *D. radiodurans* is dependent on homology, indicating that SSA (SSA, Chapter 7) might have a role. In fact, a two-stage DNA repair process has been described which involves a mechanism for fragment reassembly called synthesis-dependent strand annealing (SDSA) which is followed and completed by crossovers. At least two genome copies and random DNA breakage are required for effective SDSA. In SDSA, chromosomal fragments with overlapping homologies are used both as primers and as templates for massive synthesis of complementary single strands in a manner similar to what occurs in a single-round multiplex polymerase chain reaction. This synthesis depends on DNA polymerase I and incorporates more nucleotides than does normal replication in intact cells. Newly synthesized complementary single-stranded extensions become "sticky ends" that anneal with high precision, joining together contiguous DNA fragments into long, linear, double-stranded intermediates. (SDSA is shown schematically in Figure 17.10.) The

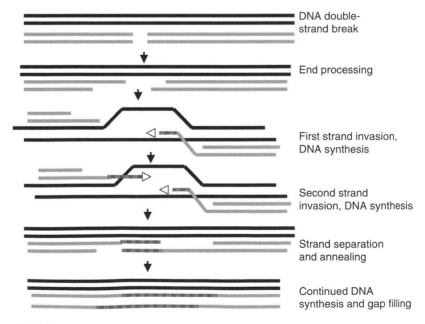

FIGURE 17.10
SDSA, a mechanism of error-free double-strand break (DSB) repair that is initiated by creating 3' overhangs from the ends of the broken DNA duplex (gray in the figure). One of these 3' ends invades a homologous region on an undamaged sister duplex (black in the figure), priming DNA synthesis and creating a D-loop that acts as a template or DNA synthesis primed by the other 3' end. If displaced, the newly synthesized DNA can anneal, closing the DSB. Newly synthesized DNA is shown as ▬▬▬▬▬.

intermediates require RecA-dependent crossovers to mature into circular chromosomes that comprise double-stranded patchworks of numerous DNA blocks synthesized before radiation, connected by DNA blocks synthesized after radiation [36].

17.5.4.4 Recombinational DNA Repair

A number of proteins with known functions in bacterial recombinational DNA repair which have been identified in *D. radiodurans*. These include RecA, single-strand-binding (SSB) protein, recombinase D (RecD), DNA polymerase I, recombinase R (RecR), and recombinase O (RecO).

The *D. radiodurans* RecA protein is 57% identical to the *E. coli* RecA protein. In vitro, the protein promotes all the key recombinogenic activities of RecA-class recombinases. It forms filaments on DNA, hydrolyses ATP and dATP in a DNA-dependent fashion, and promotes DNA-strand exchange. However, the *D. radiodurans* RecA protein differs in one respect. The DNA-strand-exchange reactions of the *E. coli* RecA protein are ordered so that the single-stranded DNA is bound before the double-stranded DNA. This is logical as the RecA protein must be targeted to single-strand gaps at stalled replication forks and other damaged DNA sites. In contrast, the *D. radiodurans* RecA protein promotes an obligate inverse DNA-strand-exchange reaction binding the duplex DNA first before the homologous single-stranded DNA.

It has been suggested that *D. radiodurans* RecA has a primarily regulatory role in DSB repair rather than a recombination function. *D. radiodurans* strains harboring the E142K RecA mutation apparently retain resistance to γ irradiation. However, the E142K mutation also retains significant DNA-strand-exchange activity in some assays, and it seems that the function of RecA in *D. radiodurans* has yet to be completely defined. Assuming that *D. radiodurans* must locate and splice together overlapping segments of its chromosomes to reconstruct a functional genome, a DNA-pairing function of RecA must be involved.

Another recombinational repair protein that is expressed in *D. radiodurans* to a much greater extent than in other bacteria is the SSB protein, which is efficient at stimulating the DNA-strand exchange that is promoted by RecA proteins from both *E. coli* and *D. radiodurans*. The concentration of the *D. radiodurans* SSB is 10-fold higher than the normal in vivo levels of the SSB protein in *E. coli*.

17.5.5 Double-Strand Break Tolerance

It has been suggested by Battista and Cox [32] that a significant contribution to the observed tolerance of ionizing radiation in *D. radiodurans* would result if many of the measured DSBs that occur are cryptic in vivo. According to this hypothesis, some fraction of the measured DSBs would be held together so that the actual separation of the DNA ends would not occur in the cell. This would require that the organism has a mechanism to stabilize opposed breaks, constraining the intervening base pairs so that actual separation of the two DNA ends does not occur. The cryptic breaks would be scored as DSBs during the processing and analysis of genomic DNA because the conditions necessary to stabilize the paired single-strand breaks are lost when the cell is disrupted.

By this mechanism, the linear continuity of the genome sequence would be preserved at many potential sites of DSBs in a manner that is error-free, and experimentally, this process would be indistinguishable from a non-mutagenic type of NHEJ. If such a system exists, the repair would be accurate and RecA-independent, passively reducing the cell's dependence on recombinational DNA repair and the accompanying homology search at any break site that is stabilized in this manner.

Two alternative, but not mutually exclusive, mechanisms might be used to stabilize base pairing between opposed single-strand breaks. First, *D. radiodurans* might encode proteins that hold the DNA together. The proteins involved could be functionally analogous to the structural maintenance of chromosomes proteins that are present in many eukaryotic and prokaryotic species. In eukaryotes, these proteins are known as cohesins and condensins, and the significance of their role in genome stabilization and DNA repair is becoming recognized. A second possibility is that the intracellular ionic composition could be sufficient to physically limit dissociation of DNA base pairs. The stability of annealed complementary DNA is dependent on the ionic strength of the medium in which the DNA is dissolved, again implicating increased intracellular Mn(II) concentrations which might help to hold DNA that contains several single-strand breaks together.

17.5.6 An Economic Niche for *D. radiodurans*

Unlike *C. elegans* and zebrafish, *D. radiodurans* has not been recruited by NASA although some of the more far-fetched theories as to its origins had it arriving from outer space. In fact, the talents of *D. radiodurans* are being used to address important problems on Earth. Using genetic engineering, *Deinococcus* has been equipped for bioremediation to consume and digest solvents and heavy metals on sites which are highly radioactive. The bacterial mercuric reductase gene has been cloned from *E. coli* into *Deinococcus* to effectively reduce Hg (II), frequently found in radioactive waste generated from nuclear weapons manufacture, to the less toxic volatile elemental mercury. A strain of *Deinococcus* has been developed that could detoxify both mercury and toluene in mixed radioactive wastes [37].

References

1. Goffeau A, et al. Life with 6000 genes. *Science* 274: 563–567, 1996.
2. Game JC. DNA double strand breaks and the *RAD50-RAD57* genes in *Saccharomyces. Semin. Cancer Biol.* 4: 73–82, 1993.

3. Game JC. The *Saccharomyces* repair genes at the end of the century. *Mutat. Res.* 451: 277–293, 2000.

4. Jablonovich Z, Liefshitz B, Steinlauf R, and Kupiec M. Characterization of the role played by the *RAD59* gene of *Saccharomyces cerevisiae* in ectopic recombination. *Curr. Genet.* 36: 13–20, 1999.

5. Game JC and Mortimer RK. A genetic study of X-ray sensitive mutants in yeast. *Mutat. Res.* 24: 281–292, 1974.

6. Feldmann H and Winnacker EL. A putative homologue of the human autoantigen Ku from *Saccharomyces cerevisiae*. *J. Biol. Chem.* 268: 12895–12900, 1993.

7 Boulton SJ and Jackson SP. Identification of a *Saccharomyces cerevisiae* Ku80 homologue: Roles in DNA double strand break rejoining and in telomeric maintenance. *Nucleic Acids Res.* 24: 4639–4648, 1996.

8. Carr AM. Radiation checkpoints in model systems. *Int. J. Radiat. Biol.* 66: S133–S139, 1994.

9. Lowndes NF and Murguia JR. Sensing and responding to DNA damage. *Curr. Opin. Genet. Dev.* 10: 17–25, 2000.

10 Humphrey T. DNA damage and cell cycle control in *Schizosaccharomyces pombe*. *Mutat. Res.* 451: 211–226, 2000.

11. Bennett CB, Lewis LK, Kathekeyan G, Lobachev K, Jin YH, Sterling JF, Snipe JR, and Resnick MA. Genes required for ionizing radiation resistance in yeast. *Nat. Genet.* 29: 426–434, 2001.

12. Game JC, Birrell GW, Brown JA, Shibata T, Baccari C, Chu AM, Williamson MS, and Brown JM. Use of a genome-wide approach to identify new genes that control resistance of *Saccharomyces cerevisiae* to ionizing radiation. *Radiat. Res.* 160: 14–24, 2003.

13. The *C. elegans* sequencing consortium. Genome sequence of the nematode *C. elegans*: A platform for investigating biology. *Science* 282: 2012–2018, 1998.

14. Schumacher B, Hofmann K, Boulton S, and Gartner A. The *C. elegans* homolog of the p53 tumor suppressor is required for DNA damage-induced apoptosis. *Curr. Biol.* 11: 1722–1727, 2001.

15. Gartner A, Milstein S, Ahmed S, Hodgkin J, and Hengartner MO. A conserved checkpoint pathway mediates DNA damage-induced apoptosis and cell cycle arrest in *C. elegans*. *Mol. Cell* 5: 435–443, 2000.

16. Stergiou L and Hengartner MO. Death and more: DNA damage response pathways in the nematode *C. elegans*. *Cell Death Differ.* 11: 21–28, 2004.

17. van Haaften G, Romeijn R, Pothof J, Koole W, Mullenders LH, Pastink A, Plasterk RH, and Tijsterman M. Identification of conserved pathways of DNA-damage response and radiation protection by genome-wide RNAi. *Curr. Biol.* 16: 1344–1350, 2006.

18. Rosenbluth RE, Cuddeford C, and Baillie DL. Mutagenesis in *Caenorhabditis elegans* I. A rapid mutagenic system using the reciprocal translocation eT1 (III;V). *Mutat. Res.* 110: 39–48, 1983.

19. Zhao Y, Johnsen R, Baillie D, and Rose A. Worms in space? A model biological dosimeter. *Gravit. Space Biol. Bull.* 18: 11–16, 2005.

20. Johnson TE and Nelson GA. *Caenorhabditis elegans*: A model system for space biology studies. *Exp. Gerontol.* 26: 299–309, 1991.

21. Nelson GA, Schubert WW, Kazarians GA, Richards GF, Benton EV, Benton ER, and Henke R. Radiation effects in nematodes: Results from IML-1 experiments. *Adv. Space Res.* 14: 87–91, 1994.

22. Nelson GA, Jones TA, Chesnut A, and Smith AL. Radiation-induced gene expression in the nematode *Caenorhabditis elegans*. *J. Radiat. Res. (Tokyo)* 43: 199–203, 2002.
23. Walker C and Streisinger G. Induction of mutations by gamma rays in pregonial germ cells of zebrafish embryos. *Genetics* 103: 125–126, 1983.
24. McAleer MF, Davidson C, Davidson WR, Yentzer B, Farber SA, Rodeck U, and Dicker AP. Novel use of zebrafish as a vertebrate model to screen radiation protectors and sensitizers. *Int. J. Radiat. Oncol. Biol. Phys.* 61: 10–13, 2005.
25. Geiger GA, Parker SE, Beothy AP, Tucker JA, Mullins MC, and Kao GD. Zebrafish as a "Biosensor"? Effects of ionizing radiation and amifostine on embryonic viability and development. *Cancer Res.* 66: 8172–8181, 2006.
26. Bladen CL, Lam WK, Dynan WS, and Kozlowski DJ. DNA damage response and Ku80 function in the vertebrate embryo. *Nucleic Acids Res.* 33: 3002–3010, 2005.
27. Imamura S and Kishi S. Molecular cloning and functional characterization of zebrafish ATM. *Int. J. Biochem. Cell Biol.* 37: 1105–1116, 2005.
28. Traver D, Winzeler A, Stern HM, Mayhall EA, Langenau DM, Kutok JL, Look AT, and Zon LI. Effects of lethal irradiation in zebrafish and rescue by hematopoietic cell transplantation. *Blood* 104: 1298–1305, 2004.
29. White O, et al. Genome sequence of the radioresistant bacterium *Deinococcus radiodurans*. *Science* 286: 1571–1577, 1999.
30. Krasin F and Hutchinson F. Repair of DNA double-strand breaks in *Escherichia coli*, which requires recA function and the presence of a duplicate genome. *J. Mol. Biol.* 116: 81–98, 1977.
31. Burrell AD, Feldschreiber P, and Dean CJ. DNA-membrane association and the repair of double breaks in x-irradiated *Micrococcus radiodurans*. *Biochim. Biophys. Acta* 247: 38–53, 1971.
32. Cox MM and Battista JR. *Deinococcus radiodurans*—the consummate survivor. *Nat. Rev. Microbiol.* 3: 882–892, 2005.
33. Levin-Zaidman S, Englander J, Shimoni E, Sharma AK, Minton KW, and Minsky A. Ringlike structure of the *Deinococcus radiodurans* genome: A key to radioresistance? *Science* 299: 254–256, 2003.
34. Daly MJ, Gaidamakova EK, Matrosova VY, Vasilenko A, Zhai M, Venkateswaran A, Hess M, Omelchenko MV, Kostandarithes HM, Makarova KS, Wackett LP, Fredrickson JK, and Ghosal D. Accumulation of Mn(II) in *Deinococcus radiodurans* facilitates gamma-radiation resistance. *Science* 306: 1025–1028, 2004.
35. Tanaka M, Earl AM, Howell HA, Park MJ, Eisen JA, Peterson SN, and Battista JR. Analysis of *Deinococcus radiodurans*'s transcriptional response to ionizing radiation and desiccation reveals novel proteins that contribute to extreme radioresistance. *Genetics* 168: 21–33, 2004.
36. Zahradka K, Slade D, Bailone A, Sommer S, Averbeck D, Petranovic M, Lindner AB, and Radman M. Reassembly of shattered chromosomes in *Deinococcus radiodurans*. *Nature* 443: 569–573, 2006.
37. Brim H, McFarlan SC, Fredrickson JK, Minton KW, Zhai M, Wackett LP, and Daly MJ. Engineering *Deinococcus radiodurans* for metal remediation in radioactive mixed waste environments. *Nat. Biotechnol.* 18: 85–90, 2000.

Glossary

14.3.3 Proteins: A family of small, widely expressed, highly conserved cytosolic proteins which bind to and influence the activities of a diverse group of molecules involved in signal transduction, cell cycle regulation, and apoptosis.

Abasic, apurinic, apyrimidinic: Referring to the product of depurination or depyrimidination in which the glycosidic bond linking a deoxyribose or ribose to a purine or pyrimidine base is broken, leaving behind only the deoxyribose or ribose in the DNA or RNA, respectively.

Abscopal effect: A significant response to radiation in tissues which are clearly separated from the radiation-exposed area with a measurable response occurring at a distance from the portals of irradiation large enough to rule out possible effects of scattered radiation.

Accelerator (linear): A machine, often called a linac that produces high-energy x-rays for the treatment of cancer.

Acentric: A fragment of a chromosome lacking a centromere.

Acetylation: Covalent attachment of an acetyl group to a second molecule such as protein.

Adaptive response: The ability of a cell, tissue, or organism to better resist stress damage by prior exposure to a lesser amount of stress.

Adaptor protein: A protein that may act in signal transduction in cells by facilitating the association between other molecular components of signaling pathways. The protein often lacks an intrinsic catalytic activity (e.g., kinase activity), and possesses distinct protein- or phosphoprotein-binding domains, thus bridging proteins with catalytic activity.

Adduct: The novel molecular structure arising after covalent linkage of a mutagen with one or another portion of a DNA molecule.

Adenocarcinoma: Tumor derived from secretory epithelial cells.

Adenovirus: A group of viruses responsible for infections in mammals and birds. Adenoviruses are useful as a vector in gene therapy because they infect both dividing and nondividing cells, but the disadvantage is that they evoke an immune response, which makes their repeated use difficult.

Aerobic: Describes organisms that use oxygen as a final electron acceptor in respiration in order to generate energy (ATP).

Akt **family:** Comprises three genes in humans: the *Akt1*, *Akt2*, and *Akt3*. Akt enzymes are serine/threonine-specific protein kinases. Akt1 is involved in cellular survival pathways, by inhibiting apoptotic processes.

Alkylating: Capable of attaching an alkyl group or similarly structured chemical group to a substrate such as a DNA base.

Allele: One alternative among different versions of a gene that may be defined by the phenototype it creates, by the protein it specifies, or by its nucleotide sequence.

Alpha particle (α-particle): A positively charged particle emitted by radioactive materials. It consists of two neutrons and two protons bound together. α-Particles are ejected from a nucleus during the decay of some radioactive elements.

Amino acids: The 20 basic building blocks of proteins.

Anaphase: Third subphase of mitosis, during which the paired chromatids are segregated to opposite poles of the cell.

Aneuploid: Describes a karyotype that deviates from diploid because of increases or decreases in the numbers of certain chromosomes.

Angiogenesis: Process by which new blood vessels are formed.

Angstrom (Å): A unit of length equal to 10^{-10} m.

Anion: A negatively charged ion, which has more electrons in its electron shells than it has protons in its nuclei. The anion is attracted to the anode.

Anneal: To pair complementary DNA or RNA sequences via hydrogen bonding, to form a double-stranded polynucleotide.

Anoikis: Form of apoptosis that is triggered by the failure of a cell to establish anchorage to a solid substrate, such as the extracellular matrix (ECM), or by loss of such anchorage.

Antibody: A soluble protein produced by plasma cell of the immune system that is capable of recognizing and binding particular antigens with high specificity.

Antigen: A molecule or portion of a molecule that can be specifically recognized by an antibody or a T-cell receptor or that provokes the production of an antibody.

Antioxidants: Molecules that slow or prevent the oxidation of other chemicals. Antioxidants terminate chain reactions involving free radicals by removing radical intermediates and can inhibit other oxidation reactions by being oxidized themselves. Antioxidants are often reducing agents, such as thiols or phenols.

Antisense RNA: An RNA molecule with a sequence that is complementary to the messenger RNA (mRNA) sequences expressed from a target gene. The resultant double-stranded RNA molecule inhibits the translation of the target protein.

Apoptosis: Complex program of cellular self-destruction, triggered by a variety of stimuli and involving the activation of caspase enzymes, that results in quick fragmentation and phagocytosis of a cell.

Aromatic: Referring to an organic molecule that contains one or more six carbon rings.

Asters: Radial microtubule arrays in animal cells which are located around each pair of centrioles.

Ataxia telangiectasia (AT): A clinical syndrome in which patients have a variety of symptoms, including ataxia (unstable gait) and telangiectasia (prominent and tortuous blood vessels). Cells from such individuals are sensitive to ionizing radiation and defective in repair of DNA damage. AT patients have a high incidence of lymphoma.

Atomic number (Z): The number of protons in the nucleus of an atom and also its positive charge.

Autocrine: Referring to the signaling path of a hormone or factor that is released by a cell and proceeds to act upon the same cell (or same cell type) that has released it.

Autophagy: Program of cellular responses to nutrient deprivation involving the digestion of a cell's organelles within its own lysosomes.

Autoradiography: A technique to identify where a radioactive isotope is localized in cells or subcellular components. The process involves covering biological material with photographic film or emulsion. The radioactivity produced by the isotope causes local exposure of the overlying film, emulsion (or detector), which upon development can be detected as dark grains close to the location of the isotope.

Autosome: A chromosome that is not a sex chromosome (i.e., neither an X nor a Y chromosome).

Bacteriophage: A virus that infects bacteria. Frequently used as carriers of cloned genes.

Base excision repair: A mechanism of DNA repair, whereby one damaged base is removed from the DNA.

Base pair: A pair of complementary nitrogenous bases in a DNA molecule. The number of base pairs is a measure of the length of DNA.

bcl-2/bax: Members of a family of genes involved in the control of apoptosis. Their products form dimers. Increased *bcl*-2 expression is anti-apoptotic increased expression of *bax* is pro-apoptotic.

Beta particle (β-particle): An elementary particle emitted from a nucleus during radioactive decay, with a single electrical charge and a mass equal to 1/1837 that of a proton. A negatively charged β-particle is identical to an electron.

Betatron: A particle accelerator developed to accelerate electrons. If the electron beam is directed at a metal plate, the betatron can be used as a source of energetic x-rays or γ-rays.

Biodosimetry: The estimation of received doses by determining the frequency of biological markers. In the case of radiation exposure, the most commonly used markers are radiation-induced chromosome aberrations.

Bioinformatics: The science of using computational methods for analyzing biological information.

Biopsy: Removal of a small portion of a tumor for examination by a pathologist for diagnosis.

Bioreductive drugs: Inactive prodrugs that are converted into potent cytotoxins under conditions of either low oxygen tension or in the presence of high levels of specific reductases.

Bloom's syndrome: A rare autosomal recessive disorder characterized by telangiectases, photosensitivity, growth deficiency, immunodeficiency, and increased susceptibility to neoplasms. The syndrome is caused by a mutation in the gene designated *BLM*.

Bragg peak: The maximum rate of energy loss which occurs just before a particle passing through matter comes to rest.

BRCA1 and BRCA2: Tumor suppressor genes in which mutations are associated with a high incidence of breast and ovarian cancer. Their protein products appear to be involved in DNA repair.

Bremsstrahlung x-rays: x-Rays resulting from an interaction of the projectile electron with a target nucleus; braking radiation.

Bromodeoxyuridine (BrdU or BrdUrd): A mutagenically active analogue of thymidine in which the methyl group at the 6′ position in thymine is replaced by bromine.

Bystander effect: An effect whereby therapeutic agents that target specific types of cells may influence the viability or other properties of nontargeted neighboring cells.

Calcium phosphate precipitation: The classic method for introducing DNA into cells. Calcium phosphate precipitates DNA in large aggregates which are taken up by cells.

Cap-dependent mRNA translation: Part of the processing of eukaryotic RNAs before release into the cytosol involving attachment of a cap of 7-methylguanosine triphosphate to the 5′ end.

Carbogen: Inhalant consisting of hyperoxic gas (95%–98% oxygen and 2%–5% carbon dioxide) with radiosensitizing properties.

Carcinogen (adj., carcinogenic): A physical or chemical agent capable of causing cancer.

CArG box: Alternate name for the serum response element (SRE) at least five copies of which are found in the promoter sequence of the human *Egr-1* gene.

Caspases: A family of cysteine aspartyl-specific proteases that effect programmed cell death or apoptosis.

Catalase: Haem enzyme of peroxisomes of many eukaryotic cells. Converts hydrogen peroxide produced by certain dehydrogenases, and oxidases to water and oxygen.

Catalyst: A substance that promotes a chemical reaction by lowering the activation energy of a chemical reaction, but itself remains unaltered at the end of the reaction.

Cataract: An opacification in the normally transparent lens of the eye.

Cathepsins: A group of proteolytic enzymes occurring in lysosomes.

Cation: A positively charged ion, which has fewer electrons than protons. The cation is attracted to the cathode.

Caveolae: Invaginated ~50 nm vesicles on the cell membrane surface bearing anchored proteins.

Cell adhesion molecules (CAMs): Molecules that are expressed on the surface of cells and that mediate the attachment of cells to the ECM and to other cells.

Cell cycle: The sequence of changes in a cell from the moment when it is created by cell division, continuing through a period in which its contents, including chromosomal DNA are doubled, and ending with the subsequent cell division and formation of daughter cells.

Centromere: Region of a chromosome that holds the two chromatids together and that binds, via a kinetochore, with mitotic or meiotic spindle fibers.

Ceramide: One of the component lipids of sphingomyelin, a major component of the lipid bilayer of the cell membrane. It can be released from the cell membrane by enzymes and act as signaling molecule.

Checkpoint: Control mechanism that ensures that the next step in the cell cycle does not proceed until a series of preconditions have been fulfilled including the completion of all previous steps.

Chimeric gene (fusion gene): Artificial construct composed of DNA fragments of different genes. The chimeric gene encodes a chimeric protein.

Chk1 and Chk2: Serine/threonine kinases involved in the induction of cell-cycle checkpoints in response to DNA damage and replicative stress.

Chromatid: Each of the two progeny strands of a duplicated chromosome joined at the centromere during mitosis and meiosis.

Chromatin: Complex of DNA, RNA, and proteins that constitutes a chromosome.

Chromosomal aberration: Any change resulting in the duplication, deletion, or rearrangement of chromosomal material.

Chromosome: In prokaryotes, an intact DNA molecule containing the genome; in eukaryotes, a DNA molecule complexed with RNA and proteins to form a threadlike structure containing genetic information arranged in a linear sequence.

Clastogenic: Describes factors causing damage to chromosomes, such as breaks or changes in the amount of proteins.

Clonogenic assay: An experimental method that assesses the probability of survival of colony-forming (i.e., clonogenic) cells after some form of treatment, such as radiation or anticancer drugs.

***C-met* proto-oncogene:** Encodes the tyrosine kinase receptor for hepatocyte growth factor (HGF), a potent mitogen and motogen for epithelial cells.

Codon: A group of three DNA or mRNA bases that code for a given amino acid. Codons thus form the "words" of the genetic code.

Collagen: Major fibrous (structural) protein of connective tissue, and occurs as fibers produced by fibroblasts.

Comet assay: Single-cell gel electrophoresis.

Complementarity: Chemical affinity between nitrogenous bases as a result of hydrogen bonding, responsible for the base pairing between the strands of the DNA double helix.

Complementation: A technique that can assist in identifying a defective gene. Two types of cells, containing different genetic defects, are fused and complementation is said to occur if the hybrid cell lacks these genetic defects.

Compton effect: Scattering of x-rays resulting in ionization and loss of energy. The energy lost by the photon is given to the ejected electron as kinetic energy.

Constitutive: Describing a state of activity that occurs at a constant level and is therefore not responsive to modulation by physiologic regulators, or a type of control that yields such a constant output.

Covalent bond: A nonionic chemical bond formed by the sharing of electrons.

Cyclin: A protein that associates with a cyclin-dependent kinase (CDK) and serves as a regulatory subunit of this kinase by activating its catalytic activity and directing it to appropriate substrates.

Cyclin-dependent kinase: Proteins whose activity varies around the cell cycle. Bind to cyclin-dependent kinases (CDKs) and activation of cyclins in these complexes is associated with progression of cells from one cell cycle phase to the next.

Cyclotron: A particle accelerator which uses the magnetic force on a moving charge to bend moving charges into a semicircular path between accelerations by an applied electric field. The field is reversed at the cyclotron frequency to accelerate the electrons back across the gap.

Cytochromes: System of electron-transferring proteins, often regarded as enzymes, with iron–porphyrin or (in cytochrome c) copper–porphyrin as prosthetic groups.

Cytogenetics: Study that relates the appearance and behavior of chromosomes to genetic phenomena.

Cytokines: Polypeptides originally defined as being released from lymphocytes and involved in maintenance of the immune system. They have pleiotropic effects not only on hematopoietic cells but also on many other cell types.

Cytokinesis: Last step of M phase during which the cytoplasm divides and daughter cells separate.

Cytoplasm: All cell contents, including the plasma membrane, but excluding any nuclei. Comprises cytoplasmic matrix, or cytosol, in which organelles are suspended plus crystalline or otherwise insoluble granules of various kinds.

Cytoskeleton: Molecules that provide physical structure and define the form and shape of cells.

Cytotoxic: Poisonous, or lethal, to cells.

D_0: A parameter in the multitarget equation: the radiation dose that reduces survival to e^{-1} (i.e., 0.37) of its previous value on the exponential portion of the survival curve.

D_{10} dose: The dose resulting in one decade of cell kill.

D_{37} dose: Dose to reduce surviving fraction to 0.37. The dose which will give an average of one hit per target or one inactivating event per cell.

Deamination: Loss of an amine group from a larger molecule such as a DNA base.

Death receptor: Receptors on cells, that when bound by ligands, stimulate signaling pathways that initiate apoptosis and lead to their death, include Fas (CD95), tumor necrosis factor (TNF) receptor type 1 (TNFR-1), and TNF-related apoptosis-inducing ligand (TRAIL).

Death-inducing factor (DIE): A nontargeted effect by which medium from clonally expanded cells which have survived ionizing radiation but manifest chromosomal instability is cytotoxic to parental nonirradiated cells.

Deletion: Loss of DNA. Deletions can be small, affecting only a small part of a single gene, or large, for example, a chromosomal deletion involving many genes.

Delta ray (δ-ray): A term sometimes used to describe any recoil particle that causes secondary ionization.

Deuterium (^{2}H, D): An isotope of hydrogen whose nucleus contains one neutron and one proton and therefore is about twice as heavy as the nucleus of normal hydrogen.

Deuteron: The nucleus of deuterium. It contains one proton and one neutron.

Dicentric chromosome: An abnormal chromosome that contains two centromeres.

Dideoxynucleotide (didN): A deoxynucleotide that lacks a 3′ hydroxyl group and is thus unable to form a 3′–5′ phosphodiester bond necessary for chain elongation. Dideoxynucleotides are used in DNA sequencing.

Dimer: A complex formed by the joining of two molecules. In a homodimer, the constituent molecules are identical; in a heterodimer, they are different.

Diploid: A genome in which all chromosomes are present in pairs, one of each pair being inherited from a father and the other from a mother, with the exception of the sex chromosomes.

DNA deoxyribophosphodiesterase: An enzyme catalyzing the hydrolytic release of 2-deoxyribose-5-phosphate from single-strand interruptions in DNA with a base-free residue on the 5′ side.

DNA polymerase: An enzyme that catalyzes the synthesis of DNA from deoxy-ribonucleotides and a template DNA molecule.

DNA PK$_{cs}$: The DNA-dependent protein kinase catalytic subunit of DNA PK.

Domain: A regionally differentiated feature for instance a sequence of amino acids forming a functional group within a protein molecule. Nuclear domains have also been recognized.

Dominant negative: A mutant allele of a gene that, when co-expressed with the wild-type allele of the gene, is able to interfere with the functioning of the latter.

Double-strand break (DSB): A lesion causing interruption of both strands (sugar–phosphate "backbone") of a DNA molecule.

D_q: Dose at which the extrapolated straight-line section of the dose–response curve crosses the dose axis, quantitatively defines the size of the shoulder (quasi-threshold dose).

Effector: An agent (such as a protein) that carries out the actual work of a biological process rather than just regulating it.

Electromagnetic radiation: Radiation consisting of associated and interacting electrical and magnetic waves that travel at the speed of light, such as light, radio waves, γ-rays, and x-rays.

Electron (e): An elementary particle with a unit negative electrical charge and a mass $1/1837$ that of the proton.

Electronvolt (eV): The amount of energy gained by a particle of charge e (-1.6×10^{-19} C) if it is accelerated by a potential difference of 1 V ($1 \text{ eV} \approx 1.6 \times 10^{-19}$ J).

Electrophile: A molecule or chemical group that is positively charged and attracts electrons.

Electrophoresis: The technique of separating charged molecules in a matrix to which an electrical field is applied.

Electroporation: Process in which high-voltage pulses of electricity are used to open pores in the cell membrane through which foreign DNA can pass.

Endocytosis: Process by which patches of plasma membrane and associated proteins are internalized into the cell cytoplasm, resulting in their forming cytoplasmic vesicles. Compare exocytosis.

Endogenous: A property intrinsic to the cell, not introduced or expressed from an exogenous source.

Endonuclease: An enzyme that can cut an intact strand of DNA or RNA at some point in the strand other than at an end.

Endoplasmic reticulum (ER): Elaborate network of membranous structures in the cytoplasm on which glycoproteins are assembled.

Endothelial cells: (1) Mesenchymal cells that form walls of capillaries or lymph ducts by assuming a tubelike shape. (2) Mesenchymal cells lining the luminal walls of larger blood vessels or lymph ducts.

Endotoxin: Glycolipids attached to cell walls of certain gram-negative bacteria, causing pathogenicity.

Enzyme: A protein catalyst produced by a cell and responsible for the high rate and specificity of one or more intracellular or extracellular biochemical reaction.

Epidemiology: Study of factors affecting the health and illness of populations.

Epigenetic: Epigenetic changes alter the expression of genes without causing permanent base change. DNA methylation or histone acetylation is the form of epigenetic change in chromatin.

Epistasis groups: Loose classification of radiation-sensitive mutants of yeast into groups based on genetic analysis and on mutant phenotypes.

Epithelium: A layer of cells that forms the lining of a cavity or duct, including the specialized epithelium that forms the skin.

Epitope: A small part of a protein molecule that can be recognized by antibodies or antigen receptors on lymphocytes and which elicits an immune response against that molecule.

Erb-B (EGFR/HER1): Membrane receptor that binds epidermal growth factor.

Erythema: Abnormal redness of the skin caused by capillary congestion. Cardinal sign of inflammation, caused by infection, allergies, sunburn, and ionizing radiation.

Erythropoietin: Growth factor that stimulates the production of red blood cells, often in response to inadequate oxygen transport by the blood.

Eukaryote: An organism whose cells possess a nucleus and other membrane-bounded vesicles.

Excitation: Raising of an electron in an atom or molecule to a higher energy level without actual ejection of an electron.

Exocytosis: Process by which cells secrete products by storing them in cytoplasmic membrane vesicles that are caused to fuse with the plasma membrane, allowing the products carried in these vesicles to be released into the extracellular space.

Exon: DNA segment of a gene that is transcribed and translated into protein.

Exonuclease: An enzyme that breaks down nucleic acid molecules by breaking the phosphodiester bonds at the 3′ or 5′ terminal nucleotides.

Extracellular matrix (ECM): Mesh of secreted proteins, largely glycoproteins and proteoglycans that surround most cells within tissues and create structure in the intercellular space.

Extravasation: Leakage. In the case of inflammation refers to the movement of white blood cells to the surrounding tissue.

Fanconi anemia (FA): Autosomal recessive disease characterized by congenital abnormalities, defective hemopoiesis and a high risk of developing acute myeloid leukemia and certain solid tumors. Defective DNA damage repair is a factor. FA can be caused by mutations in at least seven different genes.

Fas and FasL (CD95 and CD95L): Fas protein (a member of the TNF receptor superfamily) is activated by binding of the Fas ligand (Fas-L) to initiate a chain of reactions leading to the activation of caspase-8 and apoptosis.

Fenton reaction: Iron–salt-dependent decomposition of hydrogen peroxide to generate the hydroxyl radical.

Fibroblast: A precursor cell of connective tissue that is relatively easy to maintain in cell culture.

Fibrosis: Development within a tissue, often following chronic inflammation, of dense fibrous stroma that replaces normally present epithelium, resulting in loss of function of that tissue.

Filter elution: Lysis of cells containing radio-labeled DNA on a filter, the pore size of which is many times larger than the DNA filament. The flow of elution buffer carries the DNA fragments through the membrane and the rate of elution of DNA is related to the size of the DNA molecules on the membrane.

Flavoproteins: A group of conjugated proteins in which one of the flavins FAD or FMN is bound as prosthetic group. Occur as dehydrogenases in electron transport systems.

Flow cytometry: Analysis of cell suspensions in which a dilute steam of cells is passed through a laser beam. DNA content and other properties are measured by light scattering and fluorescence following staining with dyes or labeled antibodies.

Fos/Jun pair (AP1): A heterodimer between the subunits Jun and Fos, converted to the active form by phosphorylation of the Jun subunit.

Frameshift mutation: A mutational event leading to the insertion of one or more base pairs in a gene, shifting the codon reading frame in all codons following the mutational site.

Free radical: An unstable chemical species that is highly reactive due to the presence of an unpaired electron.

Gamma rays (γ-rays): High-energy, short-wavelength electromagnetic radiation. Indistinguishable from x-rays except for their source: γ-rays originate inside the nucleus, x-rays from outside.

Gap junctions: Cylindrical channel proteins (connexins) with a channel diameter of 1.5 nm, coupling cells electrically and flipping between open and closed states, their permeability changing with calcium ion concentration.

Gene family: Group of genes all of which are descended evolutionarily from a common ancestral gene. The members of a gene family often encode distinct, structurally related proteins.

Genetic background: The entire array of alleles carried in a genome with the exception of a small number of genes that form the subject of study.

Genetic instability: Increased propensity for a cell to undergo genetic alterations.

Genome: The normal complement of DNA for a given organism. In humans, the genome is comprised of 46 chromosomes within the nucleus (the DNA within the mitochondria is usually considered separately).

Genotoxic: Referring to an agent that is capable of damaging the genome (i.e., is mutagenic).

Germ line: (1) The collection of genes transmitted from one organismic generation to the next. (2) The cells within a multicellular organism that are responsible for carrying and transmitting genes from organismic generation to its offspring.

Glioblastoma: Tumor of the nonneuronal glial cells of the brain.

Glutathione peroxidase: Enzyme whose biochemical role is to reduce lipid hydroperoxides to their corresponding alcohols and reduce free hydrogen peroxide to water. In general, peroxidases protect the organism from oxidative damage.

Glutathione: A small molecule composed of three amino acids (glycine, cysteine, and glutamate) that is prevalent within cells and can bind to reactive compounds and aid in their excretion.

Glycolysis: Anaerobic degradation of glucose (usually in the form of glucose–phosphate) in the cytosol to yield pyruvate, forming initial process by which glucose is fed into aerobic phase of respiration.

Glycoprotein: A protein that has been modified post-translationally through the addition of carbohydrate side chains.

Glycosylation: Bonding of sugar residue to another organic compound.

Golgi complex: A dynamic eukaryotic organelle, comprising a system of stacked and roughly parallel interconnected sacs (cisternae) sandwiched between two complex networks of tubules (*cis* and *trans* Golgi networks), the whole situated close to, but physically separate from, the ER.

Granulocyte: Leucocyte that contains cytoplasmic granules such as a basophil, an eosinophil, or a neutrophil.

Gray (Gy): Unit of radiation dose, defined by the energy absorbed per unit mass. 1 Gray is equivalent to 1 J/kg.

H2AX, γ-H2AX: Histone H2 is one of the five main histone proteins involved in the structure of chromatin on eukaryotic cells. Phosphorylation of histone H2AX to γ-H2AX occurs rapidly after DNA damage and is important for coordination of signaling and repair activities.

Haber–Weiss reaction: Consists of two reactions; in the first Fe(III) is reduced by superoxide followed by oxidation by hydrogen peroxide. The second reaction generates hydroxyl radical (see Fenton reaction).

Haploid: Describing a genome in which all chromosomes are present in a single copy.

Helicase: An enzyme that participates in DNA replication by unwinding the double helix near the replication fork.

Hematogenous: Depending upon or facilitated by circulating blood.

Hematopoiesis (adj., -poietic): The formation of all of the cells in the blood, including its red and white cells including various cells of the immune system.

Heme oxygenase (HO): An enzyme that catalyzes the degradation of heme.

Hepatitis C: A blood-borne, infectious, viral disease caused by a hepatotropic virus.

Hepatocyte growth factor/scatter factor (HGF/SF): A mesenchymally derived, heparin-binding glycoprotein with mitogenic, motogenic, and morphogenic effects on a variety of cells. It was originally identified as two different factors which were later ascribed to the same factor.

Herceptin: Chimeric anti-HER2/Neu monoclonal antibody bearing murine antigen-combining (variable) domains and a human constant domain (also called trastuzumab).

Hermaphrodite: An organism that possesses both male and female sex organs.

Heterodimer: Molecular complex composed of two distinct subunits.

Heterozygosity: The presence of different alleles of a gene on the two copies of the same chromosome.

Hexokinase: Enzyme phosphorylating free glucose within cell, producing glucose–phosphate, which cannot pass out across the plasma membrane.

High Z energy (HZE) particles: Although not the most abundant form of ionizing radiation in space, HZE particles pose the greatest risk to humans.

Histone acetyl transferases (HATS): Enzymes, activating transcription by modifying particular amino-terminal tails of specific core histones by acetylating lysine residues in the histone tail domains. This destabilizes the nucleosome, allowing DNA-binding components of the basal transcriptional machinery better access to promoter elements.

Histone deacetylase (HDAC): HDACs remove acetyl groups from core nucleosomal proteins (histones) and thereby inhibit the transcription of genes.

Histones: Basic proteins of major importance in packaging of eukaryotic DNA. DNA and histones together comprise chromatin, forming the bulk of the eukaryotic chromosome.

Homeostasis: The maintenance of a normal physiological state. Homeostasis is often maintained through feedback systems employing signals (e.g., hormones or growth factors) that have opposing effects.

Homologous recombination: The crossing over and rejoining of corresponding (homologous) regions of DNA on opposite chromosomes that occur normally during meiosis. A similar process can occur between a segment of DNA introduced into a cell and the homologous region on one of the chromosomes, or during repair of DNA damage.

Hybrid: The offspring of two parents differing in at least one genetic characteristic.

Hybridization: (1) The fusion of two somatic cells to form a single cell. (2) The binding of complementary sequences of DNA or RNA.

Hybridoma: The term is used to describe a population of hybrid cells that produces monoclonal antibodies. Such a cell is produced by fusing an antibody-producing normal cell and a non-antibody-secreting myeloma cell.

Hydrated electron: A detached electron which has been slowed by collisions to the point where it is captured by strongly polarized molecules of water. The hydrated electron is a powerful reducing agent.

Hydrocarbon: A molecule composed of hydrogen and carbon atoms.

Hydrogen abstraction: The creation of solute radicals by the reaction of the solute molecule with either $H^\bullet$ or $^\bullet OH$ to extract a hydrogen atom from a C–H bond.

Hydrogen bond: An electrostatic attraction between a hydrogen atom bonded to a strongly electronegative atom such as oxygen or nitrogen and another atom that is electronegative or contains an unshared electron pair.

Hydrophilic: A molecule or chemical group that has high solubility in water primarily because of its polarity.

Hydrophobic: A molecule or chemical group that has low solubility in water, usually because it is nonpolar. Such molecules usually have higher solubility in lipids.

Hyperthermia: The use of heat to treat cancer.

Hypotonic: In biology, a hypotonic solution has the lower osmotic pressure of two fluids.

Hypoxia: State of lower than normal oxygen tension.

Hypoxia inducible factor-1 (HIF-1): A transcription factor that stimulates the expression of certain genes in response to a hypoxic microenvironment.

Immunoblot: Laboratory procedure, such as western blot analysis, in which proteins that have been separated by electrophoresis are transferred onto nitrocellulose sheets and are identified by their reaction with labeled antibodies.

Immunodeficient: Describing an organism lacking a fully functional immune system.

Immunoglobulin: The class of serum proteins having the properties of antibodies.

Immunohistochemistry, immunofluorescence: A histologic process whereby a colored stain is linked to an antibody (usually a monoclonal antibody) that is used on tissue sections to recognize cells that express specific proteins that bind the antibody.

Inflammation: The first response of the immune system to infection or irritation. Helps fight disease but at the cost of suspending the body's normal immune and catabolic processes which in the long term may cause progressive damage.

In situ: Occurring in the site of origin.

Intercalating agent: A compound that inserts between bases in a DNA molecule, disrupting the alignment and pairing of bases in the complementary strands (e.g., acridine dyes).

Interleukin: A growth and differentiation factor that stimulates various cellular components of the immune system.

Interleukin-1 (IL-1): Cytokine involved in the regulation of immune and inflammatory responses.

Interleukin-6 (IL-6): Cytokine that regulates B cell differentiation, T cell activation, killer cell induction, and other physiologic responses. IL-6 is induced by cytokines, ultraviolet (UV) irradiation, and other stimuli.

Interleukin-8 (IL-8): One of an extended family of cytokines that act as chemoattractants for neutrophils, T cells, and basophils.

Intermediate filaments: Insoluble fibrous proteins, providing mechanical stability in (apparently, only) animal cells.

Interphase: That portion of the cell cycle between divisions.

Interstitial: Referring to the space within a tissue that lies between cells.

Intima: The innermost layer of an artery made up of one layer of endothelial cells supported by an internal elastic lamina.

Intron: Portion of a primary RNA transcript that is deleted during the process of splicing.

In vitro: (1) Occurring in tissue culture, or in cell lysates or in purified reaction systems in the test tube. (2) Propagation of living cells in a culture vessel.

In vivo: (1) Occurring in a living organism. (2) Occurring in a living, intact cell.

Ionizing radiation: Radiation (e.g., x- or γ-rays) that is sufficiently energetic for its interactions with matter (tissue) to cause the formation of ions.

Ionophore: One of a range of small organic molecules facilitating ion movement across a cell membrane (usually the plasma membrane).

IR4, IR5, IR7: Mammalian cell mutants defective in DNA nonhomologous end joining (NHEJ).

Ischemia (adj., ischemic): Reduced blood supply.

Isogeneic (syngeneic): Having the same genotype.

Isomer: One structural form of a chemical compound that can occur naturally in a number of different structural forms.

JAK–STAT: Signaling pathway employing janus kinases (JAKs) and signal transducers and activators of transcription (STATs). The JAK–STAT pathway is involved in the regulation of cellular responses to cytokines and growth factors.

Karyotype: Arrangement of chromosomes from a particular cell according to a well-established system such that the largest chromosomes are first and the smallest ones are last.

Keratinocyte: Epithelial cell type found in tissues such as the skin.

Kilodalton: The dalton is a unit of mass equal to that of a hydrogen atom, 1.67×10^{-24}g. The mass of large molecules like proteins are typically specified in kilodaltons, or kDa (or kD) with 1kDa being equal to 1000.

Kinase: Enzyme that covalently attaches phosphate groups to substrate molecules, often proteins.

Kinetochore: Nucleoprotein complex that is associated with the centromeric DNA of a chromosome and is responsible during mitosis (or meiosis) for forming a physical connection between the chromosome and the microtubules of the spindle fibers.

Knockout (e.g., knockout mouse): Replacement of a normal gene by a specifically mutated copy. It is a valuable tool for study of phenotypic effects of mutations.

Ku70: A subunit of DNA protein kinase complex involved in repair of double-stranded DNA breaks.

Laser capture microdissection (LCM): A method for isolating pure cells of interest from specific microscopic regions of tissue sections.

Late effects (late tissue responses): Toxicity to normal tissues that becomes apparent at a time long after (months to years) the application of radiation therapy.

Leucine zipper: A secondary protein structure in which projecting leucine residues on two polypeptide changes interdigitate to form a stable dimmer.

Leukemia: Malignancy of any of a variety of hematopoietic cell types, including the lineages leading to lymphocytes and granulocytes, in which the tumor cells are nonpigmented and dispersed throughout the circulation.

Library: A collection of cells (usually bacteria or yeast) that have been transformed with recombinant vectors carrying DNA inserts from a single species.

Ligand: Molecule that binds specifically to a receptor and activates its signaling powers.

Ligase: An enzyme that covalently joins the ends of two molecules together; in the context of DNA, ligases join the 3′ end of one ssDNA to the 5′ end of the other via a phosphodiester linkage.

Linear energy transfer (LET): A measure of the density of energy deposition along the track of a given type of ionizing radiation in matter. The deposition of energy in matter by ionizing radiation occurs randomly along the particle track and in different amounts, hence LET is a quantity usually averaged over segments of the track length (track-averaged LET).

Linear quadratic equation: An equation describing biological effect as a function of dose of a cytotoxic agent, such as radiation, which contains both linear and quadratic (squared) terms of dose with constants α and β, respectively. It provides a useful model for describing the shape of radiation survival curves and for comparing isoeffective radiation treatments.

Lipid peroxidation: Oxidative degradation of lipids.

Lipid raft: A cholesterol-enriched microdomain in cell membranes.

Lipophilic: A substance that is lipid soluble. Lipophilic substances penetrate readily into cells, since they are soluble in the cell membrane.

Liposome: A small vesicle containing fluid surrounded by a lipid membrane. Liposomes may be constructed to have varying lipid content in their membranes and to contain various types of drugs or other molecules. They may be used to introduce genes into cells.

Locus: Chromosome site that can be studied genetically and is presumed to be associated with a specific gene.

Lyase: In biochemistry, a lyase is an enzyme that catalyzes the breaking of various chemical bonds by means other than hydrolysis and oxidation, often forming a new double bond or a new ring structure. Lyases differ from other enzymes in that they only require one substrate for the reaction in one direction, but two substrates for the reverse reaction.

Lymphoid cell: Referring to the lineage of hematopoietic cells that yield B and T lymphocytes as well as natural killer cells.

Lymphoma: Solid tumor of lymphoid cells.

Lysis: The destruction of the cell membrane.

Lysosome: A cytoplasmic lipid vesicle that contains degradative enzymes in a solution of low pH, allowing it to degrade various molecules that are introduced into it.

Macrophage: A type of white blood cell assisting in the body's fight against bacteria and infection by engulfing and destroying invading organisms.

Magnetic resonance imaging (MRI): An imaging technique that depends on the magnetization of tissue when a patient is placed in a strong externally applied magnetic field. The magnetic field results in the alignment of proton spins, an alignment that can be perturbed by an electromagnetic pulse. The different rates or realignment of such proton spins (relaxation) in different tissues allow an image to be generated.

Mass spectroscopy: A process used for very precise measurement of the mass of components of proteins and other biological molecules.

Meiosis (reduction division): Process whereby a nucleus divides by two divisions into four nuclei, each containing half the original number of chromosomes, in most cases forming a genetically nonuniform haploid set.

Messenger RNA (mRNA): An RNA molecule transcribed from DNA and translated into the amino acid sequence of a polypeptide.

Meta-analysis: A statistical technique for combining the results of clinical trials evaluating similar strategies. The technique facilitates the detection of small but clinically important differences between experimental and standard therapies.

Metalloproteinases: Proteolytic enzymes secreted by cells that may cause degradation of components of the extracellular matrix.

Metaphase: Second subphase of mitosis, during which chromosomes complete condensation and attach to the mitotic spindle as the nuclear membrane disappears; chromosomes are now readily seen in the light microscope.

Methylation: The addition of one or more methylation groups (CH_3) to a molecule.

Microarray: A slide coated with thousands of oligonucleotides or cDNAs that is hybridized with labeled cRNA from cells to monitor expression changes.

Microbeam: Charged-particle microbeam by which cells on a dish are individually irradiated by a predefined exact number of α particles, allowing the effects of exactly one (or more) α particle traversals to be investigated.

Microcephaly: Neurological disorder in which the head circumference is significantly smaller than average for the persons age and sex. It is associated with impaired intellectual development and may be induced by irradiation in utero.

Microdosimetry: A technique for measuring the microscopic distribution of energy deposition.

Microenvironment: The environment surrounding cells in solid tissue. The microenvironment may change quite markedly over short distances with respect to metabolic factors such as level of oxygen and pH as well as in the consistency of the extracellular matrix.

Microhomology: Short stretches (1–4 nucleotides) of complementary sequences. A minor pathway for DSB repair in mammalian cells, independent of normal NHEJ proteins has a requirement for microhomology at DNA ends.

Microinjection: Introducing DNA into a cell using a fine microcapillary pipette.

Micronucleus: A small fragment of nucleus that has its own nuclear membrane and results from certain aberrations in cell division or from damage inflicted on a cell.

Microsatellite: Short repetitive DNA sequences scattered throughout the genome.

Microsatellite instability: Instability in these sequences is associated with many sporadic and familial cancers and related to mutations in mismatch repair genes.

Microtubule: Intracellular structures containing the protein tubulin that form the mitotic spindle and provide structure and function to nerves. Some anticancer drugs, including taxanes and vinka alkaloids, bind to tubulin and inhibit the function of microtubules.

Minisatellite: Tandem repeats of a simple DNA sequence, which form minor "peaks" (on account of its peculiar base composition and hence density) after DNA fragmentation and density gradient centrifugation.

Mismatch repair: The class of DNA repair processes that depend on scanning a recently synthesized segment of DNA and removing any misincorporated bases.

Mitochondria: Cellular organelles that are responsible for the production of energy or ATP, via a complex chain of electron transfers within the mitochondrial membranes.

Mitogen: An agent that provokes cell proliferation.

Mitotic catastrophe: A type of cell death that occurs during mitosis. Response of mammalian cells to mitotic DNA damage which produces tetraploid cells with a range of different nuclear morphologies from binucleated to multi-micronucleated.

Molecular radiosensitizers: Drugs targeting signal transduction intermediates and pathways that are impacted by ionizing radiation. In addition to the vast number of potential targets, the appeal of this strategy lies in their target specificity and clinically acceptable toxicity.

Monoclonal antibody: An antibody of a single defined specificity, most commonly obtained from a single clone of antibody-producing cells or hybridoma. A monoclonal antibody binds to a specific epitope of the foreign protein it recognizes.

Monocyte: Largest of the kinds of vertebrate leucocyte. Can differentiate into macrophages and are the source of numerous growth factors, notably PDGF and IL-1.

Monolayer: A population of cells growing as a layer one cell thick.

Monosomy: Describing an aneuploid condition in which one member of a chromosome pair is missing; having a chromosome number of $2n-1$.

MRN complex: Consists of three proteins—Mre1, Rad50, and Nbs (nibrin). It plays a critical role in the response to DNA damage and telomere maintenance in mammalian systems. The primary function of the MRN complex is to sense DNA strand breaks and then to amplify the initial signal and convey it to downstream effectors.

MTT assay: A standard colorimetric assay for measuring numbers of viable cells. It is used to determine cytotoxicity of potential medicinal agents and other toxic materials.

Multiply damaged sites (MDS): Clusters of strand breaks and base damages within one or two turns of the DNA (10–20 base pairs). DSBs are the prototypical example of a MDS.

Mutagen: An agent that causes an increase in the rate of mutation.

Myc: A nuclear oncogene involved in immortalizing cells.

Myeloid: (1) Referring to the lineage of hemopoietic cell that yields granulocytes macrophages and mast cells. (2) Pertaining to or resembling bone marrow; often used as a synonym for myelogenous.

Myeloma: A malignancy of the antibody-producing cells of the bone marrow, often called multiple myeloma.

Myeloperoxidase (MPO): A peroxidase enzyme most abundant in neutrophil granulocytes. MPO produces hypochlorous acid from hydrogen peroxide and chloride anion during the respiratory burst of neutrophil. Toxic products produced by MPO are used to kill bacteria and other pathogens.

Myosin: A protein found in the majority of eukaryotic cells. At least two classes of myosin exist, a single-headed tailless variety (*myosin I*) involved in cell locomotion, and a two-headed tailed variety (*myosin II*) involved in muscle contraction.

Nano (n): A prefix that divides a basic unit by 1 billion (10^{-9}).

Nanometer (nm): A unit of length equal to 1×10^{-9} m.

Necrosis: Process of cell death involving the breakdown of a cell and its constituents through steps that are distinct from those in the apoptotic death program.

Negative π-mesons: Negatively charged particles with a mass 273 times that of the electron. They can be produced by a synchrocyclotron or linear accelerator capable of accelerating protons to energies of 400–800 MeV.

Neoplasia (adj., -plastic): (1) The state of cancerous growth. (2) Benign or malignant tumor composed of cells having an abnormal appearance and abnormal proliferation pattern.

Neurogenesis: Process by which neurons are created.

Neutron (n): An uncharged elementary particle that has a mass slightly greater than that of the proton and that is found in the nucleus of every atom heavier than hydrogen.

Nibrin: The *NBS1* gene product. Mutations in the *NBS1* gene result in Nijmegen breakage syndrome (NBS) a recessive disorder with some phenotypic similarities to AT.

Nondisjunction: An accident of cell division in which the homologous chromosomes (in meiosis) or the sister chromatids (in mitosis) fail to separate and migrate to opposite poles; responsible for defects such as monosomy and trisomy.

Nonhomologous end joining (NHEJ): Type of DNA repair consisting of fusion of two dsDNA ends in which the joining of the two ends is not informed or directed by sequences in a sister chromatid or homologous chromosome.

Northern blot: Adaptation of the Southern blotting procedure in which RNA (rather than DNA) is resolved by gel electrophoresis and transferred to a filter that is subsequently incubated with a sequence-specific, radiolabeled DNA probe.

Nuclear lamins: Intermediate filaments that contain a nuclear transport signal targeting them to the nucleus.

Nuclear magnetic resonance imaging (NMRI): See magnetic resonance imaging.

Nuclear matrix: In biology, the nuclear matrix is the network of fibers found throughout the inside of a cell nucleus.

Nuclease: An enzyme that breaks bonds in nucleic acid molecules.

Nucleoids: Mammalian cell nuclei depleted of histones and other nuclear proteins by extraction in nonionic detergent and high salt concentration.

Nucleolus: A sub-organelle of the cell nucleus. The main function of nucleolus is the production and assembly of ribosome components.

Nucleoside: A purine or pyrimidine base covalently linked to a ribose or deoxyribose sugar molecule.

Nucleosome: Protein octamer, composed of two each of histones H2A, H2B, H3, and H4, around which DNA is wrapped in chromatin.

Nucleotide excision repair (NER): A DNA repair process that involves excision of a group of nucleotides containing damaged or altered base(s) in one strand of a DNA molecule and its replacement by synthesis of new DNA using the opposite strand as a template.

Nucleotide: A building block of DNA and RNA, consisting of a nitrogenous base, a five-carbon sugar, and a phosphate group.

Nude mouse: A mouse that congenitally lacks a thymus and hence has no mature T cells. Xenografts of human tumors will often grow in such immune-deficient animals. These mice are also hairless, hence the term "nude."

Oligomer: A polymer of more than two (but not a large number of) subunits.

Oligonucleotide: A short piece of DNA or RNA usually containing a defined sequence of bases.

Oncogene: A gene whose protein product may be involved in processes leading to transformation of a normal cell to a malignant state. The gene may be known as a viral oncogene if it was detected in a transforming virus.

Oncogenesis: The progression of cytological, genetic, and cellular changes that culminate in a malignant tumor.

Oogenesis: Production of ova, involving usually both meiosis and maturation.

Ortholog: A gene in one species that is the closest relative of a gene in another species; usually orthologs represent direct counterparts of one another in the genomes of two species.

Osmosis: The net diffusion of water across a selectively permeable membrane (permeable in both directions to water, but varyingly permeable to solutes) from one solution into another of lower water potential.

Osteosarcoma: Most common type of malignant bone cancer.

Oxidation number: The charge that an element in a molecule or complex would have if all the ligands were removed along with the electron pairs that were shared with the central atom. In the nomenclature of inorganic compounds it is represented by a Roman numeral, the plus sign is omitted for positive oxidation numbers, e.g., Fe(III).

Oxidation/reduction (redox) reactions: Describes all chemical reactions in which atoms have their oxidation number (oxidation state) changed. In simple terms, oxidation describes the loss of electrons by a molecule, atom, or ion while reduction describes the gain of electrons. Oxidation and reduction more properly refers to a change in oxidation number—the actual transfer of electrons may never occur. In practice, the transfer of electrons will always cause a change in oxidation number, but there are many reactions which are classed as "redox" even though no electron transfer occurs (such as those involving covalent bonds).

Oxidative phosphorylation: Process by which energy released during electron transfer in aerobic respiration is coupled to production of ATP.

Oxidative stress: Formation of reactive oxygen species (ROS) in and outside cells, such as those resulting from the lysis of water molecules induced by ionizing radiation. This stress can not only activate several enzyme systems, but can also modify the transcription of genes. These reactions are known collectively as oxidative stress.

Oxidoreductases: Major group of enzymes; catalyze redox reactions.

p15: G_1 inhibitor induced in epithelial cells by TGF-β. Inhibits cyclin D_1–Cdk4 and cyclin D_1–Cdk6 complexes.

p16: G_1 inhibitor of epithelial cells. Inhibits cyclin D_1–Cdk4 and cyclin D_1–Cdk6 complexes. Gene is deleted in familial melanomas and other tumor types.

p21^{Waf1}: Inhibitor of Cdc2, Cdk4, and Cdk6. Induced through p53 pathway.

p27: Cell cycle inhibitor induced in epithelial cells by TGF-β. It inhibits cyclin E–Cdk2 complex.

p53: Considered the guardian of the genome. The p53 protein is involved in control of the progression of cells through the cell cycle, particularly, the transition from G_1 phase to S phase preventing cells with DNA damage from progressing into S phase. p53 is involved in the transactivation of many other genes, including p21.

p53 **gene:** A tumor suppressor gene, so named because of the molecular weight of the corresponding protein ($\sim$53 kDa).

Pachytene: The third stage of the prophase of meiosis during which the homologous chromosomes become short and thick and divide into four distinct chromatids.

Pair production: A process of energy absorption. Pair production requires higher-energy photons (>1.02 MeV, usually >5 MeV). γ-Ray photons with energy greater than 1.02 MeV may interact with a nucleus to form an electron–positron pair.

Palindrome: A word, number, verse, or sentence that reads the same backward or forward. In nucleic acids, a sequence in which the base pairs read the same on complementary strands ($5' \rightarrow 3'$).

Paracentric inversion: A chromosomal inversion that does not include the centromere.

Paracrine: Referring to the signaling path of a hormone or factor that is released by one cell and acts on a nearby cell.

Particle: A minute constituent of matter, generally one with a measurable mass. The primary particles involved in radioactivity are α-particles, β-particles, neutrons, and protons.

Pentose phosphate pathway (phosphogluconate pathway, hexose monophosphate shunt): An alternative route to glycolysis for glucose catabolism, involving initially conversion of glucose–phosphate to phosphogluconate. Some intermediates are the same as those of glycolysis. It generates extramitochondrial reducing power in the form of NADPH. Ribose–phosphate is one intermediary in the pathway, and may be used for nucleic acid synthesis.

Peptides: A compound of two or more amino acids (strictly, amino acid residues). Besides being intermediates in protein digestion and synthesis, many peptides are biologically active. Some are hormones (e.g., oxytocin, ADH, melanocyte-stimulating hormone).

Pericentric inversion: A chromosomal inversion that involves both arms of the chromosome and thus involves the centromere.

Pericytes: Cells closely related to smooth muscle cells that surround capillaries and provide the capillary walls formed by endothelial cells with tensile strength and contractility.

Perinuclear cisterna: The space separating the inner from the outer nuclear membrane.

Peripheral neuropathy: Medical term for damage to the nerves of the peripheral nervous system which may be caused by diseases of the nerve, side effects of systemic illness, or side effects of treatments for disease. Symptoms may include pain, numbness, tingling, muscle weakness, burning, and loss of feeling.

Peroxiredoxins (Prxs): A ubiquitous family of antioxidant enzymes that also control cytokine-induced peroxide levels which mediate signal transduction in mammalian cells. Prxs can be regulated by changes to phosphorylation, redox, and possibly oligomerization states.

Peroxisome: A cytoplasmic organelle that is involved in the oxidation of various substrates, notably lipids.

Phagocytosis: The process by which a cell, usually a component of the immune system, engulfs a particle (which may be another cell), internalizes this particle, and usually proceeds to degrade it.

Phenotype: (1) A measurable or observable trait of an organism. (2) The sum of all such traits of an organism.

Phosphodiester bond: In nucleic acids, the covalent bond between a phosphate group and adjacent nucleotides, extending from the 5' carbon of one pentose (ribose or deoxyribose) to the 3' carbon of the pentose in the neighboring nucleotide. Phosphodiester bonds form the backbone of nucleic acid molecules.

Phosphoglycolate: A sugar fragment frequently found at the 3' terminus of DSBs and SSBs formed in DNA by ionizing radiation and other oxidative mutagens.

Phosphoinositide 3-kinases (PI3-kinases or PI3Ks): A family of related enzymes that are capable of phosphorylating the 3 position hydroxyl group of the inositol ring of phosphatidylinositol.

Phosphoinositides: A family of molecules involved in intracellular signaling.

Phospholipids (phospholipins, phospholipoids, phosphatides): Those lipids bearing a polar phosphate end, commonly esterified to a positively charged alcohol group. Major components of cell membranes are responsible for many of their properties.

Phosphoprotein: A protein to which one or more phosphate groups have been covalently attached.

Phosphorylation: Covalent attachment of a phosphate group to a substrate, often a protein.

Photoelectric effect: Absorption of an x-ray by ionization.

Photon: The carrier of a quantum of electromagnetic energy. Photons have an effective momentum but no mass or electrical charge.

Photoproducts: The product of a photochemical reaction including UV-induced pyrimidine(6-4)pyrimidone photoproducts and cyclobutane dimmers in DNA of mammalian cells.

Pixel: A single point in a graphic image (short for picture element, using the common abbreviation "pix" for picture).

Plasma: Clear yellowish fluid of vertebrates, clotting as easily as whole blood and obtained from it by separating out suspended cells by centrifugation. It is an aqueous mixture of substances, including plasma proteins.

Plasma membrane: Lipid bilayer membrane that surrounds a eukaryotic cell and separates the aqueous environment of the cytoplasm from that in the extracellular space.

Plasmid: A circular piece of DNA that may reproduce separately from chromosomal DNA within cells, bacteria, or other organisms.

Plasminogen: The inactive proenzyme that is converted into the active plasmin protease through proteolytic cleavage.

Plating efficiency: The proportion or percentage of in vitro-plated cells that form colonies.

Pleiotropy: Ability of certain genes or proteins to concomitantly evoke a series of distinct downstream responses within a cell or organism.

Ploidy: A description of the chromosome content of the cell. Normal mammalian cells contain two copies of each chromosome (except for the sex chromosomes in males) and are diploid.

Pneumocytes: Two types of cells which line the alveoli of the lung. The type I pneumocyte is a very large, thin cell stretched over a very large area, responsible for gas exchange occurring in the alveoli. This cell does not replicate. The type II granular pneumocytes cover about 5% of the surface area of the alveoli but greatly outnumber type I cells. Type II cells are responsible for the production and secretion of surfactant; they can also replicate and will replace damaged type I pneumocytes.

Poikilothermy: Condition of any animal whose body temperature fluctuates considerably with that of its environment.

Point mutation: Substitution of a single base for another in a DNA sequence.

Polyacrylamide gel electrophoresis: Electrophoresis through a matrix composed of a synthetic polymer, used to separate small DNA or RNA molecules (up to 1000 nucleotides) or proteins.

Polyamines: Organic compounds having two or more primary amino groups—such as putrescine, cadaverine, spermidine, and spermine—that are growth factors in both eukaryotic and prokaryotic cells and are also found associated with chromatin.

Polymerase: An enzyme that catalyzes the addition of multiple subunits to a substrate molecule.

Polymerase chain reaction: A procedure that enzymatically amplifies a DNA sequence through repeated replication by DNA polymerase.

Polymorphism: An altered DNA base sequence in a gene, either between the two alleles in one individual or between different individuals that occurs naturally in the population with a frequency greater than 1%, and usually leads to little or no changes in the function of the coded protein.

Polynucleotide: Long-chain molecule formed from a large number of nucleotides (e. g., nucleic acid).

Polypeptide: A molecule made up of amino acids joined by covalent peptide bonds. This term is used to denote the amino acid chain before it assumes its functional three-dimensional configuration.

Positron (β^+): An elementary particle with the mass of an electron but positively charged. It is emitted in some radioactive disintegrations and is formed by the interaction of high-energy γ-rays with matter.

Posttranscriptional (modification): Changes occurring to some tRNA and rRNA transcripts before translation. This may involve removal of nucleotides by hydrolysis of phosphodiester bonds or chemical modification of specific nucleotides.

Potentially lethal damage: Damage to a cell that may be caused by radiation or drugs and that may or may not be repaired depending on the environment of the cell following treatment.

Premature chromosome condensation (PCC): Process occurring when an interphase cell is fused with a mitotic cell and the interphase chromatin undergoes a process of condensation by which its chromosomes become visible.

Prenylation (isoprenylation or lipidation): Addition of hydrophobic molecules to a protein. Protein prenylation involves the transfer of either a farnesyl or a geranyl–geranyl moiety to C-terminal cysteine(s) of the target protein. There are three enzymes that carry out prenylation in the cell.

Primase: An enzyme that initiates DNA synthesis by laying down a short RNA segment on the template strand; the 3'-hydroxyl end of this RNA primer then serves as the site for attachment of the initial deoxyribonucleotide by a DNA polymerase.

Primer: A DNA or RNA molecule whose 3' end serves as the initiation point of DNA syntheses by a DNA polymerase.

Priming dose: A low dose of radiation or other stressor which, given before a higher challenge dose of radiation causes the organism to better resist the challenge dose.

Probe: A single-stranded DNA (or RNA) that has been radioactively labeled and is used to identify complementary sequences.

Pro-drug: An inactive precursor of a biologically active drug.

Prokaryotic: Referring to the relatively small, nonnucleated cells of bacteria and related organisms.

Proliferating cell nuclear antigen (PCNA): Commonly used marker for proliferating cells, a 35 kD protein that associates as a trimer, interacts with DNA polymerases δ and ε and acts as an auxiliary factor for DNA repair and replication.

Promoter: A region of DNA extending 150–300 bp upstream from the transition start site that contains binding sites for RNA polymerase and a number of proteins that regulate the rate of transcription of the adjacent gene.

Prophase: First stage of mitosis and meiosis.

Protease: An enzyme that cleaves protein substrates.

Protein kinase C (PKC): A family of protein kinases involved in mitogenic signaling. Activated by second messengers, including diacylglycerol and Ca^{2+} (some isoforms). PKC be activated directly by the phorbol ester class of tumor promoters. Can induce early-response genes through *raf*.

Protein kinases: Enzymes catalyzing the transfer of a phosphate group to a protein (often also an enzyme), thereby regulating its activity.

Proteolysis (adj., -lytic): Process, usually mediated by proteases, of cleaving a polypeptide to lower molecular weight fragments, including individual amino acids.

Proteomics: Study of the structure and function of proteins.

Proton: An elementary particle that is a component of all nuclei and that has a single positive electrical charge and a mass approximately 1837 times that of the electron.

Proto-oncogene: A gene generally active in the embryo and fetus and during proliferation processes. A mutation, amplification, or rearrangement can result in the permanent activation of a proto-oncogene, which then becomes an oncogene.

PTEN **gene:** A tumor suppressor gene that is frequently mutated in human cancer.

Pulsed field gel electrophoresis (PFGE): Process whereby current is alternated between pairs of electrodes set at angles to one another to separate very large DNA molecules of up to 10 million nucleotides.

Purines: They are organic bases with carbon and nitrogen atoms in two interlocking rings, and are components of nucleic acids and other biologically active substances.

Pyrimidines: Nitrogenous bases composed of a single ring of carbon and nitrogen atoms; components of nucleic acids.

Rad: Unit of radiation corresponding to 0.01 J of absorbed radiation or 0.01 Gy.

RAD52 **group genes:** *RAD50, RAD51, RAD52, RAD54, RDH54/TID1, RAD55, RAD57, RAD59, MRE11,* and *XRS2* are genes central to the process of homologous recombination. Most of them are also required for the repair of ionizing radiation-induced DNA damage in *S. cerevisiae.*

Rad9: Evolutionarily conserved gene with roles in multiple, fundamental biological processes primarily for regulating genomic integrity. The encoded mammalian proteins participate in promoting resistance to DNA damage, cell-cycle checkpoint control, DNA repair, and apoptosis.

Radical scavengers: Compounds that react with radicals before they can reach their chemical or intracellular target. Endogenous scavengers include thiol compounds such as glutathione and cysteine which contain sulfhydryl (SH) groups that can react chemically with the free radicals as well as other antioxidants, including vitamins C and E and intracellular manganese superoxide dismutase (MnSOD).

Radioisotope: A radioactive isotope; an unstable isotope of an element that decays or disintegrates spontaneously emitting radiation.

Radiomimetic chemical: Compound having effects on living tissue similar to those produced by radiation

Radionuclide: An atom with an unstable nucleus which undergoes radioactive decay, and emits a γ-ray or subatomic particles.

Radiosensitizer: In general, any agent that increases the sensitivity of cells to radiation.

Raf: A protein kinase that is activated by GTP-bound *ras*. Acts to transduce mitogenic signaling by phosphorylation of MAP kinases.

RAG1 and RAG2 proteins: Recombination activating gene-1 and -2 (*RAG1* and *RAG2*) are lymphocyte specific enzymes which carry out the initial steps of VDJ recombination.

Ras: A family of 21 kDa proteins (H-, K-, and N-*ras*) found to be activated by point mutations at codons 12, 13, and 61 in a variety of tumors. Involved in mitogenic signaling, coupling growth signals from growth factor receptors to *raf* activation, and downstream stimulation of early-response genes.

Reactive oxygen species (ROS): Usually very small molecules which are highly reactive due to the presence of unpaired valence shell electrons. It includes oxygen ions, free radicals, and peroxides both inorganic and organic. ROS form as a natural byproduct of the normal oxygen metabolism and have important roles in cell signaling.

Receptor: A molecule inside or on the surface of cells that recognizes a specific hormone, growth factor, or other biologically active molecule. The receptor also mediates transfer of signals within the cell.

Recombinant DNA: The process of cutting and recombining DNA fragments as a means to isolate genes or to alter their structure and function.

Redox reactions: Oxidation–reduction reactions; in biology, generally catalyzed by enzymes. Involve transfer of electrons from an electron donor (reducing agent) to an electron acceptor (oxidizing agent). Sometimes hydrogen atoms are transferred, equivalent to electrons, so that dehydrogenation is equivalent to oxidation.

Relative biologic effectiveness (RBE): A factor used to compare the biologic effectiveness of different types of ionizing radiation. It is the inverse ratio of the amount of absorbed radiation required to produce a given effect to a standard (or reference) radiation required to produce the same effect.

Reoxygenation: A process by which cells in a tumor that are at low oxygen levels because of poor blood supply, and hence are resistant to radiation, gain access to oxygen following a treatment so that they become more sensitive to a subsequent radiation treatment.

Repair of potentially lethal damage (PLDR): Potentially lethal damage is that damage which can be modified (repaired) by manipulation of postirradiation conditions. PLDR can occur if cells are prevented from dividing for 6 or more hours.

Repair saturation: Explanation of the shoulder on cell survival curves on the basis of the reduced effectiveness of repair after high radiation doses.

Replication: The process of copying a double-stranded DNA molecule.

Replication fork: The structure which forms when DNA is replicating itself. It is created through the action of helicase, which breaks the hydrogen bonds holding the two DNA strands together. The resulting structure has two branching "prongs," each one made up of a single-strand DNA.

Replicon: Nucleic acid molecule containing a nucleotide sequence forming a replication origin, at which replication is initiated. It is usually one per bacterial or viral genome but often several per eukaryotic chromosome.

Resolvases: Holliday junction resolvases are endonucleases that bind and cleave four-way junction DNA molecules.

Restriction endonuclease: Nuclease that recognizes specific nucleotide sequences in a DNA molecule and cleaves or nicks the DNA at that site. Derived from a variety of microorganisms, those enzymes that cleave both strands of the DNA are used in the construction of recombinant DNA molecules.

Retinoblastoma: Tumor of the oligopotential stem cells of the retina.

Retrovirus: A class of viruses that uses a reverse transcriptase enzyme to copy its genomic RNA into DNA.

Rho family: A family of small GTP-ases, encoded by *rho* genes. Rho GTPase is a key regulator of the actin cytoskeleton.

Ribonucleoprotein (RNP): A compound that combines ribonucleic acid (RNA) and protein together. It is one of the main components of nucleolus.

Ribonucleotide: A nucleotide in which a purine or pyrimidine base is linked to a ribose molecule. The base may be adenine (A), guanine (G), cytosine (C), or uracil (U).

Ribosomes: Nonmembranous, but often membrane-bound, organelles of both pro-karyotic and eukaryotic cells, of chloroplasts and mitochondria. Sites of protein synthesis, each is a complex composed of roughly equal ratios of ribosomal RNA (rRNA) and 40 or more different types of protein.

Ribosylation: For example protein ADP-ribosylation: the covalent, posttranslational addition of an ADP-ribose moiety onto a protein.

RNA polymerase: An enzyme that catalyzes the formation of an RNA polynucleo-tide strand using the base sequence of a DNA molecule as a template.

SCID mouse: A mouse that has severe combined immunodeficiency by virtue of having no functioning T or B lymphocytes because of a mutation that prevents effective rearrangement of the immunoglobulin and T-cell receptor genes. These mice allow the growth of human tumor xenografts because of their immunodeficiency.

Semipermeable: In general membranes which are impermeable to organic solutes with large molecules, such as polysaccharides, but permeable to water and small, uncharged solutes.

Senescence: A nongrowing state of cells in which they exhibit distinctive cell phenotypes and remain viable for extended periods of time but are unable to proliferate again. Often arises after extended passaging in vitro.

SH2 and SH3 domains: The Src-homology domains are homologous to regions in src family protein kinases. They are present in many proteins involved in signal transduction and give the proteins the ability to bind to other proteins in a manner facilitating signal transduction.

Signal transduction: A process by which information is transmitted from the sur-face of the cell to the nucleus. It involves a series of molecular components (proteins), that activate subsequent members of the cascade (usually by phosphorylation), resulting in activation of transcription factors.

Single-strand annealing (SSA): Like homologous recombination SSA is a process for rejoining DSBs using homology between the ends of the joined sequences. SSA relies on regions of homology with which to align the strands of DNA to be rejoined. DNA strands are first resected to generate SBB tails and when this process has proceeded far enough to reveal complementary sequences the two DNAs are annealed and then ligated. The genes which define SSA belong to the RAD52 epistasis group of HR.

Singlet oxygen: The common name used for the two metastable states of molecular oxygen (O_2) with slightly higher energy than the ground state triplet oxygen.

siRNA: A short inhibitory double-stranded RNA molecule that causes target cellu-lar RNA to be degraded.

Sister chromatid exchange (SCE): A crossing over event that can occur in meiotic and mitotic cells. It involves the reciprocal exchange of chromosomal material between sister chromatids (joined by a common centromere). Such exchanges can be detected cytologically after BrdUrd incorporation into the replicating chromosomes.

Site-directed mutagenesis: A synthetic oligonucleotide is used to achieve gene conversion in a plasmid before introduction into its *Escherichia coli* host. Alternatively, wild-type sequences can be removed from a plasmid and the desired mutant sequence (cassette) ligated instead.

Somatic: Pertaining to the body; pertaining to all cells except the germ cells.

Southern blot analysis: A technique used for detecting specific DNA sequences in cells. DNA is extracted from cells and cut with one or more restriction enzymes. The DNA fragments are separated by gel electrophoresis and blotted onto nitrocellulose paper or a nylon membrane. The DNA is then hybridized using a labeled DNA probe with a sequence complementary to the specific sequence to be detected. The DNA fragments that hybridize with the probe can be detected by techniques such as autoradiography, phosphoimaging, or chemiluminescence.

Spectral karyotyping (SKY): A technique that allows identification of chromosomes by application of fluorescent chromosome-specific paints.

Spermatogonia: Primitive differentiated male germ cells which give rise to primary spermatocytes, through the first meiotic division.

Sphingolipids: Any of a group of lipids, containing the basic sphingosine instead of glycerol, including sphingomyelins and cerebrosides.

Sphingomyelin: Type of sphingolipid found in animal cell membranes, especially the membranous myelin sheath which surrounds some nerve cell axons. It usually consists of ceramide and phosphorylcholine.

Splicing (RNA): The process in which the initial RNA copy made from DNA during transcription is modified to remove certain sections (such as introns) before use as a template for translation. Many genes can undergo alternative splicing, which can remove some exons (as well as the introns), resulting in mRNAs of different length that may be translated to produce proteins of different sizes.

Sporulation: The developmental process by which a fungal cell, amoeba, bacteria, or protozoan becomes a spore.

Stressor: An agent that causes some type of physiologic stress.

Stroma: The mesenchymal components of epithelial and hematopoietic tissues and tumors, which may include fibroblasts, adipocytes, endothelial cells, and various immunocytes as well as associated extracellular matrix.

Sublethal damage (SLD): Nonlethal cellular injury that can be repaired or accumulated with further dose to become lethal. Classically, repair of sublethal damage is revealed by giving two treatments separated by a variable time interval.

Superoxide: The anion $O_2^{\bullet-}$, with one unpaired electron it is a free radical.

Superoxide dismutase (SOD): An enzyme that removes the superoxide ($O_2^{\bullet-}$) radical.

Surviving fraction: The fraction of cells that retain long-term proliferative potential (i.e., usually clonogenic cells) following treatment with a cytoxic agent.

Syngeneic: Refers to organisms or cells which share an identical genetic background.

Tandem repeats: Occurs in DNA when a pattern of two or more nucleotides are repeated. A short tandem repeat (STR) in DNA is a class of polymorphism that occurs when a pattern of two or more nucleotides are repeated and the repeated sequences are directly adjacent to each other. The pattern can range in length from 2 to 10 base pairs (bp) for example and is typically in the noncoding intron region.

TATA box: Short nucleotide consensus sequence in eukaryote promoter sequences bound by RNA polymerases II and III, about 25–30 bp upstream of transcribed sequence. An AT-rich region, commonly including the interrupted sequence TATAT...AAT...A.

Tautomeric: Chemical isomerism characterized by relatively easy interconversion of isomeric forms in equilibrium.

Telomerase: A reverse transcriptase that polymerizes TTAGGG repeats to offset the degradation of chromosome ends that occurs with successive cell divisions.

Thiol: In organic chemistry, a compound which contains the functional group composed of a sulfur atom and a hydrogen atom (–SH).

Thrombus (pl., thrombi): A blood clot.

Tissue culture: Procedure of propagating cells outside of living tissues in various types of flasks and dishes.

Tissue microarray: A technique by which hundreds of tissue cores taken from the original paraffin blocks are combined in a tissue microarray block. When sections are cut from these blocks immunohistochemical studies can be performed on hundreds of tumor samples at the same time.

Topoisomerases: Enzymes that allow breakage of one or both DNA strands, unwinding of DNA, and resealing of the strands. The enzymes are required for DNA and RNA synthesis and are important for the action of some anticancer drugs.

Transactivation: The process by which the transcription of a gene is increased by proteins that interact with the promoter DNA of the transcribed gene.

Transcription: Transfer of genetic information from DNA by the synthesis of an RNA molecule copied from a DNA template.

Transcription-coupled repair (TCR): A DNA repair mechanism which operates in tandem with transcription.

Transcription factor: Proteins that can bind to DNA (often after their association to form dimers) and are involved in regulating the transcription of a gene, often by associating with sequences in the promoter region of the gene.

Transduction: (1) Process whereby a signaling element, such as a protein, receives a signal and, in response, emits another signal. (2) Process by which a gene is introduced into a cell, usually by a vector such as a viral vector.

Transfection: Procedure of introducing DNA into mammalian cells.

Transferrin: A blood plasma protein for iron ion delivery.

transfer RNA (tRNA): A small ribonucleic acid molecule that contains a three-base segment (anticodon) that recognizes a codon in mRNA, a binding site for a specific amino acid, and recognition sites for interaction with the ribosome and the enzyme that links it to its specific amino acid.

Transformation: In higher eukaryotes, the conversion of cultured cells to a malignant phenotype.

Transforming growth factor alpha (TGF-α): Functional and structural analogue of epidermal growth factor. It induces the growth of epithelial cells as well as fibroblasts and keratinocytes.

Transforming growth factor beta (TGF-β): A cytokine that regulates many of the biologic processes essential for embryo development and tissue homeostasis and which therefore plays a role in the healing of a tissue and carcinogenesis.

Transition: Point mutation in which one purine base replaces the other, or in which one pyrimidine base replaces the other.

Translocation: The displacement of one part of a chromosome to a different chromosome or to a different part of the same chromosome.

Transphosphorylation: Phosphorylation of one protein molecule by another, such as the phosphorylation of one receptor subunit by the kinase carried by another.

Transrepression: The opposite mechanism to transactivation. The activated receptor interacts with specific transcription factors and prevents the transcription of targeted genes.

Transversion: Point mutation in which a purine base replaces a pyrimidine or vice versa.

Triradial: A chromosome aberration which involves deletion and rejoining of two terminal fragments from one chromosome and one from another.

Trisomy: The condition in which a cell or organism possesses two copies of each chromosome, except for one, which is present in three copies.

Tritium: A radioactive isotope of hydrogen.

Tubulin: Member of a family of small globular proteins, the most common being α and β tubulin, dimers of which make up microtubules.

TUNEL assay: A technique (terminal deoxynucleotidyl transferase dUDP nick end labeling) widely used to detect cells that are undergoing apoptosis.

Two-component model: (also called multitarget model). Describes a survival curve which is characterized by an initial slope (D_1) and final slope (D_0) and a parameter which describes the width of the shoulder, either n or D_q.

Ubiquitylation: Process by which one or more ubiquitin molecules are attached to a protein substrate molecule, which often results in the degradation of the tagged protein.

Ultrafractionation: In radiotherapy, the application of multiple fractions of 0.5 Gy/-day to the same total dose and within the same treatment time as that used for conventional fractionation.

Unsaturated: Term used in organic chemistry to describe any carbon structure which contains double or occasionally triple bonds.

V(D)J recombination: A mechanism of DNA recombination that generates diverse T-cell receptor (TCR) and immunoglobulin (Ig) molecules that are necessary for the recognition of the greatly diverse foreign antigens.

Van de Graaff generator: An electrostatic machine which uses a moving belt to accumulate very high voltages on a hollow metal globe. The potential differences achieved in modern Van de Graaff generators can reach 5 MV. Applications for these high voltage generators include driving x-ray tubes and accelerating electrons and protons for various purposes.

Vasculature: Network of blood vessels.

Vector: A short piece of DNA or RNA, such as a DNA plasmid or RNA virus, into which genetic material of interest is incorporated and which is used to transfer this genetic material into a cell for either transient or long-term expression.

Werner syndrome: A very rare, autosomal recessive disorder the most recognizable characteristic of which is premature aging. The defect is on a gene that codes DNA helicase and as a result DNA replication is impaired in this syndrome.

Western blot analysis: A procedure analogous to Southern and Northern blot analyses that allow the detection of specific proteins. Proteins are separated by electrophoresis and transferred onto a membrane. They are usually detected following binding with labeled antibodies.

Xanthine oxidase: Enzyme catalyzing the oxidation of hypoxanthine to xanthine and the further oxidation of xanthine to uric acid.

Xenobiotic: A biologically active compound that originates outside of the body and is foreign to its normal metabolism.

Xenografts: Transplants between species; usually applied to the transplantation of human tumors into immune-deficient mice and rats.

Xeroderma pigmentosum: A human genetic disease characterized by extreme sensitivity to sunlight and the early onset of skin cancers. The genetic defect lies in the inability of the individual's cells to repair DNA damage (by NER) caused by UV light.

XPA: One of the nine major proteins involved in NER in mammalian cells. The names of the proteins come from the diseases which are associated with their deficiency. XPA, XPB, XPC, XPD, XPE, XPF, XPG from xeroderma pigmentosum, and CSA and CSB from Cockayne syndrome.

XRCC genes: (x-ray cross-complementing). Human genes, which were mostly identified through their ability to correct DNA damage hypersensitivity in rodent cell lines, were assigned to the XRCC nomenclature. They represent components of several different repair pathways including base-excision repair, NHEJ, and homologous recombination.

Zinc-finger domain: The region of a protein that is formed into a finger-like projection by binding some of the amino acids (cysteines or histidines) to a zinc ion. This configuration usually provides a DNA-binding region and is often found in transcription factors.

Zymogen: An inactive precursor form of an active enzyme.

Index

A

Access, repair, restore (ARR)
  model, 218
Acid sphingomyelinase (ASMase), 291
Actin filaments, 22, 28
Activating protein-1 (AP-1) transcription
  factor, 312
Activating transcription factor-like
  (ATF) proteins, 314
Activation-induced cell death
  (AICD), 289
Adaptive response (AR), 350
  interactions of, 357–358
  mechanisms of, 356–357
Addition reactions, 15, 19
Adenine, 31–32, 128
Aerobic/anaerobic glycolysis, 402–405
Agarose gel electrophoresis, 280
AG1478, EGFR tyrosine kinase
  inhibitor, 458
Alkylphosphotriesters repair, 142
Amifostine, 457–458
Amino acids
  for protein synthesis, 44
  in RNA, 42–43, 49
Anaphase bridge, 231–232
Anti-angiogenic molecules, 399
Anti-apoptotic factors/proteins,
  258, 283
Anti-apoptotic signaling, 262, 295
Antisense oligonucleotides, 85
Anti-vascular endothelial growth factor
  (anti-VEGF) therapies, 415
Apoptosis, 26, 55, 116, 175, 199–200
  activation of, 289
  biochemical and morphologic
    manifestations of, 291
  biochemical features of, 293
  cell death by, 423
  characterization of, 280

dysregulation of, 339
extrinsic signaling in, 287
mechanism of, 279–280, 287
as mode of cell death, 339, 342
tumor susceptibility to, 343
Apoptosis, in *C. elegans*
  cell death, developmental, 446–447
  CEP-1, p53 homologue, 447
  radiation-induced, 447–448
Apoptosis-resistant cells, 405
Apoptotic signaling pathways, 282
Apurinic site, DNA damage by ionizing
  radiation, 128–129, 144
Apyrimidinic (AP) site, 128–129, 136,
  140, 145, 170
Artemis, in DSBs DNA repair, 161–162
Assays of radiation–dose relationships,
  cell survival measurement and,
  64–66
Ataxia telangiectasia (AT), 312
  protein, 177, 201
Ataxia telangiectasia-like disorder
  (ATLD), 167–168, 171,
  182–183, 193
Ataxia telangiectasia mutated (ATM)
  protein kinase, 168–169, 312, 353
A-T-like disorder, *see* Ataxia
  telangiectasia-like disorder
  (ATLD)
ATM protein
  functions, 178–180
  MRN complex and, 182–183
  response to DNA damage, 180–181,
    200–201
  role in DNA repair, 181–182
ATM signaling, 216–218
  from Chromatin, 217
Autophagic cells, morphology of, 340
Autophagy, 340
Azotobacter vinelandii, 464